AF670111

Advancing Cochlear Implants

Behnam Molaee-Ardekani
Manuel Segovia-Martinez
Editors

Advancing Cochlear Implants

Current and Emerging Methods and Technologies

Editors
Behnam Molaee-Ardekani
Cochlear Ltd, Advanced Innovation
Vallauris, France

Manuel Segovia-Martinez
Oticon Medical, Research and Technology
Vallauris, France

ISBN 978-3-032-15718-8 ISBN 978-3-032-15719-5 (eBook)
https://doi.org/10.1007/978-3-032-15719-5

This Springer imprint is published by the registered company Springer Nature Switzerland AG
The registered company address is: Gewerbestrasse 11, 6330 Cham, Switzerland

Preface

When we first entered the field of cochlear implants, we sought a comprehensive resource that not only covered the foundational engineering and system-level principles of these devices but also meaningfully connected them to their clinical and surgical applications. We were eager to explore how the design of cochlear implant systems interacts with various clinical aspects of hearing loss, treatment strategies, and auditory rehabilitation.

Although topics such as sound processor architecture, implant mechanics, biomaterials, and signal processing have been discussed in various specialized publications, we were unable to identify a single volume that brings together these technical elements alongside clinical perspectives and emerging innovations. This book aims to contribute toward addressing that need.

Cochlear implants represent one of the most remarkable achievements at the intersection of engineering, medicine, and neuroscience, restoring hearing to over a million individuals worldwide. Yet, despite their widespread success, outcomes vary significantly across users. This variability has catalyzed a multidisciplinary effort—spanning engineering, audiology, neuroscience, and surgery—to better understand, personalize, and optimize cochlear implant performance.

In this collection, we bring together diverse perspectives from leading academic and industrial research teams. The chapters explore current developments and findings while outlining a roadmap for future innovation. Over the past decade, cochlear implant systems have undergone rapid transformation, driven by advances in materials science, signal processing, neurophysiology, and artificial intelligence. Today, the field is vibrant with new ideas and technologies that promise to enhance the precision, adaptability, and effectiveness of these devices.

Our goal with this book is to provide a detailed and up-to-date account of these advancements—bridging the gap between engineering fundamentals and clinical realities, and inspiring continued progress in this dynamic and impactful domain. This volume is organized into three parts.

Part I: Foundations of Cochlear Implant Systems

We begin with a look back at the origins of cochlear implant technology and the foundational principles that shaped its evolution. Early innovations laid the groundwork for today's sophisticated systems, and the architecture of modern implants reflects the needs of diverse stakeholders—patients, clinicians, surgeons, and audiologists. From the physics of electrical stimulation to the design of user subsystems, this historical and technical narrative sets the stage for what cochlear implants are—and what they might become.

The focus then shifts to the core of cochlear implant functionality: sound coding. We examine the intricate process of transforming acoustic signals into electrical patterns that stimulate the auditory nerve. Despite decades of progress, challenges remain—biological constraints, hardware limitations, and the coarse nature of electric stimulation all impact user experience. Emerging technologies such as AI-driven coding strategies, advanced imaging, adaptive electronics, and other innovations offer promising avenues to personalize stimulation and enhance various aspects of auditory perception, including speech understanding, music appreciation, and sound localization.

Next, we explore the clinical dimension of cochlear implants. Performance is evaluated not only in controlled environments but also in the unpredictable reality of everyday life. Traditional audiologic assessments are contrasted with newer approaches that emphasize ecological validity, including wearable technologies and ecological momentary assessment. Broader outcomes—mental health, social connectedness, and quality of life, especially for elderly and pediatric populations—are also considered, reminding us that cochlear implant success is not just about hearing but about living.

Finally, we delve into the materials and mechanics that make cochlear implants durable, biocompatible, and effective. From titanium casings to ceramic components, from brazing techniques to electrode array design, we explore the engineering innovations that ensure long-term reliability and safe neural stimulation—forming the physical backbone of cochlear implant systems.

Part II: Objective Assessment and Performance Optimization

With the foundational concepts established, we now turn to the tools used to measure, monitor, and enhance cochlear implant performance. This section connects clinical practice with advanced research, presenting a broad set of objective methods that help us understand how the implant interacts with the auditory nerve, assess hearing function, and customize treatment to meet individual needs.

We begin with impedance measurements—a simple electrical property that reveals complex interactions between the implant and living tissue. Through biophysical modeling and spectral analysis, impedance becomes a window into surgical outcomes, device integrity, and tissue response.

Intracochlear electrocochleography (eCochG) follows a technique that has begun to influence cochlear implant surgery. Once confined to animal models, eCochG now enables real-time monitoring of cochlear function during implantation, helping preserve hearing and optimize electrode placement.

Electrically Evoked Compound Action Potentials (ECAPs) continue this theme, offering direct insight into auditory nerve responsiveness. The same electrodes that stimulate hearing can record neural activity, allowing clinicians to fine-tune programming and researchers to model the electrode-neuron interface.

We then expand to EEG and eABR—two powerful electrophysiological tools that assess auditory and cognitive function. These techniques reveal how CI users process speech, music, and effort, and how the brain adapts to electric stimulation. Despite challenges such as artifact interference, EEG and eABR are unlocking new dimensions of understanding.

The brain's remarkable plasticity and capacity for perceptual learning are also explored. Whether through passive exposure or active auditory training, CI users reshape their neural pathways to make sense of electric sound. This chapter highlights the importance of rehabilitation and the role of acoustic hearing in supporting adaptation.

Finally, we enter the realm of artificial intelligence. From predicting hearing outcomes to automating parameter fitting, AI is poised to revolutionize cochlear implant care. Though challenges remain—data standardization, generalization, clinical adoption—the potential is undeniable. AI offers a glimpse into a future where cochlear implants learn from users and adapt in real time.

Part III: Emerging Technologies and Future Directions

The final part of our journey ventures into the frontier—where computational models, surgical robotics, and optogenetics are reshaping what cochlear implants can do.

We begin with physical modeling of the cochlea, using finite element methods and anatomical reconstructions to simulate electrical stimulation. These models help us understand current flow, off-target effects, and optimize electrode design. They are virtual laboratories, where hypotheses become simulations and simulations guide practice.

Biophysical modeling takes us deeper, into the neural encoding of sound. From single-fiber responses to population-level neurograms, we explore how electric stimulation compares to natural hearing. These models help us evaluate speech processors, predict intelligibility, and bridge the gap between stimulus and perception. It's a story of abstraction made actionable.

Medical imaging and artificial intelligence converge next, transforming how we visualize and interpret the cochlea. From CT and MRI to synchrotron imaging, we explore the tools that map anatomy with precision. AI enhances these processes—segmenting structures, detecting malformations, and guiding electrode placement. Imaging becomes intelligent, and intelligence becomes integral.

Surgical innovation follows, with robotics, augmented reality, and virtual reality redefining the operating room. These technologies improve precision, reduce invasiveness, and elevate training. Though still emerging, they represent a paradigm shift in cochlear implant surgery—where digital tools enhance human skill.

The journey concludes with a glimpse into the future: optical cochlear implants. By replacing electrical stimulation with light, optical cochlear implants promise finer spatial resolution and more natural hearing. Through optogene therapy and targeted neural activation, this technology may overcome the limitations of current implants, especially in noisy environments. It's a vision of hearing restored not just functionally but beautifully.

We hope this book serves not only as a reference but also as a source of inspiration for those seeking to advance the science and technology of cochlear implants.

Vallauris, France

Behnam Molaee-Ardekani
Manuel Segovia-Martinez

Acknowledgements We, Behnam Molaee-Ardekani (Editor) and Manuel Segovia-Martinez (Co-Editor) express our deepest gratitude to the contributors of this volume. Their expertise and dedication have made it possible to share the interdisciplinary knowledge brought together in this work with a new generation of engineers, clinicians, and researchers.

Contents

About the Editors

Behnam Molaee-Ardekani, Ph.D., is a Senior Scientist in the Advanced Innovation Department at Cochlear Ltd. Since 2016, he has led research initiatives on cochlear implants, with a focus on audio processing, sound-coding strategies, objective measures, and the integration of machine learning into implant fitting and cochlear imaging workflows. His contributions have resulted in multiple patents and peer-reviewed scientific publications. He previously held postdoctoral positions across multiple European research projects, investigating biomarkers and brain activity dynamics in anesthesia, transcranial stimulation, epilepsy, and neurodegenerative disorders. He completed his Ph.D. in Biomedical Engineering in 2008 through a joint doctoral program between the University of Rennes 1 in France and Aryamehr (Sharif) University of Technology in Iran.

Manuel Segovia-Martinez, Ph.D., is an Audiology and Signal Processing scientist with over 20 years of experience and a demonstrated track record in the medical device industry. He earned his Ph.D. in 3D medical image processing for the diagnosis of Alzheimer's at the University of Surrey in 2001, under the supervision of the late Prof. Maria Petrou. He is currently working on the development of assisting technology for the Neurodivergent Community. Previously, he led the Oticon Medical Research and Technology and Embedded Systems departments, with a track record of releasing several generations of implantable medical devices. He has been involved in a series of intellectual developments and is the author or co-author of several patents that form the basis of Oticon Medical's CI audiology strategy.

Part I

Foundations of Cochlear Implant Systems

Cochlear Implant Development: From Early Concepts to Future Directions

1

Manuel Segovia-Martinez, Filiep Vanpoucke, David Landsberger, and Behnam Molaee-Ardekani

Abstract

This chapter introduces the enabling technologies behind cochlear implants. It begins with a retrospective overview of the key technological contributions that have significantly influenced the architectural design of current cochlear implant systems. To support this, a comprehensive literature review and a series of interviews were conducted. Readers are encouraged to consult the recommended bibliography for a deeper understanding of the history of cochlear implants. Following this historical context, the chapter presents fundamental principles of electrical stimulation in cochlear implants. This includes a brief overview of various configurations and stimulation parameters, and how these affect the auditory nerve. Building on this foundation, the current architecture of most cochlear implants is examined, analyzed through the lens of different user subsystems—namely, the patient, clinician, surgeon, and audiologist. The chapter

M. Segovia-Martinez (✉)
Research and Technology Department, Oticon Medical, Vallauris, France
e-mail: manuel.segovia@origenes.tech

F. Vanpoucke
Advanced Innovation Department, Cochlear Ltd, Mechelen, Belgium
e-mail: fvanpoucke@cochlear.com

D. Landsberger
Department of Otolaryngology-Head and Neck Surgery, New York University, New York, NY, USA
e-mail: david.landsberger@nyulangone.org

B. Molaee-Ardekani
Research and Technology Department, Oticon Medical, Vallauris, France

Advanced Innovation Department, Cochlear Ltd, Vallauris, France
e-mail: bmolaee@cochlear.com

B. Molaee-Ardekani, M. Segovia-Martinez (eds.), *Advancing Cochlear Implants*,
https://doi.org/10.1007/978-3-032-15719-5_1

concludes with a forward-looking discussion on how cochlear implant technologies may evolve in the coming years.

Keywords

Cochlear implant · Technology history · Stimulation configurations · User subsystems · Clinicians · Surgeons · Audiologists

1.1 Introduction

Cochlear implants (CI) are today one of the most successful implantable medical devices with more than 1 million users worldwide (Zeng 2022). These devices are designed to replicate the function of the natural ear by directly stimulating the auditory nerve, effectively bypassing the external, middle, and inner ear. This direct neural stimulation enables individuals with severe to profound hearing loss to perceive sound and understand speech, restoring auditory access and information flow in cases where conventional hearing aids are insufficient.

Over time, the architecture of cochlear implants has become standardized across commercial manufacturers, making it relatively easy to identify their core components (e.g., Fig. 1.1). This consistency reflects decades of technological refinement and clinical experience. In contrast, early cochlear implant systems featured a wide variety of experimental designs and concepts before converging on the modern configuration.

This chapter provides an overview of cochlear implant devices and their associated systems. It begins by outlining the architecture and functionality of contemporary cochlear implant technologies. A dedicated section follows, exploring the rich and formative history of these devices—an essential context for understanding the advancements that define today's solutions.

Following this, the chapter explores the fundamental technological principles that enable cochlear implant functionality, with a particular focus on electrical

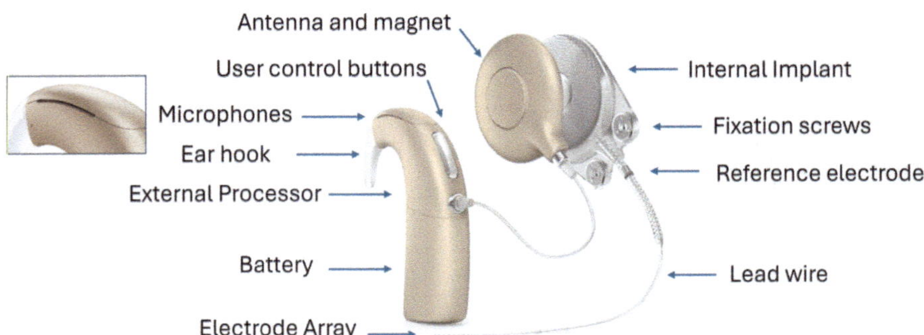

Fig. 1.1 The Neuro2 cochlear implant system from Oticon Medical Company. © Cochlear Limited. (Image courtesy of Cochlear Limited)

stimulation techniques. It then expands the scope from the device itself to the broader cochlear implant healthcare ecosystem, emphasizing the importance of a multidisciplinary approach to achieving optimal hearing outcomes.

This ecosystem is described through several interconnected subsystems, including the user's system (i.e., the cochlear implant device used directly by the individual), the clinical system, the surgical system, and the audiology system. The chapter concludes with a forward-looking perspective on the future of cochlear implant technology, exploring potential advancements within each of these subsystems.

1.2 Contemporary Cochlear Implant Systems

The basic configuration of a cochlear implant (CI) system consists of two main components: an external sound processor and an internal implant (Fig. 1.1). In addition to these core elements, users often rely on various accessories—such as chargers, remote microphones, wearing options, and digital tools—to support daily use and enhance performance in different listening environments.

The implant is the internal component, surgically placed under the skin and into the cochlea. It includes an electrode array, typically containing 12 to 22 stimulation contacts, which is inserted into the cochlea to directly stimulate the auditory nerve fibers. The electrode is connected to a hermetic receiver-stimulator. The heart of the receiver-stimulator is an electronic assembly having programmable current sources. The stimulation commands from the external sound processor are received by the receiver coil, demodulated and converted into stimulation impulses. Electrical pulses are transmitted to the electrode array, stimulating the auditory nerve and restoring the sense of audition. The receiver-stimulator is powered by the Radio Frequency (RF) signals transmitted from the external sound processor. The power supply component converts the RF energy into electrical power, which is used to operate the internal components of the receiver-stimulator. Implants also have recording circuitry. They can capture both large technical potentials, such as induced intracochlear voltage or impedance, and tiny biological potentials such as neural responses to the electrical stimulation. The RF link is bidirectional, allowing the implant to send back information from the implant to the sound processor, a function called back-telemetry.

The sound processor is the external device that captures sound from the environment through one or more microphones or other input sources, such as wireless connectivity or broadcast. It processes the sound by converting it into stimulation commands, which are then transmitted to the implant through a transmitter coil, aligned with the internal coil using a magnet. The sound processor is the brain of current CI system, the primary device directly controlling the actions of the implant. It is the intelligent part, sensing the sound environment, cleaning up the sound inputs, and converting the sound into stimulation sequences. The intelligence, i.e., the feature set, is implemented in firmware and constrained by the computational capabilities of the hardware platform. System power consumption determines battery size and therefore device size, user comfort, and hours of use. Current systems therefore use advanced, specialized, low-power digital, and analogue chips and highly efficient radio links.

The standard input source is the sound processor microphones. The microphones are normally a pair of matched microphones. These microphones are matched on their frequency response to enable spatial filtering (Chung 2004). Both hearing aids and cochlear implants have moved recently to low-power micro-electro-mechanical systems (MEMS) microphones. A typical sensitivity response is flat in the frequency range from approximately 100 Hz to 8 kHz. Some microphones offer an extended flat response to up to 10 kHz.

An important evolution in the recent past, requiring considerable firmware and hardware, is connectivity. Wireless connectivity enables patients to participate in phone calls, video calls, and virtual meetings with greater ease. Users can directly stream sound from various devices, such as smartphones, TVs, and computers, to their hearing aids or implants. This provides a clearer and more immersive listening experience, eliminating the signal degradations coming from room acoustics and background sounds. The usage of remote microphones and streaming in environments such as schools provides improved speech understanding (Fitzpatrick et al. 2009). It also offers the capacity to optimize the characteristics of the streamed signal to the needs of the user. For example, a music signal can be simplified to increase the enjoyment or access to CI users (Tahmasebi et al. 2023). In the past, wireless links relied on company-specific proprietary protocols due to strict low-power constraints. However, there is now a shift toward industry standards such as Bluetooth® Low Energy Audio (Sharma and Shrivastava 2024), offering broader accessibility, enhanced functionality—such as Auracast™—and greater user convenience.

Access to direct streaming for mobile communication has become an essential feature in CI systems. Earlier, speech compression standards used on telecommunications presented several difficulties in terms of speech understanding for CI users (Guignard et al. 2019). In contrast, current streaming technologies are better adapted to the needs of CI users (Wolfe et al. 2015; Plyler et al. 2025).

In terms of the firmware running on the sound processor, the first signal processing step is sound enhancement of the target source, by reducing background noise and dealing with signal degradations, e.g., due to reverberation. For a dual microphone, a powerful option is beamforming. How the two microphone signals are best combined—with fixed or adaptive directionality—depends on the acoustic situation. Very narrow beamformers can be formed if the left and right microphones can be combined. This requires an ear-to-ear audio link, a difficult feature in cochlear implants due to the presence of strong RF implant links. To dynamically select the best spatial pattern, modern processors have in-built acoustic scene classification capability and automation algorithms. The next step in the processing is amplification and compression, mapping the input signal optimally in the electrical dynamic range of the patient. Amplification in cochlear implants is similar to hearing aids, typically based on a wide-dynamic range compressor (WDRC) algorithm. Stimulation levels are unique to each implanted ear, depending on the position of the electrode array in the cochlea and the health of the neural tissue. The electrical stimulation window (upper and lower stimulation level) is decided during fitting, and parameters are stored in nonvolatile memory.

The sound processor devices themselves continue to move forward toward better aesthetics, robustness, and user comfort. They evolved from bulky body-worn devices to more elegant and compact behind-the-ear (BTE) devices, suitable for all age groups. Off-the-ear (OTE) models, whereby the sound processor electronic assembly and the antenna are integrated in a single package, have become a popular choice.

Accessories for cochlear implant systems include remote microphone devices, which help users cope with the effects of room acoustics by transmitting sound directly to the sound processor. Additionally, user control systems have evolved from dedicated remote controls to smartphone apps, enabling new functionalities such as device setting personalization, audio streaming, and remote rehabilitation. The move from dedicated hardware accessories to digital apps is transformative. Whereas previous user accessories provided users with a simple but limited feature set to control and configure their hearing devices, such as changing volume or switching maps, digital apps on a smartphone empower a digitally literate user with more convenience in conducting standard tasks, and, more importantly, unlock a rich set of new possibilities. They can already now participate more fully in managing their hearing. And tasks that were traditionally confined to the clinic can now be conducted in the safety and comfort of their own home, offering, e.g., hearing rehabilitation through audio streaming.

1.3 Brief History of Cochlear Implant Devices

The history of cochlear implant (CI) development is often presented with a certain bias, highlighting the contributions of specific inventors, groups, or countries. The authors of this chapter have observed that various parties have claimed on occasions to be the "father" of the cochlear implant. As often with breakthrough technologies, it is worthwhile to look at the pioneering days from multiple perspectives. A more thorough exploration reveals a rich and geographically diverse history (see Fig. 1.2). In this brief introduction, we aim to highlight some of the most significant contributions and their impact on the scientific and industrial communities in shaping the modern cochlear implant. We encourage readers to delve into various sources and trace the path of this historical development for themselves.

It is important though to acknowledge some of the previous authors who already have done excellent research on the origins of the cochlear implant, including (Blume 1999; Seitz 2002; Eshraghi et al. 2012). Blume's work not only provides an unbiased account of CI development from the first electrical stimulation to the emergence of successful commercial houses but also presents the reactions and issues faced by the deaf community due to the introduction of CI technology in France and the Netherlands (Blume 1999). Eshraghi et al. is also an excellent introduction to the history of the cochlear implant (Eshraghi et al. 2012). Their work meticulously documents the international efforts that led from the first direct

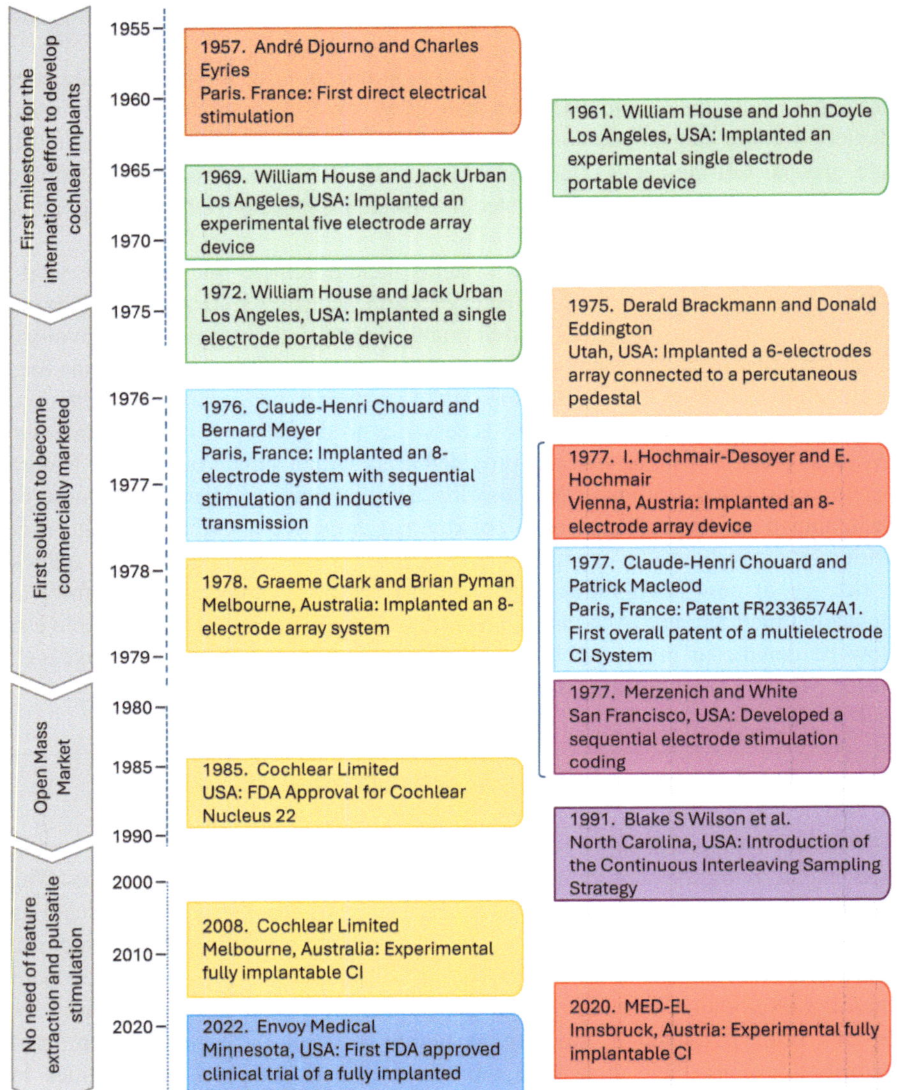

Fig. 1.2 Some key dates for CI development

electrical stimulation of the auditory nerve to the development of modern cochlear implants.

The following subsections provide a concise historical overview of the evolution of CI design and manufacturing—from the earliest devices featuring a single stimulating electrode to the pioneering advancements that led to biocompatible, multielectrode systems. A few key breakthrough projects are highlighted for their transformative impact on the research and development landscape of CI technology.

Finally, a brief summary of the major current manufacturers shaping the CI industry is presented.

1.3.1 Single-Electrode Devices

In the 1950s, the technology allowing direct peripheral nerve stimulation was under development (Seitz 2002). The use case of the auditory nerve was several times hinted on congresses and seminars. It is then that in Paris the first direct electrical stimulation of the auditory nerve was performed in 1957 (Djourno and Eyries 1957).

The pioneer work from André Djourno and Charles Eyries allowed a first medical case where some restorations of hearing by electrical stimulation were possible (Blume 1999; Seitz 2002). Unfortunately, several technical issues forced the first patient to have a second device implanted that also failed soon. Djourno continued working until around 1959 and resolved many of the technical and material issues needed for the development of cochlear implants (Djourno and Kayser 1958). This preliminary effort inspired William House to pursue his own line of development (Eshraghi et al. 2012).

In 1961, House and Doyle implanted an experimental single-electrode portable device. Thereafter, Doyle continued his own path. It is important to note that these devices showed for the first time improvement in speech performance on a closed set. However, they were suffering from biocompatibility issues, and many of them failed and had to be removed soon after implantation.

Then, William House and Jack Urban developed, starting in 1967, a biocompatible, stable, and industrial device. The first implantation of this new device came in 1972 (House and Urban 1973). We can comfortably claim that the House/Urban device is the first commercially available cochlear implant, and it was commercialized as the 3M/House single-electrode device (House 1985; Eisenberg 2015).

1.3.2 Multi-electrode Devices

Following the success of the single-electrode device, several research groups began developing implantable multielectrode arrays. They continued the effort to refine the design, implementation, and surgical procedures of a multielectrode device. Some sources report that experimental multielectrode devices were tested as early as 1957 by André Djourno and Charles Eyries, and later in 1961 by William House, John Doyle, and Jim Doyle (Blume 1999; Seitz 2002). In a notable advancement, F. Blair Simmons inserted a six-element electrode array into the modiolus near the basal turn, with each electrode individually connected via its own separate cable. The volunteer reported frequency differentiation (Simmons 1966). In 1969, House and Urban implanted an experimental five-contact electrode array that was terminated in an external connector (House and Urban 1973; Dorman and Parkin 2015). They also

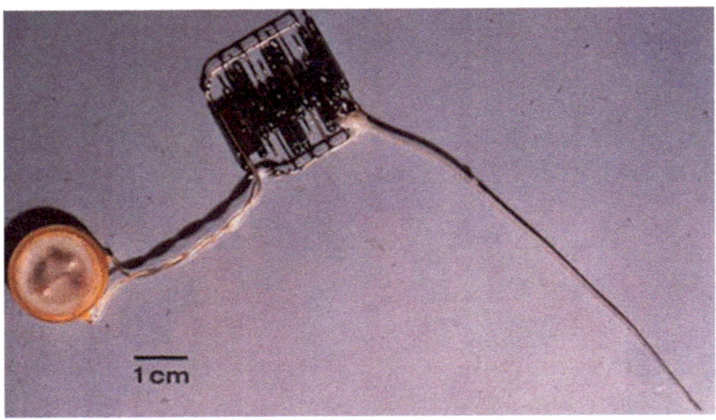

Fig. 1.3 The 1978 Vienna device, a six-electrode variant. © MED-EL GmbH. (Image courtesy of MED-EL Corporation)

investigated pitch perception enabled by these multielectrode devices (House and Urban 1973).

On September 22, 1976, Chouard and Meyer implanted an eight-electrode device in Paris (Chouard 2015). A review of their 1977 patent reveals that many fundamental elements of modern CI systems were already present in this first patent for a multielectrode implanted device (Ricard et al. 1977). Their innovations included sequential stimulation, inductive transmission between external and internal elements, and frequency-specific electrode assignment in an ascending order. However, a significant drawback was that each electrode required a separate cochleostomy, making the surgical procedure particularly lengthy and complex.

Soon after, in December 1977, a group from the Technical University of Vienna implanted another eight-electrode device (Hochmair 2013). Their patent disclosed a very compact electronic design and a flexible single-body electrode array (Hochmair and Hochmair 1979). In 1978, they developed additional six- and four-channel variants to better accommodate individual patient needs. Figure 1.3 shows an implant with six-electrode contacts. Despite these advancements, their coding strategy still did not fully resolve the issue of cross-channel interaction (Hochmair and Hochmair-Desoyer 1983). This device later evolved into the solution proposed by MED-EL Corporation.

Another historically significant multielectrode implantation was performed by the University of Melbourne on August 1, 1978 (Clark 2003). Similar to the Vienna device, it featured a single-body and flexible electrode array. This design is the precedent of what became the Cochlear Limited implant devices.

1.3.3 The Development of Electrophysiological Principles: Utah Artificial Ear Project

This subsection highlights a pivotal historical initiative in the field of CI that played a crucial role in the development and testing of the electrophysiological principles of cochlear implant systems. The Utah Artificial Ear Project, led by Dr. Donald Eddington at the University of Utah with support from House Institute, began around

1970. It is highly likely that the Utah team used the experimental five-electrode percutaneous device developed by House for initial multielectrode stimulation experiments. Insights gained from these early studies contributed to the development of a research-grade percutaneous CI between 1970 and 1974 (Dorman and Parkin 2015). This effort culminated in the implantation of a six-electrode array in 1975.

In 1975, the Utah project implanted four volunteers with a six-electrode array connected via a percutaneous connector (Eddington et al. 1978; Dorman and Parkin 2015). This research device enabled a series of psychoacoustic investigations that formed the basis of our understanding of electrical stimulation audiology (Eddington et al. 1978).

The experiments focused on the effects of stimulation parameters on auditory threshold, impedance, and electrode interaction. Researchers also documented the perception of pitch and loudness as functions of electrode placement, electrical charge, and stimulation rate. Additionally, the long-term stability of subjective auditory sensations was evaluated. These studies demonstrated the feasibility of electrical stimulation as a method for hearing restoration.

This research device eventually evolved into the Ineraid system—a percutaneous cochlear implant that was approved as an FDA investigational device in 1985. It was implanted in approximately 100 patients, with early speech performance results showing promising outcomes (Dorman and Parkin 2015).

Several features of the Ineraid device are noteworthy, both as a platform for research and development and as a potential commercial cochlear implant.

1. Percutaneous Connector for Flexible Stimulation: The percutaneous plug allowed the delivery of arbitrary stimulation waveforms. This design made the external component easily upgradeable, allowing for the integration of new functionalities without hardware limitations. A notable example occurred after Ineraid ceased operations, when MED-El adapted their sound processor for humanitarian use with Ineraid users, implementing continuous interleaved sampling (CIS) stimulation.
2. Multielectrode Design for Spectral Encoding: The inclusion of multiple electrodes enabled the encoding of spectral information, making Ineraid a preferred tool for developing and testing new signal processing strategies. It also allowed within-subject comparisons of different stimulation modes.
3. Simplicity and Longevity of the Implantable Component: The implantable portion of the device contained no electronic components and featured a compact percutaneous connector—comparable in size to modern bone conduction abutments. This simplicity contributed to its durability, with several patients documented to have functioning devices for over 40 years (Dorman and Parkin 2015).
4. MRI Compatibility and Research Applications: While current CI designs are compatible with 3.0 Tesla MRI scanners, they still produce significant artifacts that can limit brain imaging (Todt et al. 2020). In contrast, the Ineraid device has been successfully used in advanced research involving functional Magnetic Resonance Imaging (fMRI) (Lazeyras et al. 2002).

When first introduced, the Ineraid device employed a simultaneous analogue approach (Loizou 1998). While early speech performance results were promising, it became evident that channel interaction was causing significant performance issues (Tyler et al. 1995). Despite this, Ineraid played a crucial role in the development of new coding strategies, including CIS, and other modern strategies (Dorman and Parkin 2015). Its design allowed for within-subject comparisons of different stimulation modalities and coding strategies.

The device ultimately did not receive FDA approval, primarily due to the concerns about risk of skin infections associated with the percutaneous connector (Dorman and Parkin 2015)—a genuine issue for many users. Nevertheless, many users were able to use their Ineraid devices successfully over extended periods.

In summary, with a modern perspective, Ineraid offered a lot of value to the current CI landscape and could serve as inspiration for future designs.

1.3.4 Current Commercial Devices

In 1985, the FDA approval of Nucleus 22 from Cochlear Ltd. marked the beginning of the mass market success of cochlear implants. Today, several manufacturers are offering their solutions to the market. Cochlear Ltd.,[1] MED-EL GmbH,[2] and Advanced Bionics LLC[3] can trace their history to the beginning of the modern cochlear implants in the late 1970s. Another traditional manufacturer familiar to the academic and scientific community is Nurotron Ltd.[4] Unfortunately, several other pioneering manufacturers such as House/3M, Ineraid, Antwerp Bionic Systems/LAURA, and MXM/Neurelec/Oticon Medical are no longer manufacturing cochlear implants.

While writing this chapter, two new CI manufacturers are organizing clinical trials. TODOC Co. started clinical trials for a 32-channel CI device at the end of 2022 (Shin et al. 2021). Envoy Medical has received approval for their Acclaim CI system in an FDA-approved trial (Dornhoffer et al. 2023).

1.4 Evolution of CI Stimulation Technologies

The history of CI stimulation technologies is marked by continuous innovation aimed at improving the quality of sound perception for the patients. Two cornerstones have guided this evolution: the coding strategy, which determines how acoustic signals are transformed into electrical stimuli, and the electrode-neuron interface, which governs how these stimuli are delivered to the auditory

[1] https://www.cochlear.com

[2] https://www.medel.com

[3] https://www.advancedbionics.com

[4] https://en.nurotron.com

nerve. Together, these elements have shaped the effectiveness, efficiency, and user experience of cochlear implants over time.

1.4.1 Development of Coding Strategies

The transformation of sound into meaningful electrical signals, so-called coding strategy, has undergone significant refinement since the early days of cochlear implants. In the big picture, today's different emerged sound coding strategies may stem from one of the two following influential coding strategies (refer to Chap. 2 for more information on coding strategies):

- Continuous Interleaved Sampling (CIS): Introduced in the late 1980s, CIS became a cornerstone of modern CI systems. It uses fixed-rate, nonoverlapping pulses across multiple electrodes to preserve temporal and spectral cues while minimizing channel interaction.
- Fine Structure Processing (FSP): Developed in the early 2000s, FSP-based strategies aim to enhance pitch perception by encoding the temporal fine structure (TFS) of low-frequency sounds, improving music appreciation and tonal language understanding.

A major milestone in the evolution of coding strategies was the introduction of the CIS strategy published by Wilson et al. (1991). This approach marked a turning point in CI performance, as it effectively addressed the issue of channel interaction, a phenomenon where simultaneous stimulation of adjacent electrodes leads to overlapping neural responses and degraded sound quality (Loizou 1998).

CIS operates by representing the envelope of narrow frequency bands on individual electrodes using interleaved sequential biphasic pulses. While these features were present in earlier strategies, CIS unified them into a robust and clinically viable framework. Notably, the concept of sequential stimulation was independently developed in the late 1970s by two research teams: Merzenich and White at UCSF (Merzenich and White 1977; Wilson et al. 1988) and Ricard, Chouard, and Macleod at CHU Paris Saint Antoine (Ricard et al. 1977). Both groups published foundational work in 1977. The French team's patent described a sequential delivery mechanism, including claims (particularly claims 1 and 5) for tonotopically independent stimulation and a selector system for sequential electrode activation. Figure 1.4 shows the strategy developed by the UCSF team.

Initial evaluation of CIS demonstrated clear performance advantages over earlier strategies such as compressed analogue (CA) (Wilson et al. 1991). However, further studies by Eddington revealed that, with extensive and individualized fitting (on the same patients), CA could achieve outcomes comparable to those of CIS (Eddington, personal communication via Bob Shannon).

CIS gained widespread adoption due to its use of sequential pulsatile biphasic stimulation, which significantly reduced the effects of cross-channel interaction. The CIS strategy did not require the extraction of speech-specific feature values, making

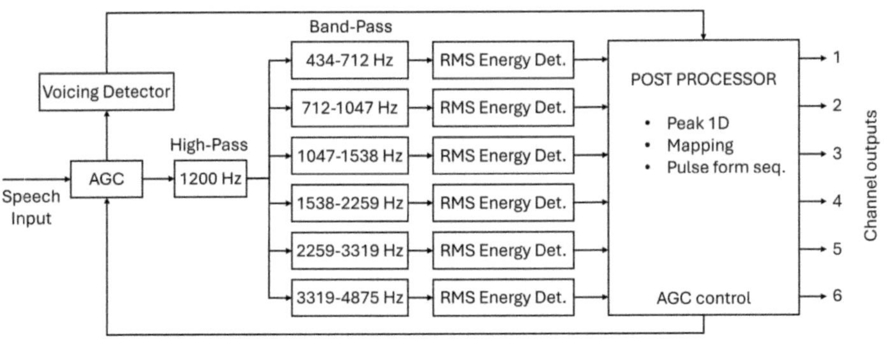

Fig. 1.4 Sequential pulse-delivering strategy. (Redrawn from Merzenich and White 1977)

it a robust strategy for any acoustic environment. Other advantages that likely contributed to the popularization of CIS included reduced power consumption and a simplified fitting process. These benefits enabled the development of a single-body behind-the-ear (BTE) processor and contributed to the commercial success of cochlear implants.

Coding strategies based on TFS, such as FSP and FS4 (Riss et al. 2014), represent advanced sound processing methods designed to improve pitch perception, music appreciation, and speech understanding in noise for CI users. TFS refers to the rapid oscillations within a sound wave and their corresponding zero-crossing information, which are essential for conveying fine temporal cues.

In contrast to traditional strategies like CIS, advanced combination encoder (ACE) (Kiefer et al. 2001), or Crystalis (Langner et al. 2020), which primarily encode the envelope (i.e., slow amplitude fluctuations), TFS-based strategies aim to present the fine timing details of the waveform. They do so by modulating the timing of stimulation pulses—particularly in the apical (low-frequency) electrodes—to synchronize with the TFS cues of the input signal.

While envelope-based strategies typically extract features using spectral or temporal methods, TFS-based strategies often rely on time-domain filterbanks to directly capture TFS cues from the temporal waveform. This approach simplifies the extraction process and enhances temporal fidelity. A novel development in this domain is the spectral event extraction (SEE) strategy, which builds upon a fast Fourier transform (FFT)-based framework. SEE enhances a predecessor strategy called spectral feature extraction (SFE) by combining spectral sharpness with temporal event tracking. It models spectral peaks using synthetic tones characterized by instantaneous frequency, phase, and energy and then tracks these peaks across frames to extract reliable temporal cues aligned with stimulation pulses (Molaee-Ardekani et al. 2025).

TFS-based coding strategies remain an evolving area in CI technology. While they aim to enhance pitch perception and music appreciation, clinical outcomes have shown considerable variability across individuals. Some studies report that TFS-based strategies do not consistently outperform envelope-based strategies in real-world listening tasks (Magnusson 2011; Riss et al. 2011). Conversely, other studies indicate a preference for TFS-based coding for vowel monosyllables, speech, and music (Müller et al. 2012; Riss et al. 2016).

1.4.2 Advancements in the Electrode-Neuron Interface

In parallel with advancements in coding strategies, researchers have also concentrated on optimizing the electrode-neuron interface—the physical and functional connection between the cochlear implant and the auditory nerve. Key areas of development include (i) electrode design, (ii) stimulation configuration, and (iii) stimulation pulse shapes. Each of these elements plays a critical role in determining how electrical stimulation activates auditory nerve fibers.

Electrode design from single-electrode to multielectrode arrays enabled place-specific stimulation, better replicating the cochlea's natural tonotopic organization. Multielectrode systems allow each electrode to target neurons located in specific cochlear regions. However, there remains ongoing debate regarding the necessity of matching frequency allocation map (FAM) of the limited number of electrodes in a CI device with cochlear tonotopy. Some studies indicate that patients with a larger mismatch between FAM and anatomically assigned frequency mapping experience poorer hearing performance and slower adaptation to a CI (Dessard et al. 2024). Others argue that the place-pitch tonotopically organized auditory neural pathway is not hardwired. The adult auditory system preserves a sufficient level of plasticity to swiftly and concurrently learn two alternating distinct mappings for speech understanding during rehabilitation (van Hoof et al. 2025). Refer to Chap. 9 for further insights into plasticity and perceptual learning in CI patients.

Researchers have sought to optimize the shape and positioning of electrode arrays within the scala tympani to enhance their proximity to auditory neurons and refine the resulting excitation patterns. In particular, perimodiolar electrode arrays (refer to Chap. 4) are designed to lie closer to the modiolus compared to straight arrays, thereby enabling more focused and spatially selective neural stimulation.

Stimulation configuration refers to the path of electrical current determined by the arrangement and activation of electrodes. Figure 1.5a1–a9 illustrates some stimulation configurations used in commercial devices, clinical settings, and animal studies (refer to Chap. 5).

- Monopolar (MP) stimulation uses an active electrode (source) within the cochlea, with the return current flowing to an external reference electrode (sink).
- Common ground (CG) or all-polar stimulation distributes the return current passively across all intracochlear electrodes.
- Distributed all-polar (DAP) combines MP and CG configurations.
- Bipolar (BP) stimulation routes current between two adjacent intracochlear electrodes, while BP+n refers to configurations where the stimulating electrodes are separated by n contacts.

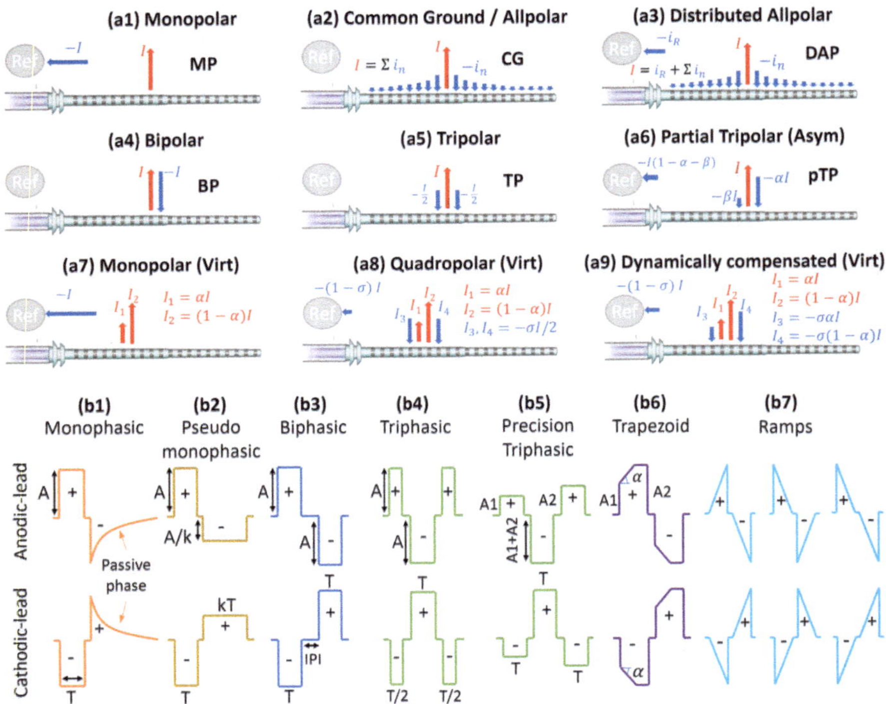

Fig. 1.5 (**a1–a9**) Various electrode stimulation configurations. (**b1–b7**) Various pulse shapes and waveforms used in CI stimulations

- Tripolar (TP) and TP+n configurations use two electrodes as return paths for a central active electrode, with the possibility that return current divided unevenly.
- Partial tripolar (pTP) includes an external reference electrode that shares part of the return current.
- Virtual electrode configurations use more than one active electrode (usually two) to steer the current in a specific location in the cochlea.

Each stimulation configuration produces a distinct current spread, resulting in a unique pattern of neural excitation. Studies have shown that tripolar (TP), bipolar (BP), and monopolar (MP) configurations generate progressively broader spreads of excitation (SOE), with TP being the most focused and MP the most diffuse (Spitzer and Hughes 2017). Spatial focusing of stimulation can be achieved using asymmetric or partial configurations to direct current toward cochlear regions beyond the reach of electrode contacts (Wu and Luo 2013). Another approach known as current steering involves simultaneous stimulation of multiple electrodes, where current superposition creates a new distribution pattern, potentially shifting the peak excitation to a location between the electrodes (Koch et al. 2007; Nogueira et al. 2009; Wu and Luo 2013; Luo et al. 2021).

Research indicates that the choice of stimulation mode in CIs should be individualized based on the recipient's neural health. Focused stimulation is not universally beneficial across all users or electrode locations. For individuals with auditory nerve fiber (ANF) degeneration, focused stimulation often requires higher current levels to activate damaged regions. This increased current can inadvertently stimulate adjacent cochlear areas, leading to mismatched or overlapping frequency-to-place maps (Zhu et al. 2012). Optical stimulation and optogenetics also offer promising technologies for achieving focused neural activation; refer to Chap. 15 for more details on this emerging approach.

Recent computational simulations using anatomically precise cochlear and ANF models suggest that different stimulation configurations (e.g., MP and BP) may be optimal for different cochlear regions within the same patient to achieve both selective activation and efficient thresholding (Cakmak et al. 2023). Refer to Chaps. 11 and 12 for more details about physical and biophysical computational modeling.

Stimulation pulse shapes are typically categorized by the number of active phases: monophasic (one phase), biphasic (two phases), and triphasic (three phases). Each phase can be either anodic or cathodic, indicating the direction of current flow from the active to the reference electrode(s) (see Fig. 1.5b1–b7). Furthermore, each phase may exhibit different waveform profiles, such as rectangular (most common), ramped (Navntoft et al. 2020), or trapezoidal (Ballestero et al. 2015). The overall polarity of a pulse is typically characterized by the polarity of its first phase (e.g., anodic-leading pulse). In the case of triphasic pulses, the polarity of the dominant phase—usually the second phase—may be used. It has been shown that both the polarity and shape of biphasic pulses significantly influence neuronal activation as measured by electrically evoked compound action potential (ECAP) (Herrmann et al. 2021).

Charge-balanced biphasic or triphasic configuration—where anodic and cathodic phases carry equal charge—provides greater assurance of preventing charge accumulation at the tissue compared to monophasic configurations, thereby reducing the risk of tissue damage. However, offset predictions for biphasic charge-balanced stimulation indicate that the voltage offset is not necessarily zero due to the capacitive properties of the electrode-tissue interface (Woods et al. 2011). Therefore, accounting for passive discharge mechanisms is considered good practice.

Today, most of CI manufacturers (e.g., Cochlear Ltd., MED-EL, and Advanced Bionics) use cathodic-leading symmetrical biphasic stimulations. Oticon Medical has used anodic-leading monophasic pulses. More recent research and developments have included asymmetric pseudo-monophasic or triphasic pulses aimed at enhancing spatial selectivity and reducing channel interaction. The choice of pulse shape and polarity affects stimulation efficiency, power consumption, and spatial selectivity and is often aligned with each manufacturer's design philosophy.

Optimizing stimulation effectiveness while minimizing energy consumption and side effects—such as facial nerve stimulation (FNS)—remains a key research challenge. For improved hearing outcomes and pitch perception, researchers aim to achieve sharp spatial selectivity by minimizing SOE along the cochlea through optimized stimulation polarity and waveform. SOE can be assessed using distributed

ECAP mapping (Rader et al. 2023), in vivo recordings (Quass and Kral 2024), and psychoacoustic measurements (Zhu et al. 2012) (refer to Chap. 7 for more information on ECAP and SOE).

It is important to note that findings from animal studies on stimulation pulses should be cautiously extrapolated to humans. While cathodic stimulation is generally more effective in animals, human studies report greater auditory nerve activation with anodic stimulation (Skidmore et al. 2022). A clinical study by Spitzer and Hughes (2017) found that anodic-leading symmetric biphasic MP stimulation produced ECAPs with larger average amplitudes—indicating stronger and more synchronous neural activation—and slightly broader SOE compared to cathodic-leading pulses. This effect is likely due to preferential stimulation of the central ANF axon by anodic-leading pulses.

Although most commercial CI devices currently use rectangular waveforms, ongoing research aims to optimize waveform design for improved neural selectivity and reduced power consumption. Ramped waveforms have gained attention for their alignment with the biophysical properties of spiral ganglion neurons (SGNs). In vitro studies using anodic-leading biphasic pulses suggest that ramped waveforms produce narrower SOE profiles than rectangular ones (Ballestero et al. 2015), offering better control over dynamic range both near and far from the stimulation site. However, animal studies by Navntoft et al. (2020) did not confirm improved dynamic range, instead reported steeper growth function slopes (measured via eABR wave II) for ramped waveforms. Further investigation is needed to fully understand the implications of ramped pulses on SOE and dynamic range. Finally, animal studies have shown that ramped waveforms require less charge to elicit neural excitation, as measured in both the brainstem and auditory cortex (Navntoft et al. 2020; Partouche et al. 2023), suggesting potential for reduced battery consumption in future CI designs.

To elicit different loudness perceptions, most CI manufacturers—including Cochlear Ltd., MED-EL, Advanced Bionics, and Nurotron—modulate the current amplitude of the stimulation pulse, a method known as pulse amplitude (PA) modulation. In contrast, Oticon Medical devices modulate the pulse duration (PD) per phase to achieve variations in loudness. A hybrid approach combining both PA and PD modulation is also conceptually feasible.

Research has shown that PA and PD modulations affect loudness growth and dynamic range differently. Specifically, PD modulation yields a significantly larger dynamic range—approximately 1.4 to 1.7 times greater—compared to PA modulation (Zhou et al. 2020; Saravanan et al. 2024). However, loudness growth functions are generally more consistent across patients with PA modulation than with PD. While both methods produce similar loudness growth at low sound levels, PD requires more charge than PA at medium and high levels. Moreover, patients with longer durations of deafness (i.e., extended auditory deprivation) tend to exhibit slower loudness growth with PD, making it a less charge-efficient strategy for this population.

1.5 Cochlear Implant Healthcare Ecosystem

Hearing loss is a chronic condition that affects millions of individuals worldwide, significantly impacting their quality of life. It is not merely a medical issue but a social and emotional one as well, necessitating a comprehensive approach to care. Because of its chronic nature, the importance of patient centricity in managing hearing loss cannot be overstated. Patient-centric care models prioritize the needs, preferences, and values of individuals, ensuring that they are active participants in their own healthcare journey (Gotowiec et al. 2022). This approach leads to better adherence to treatment plans, improved outcomes, and higher patient satisfaction.

A recent innovation in medicine and treatment of chronic conditions is the Internet of Medical Things (IoMT). It refers to a connected infrastructure of medical devices, software applications, and health systems and services (Huang et al. 2023). This network leverages advancements in wireless technology, miniaturization, and computing power to create an ecosystem where medical devices can generate, collect, analyze, and send data. Essentially, IoMT integrates medical devices with digital technology, enabling them to communicate with each other and with external systems in real time.

The advantages of IoMT are transformative for health care. First, it enhances patient care by enabling continuous monitoring and real-time data collection, which allows for prompt interventions and personalized treatment plans.

Second, IoMT improves the efficiency of healthcare delivery. By automating data collection and analysis, it reduces the workload on healthcare professionals, allowing them to focus more on patient care rather than administrative tasks. This is particularly beneficial in managing chronic conditions like hearing loss, where continuous monitoring and regular adjustments are crucial.

Moreover, IoMT facilitates remote care, reducing the need for frequent clinic visits. Patients can receive care and support from the comfort of their homes, which is especially helpful for those with mobility issues or those living in remote areas. This not only enhances patient convenience but also reduces healthcare costs by minimizing the need for in-person consultations and hospital stays.

These arguments also apply to hearing implants as a solution for patients with a moderately severe to profound hearing loss. A first main innovation driver is further improving hearing outcomes in various daily situations (Wilson and Dorman 2008; Zeng 2022). A second driver is creating more consistency and efficiencies in the aftercare, reducing clinician workload and reducing patient's burden (Nassiri et al. 2022). Both require that the system architecture of cochlear implants evolves. The overall system consists of the user subsystem and the clinician subsystem, where the main actors are the surgeon and the audiologist.

1.5.1 The User Subsystem

The user subsystem comprises the implant, sound processor, and hardware and digital accessories. Its architecture and functionality were discussed earlier in this chapter. It serves as the primary interface through which users experience restored hearing. This subsystem encompasses the entire signal pathway—from sound acquisition via external microphones to the delivery of electrical stimulation through the electrode array. It is directly responsible for compensating for lost auditory function and plays a central role in hearing restoration. Additionally, it manages both manual and automatic adjustments to optimize performance and listening comfort, adapting to the user's environment and individual hearing needs.

1.5.2 The Clinician Subsystem

The hearing journey is intense before and after the moment implantation. Candidates need to come to the clinic prior to surgery for various assessments, and once implanted, they have multiple sessions with the audiologist in the first year (Vaerenberg et al. 2014). In these sessions, the device configuration is fine-tuned (device fitting) and the users receive counseling and rehabilitation (refer to Chap. 3). Their hearing performance is also assessed through tonal and speech audiometry. Since hearing loss is a chronic disease, users need life-long support, traditionally consisting of an annual follow-up. The hearing healthcare professions involved therefore lead busy professional lives, taking care of new patients while monitoring an ever-growing number of existing patients.

1.5.3 The Surgical Subsystem

Surgeons have only one opportunity to optimally place the electrode, creating the best possible electrode-neural interface. Cochlear anatomies vary significantly, and during electrode insertion, surgeons are essentially blind to the behavior of the electrode tip within the scala tympani. Traditionally, they have relied on mechanical surgical accessories, such as device templates, to guide the procedure. At the end of surgery, the position of the electrode array and the responsiveness of the auditory nerve could be partially verified using objective measures like impedance testing and neural responses, including evoked compound action potentials.

Today, advancements in precision diagnostics and digital medical decision support tools offer enhanced capabilities. These include improved surgical planning, real-time monitoring of electrode insertion, and comprehensive post-operative verification, marking a significant evolution in cochlear implant surgery.

A first novel precision technique, involving AI-based image technologies, is automatic segmentation and analysis of the micro-anatomy of the cochlea based on computer-analysis (using AI techniques) of a CT image of the pre-implanted cochlea (Zhao et al. 2018) (refer to Chap. 13). Given the computation and memory involved, it requires uploading the image stack to a cloud application. CT images are not taken systematically in all implantation centers. If they are, a detailed automatic analysis can help surgeons understand the unique anatomical features of each patient's cochlea, allowing them to select the most suitable electrode and find the best insertion trajectory. These functionalities are still in research phase.

Objective measures, taken through the implant during implantation, are a second enabling technology. Insertion monitoring during the implantation provides real-time feedback to the surgeon about the behavior of the electrode inside the cochlear lumen. They require that the implant is powered, capable of recording intracochlear voltages. The intracochlear voltages induced by stimulation are large, on the order of volts. Typically, they are reported as impedances, which are normalized voltages. With enough pre-amplification, the implant is also capable of capturing the electrically induced compound action potentials (ECAPs), the biological response to the stimulation. These signals are on the order of 1–100 μV. Several objective measure features are interesting during implantation. Impedances come down as soon as an electrode contact is immersed in the perilymph. A generalized set of impedances, called transimpedances, allow real-time detection of anomalies, such as a tip folding, or real-time estimation of the insertion angle. For residual hearing cases, monitoring the outer hair cell function through electrocochleography (eCochG) informs the surgeon that the electrode may touch the basilar membrane or cause other forms of damage to the delicate inner ear tissues (Campbell et al. 2015) (refer to Chap. 6).

Finally, once the electrode is in its final position, surgical verification can ensure that the electrode is fully functional and well placed. This involves checking the clinical impedances, which helps verify the integrity of the electrode contacts. Evoked compound action potential measurements provide the confidence that all contacts are well placed and eliciting a response in the auditory nerve. Additionally, anomaly detection techniques are employed to find any issues that may have arisen during the surgery, allowing for immediate correction if needed (Leblans et al. 2023).

1.5.4 The Audiological Subsystem

Multiple hearing healthcare professionals take part in the aftercare. The audiologists are the primary hearing coaches of the user. Their main task is to configure the device settings, typically called the fitting process. They also evaluate the user's hearing performance. Speech and language therapists may be available to help the user in getting used to and make the most of their new sound.

The aftercare process iterates between device fitting and hearing assessments, such as tonal audiometry and speech audiometry in quiet and in noise. The latter guide the device fitting process. It involves selecting a stimulation strategy and its parameters, such as setting the stimulation rate, frequency allocation, stimulation levels, and speech enhancement parameters. The task that requires most attention in the early aftercare phase is setting the electrical stimulation window (de Quillettes et al. 2024). For every electrode or stimulation channel, a lower and upper stimulation level must be set such that the user has full access to a normal loudness range, from soft to loud. The stimulation levels are different in every implanted ear since they depend on the exact electrode position and the health status of the auditory nerve. A gradual buildup is needed for the brain to adapt to the new sound, a process called acclimatization. Therefore, the fitting process consists of a series of ±4 sessions, spread over the first 3 months.

In every session, the audiologist uses fitting software, switching between psychoacoustic and live speech mode. In psychoacoustic mode, direct electrical stimulation is used. Every channel is individually activated with a pulse train of variable amplitude. Its amplitude is varied to find the lower and upper stimulation levels. Afterward, the audiologist switches to live mode, in which the microphone signal is processed, and all channels are active, making further adjustments. Many variations on this basic process exist, including profile-based fitting methods.

This fitting process is called a behavioral one, as the user provides feedback on the perceived loudness and sound quality. When this is not possible, e.g., with congenitally deaf children, objective measures provide an alternative approach. Therefore, fitting software holds an objective measures module. Impedance measurements allow verifying the integrity of the implanted electrode array. To guide stimulation levels, the most common method is the neural response (ECAP) method. The electrically evoked stapedius reflex threshold (eSRT) and brainstem response (eABR) methods can also be used. They are somewhat more complicated, as they require additional equipment (refer to Chaps. 7 and 8 for further insights into ECAP and eABR in CI patients).

Until a few years ago, fitting functionality was typically implemented as a professional desktop software application, only available to the audiologist in the clinic. Users had to travel to the clinic to adjust their hearing. This obviously lacks patient centricity. The pandemic accelerated an evolution toward remote fitting and self-fitting (Ramos et al. 2009). In remote fitting, the audiologist is in the office and the user at home. They are connected through cloud and mobile technologies. The audiologist can adjust the settings remotely and the user can perform hearing assessments through audio streaming. To further allow users to self-tune their sound, manufacturers start to provide mobile app functionality to further self-tune settings, within certain limits.

1.6 Future of the Cochlear Implant Systems

Looking ahead to the coming decade, significant changes can be expected to cochlear implant systems. Surfing on general trends, such as artificial intelligence and personalized medicine, all components are expected to become much smarter, driven by further evolutions in computational power, artificial intelligence, and implantable hardware.

1.6.1 The User Subsystem

The terminology "conventional cochlear implant" (COCI) refers to the current architecture consisting of a sound processor and an implant. The implant is designed to last for a minimum of 75 years. With current electronic platforms, during this period the implant functionality is fixed, in essence restricted to processing stimulation and recording commands. Hearing performance and other functional enhancements come through upgrades of the sound processor firmware. In the last decade, users have benefited from better noise reduction and connectivity to smartphone connectivity. This evolution is expected to continue and accelerate. Constantly, sound processors are integrating new technologies.

In the world of artificial intelligence, deep neural networks (DNNs) are revolutionizing speech processing by significantly improving the accuracy and efficiency of tasks such as speech recognition, synthesis, and translation (Mehrish et al. 2023) (refer to Chaps. 2, 10, 13 for different applications of artificial intelligence in CI). They excel at capturing complex patterns in audio data. Where classic signal processing struggles in a cocktail party setting, DNNs can analyze a complex acoustic scene and extract a target signal from a loud and nonstationary background. The downside is the complexity. Functionality scales with size, memory, and computation. The most challenging applications, such as quasi real-time automatic recognition of conversational speech or natural speech synthesis, require computing resources only available on cloud computing platforms. Within the constraints of a desktop or smartphone platform, significant speech enhancement is still possible. The further migration of DNN functionality into a hearing aid or sound processor device is called artificial intelligence on the edge (edge AI). Chip manufacturers start to provide low-power, dedicated neural network chips with specialized instruction sets and large and optimized memory modules, compatible with the constraints of a battery-powered ear-level device. In 2024, the first hearing aid with a powerful neural network chip was launched. It is expected that future sound processors will also integrate this functionality, given that speech understanding in background noise remains the number one need of CI users.

A second evolution is the trend to smart implants, containing firmware. Like sound processors, they have powerful digital signal processing and memory chips, in addition to the standard stimulation and recording circuitry. Because of the firmware, their functionality is upgradable across the implant lifetime. The computational load can be shared between the sound processor and the implant, e.g., to minimize the data rate over the RF link. It is possible to only partially process the audio, send the data to the implant, and finalize the calculation of the stimulation instructions in the implant. In the backward direction, objective measures can be preprocessed in the implant, e.g., averaging, before sending them back to the sound processor.

Smart implants are not fully autonomous CI systems. They do not have a microphone, and they rely on external power. Subcutaneous microphones, capturing sound under the skin, have been researched for decades. Their sensitivity is somewhat worse than external microphones, and they are susceptible to body noise. Implantable batteries are also known, e.g., from the cardiology field. When adding a microphone and a battery to a smart implant, a totally implantable cochlear implant (TICI) device is obtained (Briggs et al. 2008; Lefebvre et al. 2025). TICI users have 24/7 hearing and can listen in invisible mode, not needing an external sound processor. This device aims to provide users with a truly seamless hearing restoration experience while significantly reducing the social stigma often faced by cochlear implant users (Warner-Czyz et al. 2018; Feijóo et al. 2021).

Finally, major advances are also expected for binaural hearing systems, either consisting of two implants (bilateral CI) or one CI combined with a contralateral hearing aid (bimodal). Wireless parametric links between the two devices can coordinate the setting of the sound processing path, resulting in improved spatial hearing. And audio links will allow the better signal to be sent to the best ear.

1.6.2 The Surgical Subsystem

The electrode-neural interface is the key to better hearing outcomes. Electrode array technology is expected to evolve significantly over the next decade. Further mechanical improvements to the shape and materials are expected to allow for better and consistent positioning, better preservation of residual hearing, and easier insertion. A key evolution is drug elution. Research has demonstrated the benefits of drug elution (Fleet et al. 2024). Molecules can be embedded in the silicone surface of the electrode array, augmenting its role from a static carrier to an active therapeutic element. Chronic elution of dexamethasone is very promising for reducing the inflammatory process, encapsulating the electrode in tissue sheath, reducing impedances, and preserving low-frequency residual hearing. Other molecules may follow.

In terms of surgical techniques, in many medical disciplines, robotic surgery is often considered the ultimate approach for accuracy, precision, and minimizing trauma (refer to Chap. 14). In CI surgery, more than the drilling of the mastoid and facial recess, the insertion of the electrode is considered the most critical part, especially in residual hearing cases or complex anatomies. The insertion depth, insertion angle, and trajectory must be carefully controlled, and insertion forces must be minimized. Whether a robot arm or a human surgeon is inserting the electrode, the underlying challenges are the same.

To plan the insertion trajectory, the specific anatomy must be analyzed with great care. It is expected that the current CT image analysis techniques will evolve into powerful AI-based planning tools, simulating various electrode types and trajectories, to select the optimal one with the best chance for an optimal electrode-neural interface while minimizing the risk for cochlear tissue trauma and residual hearing.

The precision of the insertion monitoring tools will also continue to evolve toward tighter, more precise control of the actual versus the intended trajectory. Augmented reality techniques, overlaying the intended trajectory with the actual surgical microscope image, may inform surgeons better. Further development of the interpretation algorithms acting on eCochG and other objective measures will lead to better guidance (Kashani et al. 2024). As an example, eCochG still lacks specificity. Multiple phenomena, including the electrode touching the basilar membrane, can cause a drop of the eCochG signal. Research is ongoing to better discriminate between problematic and non-problematic eCochG amplitude drops (Lenarz et al. 2022; Dalbert et al. 2025) (refer to Chap. 6).

Overall, with these technological advancements, surgeons will be better informed. They will be more capable of consistently placing the electrode in the best position across a variety of cochlear shapes. Robotic insertion may be enabled as well.

1.6.3 The Audiological Subsystem

Audiologists face challenges to consistently fit well the device parameters for each individual implanted ear. Evidence-based prescription rules for setting stimulation parameters are available for hearing aids, but not yet for hearing implants. This is hampered by lack of precise knowledge of the relevant aspects of the electrode-neural interface. Most implanting centers do not perform a post-operative CT. Therefore, the exact intracochlear position of the electrode array is not known. The second factor is the neural health, the nerve's capacity to respond to the electrical stimulation and transfer the information to the auditory brainstem. The neural health may vary along the implanted ear, and this is not well understood. Current ECAP techniques are not precise enough to assess neural health in all its aspects. Research on objective measures is ongoing to further unveil the nerve's excitability for each stimulation channel (Dawson et al. 2025; Garcia et al. 2025; Konerding et al. 2025). If clinical precision diagnostics can be developed and collected in large numbers, objective-measured based prescription rules may be developed, resulting in more consistent and better hearing outcomes.

The implanted cochlea is just the periphery, the input into the central auditory pathways. If an implant would be able to sense the brain response to electrical stimulation, it may self-adjust in an autonomous fashion. The theoretical concept of such a closed-loop cochlear implant has been proposed in the literature (Mc Laughlin et al. 2012). It would require that brain responses are captured, either through sensing electrodes placed on the temporal lobe or through the intracochlear electrode array. The basic feasibility of capturing brain potentials through the electrode using external equipment has already been demonstrated (Somers et al. 2021). The presence of large artifacts requires nontrivial signal processing to extract the tiny brain potentials. This is a fascinating vision, but the technology development to record successfully through an implant system pushes this approach out to longer time horizons.

In the near term, the evolution of a smarter and more distributed care model will continue. In-clinic visits to the audiologist will be intertwined with remote sessions. To drive efficiency and consistency, smarter software tools will be available to aid professionals in their daily hearing care decisions. In nonstandard cases, device fitting may be complex, with many factors to consider, requiring extensive experience. Like in other parts of medicine, it is expected that future AI technologies can help audiologists. The machine learning techniques are known. The development of powerful AI tools is essentially limited by the availability of large amounts of the relevant data. With CI devices being more connected, the data pool will grow. Data logging is part of CI sound processors already, monitoring device health, sound environments and usage patterns. Continuous impedance monitoring is being researched (Mushtaq et al. 2024; Di Lella et al. 2020). Once these data pools are anonymized and aggregated over many CI users, machine learning techniques can be used to uncover complex relations and offer new insights (refer to Chap. 10 for the application of machine learning in CI fitting and performance estimation). This area of computational audiology is expected to thrive further (Mellor et al. 2018).

The same toolset can empower users to perform selected functionalities in a safe manner, such as fine-tuning their settings themselves. In the rehabilitation space, CI users will receive help from powerful digital tools based on the recent advances in advanced real-time speech recognition, natural computer voices, large language model-based dialogue engines, and video.

1.7 Conclusion

In conclusion, the history of cochlear implant development is an inspiring example of how technology can emerge from parallel innovations around the world. Furthermore, the evolution of cochlear implant systems is poised to undergo significant transformations in the coming decade. With advancements in artificial intelligence, personalized medicine, and implantable hardware, these systems will become increasingly smarter and more efficient. The integration of deep neural networks and edge AI will enhance speech processing capabilities, allowing users to hear better in challenging environments. The development of smart implants and the potential for totally implantable cochlear implants (TICI) will further revolutionize the user experience, offering greater autonomy and convenience.

In the surgical realm, the adoption of robotic surgery and AI-based planning tools will enhance the precision and safety of cochlear implant procedures, ensuring optimal electrode placement and minimizing trauma. Audiologists will benefit from smarter software tools and AI-driven insights, enabling more consistent and personalized device fittings. The emergence of computational audiology will empower both professionals and users, providing valuable data and tools for better hearing outcomes.

Overall, as already stated by Wilson and Dorman (2008), cochlear implants have a remarkable past and a brilliant future. Having already passed the threshold of one million users (Zeng 2022), current and future users can expect better real-world hearing outcomes, more personalization, and more convenience living with their devices. Hearing healthcare professionals and healthcare providers can expect to be able to provide higher quality and consistency of care, with less effort per user, therefore unlocking the capacity to serve more people. As technology continues to advance, the journey of cochlear implant systems will remain a testament to the transformative power of innovation in enhancing human capabilities.

Acknowledgments The authors would like to thank Robert Shannon, John Galvin III, Mario Svirsky, Stefaan Peeters, Waldo Nogueira, and Enrique Lopez Poveda for their willingness to share their professional experiences.

The authors acknowledge the use of Microsoft Copilot for proofreading and enhancing the readability of the manuscript. All suggested changes were carefully reviewed and edited. The authors assume full responsibility for the content of the final publication.

References

Ballestero J, Recugnat M, Laudanski J et al (2015) Reducing current spread by use of a novel pulse shape for electrical stimulation of the auditory nerve. Trends Hear 19. https://doi.org/10.1177/2331216515619763

Blume SS (1999) Histories of cochlear implantation. Soc Sci Med 49:1257–1268. https://doi.org/10.1016/S0277-9536(99)00164-1

Briggs RJS, Eder HC, Seligman PM et al (2008) Initial clinical experience with a totally implantable cochlear implant research device. Otol Neurotol 29:114–119. https://doi.org/10.1097/MAO.0b013e31814b242f

Cakmak O, Pal S, Sahin M (2023) Improving the stimulation selectivity in the human cochlea by strategic selection of the current return electrode. Front Audiol Otol 1. https://doi.org/10.3389/fauot.2023.1259852

Campbell L, Kaicer A, Briggs R, O'Leary S (2015) Cochlear response telemetry. Otol Neurotol 36: 399–405. https://doi.org/10.1097/MAO.0000000000000678

Chouard CH (2015) The early days of the multi channel cochlear implant: efforts and achievement in France. Hear Res 322:47–51

Chung K (2004) Challenges and recent developments in hearing aids: part I. Speech understanding in noise, microphone technologies and noise reduction algorithms. Trends Amplif 8:83–124. https://doi.org/10.1177/108471380400800302

Clark G (2003) Cochlear implants: fundamentals and applications. In: Cochlear implants. Springer, New York, pp 58–99

Dalbert A, Bester C, Collins A et al (2025) Amplitude and phase changes in electrocochleographic real-time recordings during cochlear implantation and its relation to pre- and postoperative hearing. Otol Neurotol 46:e65–e73. https://doi.org/10.1097/MAO.0000000000004420

Dawson P, Fullerton A, Krishnamoorthi H et al (2025) A prospective, multicentre case-control trial examining factors that explain variable clinical performance in post lingual adult CI recipients. Trends Hear 29. https://doi.org/10.1177/23312165251347138

de Quillettes R, Kaandorp M, Merkus P et al (2024) Experienced adult cochlear implant users show improved speech recognition when target fitting parameters are applied. Ear Hear 45:1264–1273. https://doi.org/10.1097/AUD.0000000000001513

Dessard L, Gersdorff G, Ivanovik N et al (2024) Cochlear implant: analysis of the frequency-to-place mismatch with the table-based software OTOPLAN® and its influence on hearing performance. Audiol Neurotol 29:239–245. https://doi.org/10.1159/000535693

Di Lella FA, Parreño, M, Fernandez F, Boccio CM, Ausili SA (2020) Measuring the electrical status of the bionic ear. re-thinking the impedance in cochlear implants. Frontiers in Bioengineering and Biotechnology 8:568690

Djourno A, Eyries C (1957) Prothese auditive par excitation electrique a distance du nerf sensoriel a laide dun bobinage inclus a demeure. Presse Med 65:1417–1417

Djourno A, Kayser D (1958) Perspectives nouvelles en matière de prothèse sensorielle par action directe sur les fibres ou les centres nerveux. C R Seances Soc Biol Fil 152:1433–1434

Dorman MF, Parkin JL (2015) The role of the Utah Artificial Ear project in the development of the modern cochlear implant. Cochlear Implants Int 16:S1–S11. https://doi.org/10.1179/1467010015Z.000000000246

Dornhoffer JR, Lawlor SK, Saoji AA, Driscoll CLW (2023) Initial experiences with the envoy Acclaim® fully implanted cochlear implant. J Clin Med 12. https://doi.org/10.3390/jcm12185875

Eddington DK, Dobelle WH, Miladejovsky MG et al (1978) Auditory prostheses research with multiple channel intracochlear stimulation in man. Ann Otol Rhinol Laryngol 87(6 Pt 2):1–39

Eisenberg LS (2015) The contributions of William F. House to the field of implantable auditory devices. Hear Res 322:52–56. https://doi.org/10.1016/j.heares.2014.08.003

Eshraghi AA, Nazarian R, Telischi FF et al (2012) The cochlear implant: historical aspects and future prospects. Anat Rec 295:1967–1980. https://doi.org/10.1002/ar.22580

Fitzpatrick EM, Séguin C, Schramm DR, Armstrong S, Chénier J (2009) The benefits of remote microphone technology for adults with cochlear implants. Ear and hearing 30(5):590–599

Feijóo S, Foody M, Pichel R et al (2021) Bullying and cyberbullying among students with cochlear implants. J Deaf Stud Deaf Educ 26:130–141. https://doi.org/10.1093/deafed/enaa029

Fleet A, Nikookam Y, Radotra A et al (2024) Outcomes following cochlear implantation with eluting electrodes: a systematic review. Laryngoscope Investig Otolaryngol 9. https://doi.org/10.1002/lio2.1263

Garcia C, Sismono F, Goehring T et al (2025) A comparison of electrophysiological measures for characterizing the cochlear-implant electrode-neuron interface. JASA Express Lett 5. https://doi.org/10.1121/10.0038746

Gotowiec S, Larsson J, Incerti P et al (2022) Understanding patient empowerment along the hearing health journey. Int J Audiol 61:148–158. https://doi.org/10.1080/14992027.2021.1915509

Guignard J, Senn P, Koller R et al (2019) Mobile internet telephony improves speech intelligibility and quality for cochlear implant recipients. Otol Neurotol 40:E206–E214. https://doi.org/10.1097/MAO.0000000000002132

Herrmann DP, Kretzer KVA, Pieper SH, Bahmer A (2021) Effects of electrical pulse polarity shape on intra cochlear neural responses in humans: triphasic pulses with anodic and cathodic second phase. Hear Res 412. https://doi.org/10.1016/j.heares.2021.108375

Hochmair I (2013) The importance of being flexible. Nat Med 19:1240–1244. https://doi.org/10.1038/nm.3341

Hochmair IJ, Hochmair ES (1979) Multi-frequency system and method for enhancing auditory stimulation and the like. US Patent US4284856A

Hochmair ES, Hochmair-Desoyer IJ (1983) Percepts elicited by different speech-coding strategies. Ann N Y Acad Sci 405:268–279. https://doi.org/10.1111/j.1749-6632.1983.tb31640.x

House WF, Urban J (1973) Long term results of electrode implantation and electronic stimulation of the cochlea in man. Ann Otol Rhinol Laryngol 82:504–517. https://doi.org/10.1177/000348947308200408

House W (1985) A personal perspective on cochlear implants. In: Cochlear implants. Raven Press, New York, pp 13–16

Huang C, Wang J, Wang S, Zhang Y (2023) Internet of medical things: a systematic review. Neurocomputing 557:126719. https://doi.org/10.1016/j.neucom.2023.126719

Kashani RG, Kocharyan A, Bennion DM et al (2024) Combining intraoperative electrocochleography with robotics-assisted electrode array insertion. Otol Neurotol 45:143–149. https://doi.org/10.1097/MAO.0000000000004094

Kiefer J, Hohl S, Stürzebecher E et al (2001) Comparison of speech recognition with different speech coding strategies (SPEAK, CIS, and ACE) and their relationship to telemetric measures of compound action potentials in the nucleus CI 24M cochlear implant system. Int J Audiol 40. https://doi.org/10.3109/00206090109073098

Koch DB, Downing M, Osberger MJ et al (2007) Using current steering to increase spectral resolution in CII and HiRes 90K users. Ear Hear 28. https://doi.org/10.1097/AUD.0b013e31803150de

Konerding W, Arenberg J, Kral A, Baumhoff P (2025) Cochlear health alters the polarity effect and spike-initiation sites in guinea pigs. Hear Res 465. https://doi.org/10.1016/j.heares.2025.109341

Langner F, Büchner A, Nogueira W (2020) Evaluation of an adaptive dynamic compensation system in cochlear implant listeners. Trends Hear 24. https://doi.org/10.1177/2331216520970349

Lazeyras F, Boëx C, Sigrist A et al (2002) Functional MRI of auditory cortex activated by multisite electrical stimulation of the cochlea. NeuroImage 17:1010–1017. https://doi.org/10.1006/nimg.2002.1240

Leblans M, Zarowski A, Molisz A et al (2023) Cochlear implant electrode array tip-fold over detection by electrode voltage telemetry. Cochlear Implants Int 24:95–106. https://doi.org/10.1080/14670100.2022.2148890

Lefebvre PP, Müller J, Mark G et al (2025) Rehabilitation of human hearing with a totally implantable cochlear implant: a feasibility study. Commun Med 5:10. https://doi.org/10.1038/s43856-024-00719-0

Lenarz T, Buechner A, Gantz B et al (2022) Relationship between intraoperative electrocochleography and hearing preservation. Otol Neurotol 43. https://doi.org/10.1097/MAO.0000000000003403

Loizou PC (1998) Mimicking the human ear. IEEE Signal Process Mag 15:101–130. https://doi.org/10.1109/79.708543

Luo X, Wu CC, Pulling K (2021) Combining current focusing and steering in a cochlear implant processing strategy. Int J Audiol 60. https://doi.org/10.1080/14992027.2020.1822551

Magnusson L (2011) Comparison of the fine structure processing (FSP) strategy and the CIS strategy used in the MED-EL cochlear implant system: speech intelligibility and music sound quality. Int J Audiol 50:279–287. https://doi.org/10.3109/14992027.2010.537378

Mc Laughlin M, Lu T, Dimitrijevic A, Zeng F-G (2012) Towards a closed-loop cochlear implant system: application of embedded monitoring of peripheral and central neural activity. IEEE Trans Neural Syst Rehabil Eng 20:443–454. https://doi.org/10.1109/TNSRE.2012 2186982

Mehrish A, Majumder N, Bharadwaj R et al (2023) A review of deep learning techniques for speech processing. Inf Fusion 99:101869. https://doi.org/10.1016/j.inffus.2023.101869

Mellor JC, Stone MA, Keane J (2018) Application of data mining to "big data" acquired in audiology: principles and potential. Trends Hear 22. https://doi.org/10.1177/2331216518776817

Merzenich MM, White MW (1977) Cochlear implant: the interface problem. In: Hambrecht FT, Reswick BR (eds) Biomedical engineering and instrumentation vol. 3: functional electrical stimulation: applications in neural prostheses. Eborn Books

Molaee-Ardekani B, Attili Chiea R, Zhang Y et al (2025) A novel sound coding strategy for cochlear implants based on spectral feature and temporal event extraction. Technologies (Basel) 13:318. https://doi.org/10.3390/technologies13080318

Müller J, Brill S, Hagen R et al (2012) Clinical trial results with the MED-EL fine structure processing coding strategy in experienced cochlear implant users. ORL 74:185–198. https://doi.org/10.1159/000337089

Mushtaq F, Soulby A, Boyle P et al (2024) Self-assessment of cochlear health by new cochlear implant recipients: daily impedance, electrically evoked compound action potential and electrocochleography measurements over the first three postoperative months. Otol Neurotol 45:e517–e524. https://doi.org/10.1097/MAO.0000000000004227

Nassiri AM, Saoji AA, DeJong MD et al (2022) Implementation strategy for highly-coordinated cochlear implant care with remote programming: the complete cochlear implant care model. Otol Neurotol 43:e916–e923. https://doi.org/10.1097/MAO.0000000000003644

Navntoft CA, Marozeau J, Barkat TR (2020) Ramped pulse shapes are more efficient for cochlear implant stimulation in an animal model. Sci Rep 10:3288. https://doi.org/10.1038/s41598-020-60181-5

Nogueira W, Litvak L, Edler B et al (2009) Signal processing strategies for cochlear implants using current steering. EURASIP J Adv Signal Process 2009. https://doi.org/10.1155/2009/531213

Partouche E, Adenis V, Stahl P et al (2023) What is the benefit of ramped pulse shapes for activating auditory cortex neurons? An electrophysiological study in an animal model of cochlear implant. Brain Sci 13:250. https://doi.org/10.3390/brainsci13020250

Plyler PN, Wade M, Dimlich A et al (2025) Effectiveness of television streaming on cochlear implant user performance and satisfaction in quiet and in noise. Am J Audiol 34:270–280. https://doi.org/10.1044/2024_AJA-24-00204

Quass GL, Kral A (2024) Tripolar configuration and pulse shape in cochlear implants reduce channel interactions in the temporal domain. Hear Res 443:108953. https://doi.org/10.1016/j.heares.2024.108953

Rader T, Nachtigäller P, Linke T et al (2023) Exponential fitting of spread of excitation response measurements in cochlear implants. J Neurosci Methods 391:109854. https://doi.org/10.1016/j.jneumeth.2023.109854

Ramos A, Rodríguez C, Martinez-Beneyto P et al (2009) Use of telemedicine in the remote programming of cochlear implants. Acta Otolaryngol 129:533–540. https://doi.org/10.1080/00016480802294369

Ricard C, Chouard CH, Macleod P (1977) Equipement pour prothese auditive. US patent 4,207,441

Riss D, Hamzavi J-S, Selberherr A et al (2011) Envelope versus fine structure speech coding strategy. Otol Neurotol 32:1094–1101. https://doi.org/10.1097/MAO.0b013e31822a97f4

Riss D, Hamzavi JS, Blineder M et al (2014) FS4, FS4-p, and FSP: a 4-month crossover study of 3 fine structure sound-coding strategies. Ear Hear 35. https://doi.org/10.1097/AUD.0000000000000063

Riss D, Hamzavi J-S, Blineder M et al (2016) Effects of stimulation rate with the FS4 and HDCIS coding strategies in cochlear implant recipients. Otol Neurotol 37:882–888. https://doi.org/10.1097/MAO.0000000000001107

Saravanan P, Antony PJ, Megha et al (2024) A new objective method to estimate the charge integration efficiency in cochlear implant users. Int J Audiol:1–8. https://doi.org/10.1080/14992027.2024.2417243

Seitz PR (2002) French origins of the cochlear implant. Cochlear Implants Int 3:77–86

Sharma R, Shrivastava M (2024) Aura cast: a new milestone over the existing versions of bluetooth. In: 2nd international conference series on science, engineering, and technology (ICSSET) 2022, p 030004

Shin S, Ha Y, Choi G et al (2021) Manufacturable 32-channel cochlear electrode array and preliminary assessment of its feasibility for clinical use. Micromachines (Basel) 12:778. https://doi.org/10.3390/mi12070778

Simmons FB (1966) Electrical stimulation of the auditory nerve in man. Arch Otolaryngol Head Neck Surg 84:2–54. https://doi.org/10.1001/archotol.1966.00760030004003

Skidmore J, Ramekers D, Bruce IC, He S (2022) Comparison of response properties of the electrically stimulated auditory nerve reported in human listeners and in animal models. Hear Res 426:108643

Somers B, Long CJ, Francart T (2021) EEG-based diagnostics of the auditory system using cochlear implant electrodes as sensors. Sci Rep 11:5383. https://doi.org/10.1038/s41598-021-84829-y

Spitzer ER, Hughes ML (2017) Effect of stimulus polarity on physiological spread of excitation in cochlear implants. J Am Acad Audiol 28:786–798. https://doi.org/10.3766/jaaa.16144

Tahmasebi S, Segovia-Martinez M, Nogueira W (2023) Optimization of sound coding strategies to make singing music more accessible for cochlear implant users. Trends Hear 27. https://doi.org/10.1177/23312165221148022

Todt I, Guerkov R, Gehl HB, Sudhoff H (2020) Comparison of cochlear implant magnets and their MRI artifact size. Biomed Res Int 2020. https://doi.org/10.1155/2020/5086291

Tyler R, Lowder MW, Parkinson AJ et al (1995) Performance of adult Ineraid and nucleus cochlear implant patients after 3.5 years of use. Int J Audiol 34:135–144. https://doi.org/10.3109/00206099509071907

Vaerenberg B, Smits C, De Ceulaer G et al (2014) Cochlear implant programming: a global survey on the state of the art. Sci World J 2014:1–12. https://doi.org/10.1155/2014/501738

van Hoof M, Lambriks L, van der Heijden K et al (2025) Learning to hear again with alternating cochlear frequency allocations. Sci Rep 15:317. https://doi.org/10.1038/s41598-024-83047-6

Warner-Czyz AD, Loy B, Pourchot H et al (2018) Effect of hearing loss on peer victimization in school-age children. Except Child 84:280–297. https://doi.org/10.1177/0014402918754880

Wilson BS, Dorman MF (2008) Cochlear implants: a remarkable past and a brilliant future. Hear Res 242:3–21. https://doi.org/10.1016/j.heares.2008.06.005

Wilson BS, Finley CC, Weber BA et al (1988) Comparative studies of speech processing strategies for cochlear implants. Laryngoscope 98:1069–1077. https://doi.org/10.1288/00005537-198810000-00009

Wilson BS, Finley CC, Lawson DT et al (1991) Better speech recognition with cochlear implants. Nature 352:236–238. https://doi.org/10.1038/352236a0

Wolfe J, Morais M, Schafer E (2015) Improving hearing performance for cochlear implant recipients with use of a digital, wireless, remote-microphone, audio-streaming accessory. J Am Acad Audiol 26:532–539. https://doi.org/10.3766/jaaa.15005

Woods VM, Triantis IF, Toumazou C (2011) Offset prediction for charge-balanced stimulus waveforms. J Neural Eng 8. https://doi.org/10.1088/1741-2560/8/4/046032

Wu CC, Luo X (2013) Current steering with partial tripolar stimulation mode in cochlear implants. J Assoc Res Otolaryngol 14:213–231. https://doi.org/10.1007/s10162-012-0366-8

Zeng F-G (2022) Celebrating the one millionth cochlear implant. JASA Express Lett 2. https://doi.org/10.1121/10.0012825

Zhao Y, Dawant BM, Labadie RF, Noble JH (2018) Automatic localization of closely spaced cochlear implant electrode arrays in clinical CTs. Med Phys 45:5030–5040. https://doi.org/10.1002/mp.13185

Zhou N, Dong L, Galvin JJ (2020) A behavioral method to estimate charge integration efficiency in cochlear implant users. J Neurosci Methods 342:108802. https://doi.org/10.1016/j.jneumeth.2020.108802

Zhu Z, Tang Q, Zeng F-G et al (2012) Cochlear-implant spatial selectivity with monopolar, bipolar and tripolar stimulation. Hear Res 283:45–58. https://doi.org/10.1016/j.heares.2011.11.005

2 Advancements and Challenges in Signal Processing and Sound Coding for Cochlear Implants

Waldo Nogueira

Abstract

Sound coding is a critical component of cochlear implants, responsible for transforming acoustic signals into electrical stimulation patterns that activate the auditory nerve. This process begins with sound capture via an external microphone, followed by digitization, noise reduction, and the computation of stimulation patterns delivered through intracochlear electrodes. While various sound coding strategies have been developed to enhance speech perception, progress has plateaued, and cochlear implant users still experience limited performance compared to normal-hearing individuals. These limitations stem from both biological and technological constraints, including the coarse electrode-nerve interface, current spread in the cochlear fluids, neural health variability, and hardware restrictions such as limited number of electrodes and stimulation flexibility. Additionally, the nature of artificial electric stimulation of the auditory nerve limits temporal neural processing, constraining pitch perception and binaural processing. Recent advances in medical imaging, novel impedance measurements, and electrophysiological measurements with the cochlear implant electrodes enable more precise characterization from the periphery to the central auditory system, paving the way for future individualized sound coding strategies. Furthermore, the integration of artificial intelligence in the sound coding strategy and the emergence of next generation implanted electronics including integrated digital signal processors, memory, and adaptive stimulation—promise to revolutionize cochlear implants. Future cochlear implant

W. Nogueira (✉)
Department of Otorhinolaryngology, Hannover Medical School and Cluster of Excellence Hearing4all, Hanover, Germany

Department of Microelectronics and Electronic Systems and Institute of Neuroscience, Universitat Autònoma de Barcelona, Cerdanyola del Valles, Spain
e-mail: nogueiravazquez.waldo@mh-hannover.de

B. Molaee-Ardekani, M. Segovia-Martinez (eds.), *Advancing Cochlear Implants*,
https://doi.org/10.1007/978-3-032-15719-5_2

systems will be able to dynamically interpret neural responses and deliver personalized stimulation, offering new opportunities to improve speech understanding, sound localization, and music appreciation for cochlear implant users.

Keywords

Cochlear implant · Sound coding · Front-end algorithm · Artificial intelligence · Deep learning · Electrode-nerve interface · Signal processing

2.1 Introduction

Cochlear implants (CIs) represent a remarkable technological advancement, offering a sense of hearing to individuals with severe to profound hearing loss. Central to their functionality are signal processing algorithms that preprocess the sound and sound coding strategies, which convert acoustic signals into electrical pulses that are delivered to the auditory nerve. These strategies have evolved significantly over the years, incorporating various techniques trying to enhance speech (Wohlbauer and Dillier 2025; Wouters et al. 2015) and music perception (Nogueira et al. 2018) and sound quality in general. Despite these advancements, one of the major limitations of CIs is their ineffectiveness in solving the "cocktail party problem," which refers to the difficulty of understanding speech in environments containing background noise, reverberation, and other interferers. CI users often struggle to distinguish speech from background noise due to the limited spectral and temporal resolution they obtain from electric stimulation of the auditory nerve through electrodes placed inside the cochlea. This challenge is exacerbated by the poor encoding of binaural cues, which are crucial for sound localization and separating different sound sources. As a result, even with advanced sound coding strategies, users still face significant difficulties in noisy settings, impacting their ability to engage in conversations and understand speech in social situations. Addressing this limitation remains a critical area of research to improve the auditory experience for CI recipients.

Front-end algorithms for CIs are designed to optimize the initial processing of acoustic signals before they are converted into electrical pulses for auditory nerve stimulation by the sound coding strategy. In the front end, signals from the microphone or other sources are processed. This includes Bluetooth connection or dedicated wireless transmission between microphones from other CIs, hearing aids, or external microphones (e.g., Kokkinakis et al. 2012; Büchner et al. 2014; Nogueira et al. 2015; Baumgärtel et al. 2015). Front-end processing can significantly enhance speech understanding in noisy environments for CI users. Techniques such as adaptive dynamic compression, beamforming and single-channel noise reduction, impulsive noise canceling, wind noise canceling, and sound classifiers have been pivotal in this progress. Dynamic compression in CIs is used to manage the wide range of sound intensities in the environment, ensuring that both soft and loud sounds are audible and comfortable for the user. This process involves reducing the dynamic range of the input signal so that it fits within the available electric dynamic range of each individual CI user. Beamforming uses multiple microphones to focus on the direction of the desired speech signal while suppressing background noise, thereby improving the signal-to-noise ratio. Single-channel noise reduction algorithms, like spectral subtraction and Wiener filtering, aim at enhancing speech clarity by reducing unwanted noise from a single microphone input. A challenge for front-end algorithms is determining when to activate or deactivate them for each listening situation while also accounting for the individual preferences of CI users. For this reason, algorithms are necessary to control when and how strongly each front-end algorithm may be beneficial for each user.

The goal of a sound coding strategy in CIs is to transform the acoustic signal captured by the microphone into patterns of electrical stimulation delivered through intracochlear electrodes. This transformation aims to optimize sound perception by effectively utilizing the available electrical dynamic range, the limited number of stimulation channels, and the temporal resolution of the implant while also accounting for the perceptual constraints of electric hearing. Many sound coding strategies for CIs have been proposed over the last decades, and some of them have been commercialized (e.g., Wohlbauer and Dillier 2025; Wouters et al. 2015; Nogueira et al. 2018). The sound coding strategy is restricted by capabilities of current commercially available implants and sound processors as well as the intrinsic limitations of the interface between the electrodes and the nerve fibers. Continuous Interleaved Sampling (CIS) is one of the earliest and most widely used strategies, which processes sound by dividing it into multiple frequency bands and delivering interleaved electrical pulses to the electrodes (Wilson et al. 1991). Advanced Combination Encoder (ACE) builds on CIS by selecting the most energetic bands for stimulation (Seligman and McDermott 1995). The Psychoacoustic Advanced Combinational Encoder (PACE), also known as MP3000 (Nogueira et al. 2005), aims to reduce channel interaction by selecting channels that are most perceptually relevant to the listener. This approach is based on psychoacoustic masking models similar to those used in MPEG-Layer 3 (MP3) and Advanced Audio Coding (AAC) codecs for music bitrate compression, but here it is employed to select the most perceptually significant frequency bands. Fine-Structure Processing (FS4) focuses on preserving the fine temporal structure of sound, which is crucial for music appreciation and tonal languages (Hochmair et al. 2015). HiRes Fidelity 120 (F120), part of the HiRes family, uses current steering to create virtual channels, a technique that uses simultaneous stimulation of pairs of electrodes to shift the locus of the excitation between physical electrodes (Koch et al. 2007; Nogueira et al. 2009). These strategies, along with other variants, continue to be refined and combined with modern technologies like deep learning to further enhance the auditory experience for CI users (Gajecki et al. 2023; Huang et al. 2024; Gaultier and Goehring 2024).

This chapter delves into the development, current state, and future directions of signal processing and sound coding strategies for CIs, highlighting both the achievements and ongoing challenges in this critical area of auditory technology. It examines the primary difficulties in sound encoding for CIs, addressing the limitations that users still face—most notably, the persistent challenge of speech understanding, especially in noisy environments. The chapter explores the evolution and roles of front-end algorithms, sound coding strategies, and various stimulation modes that have been designed to overcome these obstacles. Additionally, it provides an overview of the methods used to evaluate these sound coding strategies, including both objective instrumental measures and behavioral assessments with users. In discussing future prospects, the chapter critically analyzes why speech understanding remains the central problem, evaluates how much improvement has been realized through successive advances in signal processing, and offers a realistic perspective on the extent of improvement that can be anticipated from future developments.

2.2 The Cochlear Implant

A CI is comprised of two main components: the external part and the implanted internal part. The external unit includes a microphone, microcontrollers, digital signal processors, batteries, cables, and an external coil equipped with a magnet. Within this unit, the microcontrollers and digital signal processors carry out the front-end signal processing and execute the sound coding strategy. The internal component is an integrated circuit (IC) functioning as a state machine that interprets radio frequency (RF) signals received transcutaneously from the external unit and controls electrode stimulation. This internal component operates without a battery, drawing power from the RF signal. Upon receiving power, the stimulator decodes RF bitstream data and converts it into electric currents for delivery to the electrodes. Depending on the manufacturer, implants may utilize constant current or constant pulse duration sources, designed with high impedance to ensure adequate voltage compliance, all while optimizing power consumption and IC area. The IC's adaptability allows for diverse stimulation modes, pulse shapes, durations, and rates. Additionally, it includes a feedback mechanism designed to monitor voltage levels at the electrode site and transmit this information back to the external unit.

Figure 2.1 illustrates the fundamental components of a CI. The external part primarily includes the sound processor, while the internal part comprises the implanted circuitry, power transfer apparatus, decoder, and stimulator. Both power and data are transmitted via a radio frequency (RF) link, which utilizes modulation techniques to enable communication and energy delivery. Amplitude-shift keying (ASK), for instance, is frequently applied in biomedical implant systems for its efficiency and straightforward implementation (Lee 2015). The implant's direct current (DC) power is derived from a rectifier that converts the RF carrier to a usable supply voltage, supported by an energy storage device and voltage regulation circuit. Data communication is handled at the RF level through the interaction of transmitter and receiver antennas. The achievable stimulation rate of the CI depends on both the bit rate and frame rate managed by the RF transmission channel, with typical bit rates for various systems ranging between 250 and 500 kbits/s (Zeng et al. 2008). The internal implant is placed within the mastoid bone of the skull and the electrode array is placed in the scala tympani of the cochlea.

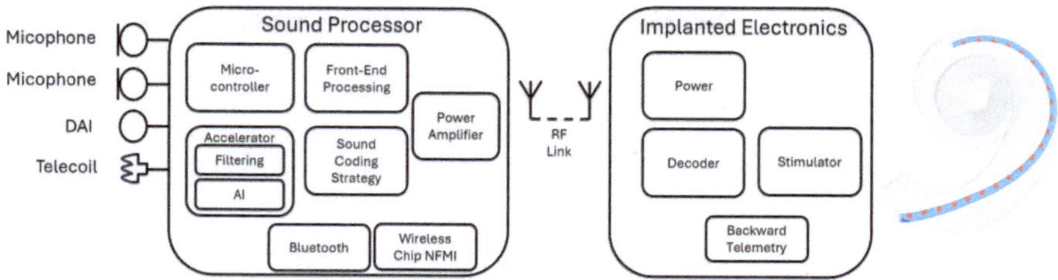

Fig. 2.1 A schematic diagram of a cochlear implant (CI). The CI consists of two main sections: an external and an internal unit. Within the external unit are components such as a power amplifier, a digital signal processor (DSP), and a radio frequency (RF) transmission module. The DSP is responsible for analyzing and processing the incoming acoustic signals, converting these sounds into digital data, which are subsequently transmitted wirelessly through the RF link. The implanted electronics decode the signals and stimulate the electrode contacts inserted in the scala tympani of the cochlea

Over the past two decades, the stimulator component within the implanted part of the CI has experienced considerable advancements. Current stimulators are capable of delivering increased current levels, in the range of 1.5–2.0 mA, and offer enhanced precision in current control, which facilitates accurate electrical stimulation of the auditory nerve (Guo et al. 2008; Ngamkham et al. 2015). These devices support higher stimulation rates of up to 10,000–20,000 pulses per second using brief pulse durations, typically around 10 μs. Additionally, current generation stimulators can operate in various stimulation configurations—such as monopolar, bipolar, and tripolar modes—and accommodate a greater number of stimulation channels and diverse pulse shapes, enabling the supply of current to as many as 12, 16, 20, 22, or 24 electrodes. Improvements in power efficiency have also been realized in modern designs.

On the sound processor side, the evolution has been marked by a significant shift from analogue technology to classical digital signal processors (DSPs) and to system-on-chip (SoC) technology. In the early days of digital sound processors, DSPs were used to process audio signals, but their large size, power consumption, and limited processing capabilities made them impractical for smaller behind-the-ear (BTE) processors.

To further enhance processing capabilities, SoCs often integrate microcontrollers, for example, based on the ARM Cortex-M3 Processor to manage system-level tasks and co-processors, such as specialized filtering and transform engines, to accelerate computationally intensive tasks like fast Fourier transforms (FFTs) and infinite and finite impulse response (FIR) filtering (e.g., Ezairo 8300[1] and NXP NXH2004[2]). The implementation of modern real-time signal processing algorithms, such as speech enhancement and adaptive beamforming, relies heavily on the availability of high-

[1] https://www.onsemi.com/products/custom-assp/audio-video-assp/audiology-dsp-systems/ezairo+8300

[2] https://www.nxp.com/products/NXH2004

performance, low-latency processing capabilities. Without advances in SoC design, it would be impossible to meet the stringent requirements of modern sound processing algorithms, which demand billions of operations per second and latency as low as 5–10 milliseconds (e.g., Stone and Moore 2005). Moreover, in recent years, the CI sound processor is becoming very similar to multimedia devices such as hearables or even smartphones. Current sound processors implement Bluetooth connectivity and RF communication links with external devices and with other CIs and hearing aids for bilateral and bimodal processing.

2.3 Why Is Speech Understanding Still the Major Problem?

CI sound coding strategies often face challenges that affect speech understanding. Since the introduction of multichannel devices, extensive research has shown a correlation between hearing performance and the number of spectral channels, with marked improvements observed as the channel count increases (e.g., Dorman and Loizou 1997; Fu et al. 1998). However, the most significant benefits are realized when advancing from 1 to 8 channels. Despite modern CIs featuring between 12 and 26 electrodes, evidence suggests that increasing stimulation sites beyond 8 does not enhance speech understanding further (Fishman et al. 1997; Friesen et al. 2001; Garnham et al. 2002). Friesen et al. (2001) demonstrated that only eight channels are sufficient for effective speech comprehension. Similarly, Croghan et al. (2017) found that while additional electrodes might contribute to improved speech understanding, performance plateaus at eight electrodes, highlighting a limitation in CI technology. This constraint implies that CIs may deliver a similar amount of information as traditional vocoders. The inherent limitations may stem from a combination of biological, technical, perceptual, and cognitive factors, emphasizing the need for advancements in sound coding strategies.

2.3.1 Electrode-Nerve Interface Limitations

The electrode-nerve interface in CIs refers to the complex interaction between the electrode array and the auditory nerve, which is responsible for transmitting electrical signals from the CI to the auditory nerve. This interface is characterized by the insertion of the electrode array into the scala tympani of the cochlea, where it comes into close proximity with the auditory nerve fibers. The electrode-nerve interface is shaped by multiple factors, including the design of the electrode array, its insertion depth and angle, the condition of the auditory nerve fibers, and the properties of the surrounding tissue near the electrode contacts. These elements collectively influence how electrical stimulation is delivered to the auditory nerve fibers and, ultimately, how sound is perceived.

While a healthy cochlea contains thousands of precisely tuned hair cells, CIs typically rely on just 12 to 24 intracochlear electrodes to stimulate broad regions of the auditory nerve. The limited number of electrodes, combined with imprecise targeting of specific nerve fibers, leads to reduced frequency resolution and sound quality. This is largely due to current spread from each electrode, which causes unintended stimulation of adjacent nerve fibers—a phenomenon known as channel interaction.

Moreover, place-frequency mismatch in CIs occurs when the frequencies delivered by the electrodes do not align with the natural tonotopic organization of the cochlea. This mismatch can lead to distorted sound perception because the electrodes may stimulate regions of the cochlea that correspond to different frequencies than intended. For example, an electrode meant to stimulate a region associated with high frequencies might inadvertently stimulate a region tuned to lower frequencies. This can affect the clarity and quality of the perceived sound, making it challenging for users to accurately perceive speech and music.

Additionally, the number and health of the surviving auditory nerve fibers vary greatly among users and across different regions of the cochlea, further limiting the achievable resolution, regardless of how sophisticated the sound coding algorithm is. Variability in cochlear anatomy and auditory nerve condition among individuals contributes to inconsistent outcomes and can degrade perceived sound quality due to channel interaction.

2.3.2 Perceptual Factors

Limitations at the level of the electrode-nerve interface cause perceptual limitations. Limited spectral resolution, which affects the ability to distinguish between different frequencies, leads to difficulties in understanding speech, especially in noisy environments. Temporal resolution is another limitation, as current strategies may not accurately convey the timing cues necessary for speech and music perception.

Spectral resolution is a crucial component of auditory processing, significantly impacting speech understanding in both quiet and noisy environments. In quiet settings, it allows for the differentiation of phonemes and identification of subtle variations in speech sounds, thereby enhancing clarity and comprehension. In noisy environments, high-spectral resolution is essential for segregating speech from background noise, facilitating better speech intelligibility and focus on a target speaker. Ultimately, robust spectral resolution improves the auditory system's capacity to parse complex acoustic signals, leading to enhanced speech perception and effective communication across diverse listening conditions.

In the temporal domain, there are limitations related to the properties of electric stimulation, which restrict the processing of temporal fine structure and result in lower temporal pitch sensitivity compared to acoustic stimulation. Human psychophysical assessments reveal a notable temporal deficit, even when bypassing speech processors. Pitch-ranking studies reveal that, for most CI listeners, pitch does not increase with increases in pulse rate above about 300 Hz, but this value varies across listeners, reaching 800–900 Hz in some cases (Kong and Carlyon 2010).

Limitations in temporal processing with electric stimulation of the auditory nerve are observed already with single-electrode activation. Activation of multiple electrodes through sound coding strategies can further impair temporal encoding, as channel interactions may distort spectral and temporal envelopes and fine-structure information (e.g., De Groote et al. 2025). The temporal envelope, representing changes in sound amplitude over time, aids in understanding speech dynamics. Sound processing algorithms in CIs must enhance these cues for better speech clarity in noise. Fundamental frequency (F0) perception is necessary for pitch and intonation recognition, helping differentiate speakers and convey meanings in tonal languages. F0 interacts with temporal fine structure (TFS) and envelope, with TFS providing phase information crucial for pitch, while the envelope assists in detecting speech rhythm. TFS is helpful in separating sound sources in noisy conditions and improves speech intelligibility.

In binaural hearing, cues such as interaural time differences (ITDs), interaural level differences (ILDs), and spectral information play a vital role in sound localization accuracy. These auditory cues are essential to obtain benefits of binaural listening, supporting phenomena including binaural summation, spatial unmasking, and the squelch effect, all of which contribute to enhanced speech understanding in complex acoustic environments. Despite these advantages, CI users often face limitations in spatial hearing and accurate sound localization, with diminished sensitivity to binaural cues like ITDs and ILDs (Kan and Litovsky 2015). In contrast to individuals with normal hearing, bilateral CI users primarily rely on ILD cues for spatial perception (Seeber and Fastl 2008), and their ability to detect ITDs declines with increasing stimulation rates, often becoming negligible above 200 pulses per second for most users. It has been speculated that fundamental challenges in the temporal processing capabilities of electrical hearing, especially at higher frequencies, may be responsible for difficulties in both pitch and ITD perception (Carlyon et al. 2025).

2.3.3 Central Factors

CI outcomes are determined not only by the technical capabilities of the device but also by the brain's capacity to interpret electrical auditory signals. In cases of prolonged auditory deprivation, optimal device performance may still result in suboptimal auditory outcomes, as neural decoding mechanisms can be compromised, leading to reduced speech comprehension and sound recognition (Lazard et al. 2012; Beyea et al. 2016). Cognitive processes—particularly neural plasticity—play a critical role in the adaptation to CI stimulation (Kral et al. 2019; Kral and Tillein 2006). The effectiveness of this neural adaptation depends on variables such as the age at implantation and the duration of prior auditory deprivation. Extended periods of deafness have been linked to impaired temporal processing in central auditory pathways (Middlebrooks 2018). Furthermore, CI users may experience progressive changes in pitch perception over time (Reiss et al. 2007). Experiments have demonstrated that recipients can adapt to discrepancies between acoustic frequencies and their spatial representation within the cochlea (Svirsky et al. 2015). Studies of auditory system development have identified both degenerative and compensatory plastic changes in the neural encoding of spatial information (e.g., Klinke et al. 1999; Kral et al. 2002).

2.3.4 Hardware Limitations

2.3.4.1 Sound Processor

Research on sound coding strategies relies on evaluations with CI users to assess the potential benefits. Experimentation necessitates the use of commercial implants familiar to users, and altering the sound coding strategy could result in poorer sound perception. Consequently, studies typically employ the user's clinical sound coding strategy as a baseline, posing a challenge to advancing sound coding approaches. Given the established commercial sound strategies as references, contributions aiming to enhance sound perception may not show improvement when examined collectively. The absence of a standardized framework for developing sound coding strategies further impedes progress, as disparate proposals arise independently. A unified platform would enable comparison and assessment across various strategies, thus fostering advancements in sound processing algorithms.

Implementing real-time speech enhancement algorithms in wearable devices presents several hardware limitations. One of the primary challenges is latency, which must be minimized to ensure seamless user experience. In bimodal hearing scenarios, where a CI is used in one ear and a hearing aid in the other, synchronization is crucial to avoid temporal mismatches. Delays exceeding 10–30 milliseconds can negatively affect speech recognition and auditory processing. For localization, even smaller delays may be required (Zirn et al. 2019). Addressing these delay tolerances is essential for optimizing hearing device configurations and enhancing communication for users. Achieving such low latency requires highly efficient algorithms and fast processing capabilities, which can be constrained by the limited computational power of wearable devices. Additionally, the number of operations that can be performed per second is restricted by the device's processing power and battery life. Wearable devices often rely on low-power DSPs that must balance performance with energy efficiency. This limitation necessitates the use of lightweight algorithms that can operate within these constraints while still providing effective noise reduction and speech enhancement.

2.3.4.2 Integrated Circuit Implant Electronics

CIs use a stimulator to generate biphasic, charge-balanced pulses for auditory nerve stimulation, emphasizing compact size and low power consumption, especially in multichannel devices. A programmable current source is crucial for adjusting stimulation amplitude to fit individual patient thresholds and clinical conditions. Current-mode stimulators, often using current mirror circuits, are favored for ease of charge control but must take into consideration trade-offs between chip area, current amplitude resolution, voltage compliance, and power consumption. Techniques like wide-swing cascodes, integrating DAC functions, and voltage-controlled resistors offer benefits but can introduce challenges such as limited voltage compliance and increased nonlinearity (e.g., Ghovanloo and Najafi 2005; Guo et al. 2008; Ngamkham et al. 2015). These constraints limit the implant's ability to generate diverse pulse shapes and stimulation modes, impacting electrode activation flexibility and current amplitude resolution. Achieving a programmable, high-resolution current source in a compact, power-efficient design remains technically challenging yet vital for effective hearing restoration.

2.3.4.3 Binaural Link

Cochlear implants operate independently, lacking true binaural processing as found in natural hearing, with each implant processing and delivering acoustic information to the auditory nerve without sharing timing or spectral cues essential for natural binaural hearing. Standard bilateral cochlear implantation utilizes two independent devices, each with separate processors and electrode arrays, leading to inconsistent ILD encoding due to varied signal levels and microphone placements between ears. The sound processors can be binaurally linked through wireless streaming functionalities, with near-field magnetic induction (NFMI) providing limited binaural linking by synchronizing features like dynamic compression and beamforming. NFMI technology (e.g., Kim and Kim 2014), similar to traditional telecoil, uses frequencies ranging from 3 to 15 MHz and allows information transmission across both sides of the head. Although the raw transmission bitrate is 596 kbit/s, the use of an Adaptive Differential Pulse Code Modulation (ADPCM) codec compresses the audio signal to just 64 kbit/s, aiding in power consumption reduction because in the wireless communication between the hearing devices the transmission energy is proportional to the number of bits; therefore, fewer bits means lower total energy consumed (Hinrichs et al. 2021). However, true binaural coordination like that in natural hearing is still difficult to achieve, because the transmission of ITDs is limited by slow sampling rates and a lack of synchronization between processors and implants, which lowers the accuracy of temporal cues.

2.4 Front-End Algorithms for Cochlear Implants

Recent years have seen significant advancements in front-end processing for CIs. The integration of adaptive noise reduction algorithms has enabled devices to more effectively suppress (nearly) steady-state background noises. Wind noise reduction and impulse or transient noise cancelers have also been introduced and refined, trying to improve sound quality and perception for different types of background noises and acoustic situations. The widespread adoption of directional and adaptive microphone arrays allows CIs to focus on speech coming from the front while attenuating noise from other directions, including multi-talker babble, restaurant noise, and environmental sounds. Additionally, automatic scene analysis systems can detect and classify acoustic environments in real time, automatically adjusting processing parameters for optimal performance without user intervention. Dynamic range optimization and adaptive gain control (AGC) have further enhanced listening comfort, ensuring that both soft and loud sounds are kept within a comfortable range for the listener. Altogether, these advancements in front-end processing must work synergistically with sound coding strategies to provide a clearer, more natural, and less effortful listening experience for CI users. These algorithms provide benefits for adult CI users; however, when fitting young children, some audiologists believe that front-end algorithms need to be disabled or only enabled in specific situations to enhance the most realistic perception of everyday noises during the auditory development of the child (e.g., Baumann et al. 2025).

2.4.1 Monaural and Binaural Beamforming

A beamformer is a signal processing technique used in hearing aids and CIs to enhance sound coming from a specific direction while reducing noise from other directions. It relies on multiple microphones to create a directional focus, improving speech understanding in noisy environments. It works by combining the signals arriving with varying delays and intensities at microphones placed in different locations. This allows the beamformer algorithm to focus on sounds from a particular direction (e.g., in front of the user) and improve the signal-to-noise ratio (e.g., Spriet et al. 2007).

The simplest beamformer is the delay-and-sum, which is a simple spatial filtering method that aligns (delays) signals received at different microphones based on the direction of arrival and then sums them. This constructive addition enhances the signals from the desired direction while attenuating signals from others due to phase mismatches. The Griffiths-Jim beamformer (Griffiths and Jim 1982) is a structure that combines a fixed beamformer (like delay-and-sum) with an adaptive noise canceler (often a Capon beamformer) to suppress noise while preserving the desired signal. The Capon beamformer, also known as minimum variance distortionless response (MVDR; Capon 1969), is a spatial filtering technique used in array signal processing to enhance a desired signal while minimizing interference and noise. It works by adjusting the beamforming weights to minimize the output power of the array, subject to the constraint that the gain in the direction of the desired signal remains unity. This ensures that the desired signal passes through undistorted while suppressing other sources. A binaural beamformer takes advantage of microphones on both ears (or devices) to create a more directional characteristic. Monaural beamforming can provide an improvement of around 5 dB in speech reception threshold (SRT) when the target speech signal is presented from the front and background and interferences are coming from other directions or the background noise is diffuse (e.g., Büchner et al. 2014; Honeder et al. 2018). The binaural beamforming can provide 1–2 dB SRT further improvement (Büchner et al. 2014; Baumgärtel et al. 2015).

2.4.2 Dynamic Compression

Dynamic compression, as part of the front end, prior to its processing through the sound coding strategy, can be implemented with adaptive gain control (AGC). CI listeners have an electric dynamic range of approximately 6–20 dB, which is considerably smaller than the 100-dB acoustic dynamic range of normally hearing listeners (e.g., Zeng and Galvin 1999; Theelen-van den Hoek et al. 2014). AGC, together with a back-end compression and the mapping block of the sound coding strategy, is used to compensate for this reduced dynamic range in CI listeners.

The AGC is completely characterized by a set of simple parameters such as the attack/release time, threshold, and compression ratio. These standard single-channel compressors are implemented in various configurations in Advanced Bionics LLC, Cochlear Ltd., and MED-EL GmbH devices (Vaerenberg et al. 2014).

Broadband AGC processes the entire frequency range of the input signal uniformly, ensuring that sudden loud noises are quickly attenuated to protect the user from discomfort. Dual-loop AGC in CIs employs two feedback circuits: a fast-acting loop that quickly responds to sudden loud sounds, providing protection from transients, and a slow-acting loop that adjusts to overall background changes, maintaining comfortable listening and consistent speech audibility (Stone et al. 1999). The fast-acting loop uses time constants in the range of 2 ms and 240 ms for the attack and recovery, respectively, with activation at 71 dB and a compression ratio of 12:1. The slow-acting AGC uses time constants of 240 ms and 1500 ms for the attack and recovery, respectively, with activation threshold at 63 dB SPL and a compression ratio of 2:1. This combination enables better management of both brief, loud noises and ongoing, varying sound levels, enhancing speech understanding and listening comfort across dynamic environments. Some advanced systems are exploring or implementing a third-loop AGC, which adds further fine-tuned control—such as optimizing loudness growth or selectively enhancing specific sound features—offering even greater adaptability to complex acoustic situations for CI users, for example, in situations with sudden and strong noises. This third loop may be related to an impulse noise canceler.

Alternatively, AGCs can be multichannel systems; this allows for more precise adjustments in time and frequency; however, care needs to be taken as it may reduce spectral contrast and may have counterproductive effects in terms of temporal modulations across different bands (Stone and Moore 2008). The AGC can also be implemented completely or in part in the back-end of the sound coding strategy. Back-end compression can provide more intuitive control in providing the desired input acoustic dynamic range into the individual electric dynamic range of the user. These back-end configurations include the adaptive dynamic range optimization (ADRO™) (Blamey 2005), available as an option in Cochlear devices, and the VoiceGuard system implemented in Oticon devices (Bozorg-Grayeli et al. 2016). Front-end compression and back-end compression provide speech understanding benefits for CI users in quiet and noisy environments (e.g., Langner et al. 2020; Boyle et al. 2009).

2.4.3 Sound Classifiers

Sound classifiers in CIs are algorithms designed to recognize different acoustic environments. These algorithms were first introduced in hearing aids (Hamacher et al. 2005). The classifiers analyze incoming sound signals and categorize them into predefined environments, such as quiet, noise, music, or speech. The classified information is then used to automatically adapt the front-end processing accordingly. The implant adjusts its processing strategies to optimize sound quality and speech intelligibility for the user. For instance, in noisy settings, the device may prioritize speech signals while suppressing background noise. Modern CIs use scene classification technology to provide seamless transitions between environments, reducing the need for manual adjustments while enhancing user comfort.

2.5 Sound Coding Strategies and Their Processing Blocks

A sound coding strategy for a CI refers to the method used to convert acoustic signals into electrical signals that stimulate the auditory nerve. This section provides an overview of different components of the sound coding as summarized in Fig. 2.2.

Sound coding determines the pattern of electrical stimulation delivered to the electrode array over time, based on a specific coding strategy. The electrodogram serves as a graphical representation of this pattern and is widely used in CI research and programming. Figure 2.4 illustrates examples of electrodograms corresponding to various sound coding strategies. These visualizations are valuable tools for analyzing, comparing, and optimizing coding strategies, enabling assessment of channel selection, temporal fine structure, stimulation rates, and the performance of compression or noise reduction algorithms in real-world CI processing.

2.5.1 Filterbank

In digital signal processing, both filterbanks and transforms (such as the Fourier Transform) are tools used to analyze and decompose signals into their frequency components. Under certain conditions, an appropriately designed filterbank can be mathematically equivalent to a transform; for example, a uniform filterbank with ideal bandpass filters effectively performs the same operation as a discrete Fourier transform. This equivalence allows engineers to choose the approach—either filterbanks for flexibility in bandwidth and time resolution or transforms for computational efficiency and global spectral analysis—that best suits their application.

In CI sound coding strategies, filterbanks are fundamental for decomposing incoming audio signals into multiple frequency bands, each corresponding to different cochlea regions along the electrode array. The filterbank determines the temporal and spectral resolution of the sound analysis and therefore determines how much temporal and spectral information is transmitted through the electric stimulation. The spectral and temporal resolution that can be transmitted through CIs is limited by the electrode-nerve interface, i.e., by the upper limit of temporal pitch, the number of

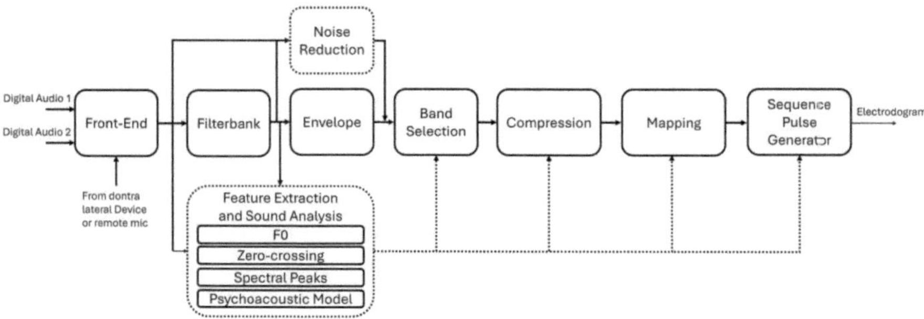

Fig. 2.2 Block diagram of a generic sound coding strategy for cochlear implants

available electrodes and channels as well as channel interactions. However, restricting both temporal and spectral resolution in the analysis will preclude that CI users may learn how to extract them and therefore may not develop sensitivity to them.

Two main approaches are commonly used to implement these filterbanks: finite/infinite impulse response (FIR/IIR) filters and the discrete Fourier transform. IIR filters are a class of digital filters that provide computational efficiency, enabling real-time processing on implant hardware with low latency and power consumption. FIR and IIR filters can be used to mimic the human cochlea's natural bandpass filtering, allowing the system to isolate and process specific frequency components before further analysis, such as envelope extraction or selection of spectral peaks. Alternatively, some strategies employ the Fourier transform for spectral analysis. The FFT approach converts the time-domain signal into its frequency-domain representation. The choice between FIR/IIR-based and FFT-based filterbanks impacts system flexibility, spectral accuracy, and computational demand and is a key design consideration in optimizing CI sound coding strategies for speech and music perception. Several alternative approaches have been explored in prior work, including adaptive filterbanks based on FFTs with varying resolutions (Albalate et al. 2006), wavelets (Nogueira et al. 2006), and computer models that simulate the basilar membrane's mechanics (e.g., Nogueira et al. 2007a; Harczos et al. 2013). Despite these innovative attempts, the inherent limitations of the electrode-nerve interface have constrained their efficacy. Ultimately, these ideas have been incorporated into CI sound coding strategies, effectively merging FFTs or filterbanks with feature extraction techniques to obtain better temporal and spectral resolution.

2.5.2 Feature Extraction

Feature extraction consists of a set of algorithms used to capture acoustic cues necessary to control timing, amplitude, or current weighting of electric pulses. Zero-crossing detection is one approach used to extract temporal fine-structure information, particularly in low-frequency channels. It identifies the time instants when the waveform crosses the zero-amplitude axis. This information is then used to synchronize stimulation pulses to these precise temporal events, transmitting timing cues that can potentially provide better pitch perception, melody recognition, and understanding tonal languages—as implemented, for example, in the FS4 or the SFE (Molaee-Ardekani et al. 2025) coding strategies. Another feature extraction method is the spectral peak locator algorithm used to analyze the FFT with higher frequency resolution. It assumes that the sound can be decomposed as a set of sinusoids which are estimated, for example, by fitting a parabola or assuming a sinusoidal model (Nogueira et al. 2009). The frequency peaks are then combined with current steering to steer the electrical field in sites positioned between pairs of simultaneously stimulated electrodes. The spectral peaks are used not only to convey spectral information but also to amplitude modulate pulses in time. Alternatively, the fundamental frequency of speech may be extracted and used to modulate the envelopes in the sound coding strategy (Milczynski et al. 2009).

The root mean square (RMS) is also a feature that is widely utilized for dynamic compression, where the signal's energy level is measured to adaptively adjust gain in the AGG, ensuring that soft sounds remain audible, while loud sounds do not become uncomfortably intense—this process is vital for preserving the natural dynamic range and optimizing the listening comfort of implant users. Other features to create psychoacoustic models can be extracted to select the most relevant bands for stimulation (Nogueira et al. 2005).

2.5.3 Envelope Detection

Envelope detection refers to the extraction of the slow-varying amplitude contour of audio signals within each frequency channel. One sophisticated method for envelope detection is the Hilbert transform, a mathematical tool that provides an analytic signal from the original waveform. When a signal is passed through a filterbank to decompose it into multiple frequency channels, the Hilbert transform can be applied to each channel output to separate the instantaneous amplitude (envelope) from the instantaneous phase (fine structure) of the signal. Specifically, the magnitude of the analytic signal resulting from the Hilbert transform yields a smooth and accurate representation of the channel envelope. This process offers advantages over simple rectification and lowpass filtering by providing a more precise tracking of the envelope. Envelope information obtained in this way is then used to modulate the current amplitude of electrical pulses delivered to different electrodes in the cochlea, enabling CI users to perceive the dynamic aspects of speech and other sounds.

2.5.4 Noise Reduction

2.5.4.1 Single-Channel Noise Reduction

Single-channel noise reduction techniques for CIs aim to enhance speech clarity in environments with background noise. These methods typically function by analyzing the signal-to-noise ratio (SNR) across various frequency bands and applying greater attenuation to those where noise is more prominent than the speech signal. This strategy is fundamentally related to the Wiener filter, a mathematical approach that minimizes the mean square error between the original and processed signals. The Wiener filter works by estimating the noise and applying a gain function based on the SNR to enhance the speech component (e.g., Scalart and Filho 1996). Such algorithms are particularly effective in improving auditory perception for CI users in settings with consistent or slowly varying background noise (Dingemanse and Goedegebure 2018; Müller et al. 2025; Stronks et al. 2022; Dawson et al. 2011). A critical challenge in designing these systems is ensuring that the speech signal remains undistorted while suppressing the background noise.

2.5.4.2 Wind Noise Reduction

Annoyance from wind noise is a common complaint from hearing aid and CI recipients. Wind noise is even more problematic with directional microphone systems compared to omnidirectional microphones. Therefore, wind noise reduction is an important feature in modern CIs, as wind turbulence across external microphones can generate low-frequency noise, making speech and environmental sounds difficult to understand. Wind noise reduction algorithms typically work by detecting the unique acoustic signature of wind—a combination of broad dispersion of low-frequency energy and a lack of harmonics—and then selectively attenuating those low-frequency signals or spatially filtering the input when wind is detected. Studies have shown that between 1 and 14 dB of noise in frequencies below 400 Hz (Chung et al. 2009; Chung and McKibben 2011; Chung 2012) and distortions to microphone directionality (Chung 2012) can be caused by wind noise. Some systems also use dual microphones to better distinguish wind from desired signals. Methods to reduce wind noise include switching from a directional to omni microphone pattern, attenuating low-frequency noise, or possibly through the use of background noise reduction technologies (Chung 2012). Low-level narrow-band compressors can also be activated to reduce the effect of wind noise in the low-frequency channels while leaving mid- and high-frequency channels generally unchanged (Goorevich et al. 2012).

This feature is now commercially available in most leading CI processors, including those from Cochlear, MED-EL, and Advanced Bionics, where it is typically activated automatically or can be configured by the audiologist.

2.5.4.3 Impulse or Transient Noise Reduction

Impulse or transient noise reduction is a digital signal processing algorithm designed to detect and suppress transient, high-amplitude noises such as clattering dishes, key jangling, or sudden loud bangs (e.g., Dyballa et al. 2015). These impulse noises create brief, intense acoustic events that can be highly uncomfortable for CI users, often causing momentary discomfort, distortion, or masking of important speech information. Transient noise cancelers typically operate by detecting the signature of impulsive sounds—characterized by their abrupt onset and high amplitude over a short duration—and then rapidly attenuating or muting the signal, or replacing it with interpolated or predicted values, to minimize perceptual disruption. Some algorithms apply nonlinear filtering or adaptive thresholding to identify and suppress impulse spikes in real time without significantly affecting speech intelligibility or naturalness.

The implementation of effective impulsive noise cancelation is especially valuable because CIs often lack the protective mechanical and neural mechanisms present in the healthy ear, making users more vulnerable to these distressing artifacts. Studies have confirmed that such algorithms can significantly improve listening comfort for implant users in real-world noisy environments. An improvement of 1.7 dB SRT in speech understanding in noise under transient noise conditions has been reported (e.g., Dyballa et al. 2015). While some CI manufacturers have started introducing impulsive noise reduction features in their latest processor generations, research and optimization of these methods remain active areas to achieve robust detection and minimal signal distortion. Its design and effectiveness need to be assessed, taking into account the compression already performed by the fast loop of the dual-loop AGC.

2.5.5 Band Selection

In CI sound coding strategies, band selection is crucial for determining which subset of frequency bands is stimulated at any given time, addressing the common issue where the number of analysis bands exceeds the available physical electrodes. Incorrect band selection can negatively impact speech performance, especially in noisy environments (Qazi et al. 2013). Traditional strategies, like N-of-M, select the N channels with the highest energy out of M potential bands to optimize the delivery of essential spectral features. More sophisticated methods enhance this selection process by incorporating psychoacoustic models inspired by human auditory perception to refine choices in the spectral (Nogueira et al. 2005; Bolner et al. 2020) and temporal domains (Kludt et al. 2021; Lamping et al. 2020). For instance, the PACE/MP3000 strategy demonstrated that reducing the number of selected bands from eight to four could maintain speech understanding performance. Kludt et al. (2021) found that incorporating temporal masking into psychoacoustic models slightly improved speech understanding compared to PACE/MP3000, while Lamping et al. (2020) reported a notable enhancement of 2.4 dB in speech understanding over a research sound coding strategy.

2.5.6 Compression

In a healthy cochlea, the auditory system can detect sounds across a wide range of intensities—from the faintest whisper to very loud noises—but CI users can typically perceive only a narrow span of electrical stimulus levels as comfortable and distinguishable. To accommodate this, compression is used to map the large dynamic range of incoming sounds into the much smaller "electric dynamic range" available to the user. This is commonly achieved by estimating the amplitude envelope of each frequency channel (often using root mean square, or RMS, analysis) and applying a nonlinear mapping to ensure soft sounds are rendered audible, while loud sounds are kept within comfortable limits. The compression function may be set differently for each channel or for a group of channels. The resulting compressed signals then control the amplitude of electrical pulses delivered to the electrodes.

2.5.7 Mapping

Mapping refers to the conversion from compressed stimulation levels into the threshold (T-level) and maximum comfortable level (C- or M-level) of each individual user. Because each user's auditory nerve, electrode placement, and perception of electrical stimulation differ, mapping is a critical step carried out by audiologists during device programming. This involves defining, for each electrode, the T-level at which sound becomes just audible, and C- or M-level, beyond which stimulation is uncomfortable or potentially painful. Proper mapping ensures that soft sounds are audible, speech is clear, and loud environmental sounds are not unpleasant, directly impacting speech intelligibility, sound quality, and overall user satisfaction. Regular remapping may be needed to accommodate physiological changes or acclimatization over time, and advanced mapping strategies may include adjustments for different sound environments or user preferences.

2.5.8 Sequence Pulse Generation

The sequence pulse generator organizes stimulation instructions into a schedule that adheres to specific parameters such as pulse rate, duration, amplitude, and order of channel activation (see Fig. 2.3). This occurs after feature extraction and mapping have determined when and how strongly each electrode should be stimulated. By controlling pulse sequences with microsecond-level precision, this block guarantees that electrical stimulation accurately represents the temporal and spectral features of the original sound as intended by the coding strategy. Most strategies, such as CIS and ACE, use nonoverlapping pulse sequences to avoid channel interaction (crosstalk) and ensure each electrode receives stimulation only at its programmed time slot.

In the pulse generation, the stimulation mode (monopolar, bipolar, multipolar, etc.) and the pulse shape (biphasic, triphasic, etc.) for each channel are determined. A more detailed description of stimulation modes and pulse shapes can be found in Chaps. 1 and 5. Finally, the order of stimulation of the channels is applied. The order of stimulation in CI sound coding strategies describes the sequence and timing by which the electrodes along the cochlear array are activated to deliver electrical pulses to the auditory nerve. This sequencing can be approached in several ways, each impacting the quality of auditory perception and the risk of channel interaction. One common scheme is base-to-apex stimulation, where pulses are delivered sequentially from the electrodes located at the basal (high-frequency) end of the cochlea toward those at the apical (low-frequency) end. Alternatively, an apex-to-base order reverses this progression. Base-to-apex stimulation is generally favored because it maps high frequencies first at the base, which aligns with the natural tonotopic organization of the cochlea.

To further minimize channel interaction and electrical crosstalk, some coding strategies implement a channel-hopping or interleaved order, where electrodes are stimulated in an arrangement that maximizes the spatial distance between

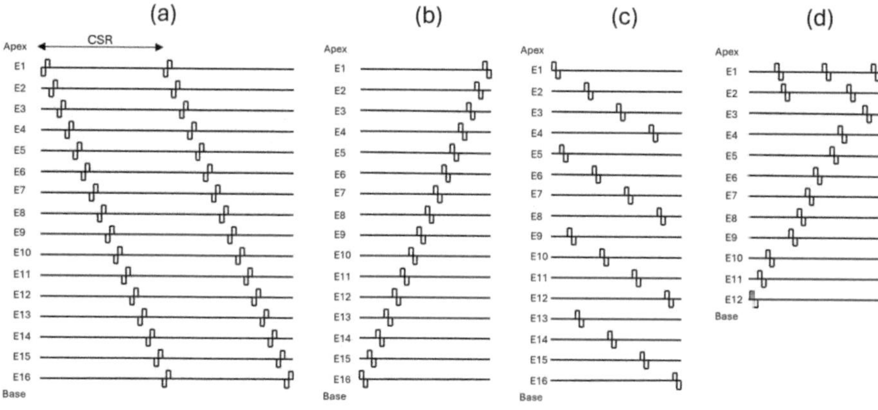

Fig. 2.3 Examples of different orders of stimulation within a stimulation cycle. A stimulation cycle is the inverse of the channel stimulation rate (CSR). The stimulation order can vary: it may proceed from base to apex, as in Cochlear's ACE sound coding strategy (**a**); from apex to base (**b**); it may jump across different electrodes (**c**), as in Advanced Bionics' Fidelity 120 sound coding strategy (note that in the figure, no current steering is depicted), or it may stimulate apical electrodes more frequently (**d**), as in MED-EL's FS4 strategy

consecutively activated channels—effectively skipping over adjacent electrodes within each stimulation cycle. By choosing an order that increases the interval between channels engaged in close temporal proximity, such strategies reduce the overlap of electric fields and subsequent neural excitation interference, leading to cleaner transmission of spectral and temporal information. The selection of stimulation order—whether sequential, grouped, or distance-maximizing—is a critical parameter in preserving speech intelligibility and sound quality for CI users and is often tailored to individual device or user-specific characteristics.

2.6 Commercially Available Sound Coding Strategies

Based on the physiology of the auditory nerve and its behavior to electric stimulation, several sound coding strategies have been developed over the last 40 years. These range from single-channel sound coding strategies to feature-extraction-based sound coding strategies, leading up to current sound coding strategies that can be considered a mixture of feature extraction and waveform coding sound coding strategies. These sound coding strategies aim to emphasize certain characteristics such as spectral, temporal, or dynamic characteristics.

In recent years, major advances in CI sound coding strategies have aimed to enhance speech intelligibility and sound quality in quiet and noisy environments as well as music perception. Some developments that have reached commercialization include fine-structure coding implemented using zero-crossing detectors in the MED-EL's FS4 or carrier synthesis implemented in the Advanced Bionics' F120 based on a spectral peak locator with the aim to improve speech on speech understanding, speech understanding in tonal languages and music perception. Another key development is current steering or virtual channels, which have been used to create additional spectral steps between physical electrodes trying to encode more detailed spectral information. Moreover, innovative N-of-M channel selection algorithms have been developed and commercialized, as in PACE/MP3000 strategy. A further contribution is back-end compression to encode more efficiently the acoustic dynamic range into the available electric dynamic range of the CI user such as in the voice-guard algorithm of the Crystalis sound coding strategy and the ADRO algorithm that can be used together with the ACE strategy (e.g., Blamey 2005).

This section presents several widely used commercial sound coding strategies—ACE (Cochlear), FS4 (MED-EL), F120 (Advanced Bionics), Crystalis (Oticon Medical), and APS (Nurotron). Figure 2.4 illustrates the electrodograms of these strategies in response to three different input signals.

2.6.1 Cochlear's Advanced Combination Encoder (ACE) and PACE/MP3000

The ACE sound coding strategy, employed in Cochlear Ltd. implants, is designed to optimize the delivery of both spectral and temporal information. It is based on its predecessor, the SPEAK sound coding strategy (Seligman and McDermott 1995). At its core, ACE combines two principal strategies: Spectral Peak (SPEAK) and CIS. The incoming sound signal is first analyzed with a bank of bandpass filters (typically 22, corresponding to the number of physical electrodes), which decompose the audio into narrow frequency bands. During each brief stimulation cycle, the ACE strategy identifies the N bands (e.g., 8–12) with the highest amplitude, representing the most salient or information-rich components at that moment. Only these N-of-M channels are stimulated in each cycle, enabling a focus on the most important spectral cues while minimizing channel interaction and reducing power consumption. The default stimulation rate for Nucleus CIs using the ACE coding strategy is 900 pps.

Stimulation pulses are delivered to the cochlea using CIS principles, providing rapid, nonoverlapping electrical stimulation. This approach aims to strike a balance between representing fine spectral details and maintaining high temporal resolution. By actively selecting the most prominent spectral peaks, ACE offers comparable speech intelligibility to the CIS sound coding strategy, also in challenging listening conditions (Skinner et al. 2002). The previously discussed PACE/MP3000 strategy (Nogueira et al. 2005; Büchner et al. 2008a) operates similarly to ACE but differs by substituting the maxima selection in the N-of-M block with a psychoacoustic model, thereby selecting the most perceptually relevant bands for stimulation.

The ACE and the PACE/MP3000 sound coding strategies can be used together with different types of beamformers such as forward-focus (Hey et al. 2019), the SNR-NR noise reduction algorithm, and the ADRO back-end compressor. These techniques are automatically activated using the Scene Classification Algorithm for Noise (SCAN).

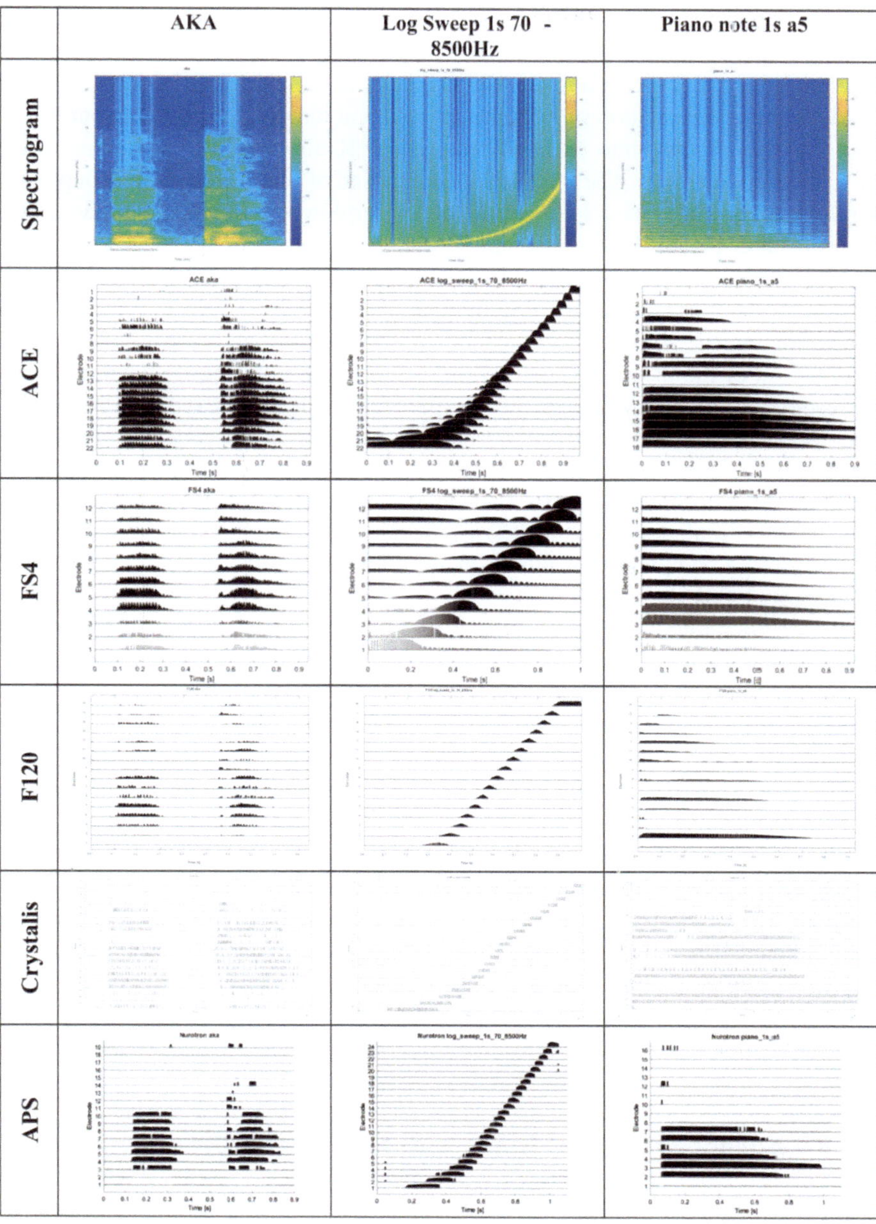

Fig. 2.4 Spectrogram and corresponding cochlear implant electrodograms of common sound coding strategies ACE (Cochlear), FS4 (MED-EL), F120 (Advanced Bionics), Crystalis (Oticon Medical), and APS (Nurotron) for the logatome "aka," a log sweep, and a piano note. X-axis in the electrodograms represents time, and the vertical Y-axis represents the electrodes being stimulated. The black bars in the electrodograms are related to amplitude of the pulses (ACE, FS4, F120, APS) or the pulse duration (Crystalis). Different tools were used to generate the electrodograms, and no calibration was considered; therefore, the input levels to each sound coding strategy may be different

2.6.2 MED-EL's Fine Structure 4 (FS4)

The fine structure 4 (FS4) sound coding strategy is used in MED-EL GmbH CIs to enhance the representation of both the spectral and temporal aspects of sound. The sound is decomposed into 12 bands using an FIR filterbank corresponding to the 12 available electrodes. The filterbank is composed of filters with short group delay for high frequencies and long group delay for low frequencies, similar to the mechanism of the basilar membrane.

FS4 extends envelope detection by transmitting the temporal fine-structure information for the four most apical (low-frequency) electrodes. This is accomplished by synchronizing the timing of stimulation pulses with the zero crossings or phase information of the incoming audio waveform in these four apical channels, which correspond to frequency regions that are particularly important for pitch perception and the understanding of tonal nuances in music and speech. The remaining electrodes, typically representing higher frequencies, continue to transmit envelope information as in conventional CIS strategies, but without the added temporal detail. The default stimulation rates in MED-EL devices are set to the highest possible rate within voltage compliance, resulting in approximately 1500–2000 pps on average (Shader et al. 2019).

By incorporating fine-structure information in the low-frequency range, FS4 aims to improve pitch discrimination, melody recognition, and speech comprehension in noise. The selective application of fine-structure coding only to the lower-frequency channels also helps balance computational demands and battery usage while targeting the frequencies most beneficial for fine temporal cues. Some clinical studies have shown that FS4 can lead to perceptual benefits in music appreciation and tonal languages (e.g., Hochmair et al. 2015; Müller et al. 2012). However, other studies have found no significant difference between fine-structure processing strategies and CIS sound coding strategies (Riss et al. 2011).

The MED-EL FS4 sound coding strategy can be used with various beamforming configurations, the single-channel noise reduction called ambient noise reduction (ANR; Müller et al. 2025), transient noise reduction (Kurz et al. 2022), and wind noise reduction algorithm (Hagen et al. 2020). These algorithms are automatically activated based on the sound scene analysis system known as *adaptive intelligence*.

2.6.3 Advanced Bionics' HiRes with Fidelity (F120)

The F120 sound coding strategy, implemented in Advanced Bionics LLC CIs, is designed to increase spectral and temporal analysis of sound and represent it in the electrical excitation patterns delivered to the electrodes through current steering. For a technical description, see Nogueira et al. (2009) and Koch et al. (2007). At its core, F120 utilizes an FFT-based spectral analysis to decompose incoming audio into 15 analysis bands and a spectral peak locator to estimate 120 possible locations. These 120 locations are encoded through current steering, which allows stimulation to be directed not only to specific electrodes but also to positions in between them by varying the relative currents delivered to adjacent electrodes. Additionally, F120 encodes not just the amplitude envelope of sound but also the fine temporal structure for frequencies up to about 1200 Hz. Default stimulation rates for Advanced Bionics range from 1800 to 2900 pps (per single electrode). Results have not shown a clear advantage for current steering sound coding strategies (e.g., Büchner et al. 2008b).

The F120 sound coding strategy can be combined with different types of monaural and binaural beamformers (Büchner et al. 2014); it uses a dual-loop AGC, and it can be used together with the single-channel noise reduction ClearVoice (Gertjan Dingemanse and Goedegebure 2018), WindBlock, and transient noise reduction (Dyballa et al. 2015, 2016). These algorithms are automatically activated through sound classifiers and actuators called AutoSense.

2.6.4 Oticon Medical's Crystalis

The Crystalis sound coding strategy is used in Oticon Medical A/S CIs. It uses a standard spectral analysis using an FFT-based filterbank, followed by envelope detection to regroup the filterbank output into 20 channels, matching the number of available electrodes. By default, eight of these regrouped channels with the highest envelope magnitude are selected, implementing an N-of-M strategy with $N = 8$. These selected channels are then stimulated using pulses with adjustable phase durations (Seligman and McDermott 1995). A compression function is used to map sound pressure amplitude in dB to pulse width in μs.

The Oticon Medical Crystalis sound coding strategy delivers stimulating pulses to the cochlea in a way that differs significantly from other sound coding strategies with conventional amplitude modulation of biphasic pulses. It modulates pulse width (duration) to reflect changes in the stimulation charge to be delivered. In other words, the pulse height remains fixed, while the width varies according to the magnitude of the envelope associated with each stimulating electrode. The stimulation rate is adjustable to seven preset values between 250 and 1000 pps, with a default clinical setting of 500 pps applied to all participants. Finally, the stimulation information is processed in the CIS-like pulse frame encoder (Wilson et al. 1991) to deliver the pulses to the electrode array.

Oticon Medical CIs (Neuro Zti and Digisonic SP) use a stimulation mode called distributed all-polar (DAP). In this mode, stimulation current is injected by one active electrode and returns through all non-stimulating electrodes, including the extra-cochlear reference electrode in proportion depending on the impedance of each electrode contact. The return electrodes do not actively shape the pulse. Implant power is only required for the first (anodic) phase. In the second (cathodic) phase, the stimulating electrode, connected in series with a capacitor (220 nF ± 10%), is grounded, resulting in exponential-like charge recovery.

The Crystalis sound coding strategy can be combined with various front-end and back-end processing tools, including different types of beamformers, the VoiceTrack noise reduction algorithm (Guevara 2016), and the VoiceGuard compressor (e.g., Langner et al. 2020).

2.6.5 Nurotron's Advanced Peak Selection (APS) and Symphony

Nurotron Co. Ltd. is a 26-electrode CI system with a 24-channel intracochlear electrode array originating at the University of California (Zeng et al. 2015). The Nurotron device can be programmed with two strategies: APS or Symphony. The APS strategy works with the principle of an N-of-M channel selection with 24 analysis channels and 6 to 8 active stimulation channels, dependent on the input energy distribution. The Symphony strategy integrates current steering to create virtual intermediate channels and peak picking. Both strategies intend to represent temporal fine structure. The APS sound coding strategy integrates the c-tone algorithm (Ping et al. 2017).

2.7 Research Platforms for Signal Processing and Sound Coding Strategies

Research and development of novel signal processing algorithms and sound coding strategies for CIs is facilitated by specialized research platforms that simulate sound coding and enable the streaming of electrical excitation patterns to CI electrodes for experimentation in humans. These interfaces are typically housed on PCs, tablets, or embedded within commercial sound processors, allowing researchers to prototype and test sound coding strategies in controlled settings using high-level programming languages such as C/C++, Matlab, Python, or graphic user interfaces (GUIs) for parameter control within the sound coding strategy. Access to these research interfaces is granted to researchers under established agreements with manufacturers, and their use requires institutional review board approval. A comprehensive review of CI research platforms can be found in Litovsky et al. (2017).

When utilizing CI research interfaces, ensuring safe electrical stimulation is paramount, which involves operating below the maximum comfortable loudness levels. Safe stimulation is achieved by adhering to biological limits and preventing tissue or neural damage through charge-balanced stimulation. Based on prior studies (e.g., Shannon 1992; McCreery et al. 1995, 1997), the FDA established guidelines with maximum allowable charge density limits of 216 $\mu C/cm^2$ for clinical sound processors and a recommended limit of 100 $\mu C/cm^2$ for research use. Additionally, pulse shapes must be configured to ensure the net charge post-pulse is zero to prevent accumulation that can damage tissues and nerves. Moreover, CIs must operate within the maximum supply voltage the implant can provide; otherwise, the stimulation is out of compliance. Out-of-compliance stimulation can lead to distortion of the stimulating pulse or waveform, and in extreme cases result in charge imbalances. Furthermore, it is mandatory to prevent uncomfortably loud stimulation that exceeds the maximum acceptable loudness of the CI user. CI manufacturers have contributed to CI sound coding research through platforms such as Nucleus Implant Communicator (NIC; Cochlear), Research Interface Box (RIB, MED-EL, Austria), HRStream/PulseWave (Advanced Bionics), and Oticon Medical Research Platform (OMRP) (see Fig. 2.5). The following subsection provides detailed insights into these research platforms, emphasizing sound coding strategies.

2.7.1 Nucleus Interface Communicator (NIC) and Nucleus Matlab Toolbox (NMT)

Cochlear Ltd. has developed the Nucleus Implant Communicator (NIC) and the Nucleus Matlab Toolbox (NMT) to support signal processing and sound coding research by facilitating real-time interaction with implant systems. These tools enable researchers to configure and control implant settings using custom applications in Matlab, Python, C++, and other languages. The NIC is a library that allows direct communication between CI systems and computer applications, enabling precise manipulation of auditory stimuli, such as stimulation amplitude, pulse duration, and rate, across its three versions: NIC2, NIC3, and NIC4.

NIC2 supports synchronous bilateral stimulation using L34 research-oriented sound processors. NIC3 uses the RF generator box and enables bilateral stimulation and synchronization through a shared clock system, but it lacks the intracochlear potential measurement capabilities of NIC2. The latest version, NIC4, was developed to replace the obsolete custom L34 sound processor setup with commercial off-the-shelf sound processors like CP910 or CP920 connected through the clinical USB POD to a PC (Fig. 2.5a). NIC4 provides low-latency, continuous streaming and includes NIC3's features, such as indefinite low-latency streaming and automatic insertion of power-up frames to maintain implant power. NIC4 also enhances functionality with descriptive errors, compliance telemetry, and checks to ensure that individual maximum stimulation levels are not exceeded. It requires researchers to ensure that the subjects' optimal comfort level is not exceeded by determining maximum acceptable loudness and verifying compliance with Cochlear's Custom Sound software used in the clinic. NIC4 supports unilateral streaming but does not yet facilitate bilateral or bimodal streaming, which NIC3 and the RF generator box allow.

The NMT provides functions for simulating CI processing strategies like ACE and CIS. It supports comprehensive testing and development of new sound processing strategies by generating electrodograms and simulating clinical sound processing strategies, ensuring sufficient accuracy for comparative analysis. The Nucleus Matlab toolbox is open source.[3]

2.7.2 Research Interface Board (RIB) and Stimgenerator Tool

The Research Interface Board version 2 (RIB2) interface is a research tool administered by the University of Innsbruck to interface with MED-EL GmbH implants. Its current version employs a DLL that interfaces with MED-EL's MAXBOX USB setup, connecting a PC to the coil linked to the implant (Fig. 2.5b). This platform has been used in many clinical studies (e.g., Neustetter et al. 2012; Imsiecke et al. 2018), and also a revision is provided in Litovsky et al.

[3] https://github.com/Cochlear-Research/nucleus-toolbox

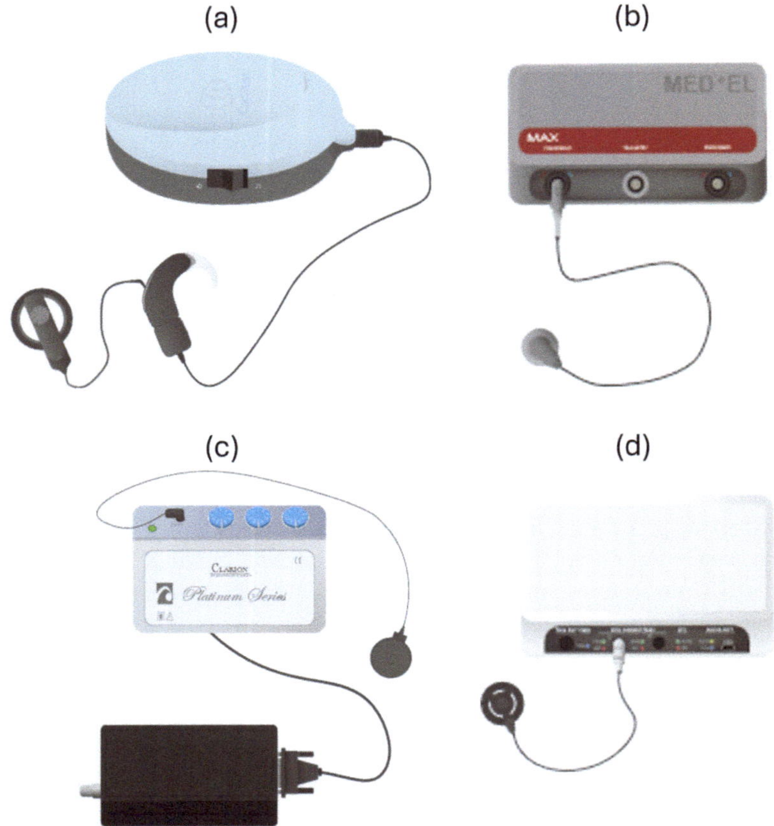

Fig. 2.5 Hardware used in different research interfaces from different CI manufacturers. (**a**) Cochlear Ltd. programming POD with speech processor used for NIC. (**b**) MAXBOX and coil used for the MED-EL RIB research interface. (**c**) Portable Speech Processor (PSP) and serial interface board (SIB) used for the Advanced Bionics HRStream and PulseWave research platform. (**d**) Interface for the Oticon Medical research platform

(2017). Using this platform, stimulation mode can be configured to deliver single or parallel stimulation channels in monopolar mode with biphasic or triphasic stimulation.

Novel sound coding strategies can be implemented using programming languages like C++, Matlab, and Python. Once the excitation patterns are computed, the RIB2 DLL facilitates their streaming through the MAXBOX to the coil and CI, enabling experimentation either with pre-calculated excitation patterns or in real time. The MAXBOX interface supports bilateral operation, providing fully synchronized stimulation.

MED-EL also offers the Stimgenerator, a tool to for generating excitation patterns based on FS4 and other methodologies like fine-structure processing (FSP) or High Definition CIS (HDCIS; e.g., Riss et al. 2016). This tool runs within Matlab on a PC, allowing for the pre-calculation and presentation of stimulus patterns derived from audio signals through RIB2.

2.7.3 BEPS+, HRStream/PulseWave, and GMT-Toolbox

Advanced Bionics LLC currently offers two research systems for sound coding studies: the Bionic Ear Programming System (BEPS+) and HRStream (recently updated as PulseWave). Both platforms support experimentation with mono- and multipolar stimulation modes using charge-balanced pulsatile waveforms of varying shapes.

BEPS+ is considered a highly advanced fitting platform, providing comprehensive access to the HiRes90K implants' functionality through an intuitive GUI. This versatile interface used in several clinical studies (e.g., Langner et al. 2017; Arenberg et al. 2018) allows for both the fitting of the implant using predefined pulse tables—sets of parameters that define the characteristics of electrical stimulation in Advanced Bionics implants—and for loading sound coding strategies into Harmony or Naída Q70 sound processors via the CPI-2 or CPI-3 boxes. Moreover, BEPS+ ensures adherence to safety limits for electrical stimulation.

Additionally, Advanced Bionics supports the design of sound coding strategies through HRStream, which in the most recent versions is referred to as the PulseWave™ platform (Fig. 2.5c). The PC is connected over USB to a streaming interface board (SIB). The SIB is connected to a standard speech processor with the clinical programming cable. The SIB connects to the portable speech processor, which transmits the signals to the implant. PulseWave enables the researcher to set any stimulus pattern on any of the 16 electrode contacts of the HiRes90K device—of course, within the limitations of the implant hardware. Stimuli are preprocessed on a standard PC and then streamed to the subject's implant. Due to the nature of PulseWave, any tests with the subjects can only be conducted acutely. The PulseWave library can be used from Matlab, Python, C#, or other languages.

Moreover, Advanced Bionics offers an open-source library for the generation of excitation patterns from F120, called the GMT tool, which is available online.[4] In principle, it is possible to generate the electrodograms from audio signals using the GMT tool and stream them to a subject using PulseWave.

2.7.4 Oticon Medical Research Platform (OMRP)

Oticon Medical A/S has developed a research interface for CIs that facilitates advanced investigations in auditory signal processing (Backus et al. 2015). As illustrated in Fig. 2.5d, the platform comprises (1) a pair of BTE devices, each containing front and rear calibrated microphones, (2) a four-channel USB analogue-to-digital converter, (3) real-time PC-based sound processing software called the Master Hearing Aid, and (4) USB-connected hardware and output coils capable of driving two implants simultaneously. This system can process input from all four microphones concurrently and generate synchronized binaural outputs for two CIs

[4] https://github.com/jabeim/GMT

(SP or XP types). Researchers can implement both audio preprocessing algorithms—such as binaural beamforming—and innovative binaural stimulation strategies, within the constraints of the implant hardware.

The platform offers highly precise control over interaural electrode timing (±1 μs) and energy levels (±2%), making it well-suited for real-time experiments involving interaural time and level differences. It also supports flexible, on-the-fly adjustments to stimulation rates, enabling researchers to explore how variations in stimulation frequency influence pitch perception. Thanks to its real-time processing capabilities, the system allows immediate modifications to signal processing strategies, facilitating the study of acute perceptual responses (Zhang et al. 2025a).

In addition, Oticon Medical provides a Matlab-based simulation environment known as the simulation chain. This tool emulates the behavior of any CI system—whether currently available or under development. It serves two main functions: (1) as a dynamic tool for system development and debugging and (2) as a verification platform for firmware implementations.

Beyond the simulation chain, the OMRP also supports integration with other programming languages (e.g., C++) and external environments such as openMHA (Kayser et al. 2022), through a dedicated communication driver. Stimulation pulses can be either sourced from predefined stimulation patterns or dynamically generated from incoming audio signals. These pulses are transmitted to the OMRP via the driver, enabling delivery to the CI for experimental purposes.

The entire system can be configured to follow clinical protocols and is capable of delivering stimulation while simultaneously recording electrophysiological and/or behavioral responses. This functionality is facilitated by Oticon Medical's experiment builder platform, OMEXP (Sulas et al. 2023), which provides researchers with a flexible and programmable framework for conducting controlled auditory experiments.

2.7.5 TU Dallas Research Platform

The CCi-MOBILE Research Platform (e.g., Ghosh et al. 2022), designed and managed by TU Dallas, is a powerful tool designed to facilitate auditory research and experimentation with CIs. Compatible with Nucleus devices from Cochlear Ltd., including CI24, CI24RE, and CI512, among others, the platform is utilized by researchers worldwide. It enables direct, real-time interaction with CI devices, allowing for the manipulation of auditory stimuli and the implementation of custom algorithms in acute experimentation with both preprocessed stimuli in Matlab and real-time processing for chronic studies. The platform supports a range of simulation and signal processing tasks, such as bilateral synchronized real-time streaming and on-the-go adjustments of sound processing parameters. Additionally, CCi-MOBILE provides programming flexibility, allowing researchers to implement sound coding algorithms as applications using Matlab, C, C++, Java, C#, and other programming languages.

2.8 Evaluation of Signal Processing Algorithms and Sound Coding Strategies

2.8.1 Evaluation in Cochlear Implant Users

Signal processing algorithms and sound coding strategies for CIs are evaluated using research platforms, such as those described in the previous section, along with CI subject evaluations (see Chap. 3 for details). These evaluations may include (1) subjective speech perception tests (e.g., the MUltiple Stimuli with Hidden Reference and Anchor (MUSHRA) test), where users report what they hear or repeat sentences presented under various listening conditions in quiet and noisy environments; (2) user-centered tools like rating scales and questionnaires (e.g., the Speech, Spatial and Qualities of Hearing Scale (SSQ) and the Listening Effort Questionnaire (LEQ)), which assess sound quality, speech clarity, music enjoyment, and overall listening comfort; and (3) objective instrumental measures (e.g., Speech Transmission Index (STI) and Speech Intelligibility Index (SII)), which provide consistent and quantifiable data on improvements in signal-to-noise ratio, spectral resolution, and speech intelligibility through prediction algorithms. Additionally, the design of CI signal processing algorithms can benefit from integrating computational models with objective instrumental metrics, offering physiologically based, consistent, and quantifiable assessments of signal processing performance that are independent of listener variability.

2.8.2 Computational Models

Computational models are very useful along with experimental methods, enhancing understanding of the factors that limit sound perception in CI users. Moreover, these models can inform the development of future audio processing strategies by predicting potential benefits for CI users, thus reducing the need for laborious and costly experiments. Moreover, computational models allow precise control over experimental variables, facilitating independent analysis of each factor's impact.

Several noteworthy models have been developed to predict CI user performance. Nogueira et al. (2007b) utilized a hidden Markov model (HMM) alongside a simple one-dimensional model of the electrode-nerve interface to compute excitation patterns derived from CI sound coding strategies, predicting speech understanding performance with varying numbers of bands selected by PACE/MP3000 and ACE. Fredelake and Hohmann (2012) developed a model for speech intelligibility in electric hearing, integrating a one-dimensional electrode-nerve interface with a leaky integrate-and-fire system, incorporating refractoriness to evoke neural spikes. These spikes are accumulated over time and tonotopic location to create an "internal representation" of speech, serving as a quantitative perceptual measure linked to the electrical stimulus. A dynamic time-warping (DTW) algorithm was employed as a pattern classifier to predict speech intelligibility. Jürgens et al. (2018) refined this model by substituting DTW with an automatic speech recognizer based on HMM using Gaussian probability distribution, within the framework for auditory discrimination experiments (FADE; Schädler et al. 2015). Alvarez et al. (2023) extended this work by incorporating a detailed 3D computational model of the electrode-nerve interface that simulates nerve degeneration, predicting effects on speech reception thresholds (SRTs) and spectral modulation thresholds (SMTs) in CI users with various stimulation techniques, including current steering and parallel or triple channel stimulation.

Instead of using back-end recognizers, such as those based on DTW or HMM, it is also possible to use measures that assess the similarity between a neurogram generated with CI stimulation and one generated with normal-hearing acoustic stimulation. Tabibi et al. (2021) introduced a framework combining a phenomenological auditory nerve model with a CI sound coding processing stage and a current spread stage. The Neurogram Similarity Index (NSIM; Hines and Harte 2012) has been employed as a comparative tool to evaluate neurograms generated by CI models against those from normal-hearing models, helping to predict melody contour identification capabilities in CI users. Computational models designed to assess the transmission of information from sound to neural spikes within the auditory pathways of CI users have demonstrated potential in enhancing predictions of speech performance with CI stimulation (Leclère et al. 2023). The Spike Activity Mutual Information Index (SAMII; Cardinale and Nogueira 2022) serves as an objective measure for predicting speech intelligibility in hearing-impaired individuals, utilizing mutual information derived between neurograms generated by normal-hearing and CI models, and can be adapted for CI application. This methodology processes target speech and speech-in-noise signals via a computational model of the peripheral auditory system, simulating neural activity within auditory nerve fibers organized into critical bands. The calculated mutual information between these integrated neural signals is averaged across analysis windows to produce SAMII, which has been further extended to binaural contexts. SAMII can be extended to predict speech performance of CI users under diverse neural health conditions and sound coding strategies, utilizing CI models of the auditory periphery for left, right, and bilateral ear configurations.

2.8.3 Vocoder Simulations

In CI research, vocoder-based simulations are commonly used to try to approximate limitations in hearing with CI stimulation in normal-hearing listeners (Cychosz et al. 2024). These models transform acoustic signals into simplified representations that try to reflect the limited spectral and temporal resolution inherent to CI processing. Vocoders offer a controlled framework for investigating how limitations in information transmission from CI electrodes to the auditory nerve influence the perception of speech and other acoustic signals. For example, sine-wave vocoders generate carriers using pure tones modulated by the amplitude envelope of each frequency band computed in the sound coding strategy (Dorman et al. 2002), offering a clean but overly simplistic representation of spectral channels. Attempts have been made to simulate current spread by mixing all envelopes across many tonal carriers (Grange et al. 2017). Noise vocoders (e.g., Shannon et al. 1995), in contrast, use band-limited noise carriers, which may better emulate the roughness and degraded pitch perception often reported by CI users. More advanced approaches, such as harmonic complex vocoders (e.g., Deeks and Carlyon 2004), employ carriers composed of harmonically related tones to simulate the pulsatile nature of electrical stimulation and preserve pitch cues more effectively. Pulse-train vocoders go a step further by using amplitude-modulated pulse trains that closely resemble the stimulation patterns delivered by CIs.

Despite their utility, vocoder simulations must be interpreted with caution. They offer only an approximation of the information transmitted from electrodes to the auditory nerve and do not fully capture the perceptual experience of CI users. The sound produced by a vocoder may differ substantially from what an implant user actually perceives, due to individual factors such as neural survival, electrode placement, and auditory adaptation. Therefore, while vocoders are indispensable tools for experimental research, their limitations must be acknowledged when drawing conclusions about CI perception. See Cychosz et al. (2024) for a review. Cochlear implantation in individuals with single-sided deafness (SSD) presents a unique opportunity to directly compare auditory percepts evoked by electrical stimulation from a CI with those elicited acoustically via the normal-hearing ear. Studies such as those by Dorman et al. (2020) and Karoui et al. (2019) have demonstrated substantial inter-individual variability in these perceptual comparisons. Furthermore, vocoder simulations based on harmonic tone complexes may yield acoustic representations that more closely resemble CI-evoked percepts than those produced by sine-wave or noise vocoders. Despite their utility, vocoders operate through fundamentally different mechanisms than CIs and should be viewed primarily as tools for modeling limitations in auditory information processing. Their most straightforward application lies in the pilot evaluation of front-end signal processing algorithms prior to clinical evaluation in CI users, where they can help assess how such algorithms affect the transmission of speech-relevant cues under simulated implant-like conditions.

2.9 Discussion

CIs have undergone significant innovation over several decades, leading to substantial enhancements in the auditory experience for users. However, despite these advancements, limited improvements in speech understanding in real acoustic environments have been observed, and speech understanding remains a major problem. This section summarizes the expected improvements and explores the reasons for this limited progress. The discussion aims to provide insights into the challenges and limitations of current sound coding strategies and to identify areas for future research and development, with the ultimate goal of improving speech understanding in noise for CI users.

2.9.1 How Much Improvement Has Been Expected from Signal Processing and Sound Coding Strategies for Cochlear Implants?

Since at least two decades, significant improvements were anticipated from advancements in front-end processing and sound coding strategies. The ultimate goal was to develop sound coding strategies that could restore speech understanding performance in noise to normal levels, without the need for front-end processing. Specifically, it was expected that the use of more accurate computational and perceptual models to compute excitation patterns for CIs would yield more natural hearing outcomes and sound coding strategies that mimic the natural auditory system have been proposed (e.g., Harczos et al. 2013; Nogueira et al. 2007a). Additionally, increasing the rate of stimulation was anticipated to improve speech understanding outcomes (e.g., Litvak et al. 2003; Runge-Samuelson 2009). The introduction of current steering also generated substantial expectations, as it potentially enabled the creation of additional stimulation sites beyond the physical number of electrodes in a CI. However, the lack of proven benefit from these advancements led to the conclusion that more focused stimulation may be required, prompting research on current focusing. To date, there is no clear evidence that focusing the current field for stimulation improves speech understanding performance. Furthermore, the emphasis on transmitting temporal fine structure or temporal modulations was expected to yield significant advances in speech performance outcomes, but these benefits have not been observed or have been modest at best (e.g., Riss et al. 2009, 2011). In practice, initial improvements in sound coding strategies were substantial, especially in providing access to speech in quiet environments. However, with successive generations of coding strategies, further improvements have tended to plateau. Today, many adult CI users can achieve 70–90% speech recognition in quiet settings, and the gains from newer coding innovations are often incremental—yielding only modest percentage increases, especially for speech in noise, music perception, and spatial hearing. Despite ongoing refinements, persistent limiting factors like current spread in the cochlea, overlap between electrode channels, and variability in neural health restrict the transmission of more complex auditory information. Front-end signal processing advances—such as automatic gain control, impulse noise reduction, and directional microphones—do increase listening comfort and yield incremental improvements but cannot achieve the dramatic leaps seen with early technological breakthroughs. For this reason, during the last two decades, considerable effort has been devoted to designing novel front-end algorithms that can improve speech understanding performance given the limitations in speech understanding performance with electric stimulation of the auditory nerve with few intra-cochlear electrodes.

Nonetheless, the development of innovative hardware enabling more flexible electrical stimulation—combined with deeper insights into cochlear anatomy, improved characterization of the auditory nerve, and the integration of emerging technologies like AI—opens up exciting possibilities for the future.

2.9.2 What Do We Learn from So Many Developments?

From decades of advances in front-end signal processing and sound coding strategies for CIs, it has become clear that technology can dramatically improve access to sound for people with severe-to-profound hearing loss, but there are also inherent biological and perceptual limits that technology alone cannot fully overcome. Developments in front-end processing have made devices more robust and comfortable in challenging listening environments, while ever more sophisticated sound coding strategies—such as N-of-M selection, current steering, fine-structure encoding, and back-end compressors—have provided modest improvements in speech understanding in noise for CI users.

However, these developments have also revealed that while each incremental improvement can yield measurable gains, the biggest barriers remain in restoring natural speech understanding in noise, music appreciation, and the benefits of true binaural hearing. This is because factors like channel interaction, current spread, limited neural survival, and the coarse "sampling" possible with the electrode interface fundamentally constrain how much information can be delivered and meaningfully processed by the brain. Additionally, performance varies widely between individuals due to biological and cognitive variability.

One of the main challenges in demonstrating the benefits of technologies like CI sound coding strategies is the substantial variability in speech understanding among users (e.g., Velde et al. 2021). This variability often far exceeds the expected gains from the coding strategy itself, which limits the ability to effectively evaluate its impact—especially when considering its implementation in commercial sound processors.

Front-end algorithms provide large improvements in speech understanding in noise. However, this is only in some specific situations. Moreover, in front-end algorithms, we have learned that care needs to be taken in adjusting these algorithms for CI users. CI users may obtain worse temporal and spectral resolution than normal-hearing subjects and hearing-impaired subjects with mild-to-severe hearing loss. A very important aspect is that these front-end algorithms cannot modify the target speech signal. Distortions introduced by these algorithms can produce detrimental effects for CI users even if the noise is reduced (e.g., Nogueira et al. 2015).

During the development of sound coding strategies, we have learned that very fast stimulation rates do not seem to provide improved speech understanding performance. Even if high stimulation rates offer better temporal sampling, can induce stochastic-like firing of auditory neurons, and can increase the electric dynamic range, no clear or consistent advantage has been shown for high rates (Shannon et al. 2011). It is worth noting that the optimal stimulation rate may vary depending on the individual user, the type of CI, and the specific sound processing strategy being used. Therefore, the stimulation rate may have to be adjusted to achieve the best possible sound perception and speech recognition outcomes for each user, but it seems that rates of around 500–600 pulses per second may be sufficient.

There is no single "best" phase duration or pulse width for CIs. Optimal settings are highly individualized and depend on the patient's specific neural health, hearing loss characteristics, and the type of CI used. While most modern CI systems use short phase durations (≤25 μs) for intensity encoding, some vary pulse duration (~10–100 μs) with fixed pulse amplitude (~0.5 mA). Clinical practice may also adjust pulse duration to address specific issues like facial nerve stimulation or auditory neuropathy. If pulse width was increased due to electrodes being out of compliance, this will likely restore adequate loudness growth necessary for maximizing patients' speech understanding. Moreover, research shows that widening pulse width can result in narrower spread of excitation (Zhou et al. 2020). This is also why increasing pulse width can be beneficial in cases of facial nerve stimulation. Increase in pulse duration will directly cause lowering of stimulation rate which, below 500 pps, may lead to decreased speech understanding performance.

In terms of temporal processing, we have learned that CI users normally are poor at using timing information to detect changes in either pitch or sound location. This deficit occurs even for listeners with good speech perception and even when the speech processor is bypassed to present simple, idealized stimuli to one or more electrodes (Carlyon et al. 2025). Therefore, the fundamental mechanisms of temporal processing in CIs need to be better understood.

Numerous attempts have been developed to reduce the detrimental effects of spread of excitation and channel interaction with CIs. Several studies used current focusing strategies based on multipolar stimulation to restrict the spread of excitation and to reduce interaction between adjacent electrode channels (Bierer and Litvak 2016; Srinivasan et al. 2013; Mens and Berenstein 2005; Berenstein et al. 2008; Arenberg et al. 2018; De Jong et al. 2019; Nogueira et al. 2017). So far, results have been mixed, and only one study found benefits for CI speech perception for most recipients by using current focusing (Srinivasan et al. 2013).

Moreover, it has been shown that increasing channel interaction—either by adding parallel channels (Langner et al. 2017) of stimulation or by blurring, i.e., driving one channel with several electrodes stimulated simultaneously (Goehring et al. 2021)—does not result in a clear worsening of speech understanding performance. These results may indicate that CI users may use a limited number of effective channels to understand speech.

Current steering in CIs, which aims at enhancing spectral resolution by stimulating multiple electrodes simultaneously, faces limitations related to current spread. It has been shown that with current steering, it is possible for CI users to discriminate between virtual channels; however, it has also been shown that with current steering, it is difficult to improve perceived spectral resolution. Spectral resolution and spectral discrimination are crucial concepts in understanding how CIs affect hearing and speech perception. Spectral resolution refers to the ability to distinguish between sounds that differ in their spectral content (frequency and intensity), while spectral discrimination specifically tests the ability to differentiate between these spectral variations (McKay 2005).

Apical electric stimulation can be achieved with longer electrode arrays. Alternatively, technics such as "Phantom" electrode stimulation or partial tripolar stimulation have been explored to virtually extend the electrode array toward the apex. It seems that encoding low frequencies starting from 70 Hz improves speech understanding for CIs. These low frequencies are mapped in principle in more apical areas of the cochlea than those achieved with short electrode arrays. However, it is still under debate whether apical electric stimulation provides improvements in terms of speech understanding performance.

Thanks to advances in clinical medical imaging and the routine use of cone beam computed tomography (CBCT) and X-ray imaging, it has been possible to analyze more accurately the location of the electrodes. Combined with models of the frequency-to-place map on the basilar membrane, there is an attempt to create frequency allocation tables in the filterbank that correspond to the tonotopic frequencies of the electrodes. This research area is providing great knowledge on the characterization of the anatomy of the cochlea as well as the relation between pitch and place of stimulation in the cochlea (Helpard et al. 2021; Landsberger et al. 2015).

The large variability in speech understanding outcomes with CIs may be partly explained by the status of the auditory nerve. Several measures have been proposed to infer the condition of the auditory nerve based on evoked compound action potentials (eCAP) measurements (Ramekers et al. 2022; Garcia et al. 2021), as well as detection threshold measures with focused stimulation (e.g., Long et al. 2014; Bierer and Litvak 2016). Analysis of these measures reveals large variability in the status of the auditory nerve. It will become very important to validate these measures with animal and computational models (Zhang et al. 2025b). Therefore, it has become clear that the characterization of the status of the auditory nerve and the electrode-nerve interface will play a key role in the future of designing and adapting sound coding strategies for CIs.

2.9.3 How Much Can Be Expected for the Future?

In the future, many developments are expected in front-end algorithms, sound coding strategies, implanted electronics, stimulators, closed-loop devices, connectivity, and sensors, as well as multimodal stimulation.

The future of signal processing algorithms for reducing background noise and reverberation in CIs is expected to be shaped by advances in adaptive noise suppression, multi-microphone array processing, and machine-learning-based approaches. Key areas with potential for significant improvement include real-time scene analysis, where algorithms dynamically classify and adapt to different acoustic environments, and the integration of deep neural networks that can better separate speech from complex noise backgrounds and reverberation. Additionally, the incorporation of spatial audio processing, user intent detection, and personalized algorithm tuning are likely to provide more natural and robust listening experiences.

As mentioned earlier, beamformers often degrade binaural cues, which are essential for maintaining spatial awareness in CI users. Therefore, preserving these cues is a critical consideration. Beamformers with cue preservation (Gößling et al. 2020), which leverage microphone arrays to emphasize sounds from specific directions while carefully controlling interaural coherence, are promising.

We are experimenting with an AI revolution that soon will be seen in signal processing for CIs. AI is a broad field that encompasses various techniques for creating intelligent machines, while deep learning is a subset of AI that focuses on the use of neural networks with multiple layers to analyze and learn from complex data. The future of deep-learning-based speech enhancement for real-time applications is highly promising, as these models continue to outperform traditional algorithms in suppressing noise and improving speech clarity. Regression models are commonly used for speech enhancement to reduce noise and improve speech quality by learning to predict clean speech signals from noisy inputs. In parallel, generative approaches, such as Generative Adversarial Networks (GANs) and diffusion model, have also been investigated for this purpose. These models aim to perform speech enhancement by learning to synthesize clean speech from noisy input signals, as demonstrated in studies by Pascual et al. (2017) for GANs and by Richter et al. (2022) for diffusion models. These models can be trained using a variety of algorithms, such as maximum likelihood or adversarial training, and have shown promise in restoring distorted speech and improving speech quality. A combination of regression and generative methods is being designed to suppress noise and restore speech, respectively. The development of foundational AI models tailored to speech enhancement is an exciting frontier. In broader tech, "foundation models" are large AI models trained on vast datasets that can be adapted to many tasks (e.g., Luong et al. 2025). However, these models are too large and computationally intensive to be deployed on resource-constrained devices and for speech enhancement. To address this challenge, researchers have developed techniques for distilling these foundational models into smaller, more efficient models using knowledge distillation (Hinton et al. 2015). This involves training smaller models to mimic the behavior of the larger model, resulting in a more compact and efficient model that can be deployed on resource-constrained devices. Advances in neural network efficiency and specialized hardware are making it increasingly feasible to deploy these complex models in real time and with low latency, even on portable and low-power devices like CIs.

2.9.3.1 Sound Coding Strategies

AI will not only be incorporated as a front-end algorithm, future deep learning sound coding strategies for CIs are moving toward end-to-end processing, where deep learning models directly map raw audio input to optimal electrical stimulation patterns for the electrode array. Unlike traditional front-end AI algorithms—which separately preprocess sound and often add latency and computational complexity—end-to-end approaches integrate all signal processing steps into a unified, efficient model (Gajecki and Nogueira 2022; Gajecki et al. 2023; Huang et al. 2024).

Future sound coding strategies for CIs will likely be highly individualized, tailoring stimulation patterns to each user's unique anatomy and neural health. By incorporating data such as the precise insertion depth of each electrode contact, the status and distribution of surviving auditory nerve fibers, characterization of the electric fields when stimulating the electrodes, and individual anamnesis factors like etiology and duration of deafness, these strategies can optimize frequency allocation, stimulation parameters, and dynamic range. This personalized approach has the potential to maximize speech intelligibility, sound quality, and overall auditory outcomes for each CI recipient. Future developments in the area of current focusing with phased array stimulation variants are expected (Van den Honert and Kelsall 2007); also the use of novel stimulation modes, for example, based on combinations of analogue (Nogueira and Büchner 2012; Stupak et al. 2018) and pulsatile stimulation or trying to steer the current more apically through Phantom and mimic benefits observed in electric acoustic stimulation (EAS) (Krüger et al. 2021) are promising.

Moreover, future sound coding strategies will have to take into account the binaural system and aim to tightly link the AGCs (Archer-Boyd and Carlyon 2019), band selection, and stimulation timing between two CIs, enabling fully synchronized delivery of interaural time and level cues. Binaural sound coding strategies can be specifically designed to enhance sound localization for CI users by accurately encoding ITDs at low pulse rates by delivering stimulation at rates that allow the auditory nerve to better phase-lock to temporal cues. The fundamental asynchronous stimulus timing (FAST) sound coding strategy developed by Cochlear (Smith et al. 2013) preserves fine temporal information essential for spatial hearing. This approach aims to improve users' ability to localize sounds while maintaining efficient power consumption. However, such strategies must face the inherent limitation in temporal processing reviewed in the previous subsection. To address these challenges, novel strategies have been proposed that involve exchanging compression and inhibition cues across ears, for instance, to mimic the action of the medial olivocochlear reflex (Lopez-Poveda et al. 2025). These interaural mechanisms, inhibition together with excitation, can also be modeled through AI algorithms by fusing the latent spaces across sides in end-to-end sound coding strategies (Gajecki and Nogueira 2024). The effect of synchronizing the bands selected on each side in N-of-M sound coding strategies has shown benefits in certain speech situations with background noise as well as sound localization through more consistent encoding of interaural time differences. Also, trends toward better speech understanding have been observed with synchronized electric stimulation (Gajecki and Nogueira 2020, 2021).

By coordinating these processing elements in real time, such strategies have the potential to significantly enhance spatial hearing and allow users to learn and utilize binaural cues more naturally. However, their clinical implementation is currently challenged by hardware limitations, including the lack of real-time, low-latency communication between devices, synchronization constraints, and increased computational and power demands. Overcoming these challenges is essential to realize the full benefits of binaural integration for CI users.

2.9.3.2 Implanted Electronics, Stimulators, and Closed-Loop Devices

Novel implanted electronics that go beyond traditional designs are already incorporating internal memory to store personalized hearing settings and digital signal processing chips. These implants will also feature adaptive configuration of the current sources as well as recording capabilities simultaneous to electric stimulation. This flexibility in future implants will allow the design of novel methods to stimulate the auditory nerve.

Moreover, new capabilities in CIs are emerging that allow the devices to record not only peripheral signals, such as ECAPs from the auditory nerve, but also cortical signals reflecting brain responses to sound (Aldag et al. 2022; Attias et al. 2022). By integrating these neural recordings, future CIs can become closed-loop devices that automatically monitor and adapt stimulation in real time based on both peripheral and central auditory system feedback. This closed-loop approach promises more precise, individualized fitting, and dynamic optimization of sound coding, leading to improved hearing outcomes and more intuitive user experiences.

Future CIs will be much smaller or even fully implanted. In this respect, the sound coding strategy has huge potential in contributing to power savings. Sound coding strategies that save power consumption in CIs typically employ techniques such as selective channel stimulation, where only the most informative spectral bands are activated in each processing cycle (e.g., N-of-M strategies), thereby reducing the number of electrical pulses delivered. Some coding strategies also use lower stimulation rates and long pulse durations to reduce the peak current needed to elicit sufficient loudness and reduce power consumption (Langner et al. 2017). Additionally, approaches like unfocused stimulation or simultaneous multielectrode stimulation can deliver current across several electrodes at once, reducing the total amount of peak current needed and therefore saving power consumption. By optimizing the number, timing, and spatial focus of pulses, these power-efficient strategies extend battery life and contribute to more compact, comfortable, and long-lasting CI systems.

2.9.3.3 Connectivity and Sensors

Major advantages expected in CIs include seamless connectivity and synchronized signal transmission between two implants, enabling binaural processing and better spatial hearing. Integration with hearing aids (bimodal systems) and remote microphones will allow users to benefit from coordinated amplification, improved speech understanding, and easier access to sound in challenging environments. The advent of Auracast™ Bluetooth technology promises universal, low-latency wireless audio streaming from public and personal devices, enhancing user experience by supporting direct, high-quality transmission from smartphones, televisions, and assistive listening systems—opening new possibilities for accessibility and convenience in everyday life as well as binaural signal processing. These advancements will also require the development of novel signal processing algorithms and coding strategies specifically designed to transmit, synchronize, and optimize audio signals across diverse devices and communication channels.

CIs are increasingly being equipped with a range of integrated sensors that will be processed in more sophisticated SoCs that will allow the collection of diverse physiological data, transforming them into platforms for continuous biological data collection that can be used to adapt sound coding strategies. Modern devices may include sensors for heart rate detection, accelerometers for movement and activity tracking, temperature sensors, and even blood oxygenation monitors. These data can be used to further individualize device settings and optimize sound processing in real time.

2.9.3.4 Multimodal Stimulation

Future advancements in CI technology may involve the combination of electric and acoustic stimulation, allowing for a more natural and nuanced sound perception. Additionally, the use of extra-cochlear electric stimulation (Krüger and Nogueira 2023), which involves stimulating the auditory nerve outside of the cochlea, may also be explored to further enhance sound processing and recognition. Furthermore, multimodal stimulation, which incorporates tactile feedback, may be a promising area of research, as it has the potential to enhance performance with CIs by providing users with a more comprehensive and immersive auditory experience (Fletcher et al. 2018; Huang et al. 2020). By combining these different stimulation modalities, researchers may be able to develop more effective and user-friendly CI systems that better mimic the natural hearing process.

What we ultimately learn is that while front-end and coding advances are essential and ongoing—optimizing speech understanding performance—the next major leaps forward may likely require fundamental innovations beyond signal processing alone. This could include breakthroughs in electrode design, neural interface technology, optogenetics, or direct brain stimulation.

Acknowledgments I am very thankful to Robert Hart from the Auditory Prosthetic Group, Department of Otolaryngology at Hannover Medical School for support in preparing some of the Electrodograms and CI research interface images. Funding to prepare this chapter was received from the European Research Council (ERC) under the European Union's Horizon-ERC programme (Grant agreement READIHEAR No. 101044753. PI: Waldo Nogueira) and the ATRAE project NEUROHEAR from the Agencia Estatal de Investigación (Grant NEURO-HEAR: ATR2023-145064, PI: Waldo Nogueira). The author acknowledges the use of Microsoft Copilot for proofreading and enhancing the readability of the manuscript. All suggested changes were carefully reviewed and edited. The author assumes full responsibility for the content of the final publication.

References

Albalate A, Nogueira W, Edler B, Buchner A (2006) Signal analysis by using adaptive filterbanks in cochlear implants. In: 2006 IEEE biomedical circuits and systems conference. IEEE, pp 210–213. https://doi.org/10.1109/BIOCAS.2006.4600345

Aldag N, Büchner A, Lenarz T, Nogueira W (2022) Towards decoding selective attention through cochlear implant electrodes as sensors in subjects with contralateral acoustic hearing. J Neural Eng 19:016023. https://doi.org/10.1088/1741-2552/ac4de6

Alvarez F, Kipping D, Nogueira W (2023) A computational model to simulate spectral modulation and speech perception experiments of cochlear implant users. Front Neuroinform 17:Article 934472. https://doi.org/10.3389/fninf.2023.934472

Archer-Boyd AW, Carlyon RP (2019) Simulations of the effect of unlinked cochlear-implant automatic gain control and head movement on interaural level differences. The Journal of the Acoustical Society of America 145:1389–1400. https://doi.org/10.1121/1.5093623

Arenberg JG et al (2018) A dynamically focusing cochlear implant strategy can improve vowel identification in noise. Ear Hear 39:1136–1145. https://doi.org/10.1097/AUD.0000000000000566

Attias J, HabibAllah S, Tarigoppula VS, Glick H, Chen C, Kanthaiah K, Litvak L (2022) Cortical auditory evoked potentials recorded directly through the cochlear implant in cochlear implant recipients: a feasibility study. Ear Hear 43(5):1426–1436. https://doi.org/10.1097/AUD.0000000000001212

Backus B, Adiloğlu K, Herzke T (2015) A binaural CI research platform for Oticon Medical SP/XP implants enabling ITD/ILD and variable rate processing. Trends Hear 19:Article 2331216515618655. https://doi.org/10.1177/2331216515618655

Baumann U, Weißgerber T, Hoppe U (2025) Anpassung von Cochleaimplantatsystemen. HNO 73: 335–356. https://doi.org/10.1007/s00106-025-01593-5

Baumgärtel RM, Hu H, Krawczyk-Becker M, Marquardt D, Herzke T, Coleman G, Adiloğlu K, Bomke K, Plotz K, Gerkmann T, Doclo S, Kollmeier B, Hohmann V, Dietz M (2015) Comparing binaural pre-processing strategies II: speech intelligibility of bilateral cochlear implant users. Trends Hear 19:Article 2331216515617917. https://doi.org/10.1177/2331216515617917

Berenstein CK, Mens LHM, Mulder JJS, Vanpoucke FJ (2008) Current steering and current focusing in cochlear implants: comparison of monopolar, tripolar, and virtual channel electrode configurations. Ear Hear 29:250–260. https://doi.org/10.1097/AUD.0b013e3181645336

Beyea JA, McMullen KP, Harris MS, Houston DM, Martin JM, Bolster VA, Adunka OF, Moberly AC (2016) Cochlear implants in adults: effects of age and duration of deafness on speech recognition. Otol Neurotol 37(9):1238–1245. https://doi.org/10.1097/MAO.0000000000001162

Blamey PJ (2005) Adaptive dynamic range optimization (ADRO): a digital amplification strategy for hearing aids and cochlear implants. Trends Amplif 9(2):77–98. PMID: 16012705; PMCID: PMC4111489. https://doi.org/10.1177/108471380500900203

Bolner F, Magits S, van Dijk B, Wouters J (2020) Precompensating for spread of excitation in a cochlear implant coding strategy. Hear Res 15(395):107977. https://doi.org/10.1016/j.heares.2020.107977

Boyle PJ, Büchner A, Stone MA, Lenarz T, Moore BCJ (2009) Comparison of dual-time-constant and fast-acting automatic gain control (AGC) systems in cochlear implants. Int J Audiol 48(4): 211–221. https://doi.org/10.1080/14992020802581982

Bozorg-Grayeli A, Guevara N, Bebear JP, Ardoint M, Saaï S, Hoen M, Gnansia D, Romanet P, Lavieille JP (2016) Clinical evaluation of the xDP output compression strategy for cochlear implants. Eur Arch Otorrinolaringol 273:2363–2371. https://doi.org/10.1007/s00405-015-3796-1

Bierer JA, Litvak L (2016) Reducing Channel Interaction Through Cochlear Implant Programming May Improve Speech Perception. Trends in Hearing 20:. https://doi.org/10.1177/2331216516653389

Büchner A, Nogueira W, Edler B, Battmer RD, Lenarz T (2008a) Results from a psychoacoustic model-based strategy for the nucleus-24 and freedom cochlear implants. Otol Neurotol 29(2): 189–192. https://doi.org/10.1097/mao.0b013e318162512c

Büchner A, Brendel M, Krüeger B, Frohne-Büchner C, Nogueira W, Edler B, Lenarz T (2008b) Current steering and results from novel speech coding strategies. Otol Neurotol 29(2):203–207. https://doi.org/10.1097/mao.0b013e318163746

Büchner A, Dyballa KH, Hehrmann P, Fredelake S, Lenarz T (2014) Advanced beamformers for cochlear implant users: acute measurement of speech perception in challenging listening conditions. PLoS One 9(4):e95542. PMID: 24755864; PMCID: PMC3995709. https://doi.org/10.1371/journal.pone.0095542

Capon J (1969) High-resolution frequency-wavenumber spectrum analysis. In: Proceedings of the IEEE, vol 57. IEEE

Cardinale FA, Nogueira W (2022) Predicting speech intelligibility using the Spike Activity Mutual Information Index. In: Proceedings of interspeech, pp 3503–3507. https://doi.org/10.21437/Interspeech.2022-10488

Carlyon RP, Deeks JM, Delgutte B et al (2025) Limitations on temporal processing by cochlear implant users: a compilation of viewpoints. Trends Hear 29. https://doi.org/10.1177/23312165251317006

Chung K (2012) Wind noise in hearing aids: II. Effect of microphone directivity. Int J Audiol 51: 29–42. https://doi.org/10.3109/14992027.2011.609184

Chung K, McKibben N (2011) Microphone directionality, pre-emphasis filter, and wind noise in cochlear implants. J Am Acad Audiol 22:586–600. https://doi.org/10.3766/jaaa.22.9.4

Chung K, Mongeau L, McKibben N (2009) Wind noise in hearing aids with directional and omnidirectional microphones: polar characteristics of behind-the-ear hearing aids. J Acoust Soc Am 125:2243–2259. https://doi.org/10.1121/1.3086268

Croghan NBH, Duran SI, Smith ZM (2017) Re-examining the relationship between number of cochlear implant channels and maximal speech intelligibility. J Acoust Soc Am 142(6):EL537. https://doi.org/10.1121/1.5016044

Cychosz M, Winn MB, Goupell MJ (2024) How to vocode: using channel vocoders for cochlear-implant research. J Acoust Soc Am 155(4):2407–2437. https://doi.org/10.1121/10.0025274

Dawson PW, Mauger SJ, Hersbach AA (2011) Clinical evaluation of signal-to-noise ratio-based noise reduction in nucleus cochlear implant recipients. Ear Hear 32:382–390. https://doi.org/10.1097/AUD.0b013e318201c200

De Groote E, Macherey O, Deeks JM, Roman S, Carlyon RP (2025) Temporal pitch perception of multi-channel stimuli by cochlear-implant users. J Assoc Res Otolaryngol 26(3):301–315. https://doi.org/10.1007/s10162-025-00983-4

De Jong MAM, Briaire JJ, Frijns JHM (2019) Dynamic current focusing: a novel approach to loudness coding in cochlear implants. Ear Hear 40:34–44. https://doi.org/10.1097/AUD.0000000000000593

Deeks JM, Carlyon RP (2004) Simulations of cochlear implant hearing using filtered harmonic complexes: implications for concurrent sound segregation. J Acoust Soc Am 115:1736–1746. https://doi.org/10.1121/1.1675814

Dingemanse GJ, Goedegebure A (2018) Optimising the effect of noise reduction algorithm ClearVoice in cochlear implant users by increasing the maximum comfort levels. Int J Audiol 57(3):230–235. https://doi.org/10.1080/14992027.2017.1390267

Dorman MF, Loizou PC (1997) Speech intelligibility as a function of the number of channels of stimulation for normal-hearing listeners and patients with cochlear implants. Am J Otolaryngol Suppl 18:S113–S114

Dorman MF, Loizou PC, Spahr AJ, Maloff E (2002) A comparison of the speech understanding provided by acoustic models of fixed-channel and channel-picking signal processors for cochlear implants. J Speech Lang Hear Res 45:783–788. https://doi.org/10.1044/1092-4388(2002/063)

Dorman MF, Natale SC, Baxter L, Zeitler DM, Carlson ML, Lorens A, Skarzynski H, Peters JPM, Torres JH, Noble JH (2020) Approximations to the voice of a cochlear implant: explorations with single-sided deaf listeners. Trends Hear 24:2331216520920079. https://doi.org/10.1177/2331216520920079

Dyballa KH, Hehrmann P, Hamacher V, Nogueira W, Lenarz T, Büchner A (2015) Evaluation of a transient noise reduction algorithm in cochlear implant users. Audiol Res 5(2):116. PMID: 26779325; PMCID: PMC4698598. https://doi.org/10.4081/audiores.2015.116

Dyballa KH, Hehrmann P, Hamacher V, Lenarz T, Büchner A (2016) Transient noise reduction in cochlear implant users: a multi-band approach. Audiol Res 6(2):154. PMID: 27942372; PMCID: PMC5134678. https://doi.org/10.4081/audiores.2016.154

Fishman K, Shannon RV, Slattery WH (1997) Speech recognition as a function of the number of electrodes used in the SPEAK cochlear implant speech processor. J Speech Hear Res 40:1201–1215. https://doi.org/10.1044/jslhr.4005.1201

Fletcher MD, Mills SR, Goehring T (2018) Vibro-tactile enhancement of speech intelligibility in multi-talker noise for simulated cochlear implant listening. Trends Hear 22:Article 2331216518797838. https://doi.org/10.1177/2331216518797838

Fredelake S, Hohmann V (2012) Factors affecting predicted speech intelligibility with cochlear implants in an auditory model for electrical stimulation. Hearing Research 287:76–90. https://doi.org/10.1016/j.heares.2012.03.005

Friesen LM, Shannon RV, Baskent D, Wang X (2001) Speech recognition in noise as a function of the number of spectral channels: comparison of acoustic hearing and cochlear implants. J Acoust Soc Am 110(2):1150–1163. PMID: 11519582. https://doi.org/10.1121/1.1381538

Fu QJ, Shannon RV, Wang X (1998) Effects of noise and spectral resolution on vowel and consonant recognition: acoustic and electric hearing. J Acoust Soc Am 104:3586–3596. https://doi.org/10.1121/1.423941

Gajecki T, Nogueira W (2020) The effect of synchronized linked band selection on speech intelligibility of bilateral cochlear implant users. Hear Res 396:Article 108051. https://doi.org/10.1016/j.heares.2020.108051

Gajecki T, Nogueira W (2021) Enhancement of interaural level differences for bilateral cochlear implant users. Hear Res 409:Article 108313. https://doi.org/10.1016/j.heares.2021.108313

Gajecki T, Nogueira W (2022) An end-to-end deep learning speech coding and denoising strategy for cochlear implants. In: ICASSP IEEE international conference on acoustics, speech and signal processing (ICASSP), Singapore, pp 3109–3113. https://doi.org/10.1109/ICASSP43922.2022.9746963

Gajecki T, Nogueira W (2024) A fused deep denoising sound coding strategy for bilateral cochlear implants. IEEE Trans Biomed Eng 71(7):2232–2242. https://doi.org/10.1109/TBME.2024.3367530

Gajecki T, Zhang Y, Nogueira W (2023) A deep denoising sound coding strategy for cochlear implants. IEEE Trans Biomed Eng 70(9):2700–2709. https://doi.org/10.1109/TBME.2023.3262677

Garcia C, Goehring T, Cosentino S, Turner RE, Deeks JM, Brochier T, Rughooputh T, Bance M, Carlyon RP (2021) The Panoramic ECAP Method: estimating patient-specific patterns of current spread and neural health in cochlear implant users. J Assoc Res Otolaryngol 22(5): 567–589. https://doi.org/10.1007/s10162-021-00795-2

Garnham C, O'Driscoll M, Ramsden R, Saeed S (2002) Speech understanding in noise with a Mel-El COMBI 40+ cochlear implant using reduced channel sets. Ear Hear 23:540–552. https://doi.org/10.1097/00003446-200212000-00005

Gaultier C, Goehring T (2024) Recovering speech intelligibility with deep learning and multiple microphones in noisy-reverberant situations for people using cochlear implants. J Acoust Soc Am 155(6):3833–3847. PMID: 38884525. https://doi.org/10.1121/10.0026218

Ghosh R, Ali H, Hansen JHL (2022) CCi-MOBILE: a portable real-time speech processing platform for cochlear implant and hearing research. IEEE Trans Biomed Eng 69(3): 1251–1263. https://doi.org/10.1109/TBME.2021.3123241

Ghovanloo M, Najafi K (2005) A compact large voltage-compliance high output-impedance programmable current source for implantable microstimulators. IEEE Trans Biomed Eng 52: 97–105. https://doi.org/10.1109/TBME.2004.839797

Goehring T, Archer-Boyd AW, Arenberg JG et al (2021) The effect of increased channel interaction on speech perception with cochlear implants. Sci Rep 11:Article 10383. https://doi.org/10.1038/s41598-021-89932-8

Goorevich M, Zakis JA, Case S, Harvey T, Holmberg P et al (2012) Effect of a wind noise reduction algorithm on cochlear implant sound processing. In: Paper presented at the 12th international conference on cochlear implants and other implantable technologies, Baltimore, MD, USA

Gößling N, Marquardt D, Doclo S (2020) Perceptual evaluation of binaural MVDR-based algorithms to preserve the interaural coherence of diffuse noise fields. Trends Hear 24:Article 2331216520919573. https://doi.org/10.1177/2331216520919573

Grange JA, Culling JF, Harris NSL, Bergfeld S (2017) Cochlear implant simulator with independent representation of the full spiral ganglion. J Acoust Soc Am 142(5):EL484. https://doi.org/10.1121/1.5009602

Griffiths L, Jim C (1982) An alternative approach to linearly constrained adaptive beamforming. IEEE Trans Antennas Propag 30:27–34

Guevara N (2016) The voice track multiband single-channel modified Wiener-filter noise reduction system for cochlear implants: patients' outcomes and subjective appraisal. Int J Audiol 55(8): 431–438. https://doi.org/10.3109/14992027.2016.1172267

Guo S, Hoi L, Loizou P (2008) A 9-Bit configurable current source with enhanced cutput resistance for cochlear stimulators. In: 2008 IEEE custom integrated circuits conference, San Jose, CA, pp 511–514. https://doi.org/10.1109/CICC.2008.4672134

Hagen R, Radeloff A, Stark T, Anderson I, Nopp P, Aschbacher E, Möltner A, Khajehnouri Y, Rak K (2020) Microphone directionality and wind noise reduction enhance speech perception in users of the MED-EL SONNET audio processor. Cochlear Implants Int 21(1):53–65. https://doi.org/10.1080/14670100.2019.1664529

Hamacher V, Chalupper J, Eggers J et al (2005) Signal processing in high-end hearing aids: state of the art, challenges, and future trends. EURASIP J Adv Signal Process 2005:152674. https://doi.org/10.1155/ASP.2005.2915

Harczos T, Chilian A, Husar P (2013) Making use of auditory models for better mimicking of normal hearing processes with cochlear implants: the SAM coding strategy. IEEE Trans Biomed Circuits Syst 7(4):414–425. https://doi.org/10.1109/TBCAS.2012.2219530

Helpard L, Li H, Rohani SA, Zhu N, Rask-Andersen H, Agrawal S, Ladak HM (2021) An approach for individualized cochlear frequency mapping determined from 3D synchrotron radiation phase-contrast imaging. IEEE Trans Biomed Eng 68(12):3602–3611. https://doi.org/10.1109/TBME.2021.3080116

Hey M, Hocke T, Böhnke B, Mauger SJ (2019) Forward Focus with cochlear implant recipients in spatially separated and fluctuating competing signals—introduction of a reference metric. Int J Audiol 58(12):869–878. https://doi.org/10.1080/14992027.2019.1638527

Hines A, Harte N (2012) Speech intelligibility prediction using a neurogram similarity index measure. Speech Comm 54(2):306–320

Hinrichs R, Gajecki T, Ostermann J, Nogueira W (2021) A subjective and objective evaluation of a codec for the electrical stimulation patterns of cochlear implants. J Acoust Soc Am 149(2): 1324–1337. https://doi.org/10.1121/10.0003571

Hinton G, Vinyals O, Dean J (2015) Distilling the knowledge in a neural network. arXiv:1503.02531

Hochmair I, Hochmair E, Nopp P, Waller M, Jolly C (2015) Deep electrode insertion and sound coding in cochlear implants. Hear Res 322:14–23. https://doi.org/10.1016/j.heares.2014.10.006

Honeder C, Liepins R, Arnoldner C, Šinkovec H, Kaider A, Vyskocil E, Riss D (2018) Fixed and adaptive beamforming improves speech perception in noise in cochlear implant recipients equipped with the MED-EL SONNET audio processor. PLoS One 13(1):e0190718. PMID: 29304186; PMCID: PMC5755914. https://doi.org/10.1371/journal.pone.0190718

Huang J, Lu T, Sheffield B, Zeng FG (2020) Electro-tactile stimulation enhances cochlear-implant melody recognition: effects of rhythm and musical training. Ear Hear 41(1):106–113. https://doi.org/10.1097/AUD.0000000000000749

Huang EH-H, Chao R, Tsao Y, Wu C-M (2024) ElectrodeNet—a deep-learning-based sound coding strategy for cochlear implants. IEEE Trans Cogn Dev Syst 16(1):346–357. https://doi.org/10.1109/TCDS.2023.3275587

Imsiecke M, Krüger B, Büchner A, Lenarz T, Nogueira W (2018) Electric-acoustic forward masking in cochlear implant users with ipsilateral residual hearing. Hear Res 364:25–37. https://doi.org/10.1016/j.heares.2018.04.003

Jürgens T, Hohmann V, Büchner A, Nogueira W (2018) The effects of electrical field spatial spread and some cognitive factors on speech-in-noise performance of individual cochlear implant users: a computer model study. PLoS One 13(4):e0193842. https://doi.org/10.1371/journal.pone.0193842

Kan A, Litovsky RY (2015) Binaural hearing with electrical stimulation. Hear Res 322:127–137. https://doi.org/10.1016/j.heares.2014.08.005

Karoui C, James C, Barone P, Bakhos D, Marx M, Macherey O (2019) Searching for the sound of a cochlear implant: evaluation of different vocoder parameters by cochlear implant users with single-sided deafness. Trends Hear 23:2331216519866029. https://doi.org/10.1177/2331216519866029

Kayser H, Herzke T, Maanen P, Zimmermann M, Grimm G, Hohmann V (2022) Open community platform for hearing aid algorithm research: open Master Hearing Aid (openMHA). SoftwareX 17:100953. https://doi.org/10.1016/j.softx.2021.100953

Kim JS, Kim CH (2014) A review of assistive listening device and digital wireless technology for hearing instruments. Korean J Audiol 18(3):105–111. PMCID: PMC4285000. https://doi.org/10.7874/kja.2014.18.3.105

Klinke R, Kral A, Heid S, Tillein J, Hartmann R (1999) Recruitment of the auditory cortex in congenitally deaf cats by long-term cochlear electrostimulation. Science 285:1729–1733. https://doi.org/10.1126/science.285.5434.1729

Kludt E, Nogueira W, Lenarz T, Büchner A (2021) A sound coding strategy based on a temporal masking model for cochlear implants. PLoS One 16(1):e0244433. https://doi.org/10.1371/journal.pone.0244433

Koch DB, Downing M, Osberger MJ, Litvak L (2007) Using current steering to increase spectral resolution in CII and HiRes 90K users. Ear Hear 28(2 Suppl):38S–41S. https://doi.org/10.1097/AUD.0b013e31803150de

Kokkinakis K, Azimi B, Hu Y, Friedland DR (2012) Single and multiple microphone noise reduction strategies in cochlear implants. Trends Amplif 16(2):102–116. PMID: 22923425; PMCID: PMC3691954. https://doi.org/10.1177/1084713812456906

Kong YY, Carlyon RP (2010) Temporal pitch perception at high rates in cochlear implants. J Acoust Soc Am 127(5):3114–3123. https://doi.org/10.1121/1.3372713

Kral A, Tillein J (2006) Brain plasticity under cochlear implant stimulation. Adv Otorhinolaryngol 64:89–108. https://doi.org/10.1159/000094647

Kral A, Hartmann R, Tillein J, Heid S, Klinke R (2002) Hearing after congenital deafness: central auditory plasticity and sensory deprivation. Cereb Cortex 12(8):797–807. https://doi.org/10.1093/cercor/12.8.797

Kral A, Dorman MF, Wilson BS (2019) Neuronal development of hearing and language: cochlear implants and critical periods. Annu Rev Neurosci 42(1):47–65. https://doi.org/10.1146/annurev-neuro-080317-061513

Krüger B, Nogueira W (2023) Towards extracochlear electric-acoustic stimulation of the auditory system. Paper presented at the 2023 conference on implantable auditory prostheses, Lake Tahoe, CA

Krüger B, Büchner A, Nogueira W (2021) Phantom stimulation for cochlear implant users with residual low-frequency hearing. Ear Hear 43(2):631–645. https://doi.org/10.1097/AUD.0000000000001121

Kurz A, Rak K, Hagen R (2022) Improved performance with automatic sound management 3 in the MED-EL SONNET 2 cochlear implant audio processor. PLoS One 17(9):e0274446. https://doi.org/10.1371/journal.pone.0274446

Lamping W, Goehring T, Marozeau J, Carlyon RP (2020) The effect of a coding strategy that removes temporally masked pulses on speech perception by cochlear implant users. Hear Res 391:Article 107969. https://doi.org/10.1016/j.heares.2020.107969

Landsberger DM, Svrakic M, Roland JT Jr, Svirsky M (2015) The relationship between insertion angles, default frequency allocations, and spiral ganglion place pitch in cochlear implants. Ear Hear 36(5):e207–e213. https://doi.org/10.1097/AUD.0000000000000163

Langner F, Saoji AA, Büchner A, Nogueira W (2017) Adding simultaneous stimulating channels to reduce power consumption in cochlear implants. Hear Res 345:96–107. https://doi.org/10.1016/j.heares.2017.01.010

Langner F, Büchner A, Nogueira W (2020) Evaluation of an adaptive dynamic compensation system in cochlear implant listeners. Trends Hear 24:Article 2331216520970349. PMID: 33356927; PMCID: PMC7887680. https://doi.org/10.1177/2331216520970349

Lazard DS, Vincent C, Venail F, Van de Heyning P, Truy E, Sterkers O, Skarzynski PH, Skarzynski H, Schauwers K, O'Leary S et al (2012) Pre-, Per- and Postoperative factors affecting performance of postlinguistically deaf adults using cochlear implants: a new conceptual model over time. PLoS One 7(11):e48739. https://doi.org/10.1371/journal.pone.0048739

Leclère T, Johannesen PT, Wijetillake A, Segovia-Martínez M, Lopez-Poveda EA (2023) A computational modelling framework for assessing information transmission with cochlear implants. Hear Res 432:Article 108744. https://doi.org/10.1016/j.heares.2023.108744

Lee EKF (2015) A discrete controlled fully integrated class E coil driver with power efficient ASK modulation for powering biomedical implants. IEEE Trans Circuits Syst I: Regul Pap 62(6): 1678–1687. PMID: 1592409; PMCID: PMC2782849. https://doi.org/10.1109/TCSI.2015.2415176

Litovsky RY, Goupell MJ, Kan A, Landsberger DM (2017) Use of research interfaces for psychophysical studies with cochlear-implant users. Trends Hear 21:Article 2331216517736464. https://doi.org/10.1177/2331216517736464

Litvak L, Delgutte B, Eddington D (2003) Improved neural representation of vowels in electric stimulation using desynchronizing pulse trains. J Acoust Soc Am 114(4 Pt 1): 2099–2111. PMID: 14587608; PMCID: PMC2275169. https://doi.org/10.1121/1.1612494

Long CJ, Holden TA, McClelland GH, Parkinson WS, Shelton C, Kelsall DC, Smith ZM (2014) Examining the electro-neural interface of cochlear implant users using psychophysics, CT scans, and speech understanding. J Assoc Res Otolaryngol 15(2):293–304. https://doi.org/10.1007/s10162-013-0437-5

Lopez-Poveda EA, Eustaquio-Martín A, Fumero MJ, et al (2025) Binaural audio frontend processing for cochlear implants inspired by the medial olivocochlear reflex. Front Neurosci 19. https://doi.org/10.3389/fnins.2025.1678288

Luong D, Heikkinen M, Drossos K, Virtanen T (2025) Knowledge distillation for speech denoising by latent representation alignment with cosine distance. arXiv:2505.03442

McCreery DB et al (1995) Relationship between stimulus amplitude, stimulus frequency and neural damage during electrical stimulation of sciatic nerve of cat. Med Biol Eng Comput 33(3): 426–429. https://doi.org/10.1007/BF02510526

McCreery DB et al (1997) A characterization of the effects on neuronal excitability due to prolonged microstimulation with chronically implanted microelectrodes. IEEE Trans Biomed Eng 44(10):931–939. https://doi.org/10.1109/10.634645

McKay CM (2005) Spectral Processing In Cochlear Implants. International Review of Neurobiology 473–509. https://doi.org/10.1016/S0074-7742(05)70014-3

Mens LHM, Berenstein CK (2005) Speech perception with mono- and quadrupolar electrode configurations: a crossover study. Otol Neurotol 26:957–964. https://doi.org/10.1097/01.mao.0000185060.74339.9d

Middlebrooks JC (2018) Chronic deafness degrades temporal acuity in the electrically stimulated auditory pathway. J Assoc Res Otolaryngol 19:541–557. https://doi.org/10.1007/s10162-018-0679-3

Milczynski M, Wouters J, van Wieringen A (2009) Improved fundamental frequency coding in cochlear implant signal processing. J Acoust Soc Am 125(4):2260–2271. https://doi.org/10.1121/1.3085642

Molaee-Ardekani B, Attili Chiea R, Zhang Y, Felding J, Wijetillake AA, Johannesen PT, Lopez-Poveda EA, Segovia-Martínez M (2025) A novel sound coding strategy for cochlear implants based on spectral feature and temporal event extraction. Technologies 13:Article 318. https://doi.org/10.3390/technologies13080318

Müller J, Brill S, Hagen R, Moeltner A, Brockmeier SJ, Stark T, Helbig S, Maurer J, Zahnert T, Zierhofer C, Nopp P, Anderson I (2012) Clinical trial results with the MED-EL fine structure processing coding strategy in experienced cochlear implant users. ORL J Otorhinolaryngol Relat Spec 74(4):185–198. https://doi.org/10.1159/000337089

Müller V, Ukuwait A, Fröẞler K, Lang-Roth R (2025) Effect of ambient noise reduction on speech recognition and listening effort in noise in MED-EL cochlear implant users. Int J Audiol. https://doi.org/10.1080/14992027.2025.2499099

Neustetter C, Zangerl M, Spitzer P, Zierhofer C (2012) In-vitro characterization of a cochlear implant system for recording of evoked compound action potentials. Biomed Eng Online 11:22. https://doi.org/10.1186/1475-925X-11-22

Ngamkham W, van Dongen MN, Serdijn WA, Bes CJ, Briaire JJ, Frijns JHM (2015) A 0.042 mm2 programmable biphasic stimulator for cochlear implants suitable for a large number of channels. ArXiv 1502.00549

Nogueira W, Büchner A (2012) Conveying low frequency information through analog electrical stimulation in cochlear implants. In: Proceedings of the 20th European signal processing conference (EUSIPCO), Bucharest, Romania, pp 509–513

Nogueira W, Büchner A, Lenarz T et al (2005) A psychoacoustic "NofM"-type speech coding strategy for cochlear implants. EURASIP J Adv Signal Process:Article 101672. https://doi.org/10.1155/ASP.2005.3044

Nogueira W, Giese A, Edler B, Buchner A (2006) Wavelet packet filterbank for speech processing strategies in cochlear implants. In: IEEE international conference on acoustics speech and signal processing proceedings, Toulouse, France. https://doi.org/10.1109/ICASSP.2006.1661227

Nogueira W, Kátai A, Harczos T, Klefenz F, Büchner A, Edler B (2007a) An auditory model based strategy for cochlear implants. In: Annual international conference of the IEEE Engineering in Medicine and Biology Society, pp 4127–4130. https://doi.org/10.1109/IEMBS.2007.4353244

Nogueira W, Harczos T, Edler B, Ostermann J, Büchner A (2007b) Automatic speech recognition with a cochlear implant front-end. In: Interspeech, vol 3. ISCA, pp 1993–1996

Nogueira W, Litvak L, Edler B et al (2009) Signal processing strategies for cochlear implants using current steering. EURASIP J Adv Signal Process:Article 531213. https://doi.org/10.1155/2009/531213

Nogueira W, Lopez M, Rode T, Doclo S, Büchner A (2015) Individualizing a monaural beamformer for cochlear implant users. In: 2015 IEEE international conference on acoustics, speech and signal processing (ICASSP). IEEE, pp 5738–5742. https://doi.org/10.1109/ICASSP.2015.7179071

Nogueira W, Litvak LM, Landsberger DM, Büchner A (2017) Loudness and pitch perception using dynamically compensated virtual channels. Hear Res 344:223–234. https://doi.org/10.1016/j.heares.2016.11.017

Nogueira W, Nagathil A, Martin R (2018) Making music more accessible for cochlear implant listeners: recent developments. IEEE Signal Process Mag 36(1):115–127. https://doi.org/10.1109/MSP.2018.2874059

Pascual S, Bonafonte A, Serrà J (2017) SEGAN: speech enhancement generative adversarial network. arXiv:1703.09452. https://doi.org/10.48550/arXiv.1703.09452

Ping L, Wang N, Tang G, Lu T, Yin L, Tu W, Fu QJ (2017) Implementation and preliminary evaluation of 'C-tone': a novel algorithm to improve lexical tone recognition in Mandarin-speaking cochlear implant users. Cochlear Implants Int 18(5):240–249. PMID: 28629258. https://doi.org/10.1080/14670100.2017.1339492

Qazi OU, van Dijk B, Moonen M, Wouters J (2013) Understanding the effect of noise on electrical stimulation sequences in cochlear implants and its impact on speech intelligibility. Hear Res 299:79–87. https://doi.org/10.1016/j.heares.2013.01.018

Ramekers D, Benav H, Klis SFL, Versnel H (2022) Changes in the electrically evoked compound action potential over time after implantation and subsequent deafening in guinea pigs. J Assoc Res Otolaryngol 23(6):721–738. https://doi.org/10.1007/s10162-022-00864-0

Reiss LAJ, Turner CW, Erenberg SR, Gantz BJ (2007) Changes in Pitch with a Cochlear Implant Over Time. JARO 8:241–257. https://doi.org/10.1007/s10162-007-0077-8

Richter J, Welker S, Lemercier JM, Lay B, Gerkmann T (2022) Speech enhancement and dereverberation with diffusion-based generative models. IEEE/ACM Trans Audio Speech Lang Process 31:2351–2364. arXiv 2208.05830

Riss D, Arnoldner C, Reiss S, Baumgartner WD (2009) One-year results using the Opus speech processor with the fine structure speech coding strategy. Acta Otolaryngol 129(9):988–991. https://doi.org/10.1080/00016480802552485

Riss D, Hamzavi JS, Selberherr A, Kaider A, Blineder M, Starlinger V, Gstoettner W, Arnoldner C (2011) Envelope versus fine structure speech coding strategy: a crossover study. Otol Neurotol 32(7):1094–1101. https://doi.org/10.1097/MAO.0b013e31822a97f4

Riss D, Hamzavi JS, Blineder M, Flak S, Baumgartner WD, Kaider A, Arnoldner C (2016) Effects of Stimulation Rate With the FS4 and HDCIS Coding Strategies in Cochlear Implant Recipients. Otol Neurotol 37(7):882–888. https://doi.org/10.1097/MAO.0000000000001107

Runge-Samuelson CL (2009) Effects of high-rate pulse trains on electrode discrimination in cochlear implant users. Trends Amplif 13(2):76–86. https://doi.org/10.1177/1084713809336739

Scalart P, Filho JV (1996) Speech enhancement based on a priori signal to noise estimation. In: IEEE international conference on acoustics, speech, and signal processing. IEEE, pp 629–632

Schädler MR, Warzybok A, Hochmuth S, Kollmeier B (2015) Matrix sentence intelligibility prediction using an automatic speech recognition system. Int J Audiol Supplement 2:100–107. https://doi.org/10.3109/14992027.2015.1061708

Seeber BU, Fastl H (2008) Localization cues with bilateral cochlear implants. J Acoust Soc Am 123:1030–1042. https://doi.org/10.1121/1.2821965

Seligman P, McDermott H (1995) Architecture of the Spectra 22 speech processor. Ann Otol Rhinol Laryngol 104:139–141

Shader MJ, Nguyen N, Cleary M, et al (2019) Effect of Stimulation Rate on Speech Understanding in Older Cochlear-Implant Users. Ear & Hearing 41:640–651. https://doi.org/10.1097/aud.0000000000000793

Shannon RV (1992) A model of safe levels for electrical stimulation. IEEE Trans Biomed Eng 39(4):424–426. https://doi.org/10.1109/10.126616

Shannon RV, Zeng F-G, Kamath V, Wygonski J, Ekelid M (1995) Speech recognition with primarily temporal cues. Science 270:303–304. https://doi.org/10.1126/science.270.5234.303

Shannon RV, Cruz RJ, Galvin JJ 3rd. (2011) Effect of stimulation rate on cochlear implant users' phoneme, word and sentence recognition in quiet and in noise. Audiol Neurotol 16(2):113–123. https://doi.org/10.1159/000315115

Skinner MW, Holden LK, Whitford LA, Plant KL, Psarros C, Holden TA (2002) Speech recognition with the nucleus 24 SPEAK, ACE, and CIS speech coding strategies in newly implanted adults. Ear Hear 23(3):207–223. https://doi.org/10.1097/00003446-200206000-00005

Smith ZM, Parkinson WS, Krishnamoorthi H (2013) Efficient coding for auditory prostheses. Paper presented at the 2013 conference on implantable auditory prostheses

Spriet A, Van Deun L, Eftaxiadis K, Laneau J, Moonen M, van Dijk B, van Wieringen A, Wouters J (2007) Speech understanding in background noise with the two-microphone adaptive beamformer BEAM in the Nucleus Freedom Cochlear Implant System. Ear Hear 28(1): 62–72. https://doi.org/10.1097/01.aud.0000252470.54246.54

Srinivasan AG, Padilla M, Shannon RV, Landsberger DM (2013) Improving speech perception in noise with current focusing in cochlear implant users. Hear Res 299:29–36. https://doi.org/10.1016/j.heares.2013.02.004

Stone MA, Moore BCJ (2005) Tolerable Hearing-Aid Delays IV: effects on subjective disturbance during speech production by hearing-impaired subjects. Ear Hear 26(2):225–235. https://doi.org/10.1097/00003446-200504000-00009

Stone MA, Moore BCJ (2008) Effects of spectro-temporal modulation changes produced by multichannel compression on intelligibility in a competing-speech task. J Acoust Soc Am 123:1063–1076. https://doi.org/10.1121/1.2821969

Stone MA, Moore BCJ, Alcántara JI, Glasberg BR (1999) Comparison of different forms of compression using wearable digital hearing aids. J Acoust Soc Am 106:3603–3619. https://doi.org/10.1121/1.428213

Stronks HC, Briaire J, Frijns J (2022) Beamforming and single-microphone noise reduction: effects on signal-to-noise ratio and speech recognition of bimodal cochlear implant users. Trends Hear 26:Article 23312165221112762. PMID: 35862265; PMCID: PMC9310275. https://doi.org/10.1177/23312165221112762

Stupak N, Padilla M, Morse RP, Landsberger DM (2018) Perceptual differences between low-frequency analog and pulsatile stimulation as shown by single- and multidimensional scaling. Trends Hear 22. https://doi.org/10.1177/2331216518807535

Sulas E, Hasan PY, Zhang Y et al (2023) Streamlining experiment design in cognitive hearing science using OpenSesame. Behav Res 55:1965–1979. https://doi.org/10.3758/s13428-022-01886-5

Svirsky MA, Talavage TM, Sinha S, Neuburger H, Azadpour M (2015) Gradual adaptation to auditory frequency mismatch. Hear Res 322:163–170. https://doi.org/10.1016/j.heares.2014.10.008

Tabibi S, Boulet J, Dillier N, Bruce IC (2021) Phenomenological model of auditory nerve population responses to cochlear implant stimulation. J Neurosci Methods 358:Article 109212. https://doi.org/10.1016/j.jneumeth.2021.109212

Theelen-van den Hoek FL, Boymans M, Stainsby T, Dreschler WA (2014) Reliability of categorical loudness scaling in the electrical domain. Int J Audiol 53:409–417. https://doi.org/10.3109/14992027.2013.879338

Vaerenberg B, Smits C et al (2014) Cochlear implant programming: a global survey on the state of the art. Sci World J 2014:501738. https://doi.org/10.1155/2014/501738

Van den Honert C, Kelsall DC (2007) Focused intracochlear electric stimulation with phased array channels. J Acoust Soc Am 121(6):3703–3716. https://doi.org/10.1121/1.2722047

Velde HM, Rademaker MM, Damen J, Smit AL, Stegeman I (2021) Prediction models for clinical outcome after cochlear implantation: a systematic review. J Clin Epidemiol 137:182–194. https://doi.org/10.1016/j.jclinepi.2021.04.005

Wilson BS, Finley CC, Lawson DT, Wolford RD, Eddington DK, Rabinowitz WM (1991) Better speech recognition with cochlear implants. Nature 352(6332):236–238. https://doi.org/10.1038/352236a0

Wohlbauer DM, Dillier N (2025) A Hundred Ways to Encode Sound Signals for Cochlear Implants. Annual Review of Biomedical Engineering 27:335–369. https://doi.org/10.1146/annurev-bioeng-102623-121249

Wouters J, McDermott HJ, Francart T (2015) Sound coding in cochlear implants: from electric pulses to hearing. IEEE Signal Process Mag 32(2):67–80. https://doi.org/10.1109/MSP.2014.2371671

Zeng FG, Galvin J (1999) Amplitude compression and phoneme recognition in cochlear implant listeners. Ear Hear 20:60–74. https://doi.org/10.1097/00003446-199902000-00006

Zeng FG, Rebscher S, Harrison W, Sun X, Feng H (2008) Cochlear implants: system design, integration, and evaluation. IEEE Rev Biomed Eng 1:115–142. Epub ahead of print. PMID: 19946565; PMCID: PMID: PMC2782849. https://doi.org/10.1109/RBME.2008.2008250

Zeng FG, Rebscher SJ, Fu QJ, Chen H, Sun X et al (2015) Development and evaluation of the Nurotron 26-electrode cochlear implant system. Hear Res 322:188–199. https://doi.org/10.1016/j.heares.2014.09.013

Zhang Y, Johannesen PT, Molaee-Ardekani B et al (2025a) Comparison of performance for cochlear-implant listeners using audio processing strategies based on short-time fast Fourier transform or spectral feature extraction. Ear Hear 46(1):163–183. https://doi.org/10.1097/AUD.0000000000001565

Zhang Y, Kipping D, Nogueira W (2025b) Evaluating electrophysiological and behavioral measures of neural health in cochlear implant users: a computational simulation study. IEEE Trans Biomed Eng. https://doi.org/10.1109/TBME.2025.3573398

Zhou N, Zhu Z, Dong L, Galvin JJ 3rd. (2020) Effect of pulse phase duration on forward masking and spread of excitation in cochlear implant listeners. PLoS One 15(7):e0236179. Published 2020 Jul 20. https://doi.org/10.1371/journal.pone.0236179

Zirn S, Angermeier J, Arndt S, Aschendorff A, Wesarg T (2019) Reducing the device delay mismatch can improve sound localization in bimodal cochlear implant/hearing-aid users. Trends Hear 23:Article 2331216519843876. PMID: 31018790; PMCID: PMC6484236. https://doi.org/10.1177/2331216519843876

Performance Evaluation Methods in Cochlear Implant Patients

3

Emily R. Spitzer, Natalia Stupak, Madeleine M. Beyer, Gabriel McDerment, Jun Zhang, and David M. Landsberger

Abstract

There are an almost infinite number of ways to evaluate performance with a cochlear implant (CI). This chapter provides an overview of recent developments for performance evaluation, using the current clinical evaluation as a baseline. Clinical evaluation includes identifying candidacy, postoperative monitoring schedules, device programming, and audiologic evaluations. Audiologic evaluation typically comprises the measurement of aided hearing thresholds and speech recognition. Emphasis is placed on tailoring the clinical evaluation for individual patient needs and special populations. However, clinical evaluations are quite limited in their ability to capture the listening difficulties patients experience in everyday life. It is important to assess outcomes, particularly speech perception, in realistic environments. These include modifications of the stimuli or listening experience to have greater ecological validity. Music and sound quality perception are also discussed. Realistic environments are not the real world and still present significant limitations for understanding performance for CI users. A recent line of work evaluates CI performance in the listener's everyday life using tools such as ecological momentary assessment (EMA) and wearable technology, providing continuous, context-dependent insights. Finally, we

E. R. Spitzer (✉) · N. Stupak · M. M. Beyer · D. M. Landsberger
Department of Otolaryngology—Head and Neck Surgery, NYU Grossman School of Medicine, New York, NY, USA
e-mail: emily.spitzer@nyulangone.org; natalia.stupak@nyulangone.org; madeleine.campbell@nyulangone.org; david.landsberger@nyulangone.org

G. McDerment · J. Zhang
New York University, New York, NY, USA
e-mail: gabrielmcderment2029@u.northwestern.edu; junzhang2030@u.northwestern.edu

J. Zhang
Northwestern University, Evanston, IL, USA

B. Molaee-Ardekani, M. Segovia-Martinez (eds.), *Advancing Cochlear Implants*,
https://doi.org/10.1007/978-3-032-15719-5_3

explore broader outcomes beyond perception, including healthcare utilization, mental health, and social connectedness, with an emphasis on elderly and pediatric populations. Outcome assessment measures for CI users reflect a growing interest in increasing ecological validity and consideration of the broader patient experience. Future studies will aim to incorporate these new metrics into clinical practice to improve patient satisfaction and performance.

Keywords

Cochlear implant · Performance outcomes · Speech perception · Realistic environments · Ecological momentary assessment (EMA) · Quality of life

3.1 Introduction

A variety of outcome measures exist for cochlear implant (CI) users, including standardized clinical measures such as speech perception, self-reported measures such as hearing-related quality-of-life, (QOF) language-based measures, academic and psychosocial measures, and other experimental and laboratory-based measures (Boisvert et al. 2020; Granberg et al. 2014; Laird et al. 2024; Neal et al. 2022). This chapter is organized as follows: Sect. 3.2 focuses on describing the standard clinical evaluation as a reference for comparison to recent advances in outcome measurements designed to increase ecological validity and measure real-world performance. Section 3.3 discusses speech perception tests designed to capture features of realistic environments. It also presents alternative assessments of sound perception. Section 3.4 moves beyond the clinic or laboratory to discuss data and experiments conducted in the patient's everyday listening environment. Finally, Sect. 3.5 provides a brief discussion of recent advances in broader outcome measures such as healthcare utilization and mental health.

3.2 Clinical Evaluation

The standard postoperative clinical evaluation of CI performance typically encompasses the measurement of aided audiometric thresholds and speech recognition. Subjective outcomes and monitoring of device usage are also recommended. Patients are typically seen at 3 and 12 months post-activation, then biannually thereafter. More frequent evaluations may be needed in special cases, such as monitoring residual hearing preservation or fluctuating performance. Evaluations typically begin with an informal discussion regarding the patient's progress and challenges. Next, the audiologist conducts an otoscopic inspection and assesses the cochlear implant site for signs of irritation or infection. Finally, the audiologist inspects the integrity of the patient's technology, including microphones, cables, and hearing aids, if applicable.

3.2.1 Audiometric Thresholds

Aided soundfield detection should be measured for warbled tones at octave frequencies between 250 and 6000 Hz. Thresholds in the range of 20–25 dB HL are desirable, as this ensures audibility of soft sounds across the speech spectrum. Thresholds lower than 20 dB HL are undesirable, as they may result in patients hearing too much noise, including circuit noise from the processor. Additionally, there is little speech information present below 15 dB HL. This also represents the noise floor of the microphone. Thresholds should be measured for each CI ear separately, using a plug and muff in the contralateral ear as necessary. The patient should be seated in the booth based on how the loudspeaker is calibrated, either 0, 45, or 90 degrees.

3.2.2 Speech Recognition

Speech recognition should be evaluated using recorded materials, as these have been shown to result in less variable scores than monitored live voice presentations (Austin et al. 2025; Uhler et al. 2016). Patients with a contralateral hearing aid should have the aid verified to ensure it is meeting prescriptive targets to ensure audibility. The audiologist should follow the guidelines of the Minimum Speech Test Battery (MSTB-3), (Dunn et al. 2024) to assess word and sentence recognition. The MSTB-3 recommends presenting one list of CNC words (Consonant-Nucleus-Consonant; Peterson and Lehiste 1962) (i.e., 50 words) in quiet to the implanted ear alone at 60 dBA, and one list of AzBio sentences (Spahr et al. 2012) (i.e., 20 sentences) at a +10 dB signal-to-noise ratio (SNR) in the patient's everyday listening condition at 65 dBA (co-located noise at 55 dBA). If additional information regarding the patient's performance is desired, the audiologist may present an additional list of CNC words in the patient's everyday listening condition (i.e., in a bilateral listening condition). AzBio sentences may also be presented in an easier (i.e., quiet) or more difficult (i.e., +5 dB SNR) condition, or to the implanted ear alone. Some patients, especially those with long durations of hearing loss prior to implantation or other disabilities, may require easier speech recognition materials. Examples of these include HINT sentences (Hearing In Noise Test; Nilsson et al. 1994) or picture-pointing tasks (Ross and Lerman 1970).

3.2.3 Other Clinical Outcomes

Some audiologists administer questionnaires to assess subjective outcomes. These questionnaires measure important elements of CI benefit not captured by speech perception tests, including social, emotional, and quality-of-life measures. Ideally, these are administered preoperatively so that a change in outcomes can be measured, and preoperative expectations can be properly established. The MSTB-3 recommends the Speech and Spatial Qualities Index (SSQ) (Gatehouse and Noble

2004) and CI Quality of Life (CI-QOL), either the 10 or 35-item versions (McRackan et al. 2019). The CI-QOL is unique, in that it was designed specifically for adult CI recipients. CI-QOL covers six domains, including communication, emotion, entertainment, environment, listening effort, and social handicap. In cases where tinnitus and/or dizziness are reported, the Tinnitus Handicap Inventory and/or Dizziness Handicap Inventory (Jacobson and Newman 1990; Newman et al. 1996) may be used.

Adequate wear time, typically over 8–10 hours a day for adults and all waking hours for children, has been shown to correlate with improved outcomes, including better speech perception and improved quality of life (Easwar et al. 2018; Gagnon et al. 2020; Holder and Gifford 2021). The audiologist monitors average daily device usage time at each programming and evaluation appointment to effectively address barriers to increasing wear time.

Clinicians can measure several different types of electrically evoked auditory responses to assist in programming and assessing outcomes. The electrically evoked stapedial reflex threshold (ESRT) has been shown to be highly correlated with upper stimulation levels in CIs, making it useful for programming (Holder et al. 2025; Molaee-Ardekani et al. 2025). At least one study has shown better performance on a speech-in-noise test for ESRT-based maps as compared to traditional psychophysical-based maps (Yiannos et al. 2023), perhaps because clinicians tend to underfit upper stimulation levels, and a wide dynamic range has been shown to relate to better speech perception (Bresnihan et al. 2001; Holder et al. 2023, 2025). Electrically evoked compound action potentials (ECAPs) represent the synchronous firing of the auditory nerve in response to an electrical sound presented on one electrode and recorded on another electrode. While ECAP thresholds have poor agreement with behavioral threshold or upper stimulation levels (Brown et al. 2000), several studies have shown that certain attributes of the ECAP response (presence, slope of the amplitude growth function, spread of excitation) are correlated to speech perception (van Eijl et al. 2017). Less commonly measured are the electrically evoked auditory brainstem response (EABR) and electrically evoked cortical auditory evoked potential (ECAEP). Atypical EABR waveform morphology or complete absence of a response has been shown to correlate with poor speech perception outcomes, especially in children with cochlear nerve deficiency or auditory neuropathy spectrum disorder (Jafari et al. 2024; Lundin et al. 2015; Walton et al. 2008). ECAEPs measured in children can demonstrate maturation with device use (Sharma and Dorman 2006), providing an objective indicator of access to soft sounds and auditory cortical development (Kosaner et al. 2018). The clinical utility of EABR and ECAEP has been limited by the need for external electrodes and significant artifact from the implant. However, several recent studies have demonstrated the capability to record these responses in commercially available CIs (Attias et al. 2022; HabibAllah et al. 2025).

3.2.4 Modifications to Clinical Protocol

There are several cases where additional or modified evaluation protocols are necessary. Patients with residual acoustic hearing in the implanted ear may be users of electroacoustic stimulation (EAS), whereby low-frequency information is presented acoustically, and high-frequency information is presented electrically via a speech processor with an acoustic component. For these patients, unaided residual hearing is monitored at each visit to ensure the acoustic component is properly programmed. The audiologist may consider evaluating speech recognition in both the best-aided condition (EAS + contralateral aid) and without ipsilateral acoustic hearing (removing the acoustic component and plugging the ear) to determine what the benefit of EAS is to the patient.

Patients with single-sided deafness or asymmetric hearing loss may also require a modified clinical protocol. These patients should have their unaided hearing in the better ear monitored at each evaluation, or more frequently if there is concern for a change in hearing. The audiologist must attempt to isolate the CI ear for speech testing, ideally by presenting test materials directly to the implant. It should be noted that presentation level cannot be verified using this method, and there may be additional signal-quality differences, especially if the signal is streamed to the implant (instead of presented via a wired connection). However, direct presentation allows for the best isolation of the test ear and is thus recommended. If this is not feasible, a plug and muff or masking noise may be used on the better ear with the caveat that contributions from that ear to speech understanding cannot be ruled out completely. The audiologist may also perform speech evaluations in spatially separated noise. The MSTB-3 recommends AzBio sentences presented at 0 dB SNR (65 dBA) with the sentences from the front and noise facing the better ear in the everyday listening condition. If additional information is desired, the audiologist can co-locate the noise or present it to the poorer ear.

Pediatric CI recipients are evaluated more frequently than adults, typically every 3 months for the first year and then every 6–12 months thereafter, depending on the child's age and progress (Messersmith et al. 2019; Shapiro and Bradham 2012). A child who is in a period of developing auditory and language skills may require more frequent follow-up evaluations (Uhler and Gifford 2014). The Pediatric Minimum Speech Test Battery (Uhler et al. 2017) recommends a clinical workflow for choosing speech recognition materials.

3.3 Advanced Behavioral Measures

The clinical evaluation provides an important but limited picture of CI outcomes. In recent years, there has been interest in measuring perception through the CI for many other types of stimuli, including speech perception in more realistic environments, music, and environmental sounds. The audiologist's test booth represents a relatively simple listening situation; however, recently developed outcome measures seek to replicate complex real-world listening while still maintaining experimental control.

3.3.1 Speech in Realistic Environments

Clinical measures of speech understanding in noise may not accurately reflect how a CI user performs in their everyday life (Hansen and Bořil 2018; Litovsky et al. 2017). There are some elements of clinical evaluations that are likely to overestimate performance (Badajoz-Davila and Buchholz 2021), and some that underestimate it.

For example, the test booth is a quiet, non-reverberant, well-lit space. Room acoustics in realistic environments are suboptimal, resulting in energetic masking effects. Reverberation poses a significant challenge, as room reflections reduce the number of modulations present in the temporal envelope of the signal or create additional overlapping maskers. Studies have shown that speech intelligibility drops significantly in small, highly reverberant rooms, particularly at long talker-to-listener distances (Kressner et al. 2018). The reduction in speech perception from reverberation can even be greater than that from adding competing talkers (Hazrati and Loizou 2012). CI users prefer environments with shorter talker-to-listener distances and benefit from early reflections, which can enhance speech perception by increasing the loudness of the original signal (Eurich et al. 2019).

Visual information is a very important cue for CI listeners and has been shown to increase speech perception scores by 30–40 percentage points (Dorman et al. 2016). Children with CIs show a significant bias toward visual cues when auditory and visual information conflict, and degrading visual cues increases reliance on auditory information (Huyse et al. 2013). Residual acoustic hearing, when combined with visual cues, significantly enhances consonant recognition for CI users (Sheffield et al. 2015). Clinical evaluations are typically done auditory only, without visual information. While realistic environments may have suboptimal visual information, either from poor lighting or long talker-to-listener distances, the complete lack of visual cues in clinical evaluations is highly unrealistic.

Almost universally, CI listeners have been shown to struggle with understanding speech in the presence of background talkers (e.g., Cullington and Zeng 2008; Stickney et al. 2004), a form of informational masking. However, these deficits are likely underreported. Clinical testing often reduces talker variability, which can have large effects on speech intelligibility for CI users listening in real-world environments (see Table 3.1).

Clinical tests often use simple stimuli, without context (i.e., single words) or minimal context (i.e., short sentences). Real-world listening involves the use of topic familiarity, context, and prediction to understand speech. CI listeners benefit from semantic context information, often more than normal-hearing listeners (Dingemanse and Goedegebure 2019; O'Neill et al. 2019). High-context sentences have also been shown to reduce listening effort (as measured with pupillometry) as compared to low-context sentences (Winn 2016). Standard tests do not allow for conversational repair tactics, such as repetition, clarification, or rephrasing (Church et al. 2017; Crowe and Dammeyer 2021; Lind et al. 2006; Toe and Paatsch 2013). Similarly, lexical difficulty influences speech perception (Kirk et al. 1995; Tamati et al. 2024), a property rarely accounted for in clinical speech perception tests (Kirk et al. 1999). Two recently developed sentence testing materials aim to address some

Table 3.1 Studies investigating how talker features affect speech intelligibility

Talker-specific feature	General findings	Example studies
Dialect (e.g., regional accent, non-native speakers)	Negatively affects speech perception for CI users	Hay-McCutcheon et al. (2018), Kapolowicz et al. (2020) and Tamati et al. (2021)
Speaking rate	Faster rates decrease intelligibility, while slower than conversational rates improve intelligibility and reduce listening effort	Dincer D'Alessandro et al. (2018), Ji et al. (2013) and Winn and Teece (2021)
Female vs. male	Most CI users can discriminate female from male talkers better than chance levels	Abdeltawwab et al. (2016) and Geers et al. (2013)
Talker discrimination (within same gender)	Generally poor, with large variability in performance	Hay-McCutcheon et al. (2018) and Li et al. (2022)
Speech level	Whispered or shouted speech decreases speech intelligibility	Hazrati et al. (2015) and Shi et al. (2018)
Talker familiarity	CI children can identify familiar talkers, but adults may not benefit from short-term talker familiarization	Tamati and Moberly (2022) and Vongpaisal et al. (2010)
Adjustment for talking in noise	Talking in background noise results in different vocal effort than talking in quiet with added noise later (Lombard Effect); CI users may be able to benefit from these cues	Hansen et al. (2020) and Lu and Cooke (2008)

of these issues and have been validated with CI users. The Perceptional Robust English Sentence Test Open set (PRESTO) is a high-variability sentence test developed to reflect talker and dialect variability (Gilbert et al. 2013; Mattingly et al. 2018; Tamati et al. 2020, 2023). The Sentence Test with Adaptive Randomized Roving levels (STARR) aims to emulate real-world listening by varying the presentation level across sentences (Boyle et al. 2013).

Furthermore, clinical testing often uses co-located targets and maskers, or at most a 45° spatial separation using two loudspeakers. However, realistic environments contain talkers and maskers from many different, often changing directions. While this can create a challenge, it also provides auditory cues the listener can take advantage of for parsing the signal. When a target and masker are physically separated in space, binaural listening strategies such as the head shadow effect, squelch, or summation can be used to localize sound. Real-world cues such as source location certainty or head turning may further enhance speech perception. Research environments have made use of multi-speaker setups to simulate more realistic listening scenarios. For example, studies have shown that CI listeners have a reduced benefit of spatial separation from target and maskers (spatial release from masking) compared to normal-hearing listeners (Litovsky 2005; Litovsky et al. 2012), but bilateral or bimodal CI users perform better than those with a unilateral CI (Misurelli and Litovsky 2012). Other experimental manipulations include roving the talker or masker (Dorman and Gifford 2017; Weissgerber et al. 2015), providing cues on talker location (Dunn et al. 2010; Gifford et al. 2018), or allowing head movements (Pastore et al. 2018).

Clinical evaluations rarely assess how modern assistive technologies—such as automatic captioning, remote microphones, or haptic feedback—affect speech outcomes. Laboratory studies suggest that they can have a significant positive effect on areas such as academic achievement (Bell and Foiret 2020) and speech understanding (Fitzpatrick et al. 2009; Rodríguez-Correa et al. 2023; Sorgini et al. 2018; Wolfe 2014; Wolfe et al. 2015).

Finally, listeners may also have competing sources for attention, for example, reading the menu or checking their phone, that decrease the available cognitive resources for understanding speech, which are not captured with a clinical evaluation. Several studies have shown attention and cognition to be mitigating factors in outcomes (e.g., Başkent et al. 2016; Giallini et al. 2023; Purdy et al. 2017; Quittner et al. 2007).

Virtual reality (VR) is a promising tool for better simulation of real-world environments in audiology research (Ahrens et al. 2019; Steadman et al. 2019; Stecker 2019). VR allows for a variety of manipulations that would be difficult in physical space. For example, maskers can be placed at different distances from the subject or at extremely small distances from each other. As detailed above, the provision of visual cues can also enhance speech perception. Furthermore, VR could be implemented in a clinical setting, allowing for testing in more realistic environments by audiologists (Hohmann et al. 2020) while providing additional participant response information, such as head turning and eye movements (Grange et al. 2018). Although still developing, VR also has the potential for improving the acceptability of spatial hearing training for CI users (Parmar et al. 2024; Valzolgher et al. 2023). Rooms with immersive sound experiences, with or without virtual reality, can enhance realism.

3.3.2 Music and Sound Quality Assessments

Speech contains a high degree of informational redundancy, allowing listeners to comprehend it even when the signal is significantly degraded by a CI. In contrast, musical appreciation—particularly the perception of pitch, timbre, and harmony—remains especially challenging due to the lower redundancy and greater spectral complexity of music. Limitations in spectral and temporal fine-structure resolution and dynamic range encoding, combined with the complexity of the musical signal as compared to speech, all contribute to poor musical perception for CI users (for reviews, see Limb and Roy 2014; McDermott 2004). Nevertheless, many CI users report enjoying music (Mitani et al. 2007; Riley et al. 2018), and assessments of music and sound quality perception are important components of the CI outcome evaluation. This is particularly relevant as music enjoyment has been shown to correlate with quality of life following implantation (Demirel and Çiprut 2025; Dritsakis et al. 2017; Fuller et al. 2013; Lassaletta et al. 2007).

Many of the acoustic features and device limitations that impair music perception also affect prosodic perception and general sound quality through the device. For example, vocal emotion is typically conveyed through pitch contour, intensity, and speaking rate, all of which can be degraded in CI users (Jiam et al. 2017). As such, tasks requiring listeners to identify emotions conveyed in semantically neutral sentences are often more difficult for CI users than normal-hearing listeners (Chatterjee et al. 2023, 2024; Luo et al. 2007).

Although objective tasks—such as pitch discrimination or timbre identification—can reveal specific deficits, the experience of music and sound quality perception through a CI is largely subjective. Two recent methods have been developed to address this issue. The first is a CI-adapted version of the Multiple Stimulus with Hidden Reference and Anchor (MUSHRA) test, known as CI-MUSHRA (Roy et al. 2012). This approach allows for systematic evaluation of subjective sound quality by presenting different versions of a stimulus, including an unaltered hidden "reference" and a highly degraded "anchor." A labeled reference is used for comparison. The subject is asked to rate each version of the stimulus between 0 and 100, such that 100 has equally good sound quality as the reference. This allows for quantification of sound quality on a relative basis and removes potentially influencing factors from other types of assessment, such as familiarity with musical terms. Using the CI-MUSHRA, a series of studies by Roy and colleagues have found that the availability of low-frequency information, dynamic range compression, insertion angle depth, and reverberation all affect sound quality (Roy et al. 2012, 2015, 2016). The second approach involves CI users with single-sided deafness (SSD). These individuals can uniquely compare the perception from their normal-hearing (and presumably undistorted) acoustic ear to the perception from their CI, essentially serving as their own control (Dorman et al. 2017, 2020; Karoui et al. 2019; Peters et al. 2018; Svirsky et al. 2021). For example, Landsberger et al. (2020) asked SSD CI users to rate their enjoyment of musical song snippets using a modified version of the CI-MUSHRA. This modification allowed for a rating scale between 0 and 200, such that a score of 200 indicated a sound was twice as good as the reference (100, sound presented only to the normal-hearing ear). Listeners rated music as more enjoyable with both ears (mean: 123), even though one was receiving a degraded signal through the CI.

3.4 Real-World Data

The research previously discussed in this chapter has focused on controlled clinical experiments and laboratory-based studies attempted to simulate more realistic environments. While the latter is often more successful at measuring the everyday listening challenges faced by CI users, it still represents a snapshot of auditory abilities in an artificial environment. A recent line of work has leveraged newer technology to collect continuous data from the patient as they go about their everyday life (Keidser et al. 2020). These methods include ecological momentary assessment (EMA), data logging of usage patterns and acoustic environments from the speech processor, and wearable technology. Often, studies combine the use of these measurements to relate real-world subjective and objective experiences.

Unlike traditional questionnaires which rely on retrospective recall and possibly introduce memory bias, EMA asks listeners/users to fill out short questionnaires the moment they receive them. This may improve the accuracy of the questionnaire and allows for time and situation-dependent data. Questionnaires can be sent multiple times a day to the subject's smartphone, for example, allowing for much finer-grained resolution than would be possible in laboratory- or clinic-based questionnaire administration (Holube et al. 2020; Timmer et al. 2017; Wu et al. 2015). EMA can be paired with data from the subject's speech processor to answer questions such as preferences for a particular sound processing strategy in different acoustic environments (Hey et al. 2022; Lawson and Hansen 2024) or how the psychological effects of social isolation during the COVID-19 pandemic relate to daily processor usage (Dunn et al. 2021). EMA can also be used to probe the factors underlying performance in clinical evaluations. For example, O'Neill et al. (2021) found that CI users who performed poorly on clinical speech perception tests reported spending more time at home, alone, and less time conversing with others than good performers.

Physiologic measures such as heart rate, skin conductance, and pupil dilation have been used to measure listening effort in the context of speech-recognition for hearing aid users and, more recently, CI users (Mackersie et al. 2015; Plain et al. 2024; Polo et al. 2025; Russo et al. 2020; Winn and Teece 2021). Long-term or remote assessment of physiologic measures is now possible with wearable monitors, such as smartwatches or in-ear devices (Huhn et al. 2022; Kondaurova et al. 2024; Masè et al. 2020). Studies with hearing aid users have shown relationships between listening environments and heart rate; for example, louder environments are related to increases in heart rates (Christensen et al. 2021). Given the interplay between hearing loss, auditory environment, physiologic stress, and health, future studies are expected to use wearable technology to investigate how these variables affect outcomes for CI users (Caduff et al. 2020; Sarangi and Johnson 2024).

3.5 Broader Outcomes: Beyond Speech Recognition

Performance evaluation of CI users extends beyond their recognition of speech, because the ability to understand speech and communicate orally has wide-ranging impacts on human experience. While some nonauditory outcomes such as education or quality of life have long been measured, newer domains such as physical and mental health, social participation and community integration, and healthcare utilization are now being explored (Bekele Okuba et al. 2023). For example, hearing loss creates barriers to healthcare utilization, leading to poor communication with providers and reduced healthcare literacy (McKee et al. 2015; Piao et al. 2023). CI use improves healthcare utilization, with several studies showing a reduction in prescription medication use and improved health literacy (Bekele Okuba et al. 2023).

Several factors have led to an increase in implantation rates for the elderly (>70) and very elderly (>80) population. CI surgery has become shorter and less invasive, and some centers are now offering local anesthesia (Connors et al. 2021; Deep et al. 2020; Shabashev et al. 2017). Furthermore, there have been numerous studies reporting the benefits of treating hearing loss for reduction in dementia risk (Lin et al. 2011; Loughrey et al. 2018; Yeo et al. 2023). While these patients see significant benefit from CIs (Bourn et al. 2022; Spitzer and Waltzman 2021), they also have higher rates of comorbidities and frailty, making them more likely to experience complications and a reduction in perceived quality of life (Aylward et al. 2022; Gordon et al. 2021; Kay-Rivest et al. 2022). On the other hand, benefits for mental health domains such as depression, anxiety, or stress are well documented in this population (Choi et al. 2016; Contrera et al. 2016). Children and adolescents with CIs may still struggle with mental health symptoms, which may be related to difficulties with social coherence (Anmyr et al. 2015; Huber et al. 2015; Huber and Kipman 2011).

Recently, a population of CI users who were implanted as very young children (<2 years of age) has reached late adolescence and adulthood. Studies investigating this group have shown educational benefits for CI users, such as mainstream placement and higher degrees of secondary education, extend beyond the initial years postimplantation, and include occupational benefits (Blanchard et al. 2016; Cejas et al. 2023; Goh et al. 2016; Gordon et al. 2023; Huber et al. 2008; Illg et al. 2017; Moura et al. 2023; Spitzer et al. 2025). Adolescents implanted earlier showed better hearing-related quality of life (Cejas et al. 2023).

3.6 Conclusions and Future Directions

Outcome assessment measures for CI users reflect a growing interest in increasing ecological validity and consideration of the broader patient experience. The clinical evaluation provides an important picture of basic device functioning but is limited by a controlled and foreign environment, unrealistic speech perception materials, and a lack of consideration for outcomes beyond speech understanding. Recent advances in outcome measurement have addressed these issues with novel speech perception materials incorporating talker variability, reverberant environments, and multiple sound sources, for example. New technology such as virtual reality has the potential for increasing ecological validity while remaining clinically implementable. Other studies take outcome measurement a step further by collecting data in the real world. Techniques such as EMA, combined with time-locked information from the acoustic environment collected on the speech processor, can provide an incredible wealth of data on how patients listen and interact with the world outside the sound booth. Finally, researchers have begun to explore a wide range of outcome measures beyond traditional speech perception or quality of life. By leveraging unique populations of CI users, there is potential to uncover the effects of CI on a wide range of physical, mental, and social health outcomes. The ultimate goal will be to translate research measures of outcomes into clinical practice in order to improve device satisfaction and performance for all patients with CIs.

Acknowledgments The authors declare that no generative AI was used in the creation of this chapter.

References

Abdeltawwab MM, Khater A, El-Anwar MW (2016) Contralateral bimodal stimulation: a way to enhance speech performance in Arabic-speaking cochlear implant patients. ORL J Otorhinolaryngol Relat Spec 78(3):126–135. https://doi.org/10.1159/000381024

Ahrens A, Lund KD, Marschall M, Dau T (2019) Sound source localization with varying amount of visual information in virtual reality. PLoS One 14(3):e0214603. https://doi.org/10.1371/journal.pone.0214603

Anmyr L, Olsson M, Freijd A, Larsson K (2015) Sense of coherence, social networks, and mental health among children with a cochlear implant. Int J Pediatr Otorhinolaryngol 79(4):610–615. https://doi.org/10.1016/j.ijporl.2015.02.009

Attias J, HabibAllah S, Aditya Tarigoppula VS, Glick H, Chen C, Kanthaiah K, Litvak L (2022) Cortical auditory evoked potentials recorded directly through the cochlear implant in cochlear implant recipients: a feasibility study. Ear Hear 43(5):1426. https://doi.org/10.1097/AUD.0000000000001212

Austin A, Ladner K, Mendel LL (2025) Recorded word recognition testing is worth the time. Am J Audiol 34(1):179–186. https://doi.org/10.1044/2024_AJA-24-00080

Aylward A, Murphy-Meyers M, Allen CM, Patel NS, Gurgel RK (2022) Frailty and quality of life after cochlear implantation in older adults. Otolaryngol Head Neck Surg 166(2):350–356. https://doi.org/10.1177/01945998211004589

Badajoz-Davila J, Buchholz JM (2021) Effect of Test Realism on Speech-in-noise Outcomes in Bilateral Cochlear Implant Users. Ear Hear 42(6):1687–1698. https://doi.org/10.1097/AUD.0000000000001061

Başkent D, Clarke J, Pals C, Benard MR, Bhargava P, Saija J, Sarampalis A, Wagner A, Gaudrain E (2016) Cognitive compensation of speech perception with hearing impairment, cochlear implants, and aging: how and to what degree can it be achieved? Trends Hear 20:2331216516670279. https://doi.org/10.1177/2331216516670279

Bekele Okuba T, Lystad RP, Boisvert I, McMaugh A, Moore RC, Walsan R, Mitchell RJ (2023) Cochlear implantation impact on health service utilisation and social outcomes: a systematic review. BMC Health Serv Res 23(1):Article 1. https://doi.org/10.1186/s12913-023-09900-y

Bell D, Foiret J (2020) A rapid review of the effect of assistive technology on the educational performance of students with impaired hearing. Disabil Rehabil Assist Technol 15(7):838–843. https://doi.org/10.1080/17483107.2020.1775317

Blanchard M, Célerier C, Parodi M, Sabban D, Prang I, Rouillon I, Frachet B, Loundon N (2016) Long term results after cochlear implantation: schooling and social insertion of teenagers and young adults. Eur Ann Otorhinolaryngol Head Neck Dis 133:S40–S43

Boisvert I, Reis M, Au A, Cowan R, Dowell RC (2020) Cochlear implantation outcomes in adults: a scoping review. PLoS One 15(5):e0232421. https://doi.org/10.1371/journal.pone.0232421

Bourn SS, Goldstein MR, Morris SA, Jacob A (2022) Cochlear implant outcomes in the very elderly. Am J Otolaryngol 43(1):103200. https://doi.org/10.1016/j.amjoto.2021.103200

Boyle PJ, Nunn TB, O'Connor AF, Moore BCJ (2013) STARR: a speech test for evaluation of the effectiveness of auditory prostheses under realistic conditions. Ear Hear 34(2):203. https://doi.org/10.1097/AUD.0b013e31826a8e82

Bresnihan M, Norman G, Scott F, Viani L (2001) Measurement of comfort levels by means of electrical stapedial reflex in children. Arch Otolaryngol Head Neck Surg 127(8):963–966. https://doi.org/10.1001/archotol.127.8.963

Brown CJ, Hughes ML, Luk B, Abbas PJ, Wolaver A, Gervais J (2000) The relationship between EAP and EABR thresholds and levels used to program the nucleus 24 speech processor: data from adults. Ear Hear 21(2):151–163. https://doi.org/10.1097/00003446-200004000-00009

Caduff A, Feldman Y, Ben Ishai P, Launer S (2020) Physiological monitoring and hearing loss: toward a more integrated and ecologically validated health mapping. Ear Hear 41:120S. https://doi.org/10.1097/AUD.0000000000000960

Cejas I, Barker DH, Petruzzello E, Sarangoulis CM, Quittner AL (2023) Cochlear implantation and educational and quality-of-life outcomes in adolescence. JAMA Otolaryngol Head Neck Surg 149(8):708–715. https://doi.org/10.1001/jamaoto.2023.1327

Chatterjee M, Feller A, Kulkarni AM, Galvin JJ (2023) Emotional prosody perception and production are linked in prelingually deaf children with cochlear implantsa. JASA Express Lett 3(12):127601. https://doi.org/10.1121/10.0023996

Chatterjee M, Gajre S, Kulkarni AM, Barrett KC, Limb CJ (2024) Predictors of emotional prosody identification by school-age children with cochlear implants and their peers with normal hearing. Ear Hear 45(2):411. https://doi.org/10.1097/AUD.0000000000001436

Choi JS, Betz J, Li L, Blake CR, Sung YK, Contrera KJ, Lin FR (2016) Association of using hearing aids or cochlear implants with changes in depressive symptoms in older adults. JAMA Otolaryngol Head Neck Surg 142(7):652–657

Christensen JH, Saunders GH, Porsbo M, Pontoppidan NH (2021) The everyday acoustic environment and its association with human heart rate: evidence from real-world data logging with hearing aids and wearables. R Soc Open Sci 8(2):201345. https://doi.org/10.1098/rsos.201345

Church A, Paatsch L, Toe D (2017) Some trouble with repair: conversations between children with cochlear implants and hearing peers. Discourse Stud 19(1):49–68. https://doi.org/10.1177/1461445616683592

Connors JR, Deep NL, Huncke TK, Roland JT Jr (2021) Cochlear implantation under local anesthesia with conscious sedation in the elderly: first 100 cases. Laryngoscope 131(3):E946–E951. https://doi.org/10.1002/lary.28853

Contrera KJ, Betz J, Li L, Blake CR, Sung YK, Choi JS, Lin FR (2016) Quality of life after intervention with a cochlear implant or hearing aid. Laryngoscope 126(9):2110–2115

Crowe K, Dammeyer J (2021) A review of the conversational pragmatic skills of children with cochlear implants. J Deaf Stud Deaf Educ 26(2):171–186

Cullington HE, Zeng F-G (2008) Speech recognition with varying numbers and types of competing talkers by normal-hearing, cochlear-implant, and implant simulation subjects. J Acoust Soc Am 123(1):450–461. https://doi.org/10.1121/1.2805617

Deep NL, Connors J, Roland JT (2020) Cochlear implantation under local anesthesia with conscious sedation in the elderly patient: focus on surgical technique. Oper Tech Otolaryngol Head Neck Surg 31(3):231–237. https://doi.org/10.1016/j.otot.2020.07.008

Demirel AK, Çiprut A (2025) Music perception and music-related quality of life in adult cochlear implant users: exploring the need for music rehabilitation. Ear Hear 46(1):265–276

Dincer D'Alessandro H, Boyle PJ, Ballantyne D, De Vincentiis M, Mancini P (2018) The role of speech rate for Italian-speaking cochlear implant users: insights for everyday speech perception. Int J Audiol 57(11):851–857. https://doi.org/10.1080/14992027.2018.1498139

Dingemanse JG, Goedegebure A (2019) The important role of contextual information in speech perception in cochlear implant users and its consequences in speech tests. Trends Hear 23: 2331216519838672. https://doi.org/10.1177/2331216519838672

Dorman MF, Gifford RH (2017) Speech understanding in complex listening environments by listeners fit with cochlear implants. J Speech Lang Hear Res 60(10):3019–3026. https://doi.org/10.1044/2017_JSLHR-H-17-0035

Dorman MF, Liss J, Wang S, Berisha V, Ludwig C, Natale SC (2016) Experiments on auditory-visual perception of sentences by users of unilateral, bimodal, and bilateral cochlear implants. J Speech Lang Hear Res 59(6):1505–1519. https://doi.org/10.1044/2016_JSLHR-H-15-0312

Dorman MF, Natale SC, Butts AM, Zeitler DM, Carlson ML (2017) The sound quality of cochlear implants: studies with single-sided deaf patients. Otol Neurotol 38(8):e268–e273. https://doi.org/10.1097/MAO.0000000000001449

Dorman MF, Natale SC, Baxter L, Zeitler DM, Carlson ML, Lorens A, Skarzynski H, Peters JPM, Torres JH, Noble JH (2020) Approximations to the voice of a cochlear implant: explorations with single-sided deaf listeners. Trends Hear 24:2331216520920079. https://doi.org/10.1177/2331216520920079

Dritsakis G, Van Besouw RM, Kitterick P, Verschuur CA (2017) A music-related quality of life measure to guide music rehabilitation for adult cochlear implant users. Am J Audiol 26(3): 268–282. https://doi.org/10.1044/2017_AJA-16-0120

Dunn CC, Noble W, Tyler RS, Kordus M, Gantz BJ, Ji H (2010) Bilateral and unilateral cochlear implant users compared on speech perception in noise. Ear Hear 31(2):296–298. https://doi.org/10.1097/AUD.0b013e3181c12383

Dunn CC, Stangl E, Oleson J, Smith M, Chipara O, Wu Y-H (2021) The influence of forced social isolation on the auditory ecology and psychosocial functions of listeners with cochlear implants during COVID-19 mitigation efforts. Ear Hear 42(1):20. https://doi.org/10.1097/AUD.0000000000000991

Dunn CC, Zwolan TA, Balkany TJ, Strader HL, Biever A, Gifford RH, Hall MW, Holcomb MA, Hill H, King ER, Larky J, Presley R, Reed M, Shapiro WH, Sydlowski SA, Wolfe J (2024) A consensus to revise the minimum speech test battery–version 3. Am J Audiol 33(3):624–647. https://doi.org/10.1044/2024_AJA-24-00008

Easwar V, Sanfilippo J, Papsin B, Gordon K (2018) Impact of consistency in daily device use on speech perception abilities in children with cochlear implants: datalogging evidence. J Am Acad Audiol 29(9):835–846. https://doi.org/10.3766/jaaa.17051

Eurich B, Klenzner T, Oehler M (2019) Impact of room acoustic parameters on speech and music perception among participants with cochlear implants. Hearing Research 377:122–132. https://doi.org/10.1016/j.heares.2019.03.012

Fitzpatrick EM, Séguin C, Schramm DR, Armstrong S, Chénier J (2009) The benefits of remote microphone technology for adults with cochlear implants. Ear Hear 30(5):590–599

Fuller C, Mallinckrodt L, Maat B, Baskent D, Free R (2013) Music and quality of life in early-deafened late-implanted adult cochlear implant users. Otol Neurotol 34(6):1041–1047

Gagnon EB, Eskridge H, Brown KD (2020) Pediatric cochlear implant wear time and early language development. Cochlear Implants Int 21(2):92–97. https://doi.org/10.1080/14670100.2019.1670487

Gatehouse S, Noble W (2004) The speech, spatial and qualities of hearing scale (SSQ). Int J Audiol 43(2):85–99. https://doi.org/10.1080/14992020400050014

Geers A, Davidson L, Uchanski R, Nicholas J (2013) Interdependence of linguistic and indexical speech perception skills in school-aged children with early cochlear implantation. Ear Hear 34(5):562–574. https://doi.org/10.1097/AUD.0b013e31828d2bd6

Giallini I, Inguscio BMS, Nicastri M, Portanova G, Ciofalo A, Pace A, Greco A, D'Alessandro HD, Mancini P (2023) Neuropsychological functions and audiological findings in elderly cochlear implant users: the role of attention in postoperative performance. Audiol Res 13(2):Article 2. https://doi.org/10.3390/audiolres13020022

Gifford RH, Loiselle L, Natale S, Sheffield SW, Sunderhaus LW, Dietrich MS, Dorman MF (2018) Speech understanding in noise for adults with cochlear implants: effects of hearing configuration, source location certainty, and head movement. J Speech Lang Hear Res 61(5):1306–1321. https://doi.org/10.1044/2018_JSLHR-H-16-0444

Gilbert JL, Tamati TN, Pisoni DB (2013) Development, reliability, and validity of PRESTO: a new high-variability sentence recognition test. J Am Acad Audiol 24(1):26–36. https://doi.org/10.3766/jaaa.24.1.4

Goh T, Bird P, Pearson J, Mustard J (2016) Educational, employment, and social participation of young adult graduates from the paediatric Southern Cochlear Implant Programme, New Zealand. Cochlear Implants Int 17(1):31–51. https://doi.org/10.1179/1754762815Y.0000000018

Gordon SA, Aylward A, Patel NS, Bowers C, Presson AP, Smith KR, Foster NL, Gurgel RK (2021) Does frailty or age increase the risk of postoperative complications following cochlear implantation? OTO Open 5(3):2473974X211044084. https://doi.org/10.1177/2473974X211044084

Gordon KA, Papsin BC, Cushing SL (2023) Long-term language, educational, and quality-of-life outcomes in adolescents after childhood cochlear implantation. JAMA Otolaryngol Head Neck Surg 149(8):715–716. https://doi.org/10.1001/jamaoto.2023.1329

Granberg S, Dahlström J, Möller C, Kähäri K, Danermark B (2014) The ICF Core Sets for hearing loss—researcher perspective. Part I: systematic review of outcome measures identified in audiological research. Int J Audiol 53(2):65–76. https://doi.org/10.3109/14992027.2013.851799

Grange JA, Culling JF, Bardsley B, Mackinney LI, Hughes SE, Backhouse SS (2018) Turn an ear to hear: how hearing-impaired listeners can exploit head orientation to enhance their speech intelligibility in noisy social settings. Trends Hear 22:2331216518802701. https://doi.org/10.1177/2331216518802701

HabibAllah S, Chen C, Attias J (2025) Direct recording of electrically evoked cortical potentials from cochlear implants demonstrates feasibility and clinical relevance in pediatric users. Sci Rep 15(1):22644. https://doi.org/10.1038/s41598-025-06652-z

Hansen JHL, Bořil H (2018) On the issues of intra-speaker variability and realism in speech, speaker, and language recognition tasks. Speech Comm 101:94–108. https://doi.org/10.1016/j.specom.2018.05.004

Hansen JHL, Lee J, Ali H, Saba JN (2020) A speech perturbation strategy based on "Lombard effect" for enhanced intelligibility for cochlear implant listeners. J Acoust Soc Am 147(3):1418–1428. https://doi.org/10.1121/10.0000690

Hay-McCutcheon MJ, Peterson NR, Pisoni DB, Kirk KI, Yang X, Parton J (2018) Performance variability on perceptual discrimination tasks in profoundly deaf adults with cochlear implants. J Commun Disord 72:122–135. https://doi.org/10.1016/j.jcomdis.2018.01.005

Hazrati O, Loizou PC (2012) The combined effects of reverberation and noise on speech intelligibility by cochlear implant listeners. Int J Audiol 51(6):437–443. https://doi.org/10.3109/14992027.2012.658972

Hazrati O, Ali H, Hansen JHL, Tobey E (2015) Evaluation and analysis of whispered speech for cochlear implant users: gender identification and intelligibility. J Acoust Soc Am 138(1):74–79. https://doi.org/10.1121/1.4922230

Hey M, Hersbach AA, Hocke T, Mauger SJ, Böhnke B, Mewes A (2022) Ecological momentary assessment to obtain signal processing technology preference in cochlear implant users. J Clin Med 11(10):Article 10. https://doi.org/10.3390/jcm11102941

Hohmann V, Paluch R, Krueger M, Meis M, Grimm G (2020) The virtual reality lab: realization and application of virtual sound environments. Ear Hear 41:31S. https://doi.org/10.1097/AUD.0000000000000945

Holder JT, Gifford RH (2021) Effect of increased daily cochlear implant use on auditory perception in adults. J Speech Lang Hear Res 64(10):4044–4055. https://doi.org/10.1044/2021_JSLHR-21-00066

Holder JT, Henry MR, MacDonald AE, Gifford RH (2023) Cochlear implant upper stimulation levels: eSRT vs. loudness scaling. Otol Neurotol 44(9):e667. https://doi.org/10.1097/MAO.0000000000003988

Holder JT, MacDonald A, Gifford RH (2025) eSRT versus population mean: setting upper stimulation levels in adult cochlear implant users. Otol Neurotol Open 5(2):e071. https://doi.org/10.1097/ONO.0000000000000071

Holube I, von Gablenz P, Bitzer J (2020) Ecological momentary assessment in hearing research: current state, challenges, and future directions. Ear Hear 41:79S. https://doi.org/10.1097/AUD.0000000000000934

Huber M, Kipman U (2011) The mental health of deaf adolescents with cochlear implants compared to their hearing peers. Int J Audiol 50(3):146–154. https://doi.org/10.3109/14992027.2010.533704

Huber M, Wolfgang H, Klaus A (2008) Education and training of young people who grew up with cochlear implants. Int J Pediatr Otorhinolaryngol 72(9):1393–1403. https://doi.org/10.1016/j.ijporl.2008.06.002

Huber M, Pletzer B, Giourgas A, Nickisch A, Kunze S, Illg A (2015) Schooling relates to mental health problems in adolescents with cochlear implants—mediation by hearing and family variables. Front Psychol 6:1889. https://doi.org/10.3389/fpsyg.2015.01889

Huhn S, Axt M, Gunga H-C, Maggioni MA, Munga S, Obor D, Sié A, Boudo V, Bunker A, Sauerborn R, Bärnighausen T, Barteit S (2022) The impact of wearable technologies in health research: scoping review. JMIR Mhealth Uhealth 10(1):e34384. https://doi.org/10.2196/34384

Huyse A, Berthommier F, Leybaert J (2013) Degradation of labial information modifies audiovisual speech perception in cochlear-implanted children. Ear Hear 34(1):110. https://doi.org/10.1097/AUD.0b013e3182670993

Illg A, Haack M, Lesinski-Schiedat A, Büchner A, Lenarz T (2017) Long-term outcomes, education, and occupational level in cochlear implant recipients who were implanted in childhood. Ear Hear 38(5):577–587. https://doi.org/10.1097/AUD.0000000000000423

Jacobson GP, Newman CW (1990) The development of the dizziness handicap inventory. Arch Otolaryngol Head Neck Surg 116(4):424–427

Jafari Z, Fitzpatrick EM, Schramm DR, Rouillon I, Koravand A (2024) Prognostic value of electrophysiological and MRI findings for pediatric cochlear implant outcomes: a systematic review. Am J Audiol 33(3):1023–1040. https://doi.org/10.1044/2024_AJA-23-00272

Ji C, Galvin JJ, Xu A, Fu Q-J (2013) Effect of speaking rate on recognition of synthetic and natural speech by normal-hearing and cochlear implant listeners. Ear Hear 34(3):313–323. https://doi.org/10.1097/AUD.0b013e31826fe79e

Jiam NT, Caldwell M, Deroche ML, Chatterjee M, Limb CJ (2017) Voice emotion perception and production in cochlear implant users. Hear Res 352:30–39. https://doi.org/10.1016/j.heares.2017.01.006

Kapolowicz MR, Montazeri V, Baese-Berk MM, Zeng F-G, Assmann PF (2020) Rapid adaptation to non-native speech is impaired in cochlear implant users. J Acoust Soc Am 148(3):EL267. https://doi.org/10.1121/10.0001941

Karoui C, James C, Barone P, Bakhos D, Marx M, Macherey O (2019) Searching for the sound of a cochlear implant: evaluation of different vocoder parameters by cochlear implant users with single-sided deafness. Trends Hear 23:2331216519866029. https://doi.org/10.1177/2331216519866029

Kay-Rivest E, Friedmann DR, McMenomey SO, Jethanamest D, Thomas Roland J, Waltzman SB (2022) The frailty phenotype in older adults undergoing cochlear implantation. Otol Neurotol 43(10):e1085–e1089. https://doi.org/10.1097/MAO.0000000000003704

Keidser G, Naylor G, Brungart DS, Caduff A, Campos J, Carlile S, Carpenter MG, Grimm G, Hohmann V, Holube I, Launer S, Lunner T, Mehra R, Rapport F, Slaney M, Smeds K (2020) The quest for ecological validity in hearing science: what it is, why it matters, and how to advance it. Ear Hear 41:5S. https://doi.org/10.1097/AUD.0000000000000944

Kirk KI, Pisoni DB, Osberger MJ (1995) Lexical effects on spoken word recognition by pediatric cochlear implant users. Ear Hear 16(5):470–481

Kirk KI, Eisenberg LS, Martinez AS, Hay-McCutcheon M (1999) Lexical neighborhood test: test-retest reliability and interlist equivalency. J Am Acad Audiol 10(3):113–123. https://doi.org/10.1055/s-0042-1748471

Kondaurova MV, Smith A, Mishra R, Zheng Q, Kondaurova I, Francis AL, Sallee E (2024) Empatica E4 assessment of child physiological measures of listening effort during remote and in-person communication. Am J Audiol 33(4):1331–1340. https://doi.org/10.1044/2024_AJA-24-00078

Kosaner J, Van Dun B, Yigit O, Gultekin M, Bayguzina S (2018) Clinically recorded cortical auditory evoked potentials from paediatric cochlear implant users fitted with electrically elicited stapedius reflex thresholds. Int J Pediatr Otorhinolaryngol 108:100–112. https://doi.org/10.1016/j.ijporl.2018.02.033

Kressner AA, Westermann A, Buchholz JM (2018) The impact of reverberation on speech intelligibility in cochlear implant recipients. J Acoust Soc Am 144(2):1113–1122. https://doi.org/10.1121/1.5051640

Laird E, Sucher C, Nakano K, Ferguson M (2024) Systematic review of patient and service outcome measures of remote digital technologies for cochlear implant and hearing aid users. Front Audiol Otol 2:1403814. https://doi.org/10.3389/fauot.2024.1403814

Landsberger DM, Vermeire K, Stupak N, Lavender A, Neukam J, Van de Heyning P, Svirsky MA (2020) Music is more enjoyable with two ears, even if one of them receives a degraded signal provided by a cochlear implant. Ear Hear 41(3):476. https://doi.org/10.1097/AUD.0000000000000771

Lassaletta L, Castro A, Bastarrica M, Pérez-Mora R, Madero R, De Sarriá J, Gavilán J (2007) Does music perception have an impact on quality of life following cochlear implantation? Acta Otolaryngol 127(7):682–686. https://doi.org/10.1080/00016480601002112

Lawson T, Hansen JHL (2024) Situational signal processing with ecological momentary assessment: leveraging environmental context for cochlear implant users. In: ICASSP 2024–2024 IEEE international conference on acoustics, speech and signal processing (ICASSP), pp 2131–2135. https://doi.org/10.1109/ICASSP48485.2024.10447591

Li MM, Moberly AC, Tamati TN (2022) Factors affecting talker discrimination ability in adult cochlear implant users. J Commun Disord 99:106255. https://doi.org/10.1016/j.jcomdis.2022.106255

Limb CJ, Roy AT (2014) Technological, biological, and acoustical constraints to music perception in cochlear implant users. Hear Res 308:13–26. https://doi.org/10.1016/j.heares.2013.04.009

Lin FR, Metter EJ, O'Brien RJ, Resnick SM, Zonderman AB, Ferrucci L (2011) Hearing loss and incident dementia. Arch Neurol 68(2):214–220

Lind C, Hickson L, Erber NP (2006) Conversation repair and adult cochlear implantation: a qualitative case study. Cochlear Implants Int 7(1):33–48. https://doi.org/10.1179/cim.2006.7.1.33

Litovsky RY (2005) Speech intelligibility and spatial release from masking in young children. J Acoust Soc Am 117(5):3091–3099. https://doi.org/10.1121/1.1873913

Litovsky RY, Goupell MJ, Godar S, Grieco-Calub T, Jones GL, Garadat SN, Agrawal S, Kan A, Todd A, Hess C, Misurelli S (2012) Studies on bilateral cochlear implants at the University of Wisconsin's Binaural Hearing and Speech Laboratory. J Am Acad Audiol 23(6):476–494. https://doi.org/10.3766/jaaa.23.6.9

Litovsky RY, Goupell MJ, Misurelli SM, Kan A (2017) Hearing with cochlear implants and hearing aids in complex auditory scenes. In: The auditory system at the cocktail party. Springer, Cham, pp 261–291. https://doi.org/10.1007/978-3-319-51662-2_10

Loughrey DG, Kelly ME, Kelley GA, Brennan S, Lawlor BA (2018) Association of age-related hearing loss with cognitive function, cognitive impairment, and dementia: a systematic review and meta-analysis. JAMA Otolaryngol Head Neck Surg 144(2):115–126

Lu Y, Cooke M (2008) Speech production modifications produced by competing talkers, babble, and stationary noise. J Acoust Soc Am 124(5):3261–3275. https://doi.org/10.1121/1.2990705

Lundin K, Stillesjö F, Rask-Andersen H (2015) Prognostic value of electrically evoked auditory brainstem responses in cochlear implantation. Cochlear Implants Int 16(5):254–261. https://doi.org/10.1179/1754762815y.0000000005

Luo X, Fu Q-J, Galvin JJ (2007) Cochlear implants special issue article: vocal emotion recognition by normal-hearing listeners and cochlear implant users. Trends Amplif 11(4):301–315. https://doi.org/10.1177/1084713807305301

Mackersie CL, MacPhee IX, Heldt EW (2015) Effects of hearing loss on heart rate variability and skin conductance measured during sentence recognition in noise. Ear Hear 36(1):145. https://doi.org/10.1097/AUD.0000000000000091

Masè M, Micarelli A, Strapazzon G (2020) Hearables: new perspectives and pitfalls of in-ear devices for physiological monitoring. A scoping review. Front Physiol 11:568886. https://doi.org/10.3389/fphys.2020.568886

Mattingly JK, Castellanos I, Moberly AC (2018) Nonverbal reasoning as a contributor to sentence recognition outcomes in adults with cochlear implants. Otol Neurotol 39(10):e956–e963

McDermott HJ (2004) Music perception with cochlear implants: a review. Trends Amplif 8(2): 49–82. https://doi.org/10.1177/108471380400800203

McKee MM, Moreland C, Atcherson SR, Zazove P (2015) Hearing loss: communicating with the patient who is deaf or hard of hearing. FP Essent 434:24–28

McRackan TR, Hand BN, Cochlear Implant Quality of Life Development Consortium, Velozo CA, Dubno JR (2019) Cochlear implant quality of life (CIQOL): development of a profile instrument (CIQOL-35 profile) and a global measure (CIQOL-10 global). J Speech Lang Hear Res 62(9): 3554–3563. https://doi.org/10.1044/2019_JSLHR-H-19-0142

Messersmith JJ, Entwisle L, Warren S, Scott M (2019) Clinical practice guidelines: cochlear implants. J Am Acad Audiol 30(10):827–844. https://doi.org/10.3766/jaaa.19088

Misurelli SM, Litovsky RY (2012) Spatial release from masking in children with normal hearing and with bilateral cochlear implants: effect of interferer asymmetry. J Acoust Soc Am 132(1): 380–391. https://doi.org/10.1121/1.4725760

Mitani C, Nakata T, Trehub SE, Kanda Y, Kumagami H, Takasaki K, Miyamoto I, Takahashi H (2007) Music recognition, music listening, and word recognition by deaf children with cochlear implants. Ear Hear 28(2):29S. https://doi.org/10.1097/AUD.0b013e318031547a

Molaee-Ardekani B, Goiana-Martins P, Segovia-Martinez M, Luís L (2025) The effect of current amplitude and multi-electrode stimulation on eSRT for auditory fitting in cochlear implants with pulse-width loudness coding. Cochlear Implants Int 26(1):30–42. https://doi.org/10.1080/14670100.2025.2457200

Moura JE, Martins JH, Alves M, Oliveira G, Ramos D, Alves H, Caiado R, Teixeira A, Silva LF, Migueis J (2023) Children then, adults now: long-term outcomes—performance at 15, 20, and 25 years of cochlear implant use. Front Rehabil Sci 4:1275808. https://doi.org/10.3389/fresc.2023.1275808

Neal K, McMahon CM, Hughes SE, Boisvert I (2022) Listening-based communication ability in adults with hearing loss: a scoping review of existing measures. Front Psychol 13. https://doi.org/10.3389/fpsyg.2022.786347

Newman CW, Jacobson GP, Spitzer JB (1996) Development of the tinnitus handicap inventory. Arch Otolaryngol Head Neck Surg 122(2):143–148

Nilsson M, Soli SD, Sullivan JA (1994) Development of the hearing in noise test for the measurement of speech reception thresholds in quiet and in noise. J Acoust Soc Am 95(2):1085–1099. https://doi.org/10.1121/1.408469

O'Neill ER, Kreft HA, Oxenham AJ (2019) Cognitive factors contribute to speech perception in cochlear-implant users and age-matched normal-hearing listeners under vocoded conditions. J Acoust Soc Am 146(1):195–210. https://doi.org/10.1121/1.5116009

O'Neill ER, Basile JD, Nelson P (2021) Individual hearing outcomes in cochlear implant users influence social engagement and listening behavior in everyday life. J Speech Lang Hear Res 64(12):4982–4999. https://doi.org/10.1044/2021_JSLHR-21-00249

Parmar BJ, Salorio-Corbetto M, Picinali L, Mahon M, Nightingale R, Somerset S, Cullington H, Driver S, Rocca C, Jiang D, Vickers D (2024) Virtual reality games for spatial hearing training in children and young people with bilateral cochlear implants: the "Both Ears (BEARS)" approach. Front Neurosci 18. https://doi.org/10.3389/fnins.2024.1491954

Pastore MT, Natale SJ, Yost WA, Dorman MF (2018) Head movements allow listeners bilaterally implanted with cochlear implants to resolve front-back confusions. Ear Hear 39(6):1224–1231

Peters JP, Wendrich AW, van Eijl RH, Rhebergen KS, Versnel H, Grolman W (2018) The sound of a cochlear implant investigated in patients with single-sided deafness and a cochlear implant. Otol Neurotol 39(6):707–714

Peterson GE, Lehiste I (1962) Revised CNC lists for auditory tests. J Speech Hear Disord 27(1): 62–70. https://doi.org/10.1044/jshd.2701.62

Piao Z, Lee H, Mun Y, Lee H, Han E (2023) Exploring the health literacy status of people with hearing impairment: a systematic review. Arch Public Health 81(1):Article 1. https://doi.org/10.1186/s13690-023-01216-x

Plain B, Pielage H, Kramer SE, Richter M, Saunders GH, Versfeld NJ, Zekveld AA, Bhuiyan TA (2024) Combining cardiovascular and pupil features using k-nearest neighbor classifiers to assess task demand, social context, and sentence accuracy during listening. Trends Hear 28: 23312165241232551. https://doi.org/10.1177/23312165241232551

Polo EM, Simeone D, Mollura M, Paglialonga A, Barbieri R (2025) An adaptive protocol to assess physiological responses as a function of task demand in speech-in-noise testing. J Neurosci Methods 415:110348. https://doi.org/10.1016/j.jneumeth.2024.110348

Purdy SC, Welch D, Giles E, Morgan CLA, Tenhagen R, Kuruvilla-Mathew A (2017) Impact of cognition and noise reduction on speech perception in adults with unilateral cochlear implants. Cochlear Implants Int 18(3):162–170. https://doi.org/10.1080/14670100.2017.1299393

Quittner AL, Barker DH, Snell C, Cruz I, McDonald L-G, Grimley ME, Botteri M, Marciel K, Investigative Team, C (2007) Improvements in visual attention in deaf infants and toddlers after cochlear implantation. Hear Balance Commun 5(4):242. https://doi.org/10.1080/16513860701745401

Riley PE, Ruhl DS, Camacho M, Tolisano AM (2018) Music appreciation after cochlear implantation in adult patients: a systematic review. Otolaryngol Head Neck Surg 158(6):1002–1010. https://doi.org/10.1177/0194599818760559

Rodríguez-Correa PA, Valencia-Arias A, Patiño-Toro ON, Oblitas Díaz Y, Teodori De La Puente R (2023) Benefits and development of assistive technologies for deaf people's communication: a systematic review. Front Educ 8:1121597. https://doi.org/10.3389/feduc.2023.1121597

Ross M, Lerman J (1970) A picture identification test for hearing-impaired children. J Speech Hear Res 13(1):44–53. https://doi.org/10.1044/jshr.1301.44

Roy AT, Jiradejvong P, Carver C, Limb CJ (2012) Assessment of sound quality perception in cochlear implant users during music listening. Otol Neurotol 33(3):319. https://doi.org/10.1097/MAO.0b013e31824296a9

Roy AT, Michelle V, Munjal T, Carver C, Jiradejvong P, Limb CJ (2015) Reverberation negatively impacts musical sound quality for cochlear implant users. Cochlear Implants Int 16(sup3): S105–S113. https://doi.org/10.1179/1467010015Z.000000000262

Roy AT, Penninger RT, Pearl MS, Wuerfel W, Jiradejvong P, Carver C, Buechner A, Limb CJ (2016) Deeper cochlear implant electrode insertion angle improves detection of musical sound quality deterioration related to bass frequency removal. Otol Neurotol 37(2):146. https://doi.org/10.1097/MAO.0000000000000932

Russo FY, Hoen M, Karoui C, Demarcy T, Ardoint M, Tuset M-P, De Seta D, Sterkers O, Lahlou G, Mosnier I (2020) Pupillometry assessment of speech recognition and listening experience in adult cochlear implant patients. Front Neurosci 14. https://doi.org/10.3389/fnins.2020.556675

Sarangi L, Johnson J (2024) Feasibility of using wearable sensors to measure listening-related stress. Perspect ASHA Spec Interest Groups 9(6):1631–1645. https://doi.org/10.1044/2024_PERSP-24-00073

Shabashev S, Fouad Y, Huncke TK, Roland JT (2017) Cochlear implantation under conscious sedation with local anesthesia; safety, efficacy, costs, and satisfaction. Cochlear Implants Int 18(6):297–303

Shapiro WH, Bradham TS (2012) Cochlear implant programming. Otolaryngol Clin N Am 45(1): 111–127

Sharma A, Dorman MF (2006) Central auditory development in children with cochlear implants: clinical implications. Adv Otorhinolaryngol 64:66–88. https://doi.org/10.1159/000094646

Sheffield BM, Schuchman G, Bernstein JGW (2015) Trimodal speech perception: how residual acoustic hearing supplements cochlear-implant consonant recognition in the presence of visual cues. Ear Hear 36(3):e99. https://doi.org/10.1097/AUD.0000000000000131

Shi Y, Peng KA, Chen B, Gong Y, Chen J, Li Y, Fu Q-J (2018) Interaction between speech variations and background noise on speech intelligibility by Mandarin-speaking cochlear implant patients. Speech Comm 104:89–94. https://doi.org/10.1016/j.specom.2018.09.007

Sorgini F, Caliò R, Carrozza MC, Oddo CM (2018) Haptic-assistive technologies for audition and vision sensory disabilities. Disabil Rehabil Assist Technol 13(4):394–421. https://doi.org/10.1080/17483107.2017.1385100

Spahr AJ, Dorman MF, Litvak LM, Van Wie S, Gifford RH, Loizou PC, Loiselle LM, Oakes T, Cook S (2012) Development and validation of the AzBio sentence lists. Ear Hear 33(1): 112–117. https://doi.org/10.1097/AUD.0b013e31822c2549

Spitzer ER, Waltzman SB (2021) Outcomes of cochlear implantation in adults over 85 years of age. Cochlear Implants Int 22(5):296–302. https://doi.org/10.1080/14670100.2021.1913331

Spitzer ER, Lichtl AJ, Waltzman SB (2025) Adults implanted as children: long-term educational, occupational, and speech perception outcomes. Audiol Neurotol. https://karger.com/aud/article-abstract/doi/10.1159/000545788/927187

Steadman MA, Kim C, Lestang J-H, Goodman DFM, Picinali L (2019) Short-term effects of sound localization training in virtual reality. Sci Rep 9(1):18284. https://doi.org/10.1038/s41598-019-54811-w

Stecker GC (2019) Using virtual reality to assess auditory performance. Hear J 72(6):20–23. https://doi.org/10.1097/01.hj.0000558464.75151.52

Stickney GS, Zeng F-G, Litovsky R, Assmann P (2004) Cochlear implant speech recognition with speech maskers. J Acoust Soc Am 116(2):1081–1091. https://doi.org/10.1121/1.1772399

Svirsky MA, Capach NH, Neukam JD, Azadpour M, Sagi E, Hight AE, Glassman EK, Lavender A, Seward KP, Miller MK (2021) Valid acoustic models of cochlear implants: one size does not fit all. Otol Neurotol 42(10S):S2–S10

Tamati TN, Moberly AC (2022) Talker adaptation and lexical difficulty impact word recognition in adults with cochlear implants. Audiol Neurootol 27(3):260–270. https://doi.org/10.1159/000518643

Tamati TN, Ray C, Vasil KJ, Pisoni DB, Moberly AC (2020) High- and low-performing adult cochlear implant users on high-variability sentence recognition: differences in auditory spectral resolution and neurocognitive functioning. J Am Acad Audiol 31(05):324–335. https://doi.org/10.3766/jaaa.18106

Tamati TN, Pisoni DB, Moberly AC (2021) The perception of regional dialects and foreign accents by cochlear implant users. J Speech Lang Hear Res 64(2):683–690. https://doi.org/10.1044/2020_JSLHR-20-00496

Tamati TN, Faulkner KF, Pisoni DB (2023) Assessment of high-variability speech recognition in adult cochlear implant users using PRESTO. J Am Acad Audiol 34(9):217–224. https://doi.org/10.1055/a-2181-2652

Tamati TN, Jebens A, Başkent D (2024) Lexical effects on talker discrimination in adult cochlear implant usersa. J Acoust Soc Am 155(3):1631–1640. https://doi.org/10.1121/10.0025011

Timmer BHB, Hickson L, Launer S (2017) Ecological momentary assessment: feasibility, construct validity, and future applications. Am J Audiol 26(3S):436–442. https://doi.org/10.1044/2017_AJA-16-0126

Toe DM, Paatsch LE (2013) The conversational skills of school-aged children with cochlear implants. Cochlear Implants Int 14(2):67–79. https://doi.org/10.1179/1754762812Y.0000000002

Uhler K, Gifford RH (2014) Current trends in pediatric cochlear implant candidate selection and postoperative follow-up. Am J Audiol 23(3):309–325. https://doi.org/10.1044/2014_AJA-13-0067

Uhler K, Biever A, Gifford RH (2016) Method of speech stimulus presentation impacts pediatric speech recognition: monitored live voice versus recorded speech. Otol Neurotol 37(2):e70–e74

Uhler K, Warner-Czyz A, Gifford R, PMSTB Working Group (2017) Pediatric minimum speech test battery. J Am Acad Audiol 28(3):232–247. https://doi.org/10.3766/jaaa.15123

Valzolgher C, Bouzaid S, Grenouillet S, Gatel J, Ratenet L, Murenu F, Verdelet G, Salemme R, Gaveau V, Coudert A, Hermann R, Truy E, Farnè A, Pavani F (2023) Training spatial hearing in unilateral cochlear implant users through reaching to sounds in virtual reality. Eur Arch Otorrinolaringol 280(8):3661–3672. https://doi.org/10.1007/s00405-023-07886-1

van Eijl RHM, Buitenhuis PJ, Stegeman I, Klis SFL, Grolman W (2017) Systematic review of compound action potentials as predictors for cochlear implant performance. Laryngoscope 127(2):476–487. https://doi.org/10.1002/lary.26154

Vongpaisal T, Trehub SE, Schellenberg EG, van Lieshout P, Papsin BC (2010) Children with cochlear implants recognize their mother's voice. Ear Hear 31(4):555. https://doi.org/10.1097/AUD.0b013e3181daae5a

Walton J, Gibson WPR, Sanli H, Prelog K (2008) Predicting cochlear implant outcomes in children with auditory neuropathy. Otol Neurotol 29(3):302. https://doi.org/10.1097/MAO.0b013e318164d0f6

Weissgerber T, Rader T, Baumann U (2015) Impact of a moving noise masker on speech perception in cochlear implant users. PLoS One 10(5):e0126133. https://doi.org/10.1371/journal.pone.0126133

Winn MB (2016) Rapid release from listening effort resulting from semantic context, and effects of spectral degradation and cochlear implants. Trends Hear 20:2331216516669723. https://doi.org/10.1177/2331216516669723

Winn MB, Teece KH (2021) Slower speaking rate reduces listening effort among listeners with cochlear implants. Ear Hear 42(3):584. https://doi.org/10.1097/AUD.0000000000000958

Wolfe J (2014) Cochlear implants and remote microphone technology. Semin Hear 35(03):177–192. https://doi.org/10.1055/s-0034-1383503

Wolfe J, Morais M, Schafer E (2015) Improving hearing performance for cochlear implant recipients with use of a digital, wireless, remote-microphone, audio-streaming accessory. J Am Acad Audiol 26(6):532–539. https://doi.org/10.3766/jaaa.15005

Wu Y-H, Stangl E, Zhang X, Bentler RA (2015) Construct validity of the ecological momentary assessment in audiology research. J Am Acad Audiol 26(10):872–884. https://doi.org/10.3766/jaaa.15034

Yeo BSY, Song HJJMD, Toh EMS, Ng LS, Ho CSH, Ho R, Merchant RA, Tan BKJ, Loh WS (2023) Association of hearing aids and cochlear implants with cognitive decline and dementia: a systematic review and meta-analysis. JAMA Neurol 80(2):134–141. https://doi.org/10.1001/jamaneurol.2022.4427

Yiannos JM, Bester CW, Zhao C, Gell B, Jayakody DMP (2023) Speech-in-noise performance in objectively determined cochlear implant maps, including the effect of cognitive state. PLoS One 18(6):e0286986. https://doi.org/10.1371/journal.pone.0286986

Implantable Technologies in Cochlear Implants: Mechanics, Materials, and Electrochemistry

4

Guillaume Tourrel, Valérie Chaumat, Marie Egert Fischer, and Jérôme Chevalier

Abstract

This chapter explores the evolution of cochlear implant (CI) technology, focusing on implant architectures, materials, and manufacturing techniques in relation to device reliability and biocompatibility. It compares two main implant designs—modular titanium-based casings and monolithic titanium–ceramic casings—highlighting differences in structure, functionality, and manufacturing complexity. Special attention is given to ceramic materials, particularly high-purity alumina and zirconia, due to their mechanical strength, biocompatibility, and radio frequency transparency. The chapter discusses recent advancements in ceramic processing methods, such as cold sintering and additive manufacturing, which offer promising improvements in implant performance and longevity. It also explores brazing techniques used in component assembly, emphasizing the importance of wetting behavior and interface adhesion. Additionally, the design and fabrication of electrode arrays, essential for effective neural stimulation, are reviewed. Despite variations in design across manufacturers, common

G. Tourrel (✉)
Advanced Innovation Department, Cochlear Ltd, Vallauris, France
e-mail: gtourrel@cochlear.com

V. Chaumat
Université Grenoble Alpes, CEA, Liten, DTCH, Grenoble, France
e-mail: valerie.chaumat@cea.fr

M. E. Fischer
ArcelorMittal, Global Research and Development, Maizières, France
e-mail: marie.egert@arcelormittal.com

J. Chevalier
INSA Lyon, Universite Claude Bernard Lyon 1, CNRS, MATEIS, UMR 5510, Villeurbanne, France
e-mail: jerome.chevalier@insa-lyon.fr

B. Molaee-Ardekani, M. Segovia-Martinez (eds.), *Advancing Cochlear Implants*,
https://doi.org/10.1007/978-3-032-15719-5_4

technologies such as platinum-based electrodes and polydimethylsiloxane (PDMS) overmolding are consistently employed. This comprehensive overview highlights the interdisciplinary innovations shaping the future of cochlear implant technology.

Keywords

Ceramic · Titanium · Brazing · Implant · Biocompatibility · Electrode array · Stimulation

4.1 Introduction

Over the past decades, cochlear implant (CI) technology has seen significant advancements in implant architecture, materials science, and manufacturing processes. Modern CIs are designed with a strong emphasis on long-term reliability, biocompatibility, and functional performance.

The development of CIs has greatly benefited from decades of innovation in active implantable medical devices (AIMDs). Among these, the pacemaker—first implanted in 1958—has contributed the most to advancements in materials and assembly techniques, particularly in areas such as batteries, leads, and hermetically sealed electronics (Beck et al. 2010; Greatbatch 2000). However, CIs present additional complexities. For example, they typically require 12 to 22 electrodes (depending on the manufacturer), far more than the bipolar or quadripolar connections used in pacemakers. This high number of connections, which must pass through compact feedthroughs, makes it impractical to reuse standard pacemaker conception. As a result, CI manufacturers have developed custom multipolar feedthroughs with enhanced precision and material technologies.

Another unique aspect of CIs is the need for continuous communication between the implant and the external sound processor to deliver uninterrupted stimulations. This communication is facilitated by two coaxial antennas, making the hermetic sealing of the implanted electronics critically important. In Sect. 4.2 of this chapter, we present the two primary CI casing architecture solutions currently employed by manufacturers. These designs enable continuous communication while also providing robust mechanical strength and reliable hermetic sealing for the implant. It is important to note that manufacturers are ultimately responsible for the safety and quality of their implants, even when using materials and components supplied by certified subcontractors with long-term guarantees. While the AIMD standard (ISO 45502-1) served as the primary reference for many years, it has now been supplemented by the CI-specific standard (ISO 45502-2-1).

A key factor in the performance of cochlear implants is the use of advanced materials—particularly ceramics—which provide critical properties such as electrical insulation, mechanical strength, and resistance to bodily fluids. Section 4.3 of this chapter examines the technological aspects of ceramic applications in CIs. In Sect. 4.4, we explore the integration of these materials and the precision manufacturing techniques—such as brazing and micromechanical fabrication—used to bond components. These bonding technologies are essential for ensuring the long-term durability and safety of the implants.

Finally, the electrode array (EA) is another important component of CIs, discussed in Sect. 4.5. The human cochlea is small and spiral-shaped, requiring the EA to be both thin and flexible to navigate the scala tympani without causing trauma. Manufacturers have developed specialized processes to produce these delicate components. Advances in micromechanical fabrication and laser technologies have enabled the creation of increasingly precise and reliable electrodes. The electrode–tissue interface is carefully engineered to maintain safe charge densities and minimize long-term degradation, in compliance with international safety standards. While designs vary—from straight to pre-curved—all manufacturers use platinum-based contacts embedded in a flexible silicone (PDMS) matrix, ensuring both mechanical integrity and ease of insertion.

4.2 Cochlear Implants: Architectures and Elements

Cochlear implants have emerged with two different design morphologies. The most common design, adopted by major manufacturers such as Cochlear, MED-EL, and Advanced Bionics, features a modular architecture with two distinct compartments: electronics house and communication module. The implant electronics are housed in a hermetically sealed titanium casing, while the communication antenna is placed externally and connected via a lead (see Fig. 4.1). This separation allows for flexible placement and has become the industry standard due to its robustness and ease of manufacturing. In contrast, a monolithic architecture was previously designed by Neurelec (Oticon Medical) with its Neuro Zti implant. This design integrated both the electronics and the communication antenna within a single hermetically sealed capsule made of titanium and ceramic. The compound of titanium–ceramic in two sides of the implant provides a solid implant (thanks to titanium) and makes it possible to establish radio frequency (RF) communication (thanks to transparency

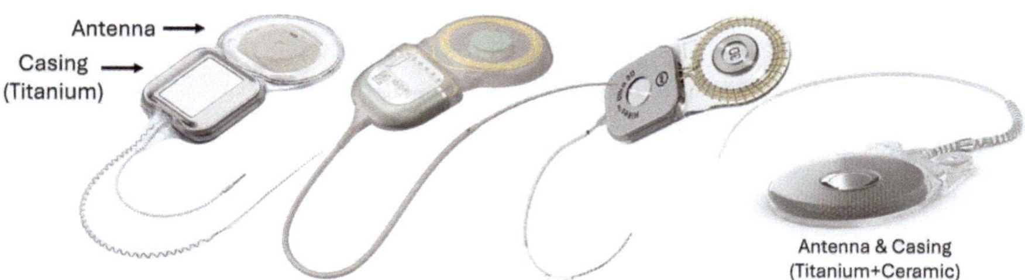

Fig. 4.1 Implant architectures featuring titanium and titanium–ceramic casings. From left to right: Cochlear Nucleus® 9 (© Cochlear Limited), MED-EL Synchrony 2 (© MED-EL GmbH), Advanced Bionics HiRes™ Ultra 3D (© 2025, Advanced Bionics LLC), and Oticon Medical Neuro Zti (© Cochlear Limited). (Implant images provided courtesy of the respective companies)

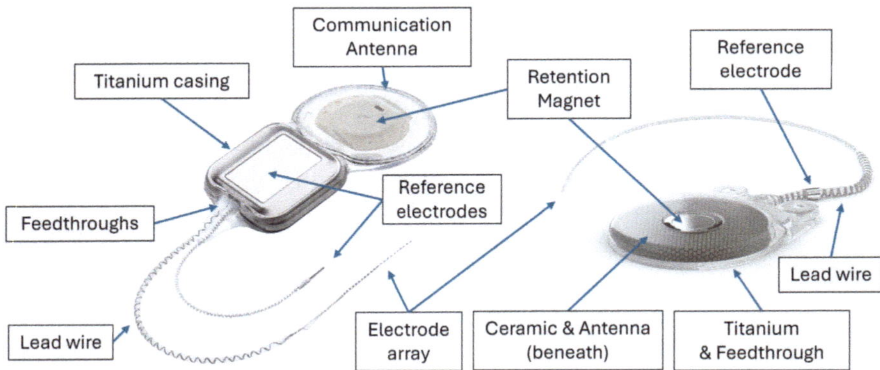

Fig. 4.2 Implant components for titanium casing design from Cochlear Ltd. (left) and titanium–ceramic casing design from Oticon Medical (right). © Cochlear Limited. (Image courtesy of Cochlear Limited)

of ceramic to RF waves). This compact configuration reduced the number of components and potential failure points but required more complex manufacturing processes, including precision ceramic brazing and laser welding. Following Cochlear's acquisition of Neurelec in 2024, this integrated design was discontinued.

Both architectures demand high standards of hermetic sealing to protect internal electronics from bodily fluids and ensure long-term reliability. The choice between these designs reflects trade-offs in size, complexity, and manufacturing flexibility, shaping the evolution of CI technology.

Overall, both implant casing architectures (i.e., titanium and compound titanium–ceramic) share the following main elements presented in Fig. 4.2.

- Casing
  The role of the casing is to safely confine the electronic board. Its function is twofold: first, to protect the patient from non-biocompatible components present on the electronic board, which could induce inflammation and poisoning, and second, to protect the electronic board from aggressive body fluids and guarantee the long-term proper functioning. Ingress of body fluid may lead to electronics corrosion, dysfunction, and the release of toxic by-products into the patient body, resulting in inflammation of surrounding tissues. It is why a robust and reliable sealed barrier is mandatory.
- Feedthroughs
  Feedthroughs are sealed conductors that connect the internal electronic board to the electrode array or to the antenna (Fig. 4.3). They provide as many independent pins as are required for wiring and contribute to the hermetic sealing. Typically hidden beneath a shielding foil (Fig. 4.2), feedthroughs are not visible on the device's exterior.

  Feedthroughs are composed of biocompatible materials, including ceramic, titanium, gold, and platinum. The ceramic serves as an insulator between the

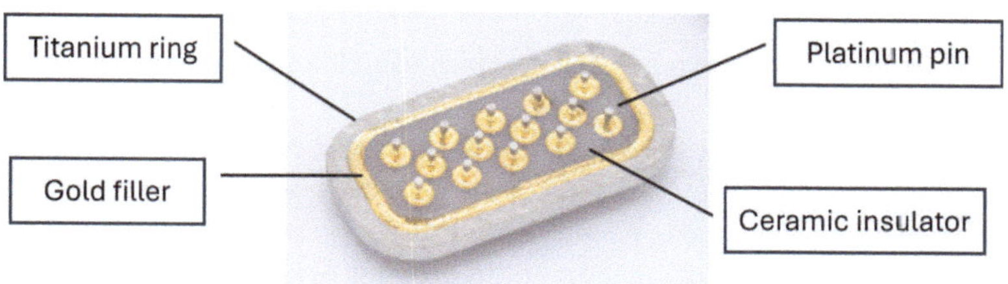

Fig. 4.3 Multipolar feedthrough fabricated by SCT, as used in Oticon Medical Neuro Zti implants. © Cochlear Limited. (Image courtesy of Cochlear Limited)

platinum pins, which may be connected to the electrode lead or antenna wires. A titanium flange surrounds the feedthrough, enabling laser welding to the titanium casing. Gold is used to bond the platinum pins and the titanium flange to the ceramic structure using brazing technology (see Sect. 4.4).

- The employed ceramic for those assembly is frequently a high-purity alumina (HPA; Al_2O_3) because it provides a good ability to brazing and good mechanical properties and is relatively easy to manufacture. Several ceramic manufacturers propose raw material, semifinished product, or customed components for the medical industry with full compliance to ISO:13485 and ISO 6474—implants for surgery—ceramic materials based on high-purity alumina. To name just a few, CoorsTek Inc. (Grand Junction, CO, USA), —Société des Céramiques Technique—SCT (Bazet, FR), and Morgan Advanced Materials (Windsor, UK) produce state-of-the art components for medical and surgical applications.
- Communication antenna
  In the titanium casing design, the communication antenna is located outside the sealed casing. The coil is connected to two feedthrough pins, allowing signal transmission between the internal electronics and the external processor. The coil itself consists of multiple loops, varying by manufacturer, and is made from wire insulated with a biocompatible enamel. To ensure durability and protection, the entire coil is encapsulated in a silicone overmolding.

  In the titanium–ceramic casing design, the antenna is positioned behind the ceramic, as the ceramic is transparent to radio frequency communication waves. Since the antenna is integrated inside the sealed casing, it can be made of enameled copper wire, which offers better electrical characteristics.
- Retention magnet
  The retention magnet ensures the self-alignment positioning of the external antenna in front of the implant. The external and internal coils must face each other to ensure good coupling and stable communication. The retention force between the external antenna and the implant antenna must be kept within an acceptable range that is comfortable for the patient and does not induce harmful skin pressure. To adapt to different skin thicknesses, several external magnet forces are provided. Various magnet materials are used, such as neodymium–

iron–boron or samarium–cobalt. As these materials are not biocompatible, they need to be sealed in a hermetic casing made of titanium, for instance.

- Lead wire
  The lead wire is composed of fine insulated wires that connect the feedthroughs to the electrode array. The lead wire must provide good flexibility to ease the surgical technique and ensure proper implant positioning on the mastoid. It must also absorb some elongation and pull stress without causing any damage to the wires, such as breaks, short circuits, or current leakage.
- Electrode array
  The electrode array is the active part that delivers electrical stimulation to the auditory nerve. This part of the device is further detailed in Sect. 4.5.
- Reference electrode
  The reference electrode serves to collect the return current delivered into the cochlea by the stimulating electrode and to maintain charge balance. It is typically made of pure platinum or a platinum alloy (e.g., platinum 90%—iridium 10%) and features a large surface area to ensure a low charge density. Depending on the manufacturer, there may be one or two reference electrodes.

4.3 Ceramic in Cochlear Implant

Ceramics are a foundational element in the design of modern CIs, offering a unique combination of biocompatibility, mechanical strength, electrical insulation, and transparency to RF waves (Vaiani et al. 2023; Guo et al. 2023). These properties make ceramics indispensable in ensuring the safety, durability, and functionality of implantable hearing devices.

As outlined in Sect. 4.2, cochlear implants can adopt different physical architectures. Regardless of the specific design, ceramics are consistently employed for critical roles. Ceramic feedthroughs are used to safely transmit electrical signals from the internal electronics to the electrode array or antenna, providing both electrical insulation and mechanical protection. Additionally, ceramics protect sensitive electronics from body fluids due to their low water vapor permeability, ensuring long-term stability in the biological environment.

In compound titanium–ceramic casing architectures, the mechanical role of ceramics becomes even more significant. While feedthroughs are typically placed on the bone-facing side—where exposure to external shocks is reduced—the ceramic must still shield internal components from mechanical stress and impact. This dual role of insulation and protection highlights the material's versatility and importance.

Among the various ceramic materials used in cochlear implants, high-purity alumina (HPA)—in accordance with ISO 6474—is the most widely adopted. Its popularity stems from its biocompatibility, brazing compatibility, and chemical stability (Piconi and Porporati 2016). In 2010, the impact resistance requirement in the CI standard (EN 45502-2-3) was raised from 1.5 to 2.5 J, placing greater emphasis on the mechanical strength of ceramic casings. In response, manufacturers have developed a broader range of ceramic materials with enhanced properties. For instance, zirconia offers superior mechanical strength and fracture toughness, making it a promising alternative for high-stress applications.

Ceramics also play a vital role in enabling wireless communication by allowing RF signals to pass through the implant casing—an essential feature in designs where the antenna is embedded within the device (see Fig. 4.2, right).

Given the critical roles ceramics play in cochlear implant systems, it is essential to understand the types of ceramics used, their material characteristics, reliability under physiological conditions, and the technological advancements shaping their future. In the following section, we explore these aspects to provide a comprehensive overview of the current landscape and future directions for ceramic materials in cochlear implant technology.

4.3.1 Ceramic Materials

This section provides an overview of various types of ceramics commonly used in medical applications, including dentistry, orthopedics, and cochlear implants. It details their key properties—such as strength, toughness, aging behavior, and hermeticity—that are critical to their performance in clinical settings.

4.3.1.1 Alumina: A Ceramic for Common Applications

Alumina (Al_2O_3) has been a key biomaterial since its introduction in the 1970s. Initially valued for its wear and chemical resistance in engineering applications, it gained success in medicine due to its biocompatibility and chemical stability. Its first major use was in orthopedic implants, particularly hip prostheses, where its high hardness and low friction—compared to metals—offered significant advantages (Piconi 2017; Ruys et al. 2022).

Alumina has also been used in implants and crowns, offering good aesthetics, sufficient strength for many applications, and resistance to bacterial adhesion. Its bioinert nature—though no material is entirely inert in the body—ensures minimal tissue reaction, making it a preferred choice for long-term restorations. However, alumina has relatively modest toughness and strength compared to modern bio-ceramics like zirconia, alumina–zirconia composites, or silicon nitride. As a result, it has been largely replaced in orthopedics and dentistry by zirconia-based ceramics, although some companies continue developing alumina femoral heads and acetabular components.

Despite these limitations, alumina remains highly relevant in applications where mechanical loads are lower. It still plays a crucial role in implanted electronic devices such as cochlear implants, serving as a protective ceramic housing for sensitive components. Its excellent electrical insulation prevents corrosion and enhances the longevity of medical devices. Additionally, advancements in powder processing, shaping, and sintering over the past 50 years have significantly improved alumina's mechanical properties. By refining grain size, increasing density, and minimizing defects, industrial-grade alumina now reaches strength of around 600 MPa, which remains sufficient for several applications.

A key advantage of alumina in electronic implants is its well-mastered brazing to titanium, ensuring highly reliable joints. However, recent standards for cochlear implants, particularly regarding impact resistance, have pushed manufacturers toward tougher ceramic alternatives.

4.3.1.2 Zirconia: A Strong Ceramic for Harsh Condition

Zirconia (ZrO_2) has emerged since the 1990s as a leading biomaterial due to its exceptional mechanical properties, particularly its superior toughness and strength compared to alumina (Piconi and Maccauro 1999; Sengupta et al. 2019). This enhanced toughness stems from transformation toughening—a unique mechanism in which stress-induced phase transformation from the tetragonal-to-monoclinic ($t \rightarrow m$) phase (Garvie et al. 1975; Mamivand et al. 2013) creates localized compressive stresses that help prevent or delay crack propagation. As a result, zirconia exhibits significantly higher fracture toughness and strength than alumina, making it a preferred choice for load-bearing applications.

The effectiveness of phase transformation toughening depends on the ability to retain metastable tetragonal grains after sintering, which can then transform under stress into the stable monoclinic form. To achieve this, zirconia must be doped with stabilizing oxides that allow retention of the tetragonal phase at room temperature. The first biomedical-grade zirconia was stabilized with magnesium oxide (Mg–PSZ, magnesia partially stabilized zirconia), but yttria-doped zirconia (Y-TZP, yttria-stabilized tetragonal zirconia polycrystal) (Kelly and Denry 2008; Tsuchida et al. 2024) quickly became the preferred choice for demanding applications due to its unprecedented combination of toughness and strength. Specifically, 3 mol.% yttria-doped zirconia (3Y-TZP) achieves toughness values exceeding 5 $MPa{\cdot}m^{1/2}$ and strength beyond 1 GPa, making it one of the strongest and toughest ceramics available.

In cochlear implants, 3Y-TZP was used as a replacement for high-purity alumina in Neuro Zti implant from Oticon Medical, which received CE marking in 2019. Following the transition from alumina in their previous implant to zirconia in the Neuro Zti implant, there were no reported cases of explantation due to ceramic-related impact. The switch to zirconia significantly enhanced the implant's durability and impact resistance, effectively addressing the issues previously associated with alumina.

In orthopedics, 3Y-TZP was initially used in femoral heads for hip replacements due to its excellent wear resistance and mechanical properties. However, concerns about low-temperature degradation (LTD)—commonly referred to as aging—have limited its widespread use. LTD occurs when zirconia is exposed to moisture over time, leading to a gradual phase transformation at the surface. This transformation can induce microcracks, increase wear, and ultimately reduce implant longevity. To counteract LTD while maintaining high toughness and strength, various strategies have been developed, including doping 3Y-TZP with small amounts of alumina and composite ceramics such as alumina-toughened zirconia (ATZ) and zirconia-toughened alumina (ZTA). These materials offer improved aging resistance and long-term durability, making them preferred or promising alternatives for biomedical applications (Zhao et al. 2013).

Zirconia has gained significant popularity in dentistry over the past two decades and is now one of the dominant ceramic materials for crowns, bridges, and implants. It exhibits low bacterial adhesion and resistance to plaque formation, potentially reducing the risk of peri-implantitis and infections—an important consideration for cochlear implants as well. Zirconia's key advantages include high strength, excellent aesthetics, and biocompatibility. Unlike metal-based restorations, zirconia provides a natural tooth-like appearance, particularly with the development of high-translucency grades. However, it is important to note that increasing yttria content enhances translucency at the expense of strength, as transformation toughening becomes less effective. Therefore, not all zirconia ceramics are the same. The term "zirconia" should be understood as referring to a class of materials—similar to steels or aluminum alloys—that share common characteristics but also exhibit significant differences. This is the responsibility of engineers and researchers to select or design the appropriate composition and microstructure for each specific application including cochlear implants. Indeed, the impact energy requirements of cochlear implants led manufacturers to evaluate and eventually transition from alumina to zirconia to achieve greater impact resistance.

Despite the advantages of zirconia, concerns remain regarding in vivo aging and long-term stability, as LTD-related failures have been reported for few orthopedic and dental implants. These failures have been linked to specific processing conditions that led to unexpected aging kinetics, but they shed light on one of the limits/challenges of biomedical zirconia.

However, it must be reminded that most zirconia implants and restorations on the market have demonstrated excellent performance. Also, today, predictive tools for aging assessment are available, allowing manufacturers to evaluate the aging resistance of zirconia before commercialization. This process is standardized under ISO 13356, which specifies requirements for biomedical-grade zirconia.

As mentioned above, the use of zirconia in cochlear implant has demonstrated a higher resistance to impact compared to alumina, and to the best of our knowledge, no failure has been reported to date. Furthermore, ongoing research aims to develop more stable zirconia formulations. These include incorporating small amounts of alumina into 3Y-TZP—already implemented in some industrial products—as well as doping with ceria or magnesia to enhance resistance to LTD. Additionally, hybrid materials and surface treatments are being explored to optimize both mechanical properties and aesthetics.

Zirconia's role in biomedical applications is thus expected to expand further with the development of new composite materials and advanced processing techniques, ensuring greater reliability in both orthopedic and dental fields. Beyond these applications, zirconia is also being increasingly used in electronic implanted devices due to its excellent electrical insulation properties. However, challenges remain in achieving the same brazing quality as with alumina. Research is ongoing to enhance brazing mechanisms and reduce the risk of long-term leakage, ensuring greater reliability for future electronic implants.

4.3.1.3 Alumina–Zirconia Composites

Both alumina (Al_2O_3) and zirconia (ZrO_2) have been widely used in biomedical applications, yet each presents inherent limitations—alumina is brittle with modest toughness, while zirconia may suffer in certain conditions from LTD. To overcome these drawbacks, alumina–zirconia composites have been developed in the medical field since the early 2000s, aiming to combine the best properties of both materials. These composites are primarily categorized into two types:

- Alumina-toughened zirconia (ATZ): Comprising approximately 75–80% yttria-doped zirconia and 20–25% alumina (e.g., De Freitas et al. 2020).
- Zirconia-toughened alumina (ZTA): Comprising approximately 75–80% alumina and 20–25% zirconia (with varying yttria contents depending on the producers) (e.g., Wang and Stevens 1989; He et al. 2008).

Through precise microstructural engineering, these composites achieve greater toughness than monolithic alumina—thanks to zirconia's transformation toughening mechanism—and improved LTD resistance compared to pure zirconia, due to alumina's stabilizing effect. Additionally, they exhibit increased wear resistance, improving bearing implants' longevity (see Table 4.1).

ZTA provides an ideal balance between alumina's strength and zirconia's toughness, associated with a high wear resistance, ensuring long-term stability for bearing implants. One of the most successful applications of ZTA in orthopedics is Biolox® Delta, introduced in the early 2000s as a next-generation femoral head material for total hip replacements. Compared to monolithic alumina (Biolox® Forte), Biolox® Delta exhibited much higher toughness and strength and, overall, a significantly lower risk of in vivo fracture, as low as 0.003% for femoral heads, compared to 0.02% with alumina ceramics. Today, Biolox® Delta dominates the orthopedic market, with over 15 million ZTA femoral heads implanted worldwide, nearly replacing monolithic alumina femoral heads. ZTA acetabular liners are also gaining traction due to their improved toughness and reliability.

ATZ offers the highest toughness and the best combination of strength and LTD resistance, making it ideal for dental implants (and a good alternative for high-load orthopedic applications). While monolithic zirconia still dominates the market for crowns and bridges because of aesthetic advantages (ATZ and ZTA are opaque),

Table 4.1 Key properties of ATZ and ZTA compared to alumina and zirconia

Material	Strength (MPa)	Toughness ($MPa{\cdot}m^{1/2}$)	Aging resistance (LTD)	Wear resistance
Alumina (Al_2O_3)	~600	~4	Excellent (no LTD)	High
3Y-TZP zirconia	~1000	~5–7	May degrade in certain conditions in vivo	Very high
ATZ (80% ZrO_2 + 20% Al_2O_3)	~1200	~9	Excellent (higher than monolithic 3Y-TZP)	Very high
ZTA (20% ZrO_2 + 80% Al_2O_3)	~900	~6	Excellent	Very high

ATZ is gaining ground in full-arch restorations and load-bearing dental implants, where fracture resistance and longevity are critical. Several companies are now investing in ATZ-based implants, offering enhanced safety and durability over traditional zirconia-only solutions.

With continuous refinements in powder processing, sintering techniques, and microstructural design, ATZ and ZTA are becoming the new gold standard for high-performance biomedical ceramics. Their ability to balance strength, toughness, and aging resistance has positioned them as the leaders in orthopedics and credible alternatives for some dental products. Looking ahead, ongoing research and development will further optimize these materials, with new hybrid composites and surface modifications enhancing both aesthetic and mechanical properties. Given their proven clinical success, ATZ and ZTA will certainly spread toward other biomedical applications and might be serious candidates for all implanted housings and components for electronic medical devices, provided their brazing ability is demonstrated.

4.3.2 Stability of Ceramics

4.3.2.1 Resistance to Corrosion

Ceramics, particularly alumina, zirconia, and silicon nitride, offer significant advantages over metallic materials in terms of chemical stability, corrosion resistance, and biocompatibility. These materials are highly inert, making them more stable when exposed to body fluids and preventing the degradation or corrosion that can occur with metals. Alumina and zirconia, for example, are not prone to oxidation or corrosion pitting, which can be a concern with certain metals, especially in the long term. This resistance to corrosion ensures that ceramics maintain their integrity and functionality within the body without the risk of ion leaching, a problem that can sometimes lead to adverse biological reactions. Additionally, silicon nitride, known for its excellent mechanical properties, also exhibits promising antibacterial qualities, which can reduce the risk of infections in medical implants. The low wear rates of ceramics further contribute to their long-term stability, ensuring minimal release of particles that could cause inflammation or other complications. Overall, the superior corrosion resistance, lack of ion leaching, and enhanced integration with body tissues make ceramics, such as alumina, zirconia, and silicon nitride, highly biocompatible and ideal materials for implants, offering better long-term safety and performance compared to many metallic alternatives.

4.3.2.2 Reliability and Durability

Despite the strong advantages outlined above, ceramics also come with their own set of considerations (no material is truly "perfect," and materials engineers must carefully weigh the pros and cons of each option). Ceramics are typically considered brittle at room temperature, as they tend to fail under high stresses but with minimal deformation, showing no signs of plasticity. Their strength also exhibits a much greater variability compared to metals. As a result, statistical methods are essential when analyzing failure, and careful control of processing defects is crucial to prevent premature fractures. Fracture mechanics is commonly applied to study the mechanical behavior of brittle materials such as glasses, ceramics, and some metals. The fundamentals of fracture mechanics are beyond the scope of this chapter. Therefore, we provide basic concepts here, and the reader is encouraged to consult several key references for further details.

Ceramics are sensitive to stress concentrations around preexisting small defects, such as pores, scratches, or cracks. Indeed, under an applied (macroscopic) tensile stress, σ, the local stresses around a preexisting defect or crack can be very high (much higher than σ). The stress field at the tip of a crack of size c can be well described by the stress intensity factor K_I, which is given by (e.g., Liu et al. 2015):

$$K_I = \sigma \times \sqrt{\pi \times c} \tag{4.1}$$

K_I represents the intensity of the local stresses at the crack tip. When K_I reaches a critical value (K_{IC}), crack propagation and failure occur. Therefore, the fracture strength σ_R of a brittle material in the presence of a defect of size c is given by:

$$\sigma_R = \frac{K_{IC}}{\sqrt{\pi \times c}} \tag{4.2}$$

Increasing the fracture strength can thus be done through an increase of the toughness (this is the advantage of zirconia or alumina–zirconia composites compared to alumina) and a decrease of defect size (this is done by improving processing methods and by a careful follow-up of manufacturing). In the case of zirconia-containing ceramics, stress concentrations occurring at the crack tip may trigger the $t \rightarrow m$ transformation, which hinders crack propagation and thus increases the toughness.

Ceramics typically show a wider statistical dispersion in their strength than metals. This is due to the absence of plasticity and the presence of preexisting defects (such as pores, cracks, or inclusions) that vary in size, orientation, and location. Therefore, reporting strength scatter is crucial in ceramic testing, as it reflects defect distribution and process reliability. The Weibull distribution is commonly used to characterize this strength variability, with its modulus m indicating how much volume affects strength. A higher Weibull modulus means less influence of volume on strength. The Weibull modulus is not inherent to the ceramic itself but is influenced by the variations in defects linked to the manufacturing process. It can range from as low as 5 to over 15 for the same 3 mol.% yttria-doped zirconia, depending on the production method. For example, additive manufacturing tends to result in lower values compared to state-of-the-art ceramic techniques.

K_{IC} provides the basis of the theories of fracture in brittle materials, since it rules the risk of immediate, brittle failure. However, it is well recognized that failure of ceramics and glasses may occur after a certain time even when stressed under K_{IC} (or σ_R). Indeed, K_{IC} represents a critical level for fast crack growth, but ceramic materials may be susceptible to slow crack propagation at K_I values under K_{IC}. This

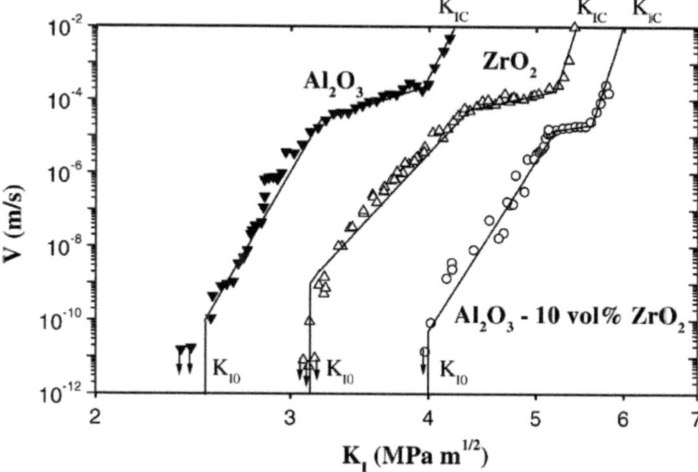

Fig. 4.4 V-K_I curves measured in biomedical grade alumina, 3Y-TZP, and ZTA (10 vol.% of pure zirconia) composites. (Figure adapted from De Aza et al. (2002) with permission from Elsevier)

phenomenon is often referred to as "subcritical crack growth" with respect to crack propagation for stress intensity factors under the toughness. This slow crack growth (SCG) is notable for its sensitivity to applied load, and it also tends to depend on concentration of environmental species, temperature, and other environmental variables. This means that under appropriate conditions, cracks keep on growing "slowly" for some time until they cause fracture.

As a general trend, slow crack growth is described based on a *V* (crack velocity) versus K_I (stress intensity factor) diagram (e.g., De Aza et al. 2002). SCG in ceramics is attributed to stress-assisted corrosion at the crack tip or any preexisting defect in the ceramic. This is indeed the combined effect of high stresses at the crack tip and the presence of water molecules that induces crack propagation in a slow manner. All oxide ceramics (e.g., alumina, zirconia, alumina–zirconia composites, glass ceramics) are prone to a reaction with the polar water molecules because of their ionic bonding nature. This is in contrast with covalent ceramics, such as SiC or Si_3N_4, which are not sensitive to stress-assisted corrosion (except if they present an oxide-containing glassy phase). Figure 4.4 shows typical V-K_I curves in alumina, zirconia (3Y-TZP), and alumina–zirconia composite, showing the better overall crack resistance of zirconia-containing materials compared to alumina. Interestingly (and fortunately), there exists a threshold stress intensity factor, K_{I0}, below which no crack growth is observed. It corresponds to the local stress intensity factor under which failure of the ceramic bonds does not break even in the presence of water species. Thus, it is strongly advised that biocomponents should be designed to work under K_{I0}, since above it, delayed fracture will occur sooner or later, generally in less than a month (a crack rate of 10^{-9} m/s seems to be low but means almost 100 microns a day). To go further in the analysis, it should even be of importance to consider that cyclic fatigue promotes SCG and thresholds should be determined under cyclic

Table 4.2 Toughness (K_{IC}), threshold stress intensity factor (K_{I0}, below which no crack growth occurs), and ratio between them for the three main structural bio-ceramics

Material	K_{IC} (MPa·m$^{1/2}$)	K_{I0} (MPa·m$^{1/2}$)	K_{I0}/K_{IC}
Alumina (Al_2O_3)	3.0–4.0	1.5–2.0	0.4–0.5
3Y-TZP zirconia	5.0–7.0	2.5–3.5	0.4–0.5
Silicon nitride	6.0–7.0	5.0–6.0	0.7–1.0

loading conditions. Table 4.2 summarizes the threshold stresses reported in biomedical grade ceramics so far and estimations of ratios between strength and threshold stress for design purposes.

4.3.2.3 Aging of Zirconia

Exposure to water—such as body fluids—can trigger the t → m phase transformation at the surface of zirconia, particularly in Y-TZP, as widely documented in the literature (Chevalier et al. 2009). Unlike the beneficial t → m transformation that occurs near cracks in the bulk—enhancing toughness through transformation toughening—surface transformation in the presence of moisture may degrade material's properties. The aging process involves moisture, destabilizing the tetragonal structure of Y-TZP, likely through the filling of oxygen vacancies with water-derived species (e.g., hydroxyl, oxygen, or hydrogen ions). This destabilization leads to internal tensile stresses due to a reduction in lattice parameters. As a result, metastable tetragonal grains at the surface gradually transform into the monoclinic phase over time. This transformation causes volume expansion, surface uplift, and large stresses, potentially leading to microcracks along grain boundaries. Aging kinetics can be monitored by measuring the monoclinic phase content on the surface over time using techniques such as X-ray diffraction or Raman spectroscopy. These kinetics typically follow sigmoidal curves described by the *Mehl–Avrami–Johnson* model, reflecting a nucleation-and-growth mechanism.

Hydrothermal aging has been recognized since the early 1980s, but its impact on the durability of biomedical implants was largely overlooked until the early 2000s. This changed following a series of fractures in zirconia orthopedic ball heads that occurred just months after implantation. These failures were attributed to accelerated aging in insufficiently densified balls, caused by a change in furnace technology (Chevalier et al. 2009). The aging mechanism and its effects on dental devices are less documented, but it has also been reported recently, though its direct impact on dental products' lifetime remains uncertain (Koenig et al. 2024; Kohal et al. 2025).

As previously mentioned, it is important to assess the aging kinetics of any zirconia-containing material before its introduction in the market, especially when new processing methods are involved. Tools are available to monitor aging, and ISO standards now include specific guidelines for this purpose. We would advise conducting accelerated tests on the surfaces of interest (not on polished sections), as aging is very sensitive to processing conditions, including surface preparation. With careful control of LTD before clinical use, and with ongoing progress in zirconia materials (including 3Y-TZP), the failures experienced in the past can largely be avoided.

4.3.3 Novelties in the Ceramic Field

Advances in the field of ceramics for medical applications, and for implantable electronic devices such as cochlear implants, may follow two main routes that can be combined, in terms of materials development and processes.

Progresses in Ceramic Materials

- *New yttria-doped zirconia with improved aging resistance:* The incorporation of alumina into zirconia has shown to enhance its resistance to hydrothermal aging, improving its long-term stability for dental and orthopedic implants. This has been implemented already with the use of E-grades from Tosoh, for example, in most of biomedical devices.
- *Ceria-doped zirconia (and other dopants):* Cerium oxide (ceria) has been explored as a stabilizing agent for zirconia, significantly enhancing its resistance to aging. The addition of ceria not only improves toughness but also prevents LTD. This has not been implemented so far but could be a credible alternative to Y-TZP in the future. Ceria-doped zirconia composites may present high toughness and ductility and to a certain extent transformation-induced plasticity resulting in less dispersion in strength data and improved fracture resistance, ideal for applications where reliability is critical, as it is the case of cochlear implants. A renewed interest is also noted for calcia-stabilized zirconia, with quite similar features.
- *Alumina–zirconia composites:* Combining alumina and zirconia creates composites that benefit from both materials' properties, as it has been described above. Such composites are already in use for orthopedic devices and dental implants. Their development for housings of cochlear implants is reasonable, provided their brazing ability with titanium is carefully studied.
- *Silicon nitride:* Known for its excellent mechanical properties, including high strength, fracture toughness, and resistance to wear and corrosion, silicon nitride is gaining traction in orthopedic and dental implants. It also offers, at least theoretically, some antibacterial properties, which would be a must for the application. As for alumina–zirconia composites, further effort will be needed to assess their viability as housings, starting again with brazing ability and durability.
- *Surface functionalization:* Surface modifications, such as coating with bioactive molecules (e.g., silanes, phosphonic acids, peptides, proteins, or growth factors), are being used to improve biocompatibility and promote cell adhesion. Techniques like plasma treatment, laser ablation, and sol–gel coatings are commonly used to tailor surface roughness and chemistry, aiding in better osseointegration and healing.

Advancements in Bioceramic Processing:

- *Cold and flash sintering:* Cold sintering, which involves densifying ceramics at significantly lower temperatures under pressure, is gaining attention for improving energy efficiency in bioceramic processing. This method enables the consolidation of materials that would typically require high sintering temperatures, thereby minimizing thermal degradation and preserving fine microstructures—such as smaller grain sizes—which are critical for maintaining desirable material properties. Flash sintering, on the other hand, uses microwave or pulsed electric fields to achieve rapid densification at reduced temperatures. This method significantly reduces processing times and energy consumption, allowing for more cost-effective production of high-performance bioceramics.

- *Co-sintering as an alternative to brazing:* Co-sintering involves sintering multiple materials together at the same time, which can be an effective alternative to brazing. This process eliminates the need for additional bonding agents or complex joining processes, ensuring material integrity and reducing the likelihood of failure at interfaces. The difficulty lies in the densification of materials with dissimilar sintering ability, by playing with size and morphology of starting powders. Cold sintering could be foreseen as a very original method along this line and could be associated to multi-materials additive manufacturing.
- *Additive manufacturing:* Stereolithography is one of the most mature 3D printing techniques for bioceramics. It allows for the creation of highly detailed, patient-specific implants with intricate geometries. It has been used for scaffold generation, for example, but is also suggested for dental products (like facets). Its development for cochlear implants would need that it presents an additional value compared to conventional technologies (i.e., pressing, densification, machining), in terms of economical cost, materials consumption, and complex geometries.

4.4 Brazing Assembly Technique for Cochlear Implant Manufacturing

4.4.1 The Principals of Brazing Process

Brazing is a well-suited method for joining metal and ceramics (Schwartz 2003; Sekulić 2013). It uses a liquid metal or alloy as a bonding material, referred to as braze, brazing alloy, or filler. During the brazing process (Fig. 4.5), the temperature is raised above the melting point of the braze (above the liquidus of the braze) but remains below the melting point of the substrates to be joined (below the solidus of each substrate), which do not melt. Thus, the braze melts, flows across the surfaces, and penetrates the gap between the substrates. Moreover, liquid–solid reactions occur between the liquid braze and the substrates. The formation of the interfaces depends on concomitant and interfering mechanisms such as wetting of the solid surfaces by the liquid braze and interfacial reactivity. During cooling, the braze solidifies to form a permanent bond between the substrates.

Brazing is typically performed at high temperatures, with a filler with a liquidus above 450 °C. It is often compared to soldering, a similar joining method, which differs primarily in the use of a filler metal (solder) with a liquidus below 450 °C.

There are two brazing configurations illustrated in Fig. 4.6 (Eustathopoulos et al. 2013):

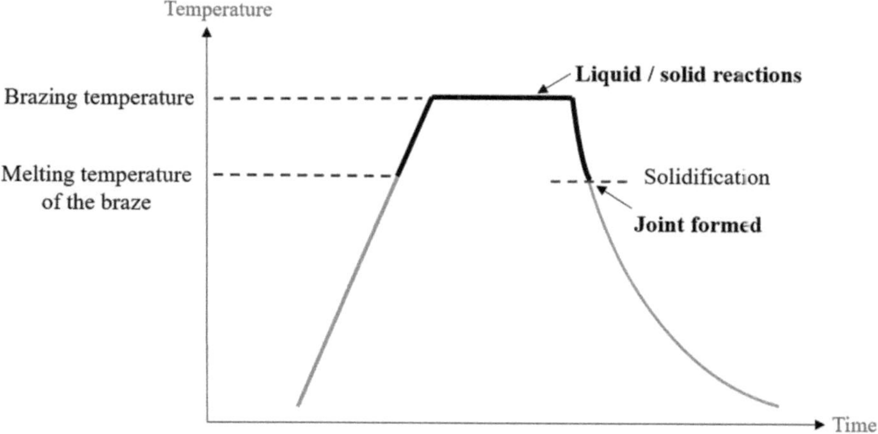

Fig. 4.5 Schematic temperature profile during brazing

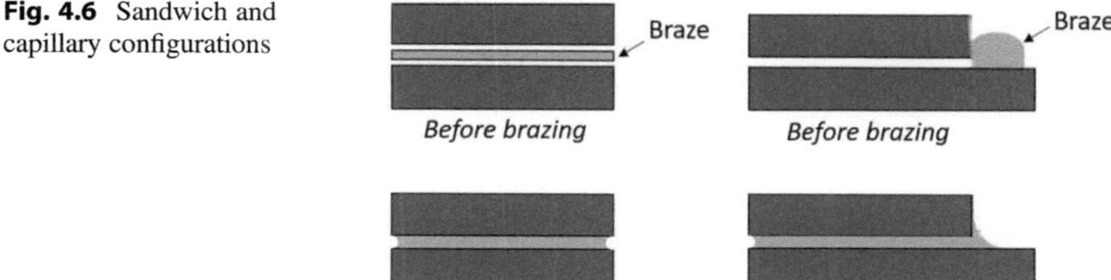

Fig. 4.6 Sandwich and capillary configurations

- Sandwich configuration (Fig. 4.6a): In this setup, the braze is preplaced between the substrates before heating begins.
- Capillary configuration (Fig. 4.6b): Here, the braze is positioned outside the gap between the substrates. During the brazing process, capillary forces draw the molten braze into the gap, enabling proper infiltration and bonding.

A key parameter in brazing is the *wetting behavior*, which is characterized by the contact angle θ at the junction of a solid (S), liquid (L), and vapor (V)—commonly referred to as the *triple line*. This angle is formed by the spreading of a nonreactive liquid droplet on a solid surface (Fig. 4.7a, d). At thermodynamic equilibrium, the contact angle is determined by the balance of surface and interfacial energies:

σ_{SV}: Surface energy of the solid with the vapor phase
σ_{SL}: Interfacial energy between the liquid and the solid
σ_{LV}: Surface energy of the liquid in equilibrium with the vapor

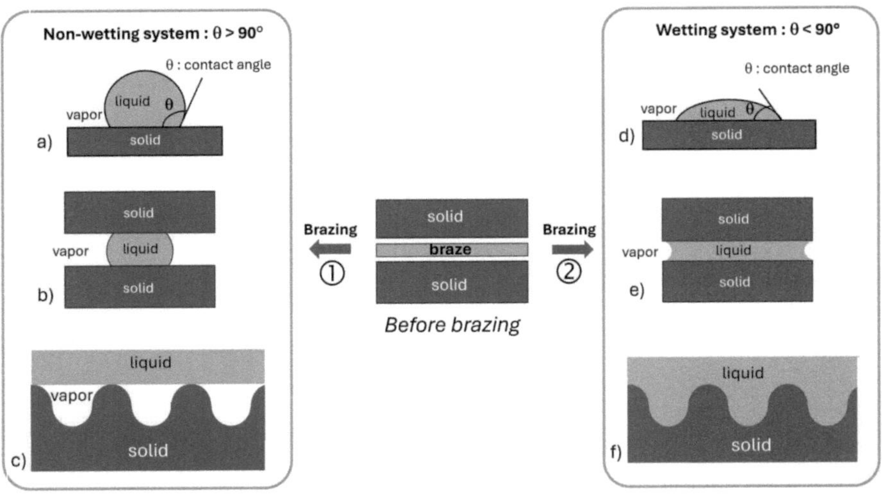

Fig. 4.7 Schematic illustration of the effect of the wetting on the brazing

These relationships are described by Young's equation (Eustathopoulos et al. 1999, 2013) as follows:

$$\cos\theta = \frac{\sigma_{SV} - \sigma_{SL}}{\sigma_{LV}}$$

Wetting behavior arises from the interplay between adhesive forces at the liquid–solid interface, which promote wetting, and cohesive forces within the liquid, which resist it. Two main configurations can be observed:

- *Non-wetting system*: When the contact angle (θ) exceeds 90° (Fig. 4.7a), the liquid braze tends to flow away from the joint or only partially fills the gap (Fig. 4.7b). In such cases, the liquid fails to penetrate surface asperities, resulting in poor effective contact (Fig. 4.7c) and weak adhesion.
- *Wetting system*: When the contact angle (θ) is less than 90° (Fig. 4.7d), a good filling of the joint by the braze is obtained (Fig. 4.7e). A high effective contact is created at the microscopic scale between the braze and the solid (Fig. 4.7f), allowing good adhesion. Therefore, good wetting is considered a necessary condition for successful brazing.

Two types of wetting can be distinguished and are detailed by Eustathopoulos et al. (1999) and (2013):

- *Nonreactive wetting:* In this case, there is no chemical reaction or dissolution of the solid substrate into the liquid braze. The spreading behavior is controlled by the viscosity of the liquid.

- *Reactive wetting:* Here, the solid substrate either dissolves into the liquid braze or reacts to form new interfacial component(s). Note that both phenomena may occur simultaneously: dissolution and the formation of new interfacial component (s). The wetting behavior then depends on the wettability of the newly formed compound(s), which may be more or less wettable than the original substrate.

Wetting is also influenced by several other factors, including temperature, duration, atmospheric conditions, surface roughness, and the presence of impurities on the solid surface.

Another key parameter of brazing is the adhesion of the solidified interfaces. During cooling from brazing temperature to room temperature, thermal stresses develop at the interfaces. To ensure a mechanically robust joint, strong interfacial bonding is essential. In case of weak interfaces, adhesive failure occurs with separation at the substrate–braze boundary. In contrast, strong interfaces can lead to two possible outcomes:

- No failure: Neither the interface nor the substrate(s) exhibit any failure.
- Cohesive failure in the substrate(s): In this case, the interface remains intact, but failure occurs within the substrate material itself, often due to mismatched properties such as the coefficient of thermal expansion.

Attention must be paid to the determination of joint designs to minimize stresses, especially for ceramic–metal joints due to the differences in the physical properties of ceramics and metals. Additionally, note that joint gap affects the capillary flow (especially in capillary configuration) and the mechanical properties of the assembly.

Liquid metals generally wet metallic substrates well due to the formation of metallic interfacial bonds, unless the surfaces are covered by oxide films. In contrast, wetting ceramics, particularly alumina and zirconia, with liquid metals (e.g., Au, Ag, Cu, Sn) or alloys (e.g., AgCu) is more difficult. However, it can be achieved through two primary methods (Fig. 4.8a, b):

- Metallization of the ceramic surfaces: This involves the direct deposition of a metal on the ceramic before brazing. The coating bonds to the ceramic and can be wetted by the liquid braze. However, the dissolution of the coating into the liquid braze must be carefully controlled to prevent de-wetting.
- Addition of an active element to the braze: Elements such as Ti can be added to the braze to form wettable reaction products during the brazing process.

In ceramic-to-metal brazing (e.g., Mishra et al. 2020), the active element can be provided by the dissolution of metallic substrate in the liquid non-active braze if the substrate contains one or more active elements (Fig. 4.8c). For example, in the case of brazing zirconia (ZrO_2) to titanium (Ti) using pure gold as a non-active braze, the active element—titanium—is supplied from the dissolution of the titanium substrate in the molten gold (Fischer et al. 2025). In the next section, the brazing of alumina and zirconia is described in more detail.

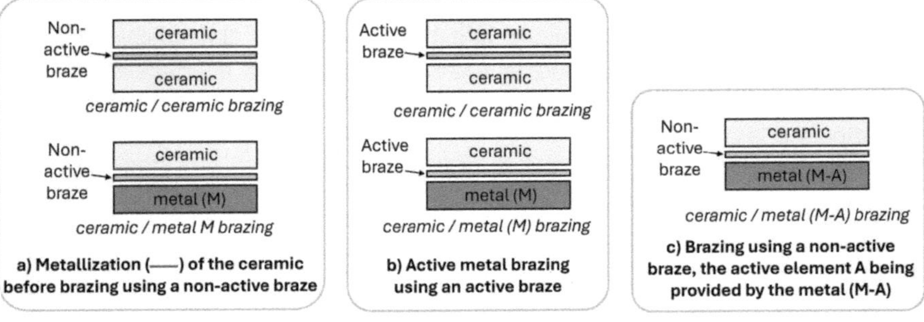

Fig. 4.8 Different approaches for ceramic/ceramic and ceramic/metal brazing

4.4.2 Brazing of Alumina and Zirconia

After World War II, the most widely used metallization process for alumina was the moly-manganese process. In this method, the ceramic surface is coated with a mixture of a glassy phase and a refractory metal—typically molybdenum (Mo) combined with manganese (Mn) or manganese dioxide (MnO_2) particles. Heating the coated ceramic at approximately 1500 °C in a wet H_2 atmosphere causes the glassy phase to flow from the ceramic grain boundaries into the capillaries between the Mo particles. Subsequently, a second layer of a metal, such as nickel, is deposited by electroplating and heat-treated at 950 °C in H_2 atmosphere. This metallized alumina can then be brazed using a non-active braze (e.g., Li et al. 2022). However, this process involves multiple steps and requires high-temperature furnace operations in H_2 atmosphere. As a result, alternative methods (e.g., Ramlow et al. 2025) such as pure metal coatings or the use of active brazes have been developed and successfully applied, particularly in CI manufacturing.

As mentioned in Sect. 4.3, in cochlear implant, the most common ceramic is high-purity alumina. It is considered an insulator in feedthroughs for both titanium and titanium–ceramic casing designs, where it is brazed to platinum pins and titanium (Fig. 4.9). In the case of titanium–ceramic casing design, the alumina part and titanium ferrules are also joined by brazing. More recently, zirconia–titanium brazing has been investigated for the new generation of implants. In all these configurations, the braze must be biocompatible, with gold (Au) being the most commonly used material.

Joining titanium and ceramic parts in implant housing typically involves two steps (Fig. 4.10): first, the ceramic is brazed to titanium ferrules; then, the brazed ceramic–titanium assembly is laser-welded to the titanium cover, resulting in a hermetically sealed enclosure for the electronic circuitry.

The next section discusses the physicochemical aspects of brazing the ceramic components of the implant, followed by an examination of joint properties such as mechanical strength.

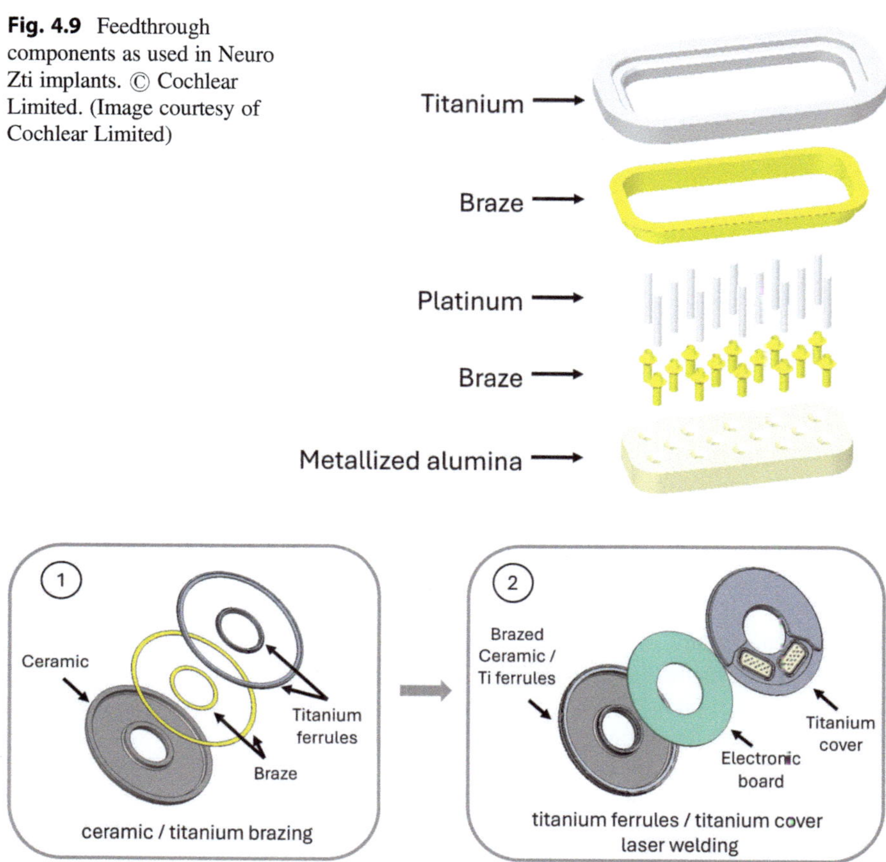

Fig. 4.9 Feedthrough components as used in Neuro Zti implants. © Cochlear Limited. (Image courtesy of Cochlear Limited)

Fig. 4.10 Two-step joining process of titanium and ceramic components in the ceramic casing design, as implemented in the Neuro Zti implant. © Cochlear Limited. (Image courtesy of Cochlear Limited)

4.4.3 Brazing Al_2O_3–Ti, ZrO_2–Ti, and Al_2O_3–Pt

Brazing between alumina and titanium has been successfully achieved following metallization of the alumina surface using direct metal deposition via physical vapor deposition (PVD). Two approaches have been reported using one and two metals:

- Niobium (Nb) coating with a thickness of 500 nm, applied by sputtering (Siddiqui and Jones 2014; Koop et al. 1980).
- Titanium–niobium (Ti–Nb) bilayer coatings; each layer is few μm in thickness (Fig. 4.11a), also applied by sputtering and proposed by (Regnier and Nguyen 2021).

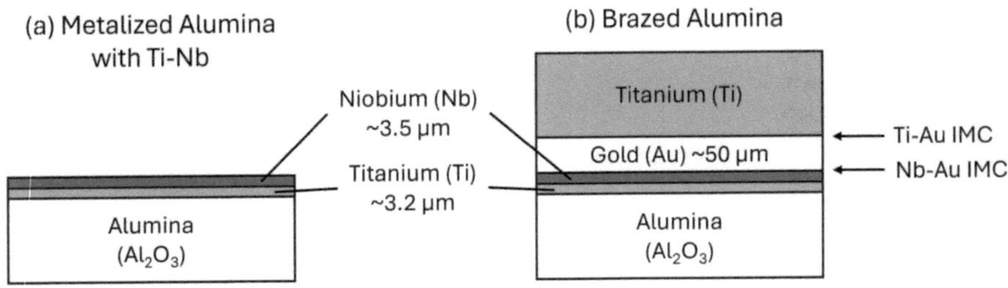

Fig. 4.11 (**a**) Metallized alumina with Ti–Nb. (**b**) Brazed alumina–Au–Ti joint

These coatings demonstrated excellent wettability with pure gold, enabling effective brazing between alumina and titanium. As a result, a Ti–Au–Al_2O_3 joint was formed by brazing titanium to alumina metallized with Ti and Nb (Fig. 4.11b). Two types of intermetallic compounds (IMCs) were identified at the interfaces:

- Ti–Au IMC at the gold–titanium interface
- Nb–Au IMC at the gold–niobium interface

In contrast, when zirconia was metallized using Ti and Nb coatings, the layers proved to be unstable during the brazing process (Fischer et al. 2023). Therefore, this method of metallizing zirconia is not recommended for forming assemblies with good mechanical properties.

Ti–Au–Al_2O_3 brazing has also been achieved without prior metallization of the alumina surface (Bian et al. 2019), the active element (Ti) being provided by the titanium substrate (partial dissolution of the titanium substrate in liquid gold). In this approach, a joint was formed at 1115 °C for 1 min using a 50 µm thick gold foil. The resulting microstructure consisted of multiple reaction layers in the following sequence: Ti | Ti_3Au | TiAu | $TiAu_2$ | $TiAu_4$ | Au with granular $TiAu_4$ | TiO_x | Al_2O_3 (Fig. 4.12). Additionally, Bian et al. (2019) investigated the effects of varying the brazing temperature (1105 °C, 1110 °C, 1120 °C, and 1125 °C) and extending the holding time at 1115 °C from 1 to 5 min on the thickness of the reaction layers. They demonstrated that the thicknesses of Ti_3Au | TiAu | $TiAu_2$ layers increase when brazing temperature or time increase. Moreover, they obtained a shear strength of 39.2 MPa after brazing at 1115 °C for 3 min.

Lei et al. (2020) evaluated Ti–Au–ZrO_2 brazing at various temperatures—1110 °C, 1130 °C, 1150 °C, 1170 °C, and 1190 °C—for 10 min using a 50 µm thick gold foil. The typical interfacial microstructure of the resulting joint consisted of the following layers: Ti | Ti_3Au | TiAu | $TiAu_2$ | $TiAu_4$ | TiO | ZrO_2. An optimal shear strength of 35 MPa was obtained at 1150 °C for 10 min.

More recently, Fischer et al. (2025) investigated the brazing of titanium to zirconia using pure gold. Their study examined the influence of several parameters, including brazing temperature (1080 °C and 1100 °C), holding time (ranging between 3 and 60 min at 1080 °C), and gold foil thickness (25, 50, 100, and

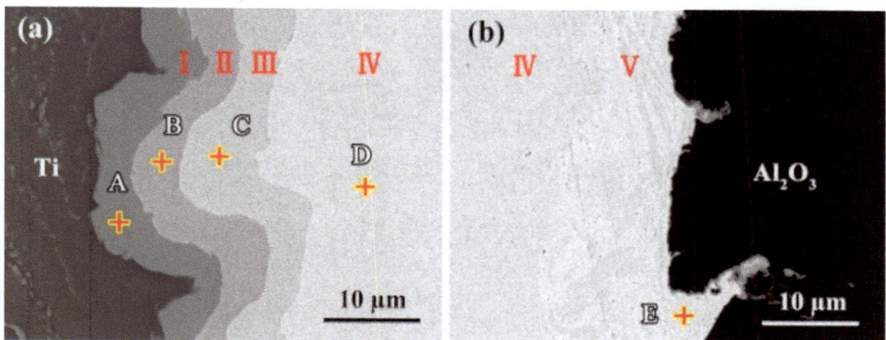

Fig. 4.12 SEM micrographs of Ti–Au–Al_2O_3 joint. (**a**) Ti–Au interface, (**b**) Au–Al_2O_3 interface. (Figure adapted from Bian et al. (2019) under CC BY license)

150 μm). At the Au/Ti interface, the formation of the four intermetallics ($TiAu_4$, $TiAu_2$, TiAu, Ti_3Au) and a Ti–Au solid solution was observed. At the Au/ZrO_2 interface, a submicron Ti_2O_3 layer was formed. The authors demonstrated that a total consumption of the liquid Au may occur with thin foils (25–50 μm), due to very fast interdiffusion in Au–Ti IMCs. In order to avoid the total consumption of liquid gold leading to isothermal solidification of the joint and obtain a thin Ti_2O_3 layer, they recommend low temperature (1080–1100 °C), thick gold foil (~100 μm), and short holding time (a few minutes).

Concerning the Al_2O_3/Pt system, successful brazing requires either metallizing the alumina substrate or using an active braze alloy such as Gold-ABA®, which is primarily gold with small additions of nickel (3%) and titanium (0.6%) by weight. Indeed, the platinum substrate, unlike titanium substrate, does not provide an active element.

4.5 Cochlear Electrode Arrays

4.5.1 Different Electrode Array Designs on the Market

Cochlear implant manufacturers offer various types of electrode arrays, each based on a distinct design concept. Depending on the company's objectives and strategy (Dhanasingh and Jolly 2017), the proposed electrode design may better address the needs of different users and patients. Electrode arrays are generally classified into two main categories:

- Straight or lateral wall electrode arrays
  A key advantage of straight or lateral wall (LW) electrode arrays (see Fig. 4.13a) is their ease of use (Hrncirik et al. 2022). These electrodes do not require specialized surgical tools for insertion into the cochlea; a single pair of forceps is typically sufficient, although each manufacturer provides their own version.

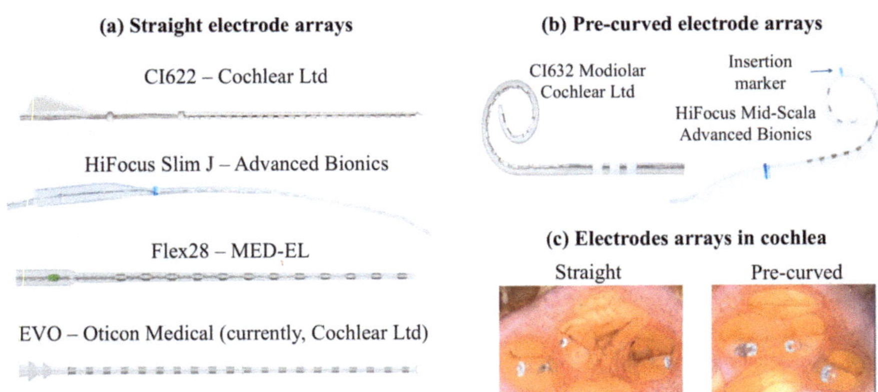

Fig. 4.13 (**a**) Straight and (**b**) pre-curved electrode arrays fabricated by different CI manufacturers. © 2025 Advanced Bionics LLC; © Cochlear Limited; and © MED-EL GmbH. (Electrode images provided courtesy of the respective companies). (**c**) Comparison of placement of electrode contacts in straight (Cochlear CI622) and pre-curved (Cochlear Contour Advance®) in scala tympani using bone histology. (Adapted from Gibson and Boyd (2016) with permission from Elsevier)

LW electrodes are designed for insertion through the round window (RW) or the extended round window (ERW) (Salcher et al. 2020). Straight electrodes are generally longer than pre-curved ones, as the external cochlear duct length (CDL) is naturally longer than the trajectory in modiolar position. Their length typically ranges from 20 to 28 mm, although longer electrode arrays—such as the 34-mm-long FLEX34 fabricated by MED-EL—are also available. These electrode arrays may cover one to two turns of the cochlea, depending on its shape and the CDL. The average CDL at the LW position is estimated to be approximately 23 mm for the basal turn (Eser et al. 2021).

Due to their flexibility, straight electrodes conform to the outer curvature of the cochlea, positioning the stimulation contacts closer to the outer hair cells of the scala media (Fig. 4.13c).

- Pre-curved electrode arrays

  The main advantage of pre-curved electrodes (see Fig. 4.13b) is their final positioning within the scala tympani (ST). The electrode contacts are closer to the cochlear modiolus and, therefore, nearer to the spiral ganglion, which can result in improved stimulation performance (Briggs et al. 2011). Pre-curved electrodes are typically shorter (less than 20 mm) than straight arrays but still cover an average of 1.5 turns. Another benefit is the reduced trauma during insertion. Unlike LW electrodes that slide along and conform to the outer cochlear wall, pre-curved electrodes regain their curved shape during insertion, minimizing friction against the lateral wall (Roland 2005; Cuda and Murri 2017; Ramos-Macías et al. 2017).

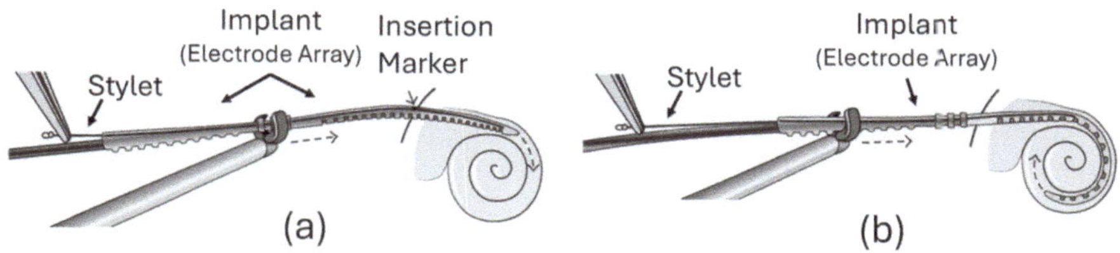

Fig. 4.14 Schematic illustration of the advanced off-stylet (AOS) technique. (**a**) The electrode array is inserted into the inner ear with the stylet inside until a marker is at the level of the cochleostomy site. (**b**) The stylet is kept stationary and the implant is advanced further into the cochlea until full insertion depth is achieved. (Figure adapted from Rau et al. (2015) under CC BY license

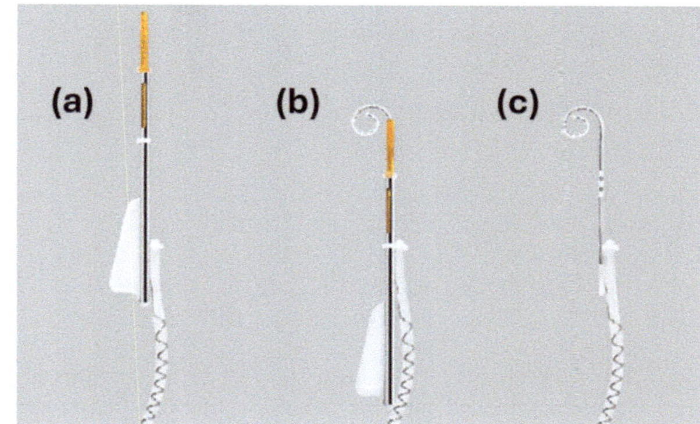

Fig. 4.15 Cochlear Ltd.—Slim Modiolar Electrode. (**a**) Electrode retracted into sheath. (**b**) Electrode deployed. (**c**) Electrode array without sheath. (Figure adopted from Cuda and Murri (2017) with permission from Springer Nature)

Unlike straight electrodes, pre-curved arrays require additional mechanisms for insertion. Due to their natural curled shape, which matches the modiolar curvature, they cannot be inserted without a dedicated insertion solution.

An early solution involved the use of a stylet integrated into the electrode array (Fig. 4.14). The stylet temporarily maintains the electrode in a straight configuration during electrode array insertion. When the electrode array is fully inserted, the stylet is gradually withdrawn (Rau et al. 2015).

A more recent approach uses a syringe-like system with a sheath (Fig. 4.15). The pre-curved electrode is retracted into the sheath before insertion. The first 5 mm of the sheath is inserted through the RW or ERW into the straight portion of the basal turn. The electrode is then pushed out of the sheath, allowing it to follow the ST and return to its curved shape, ultimately achieving a modiolar position (Beck et al. 2023).

Figure 4.16 presents a summary of the characteristics of the most commonly used electrode arrays from three cochlear implant manufacturers, including the number of electrodes, distribution of electrode contacts along the array, dimensions of the contacts at the apical (A) and basal (B) regions, and the intended purpose of each array.

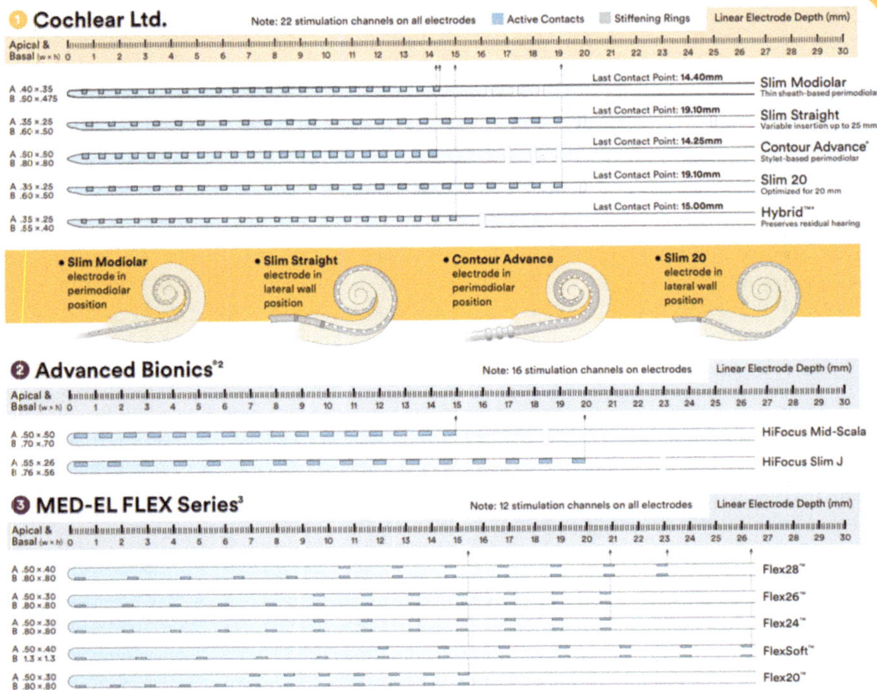

Fig. 4.16 A summary of the characteristics of the most commonly used electrode arrays from three cochlear implant manufacturers. © Cochlear Limited. (Image courtesy of Cochlear Limited)

4.5.2 Shared Technologies Among Manufacturers

In the previous section, we highlighted the various electrode array designs available on the market, each targeting different positions within the scala tympani and requiring specific surgical insertion techniques. Despite differences in final shape, number of contacts, or pitch, all manufacturers follow a common principle: electrode contacts are made of platinum or a platinum–iridium alloy. Each contact is connected to an insulated wire made of the same material, and the bundle of wires runs from the electrode array to the stimulator.

One of the key challenges in this design and manufacturing process is managing the fine wires that run along the electrode array and maintaining the precise positioning of the electrode contacts within the mold during PDMS overmolding. Each manufacturer has developed proprietary processes and precision tooling to ensure accurate overmolding and to guarantee product quality and reproducibility.

The only manufacturer that deviates from this conventional approach is TODOC Co. Their electrode array is fabricated using stacked silicone sheets and platinum foil (Shin et al. 2021; Choi et al. 2023). Instead of using multiple wires to connect the electrode contacts, electrical tracks are created directly by selective laser ablation of the platinum foil, allowing for as many tracks as needed (Fig. 4.17).

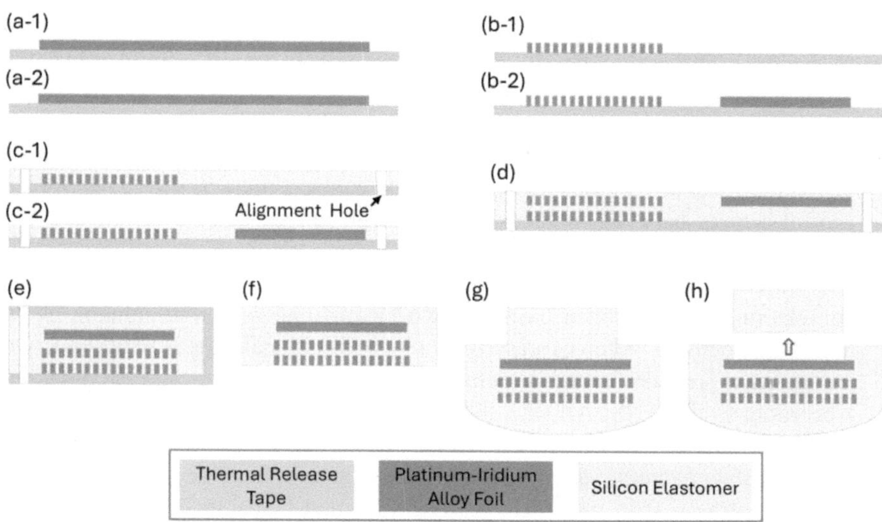

Fig. 4.17 TODOC Co. electrode manufacturing overview. The fabrication process of the 32-channel electrode array includes (**a**) attaching Pt–Ir foil to thermal release tapes; (**b**) laser micro-structuring the foil; (**c**) applying silicone molding and forming alignment holes; (**d**) bonding two 16-channel modules; (**e**) folding the array into a three-layered structure; (**f**) removing the sacrificial layer; (**g**) forming a silastic carrier and applying a pillar; and (**h**) exposing the electrode sites by removing the pillar. (Figure adapted from Shin et al. (2021) under CC BY license)

- Electrode contact fabrication

  Manufacturing miniature components such as electrode contacts is a complex task. The shape of the contact—whether round, cylindrical, rectangular, or folded—depends on the specific electrode array design.

  In micromechanics, these small parts are often produced by stamping and forming thin metal sheets. Toolmakers define the necessary steps and design the required punches accordingly. In some cases, lasers are used to cut directly from metal sheets, eliminating the need for stamping tools. Advances in laser technology, particularly femtosecond lasers, have enabled extremely precise microfabrication. These systems are widely used in industries such as watchmaking, aerospace, defense, and medical devices.

  Another method for producing such components is electrical discharge machining (EDM), which removes metal through spark erosion. In EDM, sparks are generated between a wire and the platinum foil, cutting the material with controlled roughness. This rough surface increases the specific surface area compared to a smooth geometric surface. Oticon Medical reported that using this process for its Neuro Zti implant reduced the electrical impedance of electrode contacts by a factor of five (Tourrel et al. 2018).
- Electrode–tissue interface

  These shared design principles result in a nearly identical electrode–tissue interface across manufacturers. Platinum and its alloys are among the most stable implantable materials for delivering electrical charge to living tissue, due to their low dissolution rates. Studies (Kumsa et al. 2016; Wissel et al. 2018; Shepherd et al. 2021) have shown that the main factors influencing dissolution are charge density, stimulation polarity, and mode.

Manufacturers should ensure that the electrode surface area is sufficient to maintain safe and controlled operation, adhering to standards such as EN 45502. Given the small size of the cochlea, electrode contacts must be tiny. Therefore, a careful balance between contact size and charge delivery should be made.

The Shannon limit (or Shannon level) is a safety guideline used in cochlear implant design to prevent neural or tissue damage caused by excessive electrical stimulation. It is calculated based on two factors: charge per stimulation phase and charge density. When expressed on a logarithmic scale, the sum of these two values must remain below a defined threshold to ensure safe stimulation. According to the ANSI/AAMI CI86:2017 standard, the Shannon limit is set at 1.75.

As a safety example, the smallest electrode area in the Oticon Medical EVO electrode array is 0.46 mm^2 (Molaee-Ardekani and Donahue 2023). The maximum allowed stimulation current and pulse duration per phase in the GMCI fitting software provided by the company are 2 mA and 115 μs per phase, respectively. This corresponds to a charge per phase of $2\ mA \times 115\ \mu s = 0.23\ \mu C$ per phase. The charge density is $0.23\ \mu C/0.46\ mm^2 = 50\ \mu C/cm^2$. The Shannon value is then calculated as $\log_{10}(0.23) + \log_{10}(50) = 1.06$. Since this value is smaller than 1.75, the stimulations are considered within the safe zone according to the standard.

- PDMS overmolding

Another common feature among all manufacturers is the use of silicone (PDMS) for overmolding the electrode contacts, forming the final mechanical structure of the electrode array. This overmolding process ensures cohesion between the contacts and wires, resulting in a thin, flexible, and continuous shape suitable for handling and insertion.

The shape and cross section of the electrode, along with the silicone's Shore hardness, determine the array's flexibility and mechanical behavior. Silicone hardness typically ranges from 30 to 80 Shore A (ShA) hardness and reacts differently depending on the molding and curing process. Note: Shore A is one of several hardness scales (others include Shore D, Shore 00, etc.). It measures how resistant a material is to indentation using a standardized durometer. The scale ranges from 0 (very soft) to 100 (very hard).

Two main types of PDMS are used in cochlear implant manufacturing: liquid silicone rubber (LSR) and high consistency rubber (HCR). Both are typically supplied as two-component systems in a 1:1 ratio (e.g., from NuSil Technology LLC[1]). LSR, being fluid, is easily injected into molds but may cause flash on components. HCR, with its gum-like consistency, is more viscous and requires

[1] NuSil Technology LLC, https://nusil.avantorsciences.com/us/en/

higher pressure to fill the mold cavity. Both materials demand high-precision molds to avoid flash and voids in the final product.

4.5.3 Novelties in Electrode Arrays

Research and innovation span many domains and generate numerous breakthroughs. However, long-term implantation remains highly regulated and risky. Manufacturers introducing new technologies must undergo extensive, time-consuming, and costly validation processes before gaining market approval. Below are some examples of recent innovations in the field of electrode arrays:

4.5.3.1 Drug Delivery for Inflammation Reduction

Recently introduced in clinical studies, drug delivery systems have been integrated into electrode arrays to enable direct drug release at the implantation site (scala tympani). This approach bypasses the cochlear–blood barrier, which typically limits the diffusion of drugs from systemic circulation into the inner ear.

Dexamethasone, for example, can be mixed into the silicone material used to fabricate the electrode array (Qnouch et al. 2021). This enables long-term drug release over several years (Rongthong et al. 2022), helping to reduce inflammation and tissue damage post-surgery (Toulemonde et al. 2021) and potentially improving hearing outcomes (Rahman et al. 2024). The incorporation of dexamethasone is compatible with standard electrode array manufacturing processes, requiring only minor adjustments. The drug is added during the silicone mixing phase, and the resulting compound can be used in direct injection molding without significantly altering curing time or temperature.

Once implanted and in contact with biological fluids, the surface crystals of dexamethasone begin to dissolve into the perilymph of the scala tympani. Over time, deeper layers of the drug dissolve, gradually increasing its concentration in the perilymph and leading to fibrosis limitation (Fig. 4.18).

4.5.3.2 Thin-Film Technology

As discussed earlier, traditional electrode arrays—comprising multiple platinum wires and contacts embedded in PDMS—require complex manufacturing processes and specialized expertise. An alternative approach, explored in research and short-term human or long-term animal studies, is thin-film technology. This method involves depositing multiple ultrathin layers of polymers, metals, or oxides—ranging from a few nanometers to several micrometers—onto a substrate. These layers are applied using techniques such as physical vapor deposition (PVD), chemical vapor deposition (CVD), or atomic layer deposition (ALD), depending on the material, thickness, and design requirements. Photopolymerization is then used to functionalize the layers.

By stacking conductive and insulating layers, this technique enables the creation of highly precise conductive tracks, terminals, and electrode contacts. Moreover, the fabrication process is largely automated, making it attractive for scalable medical device production.

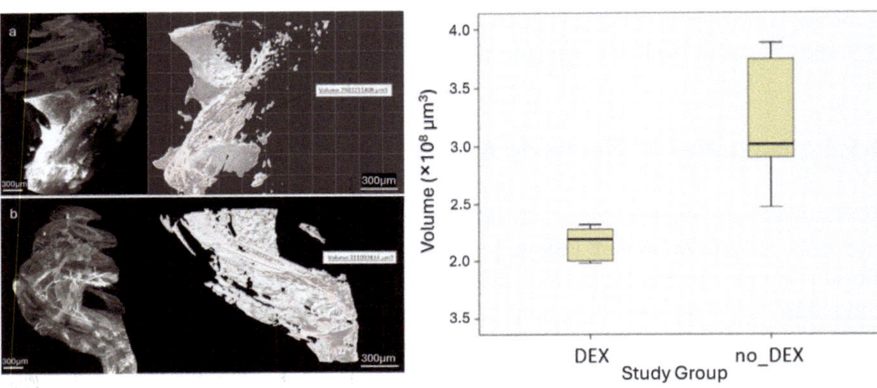

Fig. 4.18 (Left) Image of a right cochlea implanted with a non-eluting (**a**) and dexamethasone-eluting (**b**) electrode array, along with their respective fibrosis volume generated from the automatic surface creation wizard. (Right) Box plot comparing the volume of local fibrosis ($\times 10^8$ μm^3) in the dexamethasone group (DEX) and the group without dexamethasone (no_DEX). (Figure adapted from Toulemonde et al. (2021) under CC BY license)

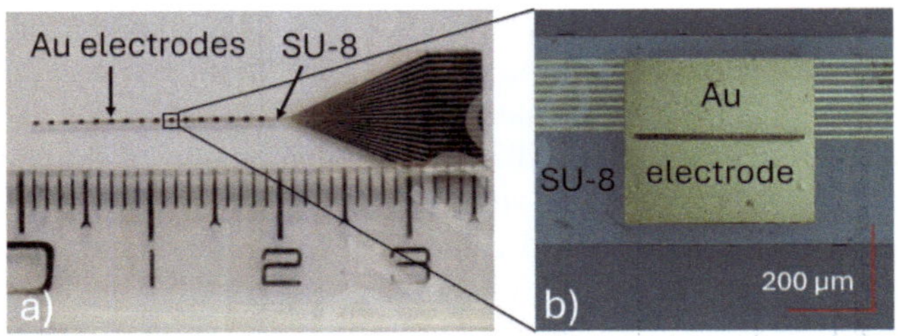

Fig. 4.19 (**a**) Au/SU-8 thin-film electrode with connection for experiment, (**b**) electrode contact detail. (Figure adapted with modification from Cattan et al. (2025). Image from author's preprint article)

In the CI field, an 8-mm-long electrode array with 15 contacts has been developed using thin-film technology (Xu et al. 2019). These electrodes exhibited an average impedance of 16 kΩ and were 97% less stiff than conventional electrode arrays (CEAs).

A thin-film electrode array (TFEA) mounted on a PDMS carrier was compared to a CEA in terms of stiffness (Askin et al. 2024). The primary advantage of thin-film technology lies in its extremely low stiffness, which helps minimize trauma to cochlear structures during insertion. In a separate study (Cattan et al. 2025), a TFEA was fabricated using SU-8 as the substrate, with gold used for the electrode contacts and tracks (Fig. 4.19). SU-8 is an epoxy-based photoresist resin commonly used in microfabrication. It is applied via spin-coating and shaped using masking techniques. Additional advantages include the potential for process automation and improved reproducibility in large-scale manufacturing. However, thin-film technology is not yet validated for long-term implantation due to several limitations. Thin films—especially those using organic materials—may lack chemical stability (Oldroyd et al. 2025). Their high flexibility and minimal barrier thickness make them susceptible to cracking and moisture ingress. Furthermore, poor adhesion between layers can lead to delamination and device failure.

Acknowledgments The authors acknowledge the use of Microsoft Copilot for proofreading and enhancing the readability of the chapter. All suggested changes were carefully reviewed and edited. The authors assume full responsibility for the content of the final publication.

References

ANSI/AAMI CI86:2017 Cochlear implant systems: requirements for safety, functional verification, labeling and reliability reporting

Aşkın GŞ, Gökçeli S, Sümer B (2024) On the design, fabrication, and characterization of a novel thin-film electrode array for use in cochlear implants. Micromachines 15:921. https://doi.org/10.3390/mi15070921

Beck H, Boden WE, Patibandla S et al (2010) 50th anniversary of the first successful permanent pacemaker implantation in the United States: historical review and future directions. Am J Cardiol 106:810–818. https://doi.org/10.1016/j.amjcard.2010.04.043

Beck R, Aschendorff A, Arndt S et al (2023) Evaluation of insertion quality of a slim perimodiolar electrode array. Eur Arch Otorrinolaringol 281:1215–1220. https://doi.org/10.1007/s00405-023-08212-5

Bian H, Song X, Hu S et al (2019) Microstructure evolution and mechanical properties of titanium/alumina brazed joints for medical implants. Meta 9:644. https://doi.org/10.3390/met9060644

Briggs RJS, Tykocinski M, Lazsig R et al (2011) Development and evaluation of the modiolar research array—multi-centre collaborative study in human temporal bones. Cochlear Implants Int 12:129–139. https://doi.org/10.1179/1754762811y0000000007

Cattan A, Borel M, Tourrel G et al (2025) Steerable thin-film electrode array for cochlear implantation: design and development for future atraumatic insertion. Soc Sci Res Netw (SSRN). https://doi.org/10.2139/ssrn.5311759

Chevalier J, Gremillard L, Virkar AV, Clarke DR (2009) The tetragonal-monoclinic transformation in zirconia: lessons learned and future trends. J Am Ceram Soc 92:1901–1920. https://doi.org/10.1111/j.1551-2916.2009.03278.x

Choi G, Ha Y, Kim D-H et al (2023) Assessing the manufacturable 32-channel cochlear electrode array: evaluation results for clinical trials. Biomed Microdevices 25:41. https://doi.org/10.1007/s10544-023-00681-z

Cuda D, Murri A (2017) Cochlear implantation with the nucleus slim modiolar electrode (CI532): a preliminary experience. Eur Arch Otorrinolaringol 274:4141–4148. https://doi.org/10.1007/s00405-017-4774-6

De Aza AH, Chevalier J, Fantozzi G et al (2002) Crack growth resistance of alumina, zirconia and zirconia toughened alumina ceramics for joint prostheses. Biomaterials 23:937–945. https://doi.org/10.1016/s0142-9612(01)00206-x

De Freitas BX, Alves MFRP, Santos C et al (2020) Mechanical properties of biocompatible Y-TZP/Al2O3 composites obtained from mechanically alloyed powders. J Braz Soc Mech Sci Eng 42:353. https://doi.org/10.1007/s40430-020-02431-1

Dhanasingh A, Jolly C (2017) An overview of cochlear implant electrode array designs. Hear Res 356:93–103. https://doi.org/10.1016/j.heares.2017.10.005

EN 45502-1: implants for surgery—active implantable medical devices—part 1: general requirements for safety, marking and for information to be provided by the manufacturer. European Committee for Electrotechnical Standardization (CENELEC), Brussels

EN 45502-2-1: active implantable medical devices—Part 2-1: particular requirements for active implantable medical devices intended to treat bradyarrhythmia (cardiac pacemakers)

EN 45502-2-3: active implantable medical devices—Part 2-3: particular requirements for cochlear and auditory brainstem implant systems

Eser MB, Atalay B, Dogan MB et al (2021) Measuring 3D cochlear duct length on MRI: is it accurate and reliable? AJNR Am J Neuroradiol 42:2016–2022. https://doi.org/10.3174/ajnr.a7287

Eustathopoulos N, Nicholas MG, Drevet B (1999) Wettability at high temperatures, Pergamon materials series, vol 3. Pergamon Press, Oxford. Series Editor: Robert W. Cahn

Eustathopoulos N, Hodaj F, Kozlova O (2013) Part I: fundamentals of brazing. In: Sekulić DP (ed) Advances in brazing: science, technology and applications. Woodhead Publishing, pp 3–35

Fischer M, Chaumat V, Godinot C, Hodaj F et al (2023) Method for assembling a zirconia part to a titanium element. US patent 2023/0191517 Al

Fischer M, Chaumat V, Flament C, Hodaj F (2025) Brazing of zirconia to titanium using pure gold: physicochemical aspects. J Eur Ceram Soc 45:117000. https://doi.org/10.1016/j.jeurceramsoc.2024.117000

Garvie RC, Hannink RH, Pascoe RT (1975) Ceramic steel? Nature 258:703–704. https://doi.org/10.1038/258703a0

Gibson P, Boyd P (2016) Optimal electrode design: straight versus perimodiolar. Eur Ann Otorhinolaryngol Head Neck Dis 133:S63–S65. https://doi.org/10.1016/j.anorl.2016.04.014

Greatbatch W (2000) The making of the pacemaker: celebrating a lifesaving invention. Prometheus Books, Amherst

Guo N, Long S, Zhu G et al (2023) Effect of TiO2 on microstructure and micro-wave dielectric properties of TiO2–CaF2 ceramics prepared by two-step sintering assisted by cold sintering. J Mater Sci Mater Electron 34:1858. https://doi.org/10.1007/s10854-023-11150-x

He X, Zhang YZ, Mansell JP et al (2008) Zirconia toughened alumina ceramic foams for potential bone graft applications: fabrication, bioactivation, and cellular responses. J Mater Sci Mater Med 19:2743–2749. https://doi.org/10.1007/s10856-008-3401-x

Hrncirik F, Roberts IV, Swords C et al (2022) Impact of scala tympani geometry on insertion forces during implantation. Biosensors 12:999. https://doi.org/10.3390/bios12110999

ISO 13356 Implants for surgery—ceramic materials based on yttria-stabilized tetragonal zirconia (Y-TZP)

ISO 6474—Implants for surgery—ceramic materials based on high purity alumina

Kelly JR, Denry I (2008) Stabilized zirconia as a structural ceramic: an overview. Dent Mater 24(3): 289–298. https://doi.org/10.1016/j.dental.2007.05.005. Epub 2007 Jul 12. PMID: 17624420

Koenig V, Douillard T, Chevalier J et al (2024) Intraoral low-temperature degradation of monolithic zirconia dental prostheses: 5-year results of a prospective clinical study with ex vivo monitoring. Dent Mater 40:198–209. https://doi.org/10.1016/j.dental.2023.11.008

Kohal R, Douillard T, Sanon C et al (2025) Signs of in-vivo aging of zirconia from explanted dental implants with porous coating after several years in function. Acta Biomater 194:498–513. https://doi.org/10.1016/j.actbio.2025.01.020

Koop JA et al (1980) Niobium coatings for joining ceramic to metal. U.S. Patent No. 4225262. United States Patent and Trademark Office

Kumsa D, Hudak EM, Montague FW et al (2016) Electrical neurostimulation with imbalanced waveform mitigates dissolution of platinum electrodes. J Neural Eng 13:054001. https://doi.org/10.1088/1741-2560/13/5/054001

Lei Y, Bian H, Fu W et al (2020) Evaluation of biomedical Ti/ZrO2 joint brazed with pure Au filler: microstructure and mechanical properties. Metals 10:526. https://doi.org/10.3390/met10040526

Li Y, Chen C, Yi R et al (2022) The brazing of Al2O3 ceramic and other materials. Int J Adv Manuf Technol 120:59–84. https://doi.org/10.1007/s00170-022-08789-x

Liu M, Gan Y, Hanaor DAH, Liu B, Chen C (2015) An improved semi-analytical solution for stress at round-tip notches. Eng Fract Mech 149:134–143. ISSN 0013-7944. https://doi.org/10.1016/j.engfracmech.2015.10.004

Mamivand M, Asle Zaeem M, El Kadiri H, Chen LQ (2013) Phase field modeling of the tetragonal-to-monoclinic phase transformation in zirconia. Acta Mater 61(14):5223–5235. https://doi.org/10.1016/J.ACTAMAT.2013.05.015

Mishra S, Sharma A, Jung DH et al (2020) Recent advances in active metal brazing of ceramics and process. Met Mater Int 26:1087–1098. https://doi.org/10.1007/s12540-019-00536-4

Molaee-Ardekani B, Donahue MJ (2023) Investigating the electrode-electrolyte interface modelling in cochlear implants. Biomed Phys Eng Express 9:055024. https://doi.org/10.1088/2057-1976/aceafb

Oldroyd P, Velasco-Bosom S, Bidinger SL et al (2025) Fabrication of thin-film electrodes and organic electrochemical transistors for neural implants. Nat Protoc. https://doi.org/10.1038/s41596-024-01116-6

Piconi C (2017) 1.5 alumina. In: Comprehensive biomaterials II. Elsevier, pp 92–121. https://doi.org/10.1016/b978-0-12-803581-8.09247-x

Piconi C, Maccauro G (1999) Zirconia as a ceramic biomaterial. Biomaterials 20(1):1–25. https://doi.org/10.1016/s0142-9612(98)00010-6. PMID: 9916767

Piconi C, Porporati AA (2016) Bioinert ceramics: zirconia and alumina. In: Anto-niac I (ed) Handbook of bioceramics and biocomposites. Springer, Cham. https://doi.org/10.1007/978-3-319-12460-5_4

Qnouch A, Solarczyk V, Verin J et al (2021) Dexamethasone-loaded cochlear implants: how to provide a desired “burst release”. Int J Pharm X 3:100088. https://doi.org/10.1016/j.ijpx.2021.100088

Rahman MT, Mostaert B, Eckard P et al (2024) Cochlear implants with dexamethasone-eluting electrode arrays reduce foreign body response in a murine model of cochlear implantation and human subjects. medRxiv. Oct 13:2024.10.11.24315311. https://doi.org/10.1101/2024.10.11.24315311

Ramlow H, Ghasemi-Tabasi H, Burn A et al (2025) Joining alumina to metals: technologies, challenges, and future prospects for high-performance structures. J Eur Ceram Soc 45:117464. https://doi.org/10.1016/j.jeurceramsoc.2025.117464

Ramos-Macías A, Borkoski-Barreiro SA, Falcón-González JC, Ramos-de Miguel A (2017) Hearing preservation with the slim modiolar electrode nucleus CI532® cochlear implant: a preliminary experience. Audiol Neurotol 22:317–325. https://doi.org/10.1159/000486409

Rau TS, Lenarz T, Majdani O (2015) Individual optimization of the insertion of a preformed cochlear implant electrode array. Int J Otolaryngol 2015:1–22. https://doi.org/10.1155/2015/724703

Regnier C, Nguyen T (2021) Implantable medical device comprising a metal/ceramics composite housing. European Patent EP 3730185 B1

Roland JT (2005) A model for cochlear implant electrode insertion and force evaluation: results with a new electrode design and insertion technique. Laryngoscope 115:1325–1339

Rongthong T, Qnouch A, Gehrke MM et al (2022) Long term behavior of dexamethasone-loaded cochlear implants: in vitro & in vivo. Int J Pharm X 4:100141. https://doi.org/10.1016/j.ijpx.2022.100141

Ruys AJ, Cowdery DJ, Soh EKL (2022) Alumina: implantable bionics and tissue scaffolds. In: Choi AH, Ben-Nissan B (eds) Innovative bioceramics in translational medicine I, Springer series in biomaterials science and engineering, vol 17. Springer, Singapore. https://doi.org/10.1007/978-981-16-7435-8_10

Salcher R, Boruchov A, Timm M et al (2020) On the intracochlear location of straight electrode arrays after cochlear implantation: how lateral are lateral wall electrodes? Otol Neurotol 42:242–250. https://doi.org/10.1097/mao.0000000000002880

Schwartz MM (2003) Brazing, 2nd edn. ASM International

Sekulić DP (2013) Advances in brazing. Science, technology and applications. Woodhead Publishing Limited. https://doi.org/10.1533/9780857096500

Sengupta P, Bhattacharjee A, Maiti HS (2019) Zirconia: a unique multifunctional ceramic material. Trans Indian Inst Metals 72:1981–1998. https://doi.org/10.1007/s12666-019-01742-9

Shepherd RK, Carter PM, Dalrymple AN et al (2021) Platinum dissolution and tissue response following long-term electrical stimulation at high charge densities. J Neural Eng 18:036021. https://doi.org/10.1088/1741-2552/abe5ba

Shin S, Ha Y, Choi G et al (2021) Manufacturable 32-channel cochlear electrode array and preliminary assessment of its feasibility for clinical use. Micromachines 12:778. https://doi.org/10.3390/mi12070778

Siddiqui MS, Jones WK (2014) Vacuum brazing of alumina to titanium for implantable feedthroughs using pure gold as the braze metal. IJMSE. https://doi.org/10.12720/ijmse.2.1.56-62

Toulemonde P, Risoud M, Lemesre PM et al (2021) Evaluation of the efficacy of dexamethasone-eluting electrode array on the post-implant cochlear fibrotic reaction by three-dimensional immunofluorescence analysis in Mongolian Gerbil cochlea. J Clin Med 10:3315. https://doi.org/10.3390/jcm10153315

Tourrel G et al (2018) Conducting polymer-based cochlear electrode. Presented at the ARO conference, San Diego, CA

Tsuchida T, Huang W, Yan J (2024) Surface conditioning of zirconia ceramic by enhanced ultrasonic vibration-assisted burnishing. Prod Eng Res Dev 18:353–366. https://doi.org/10.1007/s11740-023-01242-2

Vaiani L, Boccaccio A, Uva AE, Palumbo G, Piccininni A, Guglielmi P, Cantore S, Santacroce L, Charitos IA, Ballini A (2023) Ceramic materials for biomedical applications: an overview on properties and fabrication processes. J Funct Biomater 14(3):146. https://doi.org/10.3390/jfb14030146

Wang J, Stevens R (1989) Zirconia-toughened alumina (ZTA) ceramics. J Mater Sci 24:3421–3440. https://doi.org/10.1007/bf02385721

Wissel K, Brandes G, Pütz N et al (2018) Platinum corrosion products from electrode contacts of human cochlear implants induce cell death in cell culture models. PLoS One 13:e0196649. https://doi.org/10.1371/journal.pone.0196649

Xu Y, Luo C, Zeng F-G et al (2019) Design, fabrication, and evaluation of a parylene thin-film electrode array for cochlear implants. IEEE Trans Biomed Eng 66:573–583. https://doi.org/10.1109/tbme.2018.2850753

Zhao Y, Jiang L, Liao Y et al (2013) Low temperature degradation of alumina-toughened zirconia in artificial saliva. J Wuhan Univ Technol-Mat Sci Edit 28:844–848. https://doi.org/10.1007/s11595-013-0780-8

Part II

Objective Assessment and Performance Optimization

Impedance Measurements in Cochlear Implants: From Research to Clinical Applications

5

Sebastian Ausili and Federico Di Lella

Abstract

Electrical impedance in cochlear implants provides a unique perspective on the interface between engineering and living tissue. This chapter traces how this fundamental electrical property is used as a powerful clinical and research tool, capable of revealing critical insights into cochlear implant function, tissue response, and surgical outcomes. We begin by outlining the biophysical bases of impedance, including the roles of resistive and reactive components, and how they influence signal behavior across time and frequency domains. These principles are then applied to the context of the cochlea, where impedance measurements reflect the dynamic relationships between the electrode array and the surrounding biological environment. We explore how impedance telemetry has become central to monitoring electrode status, detecting device malfunctions, and guiding programming strategies. Through modeling approaches and spectral analyses, we highlight how impedance data can provide a deeper understanding of the electrode-tissue interface. The chapter also reviews current clinical protocols across the main manufacturers, including the use of electrode electrical field potentials for spatial mapping. Finally, we examine emerging techniques that are extending the clinical utility of impedance measurements. Together, their advancements are transforming impedance telemetry from a basic safety check into a rich source of diagnostic and prognostic information in cochlear implantation.

S. Ausili (✉)
Department of Otolaryngology, University of Miami, Miami, FL, USA
e-mail: s.ausili@miami.edu

F. Di Lella
Department of Otolaryngology, Hospital Italiano de Buenos Aires, Buenos Aires, Argentina
e-mail: federico.dilella@hospitalitaliano.org.ar

B. Molaee-Ardekani, M. Segovia-Martinez (eds.), *Advancing Cochlear Implants*,
https://doi.org/10.1007/978-3-032-15719-5_5

Keywords

Cochlear implants · Impedance · Electrode-tissue interface · Cochlear health

5.1 Introduction

Cochlear implants (CIs) represent a groundbreaking innovation in neuroprosthetic technology, restoring hearing to individuals with profound hearing loss. As one of the most successful examples of biomedical engineering, CIs rely on the precise interaction between electrodes and neural tissue to bypass damaged auditory pathways and directly stimulate the auditory nerve (Wilson 2008). This capability makes CIs a transformative tool in addressing sensorineural hearing loss, but it also introduces a range of technical and biological challenges that must be managed for optimal performance (Clark 2003). Central to the success of CIs is the electrode-tissue interface, a dynamic and complex system where electrical stimulation meets biological tissue (Shannon 1992). Understanding and managing this interface, as well as the impedance associated with it, is critical for the safety, efficacy, and longevity of these devices (Messersmith et al. 2019).

This chapter explores the critical role of impedance in CI technology, with a specific focus on its measurement and utility in surgery and clinical fitting. We will begin by defining impedance and its relevance to the electrode-tissue interface. Following this, we will examine the importance of impedance in monitoring CI function, addressing its use in both intraoperative and postoperative settings. The chapter also explores the complexities of the electrode-tissue interface, highlighting its dynamic nature and the challenges posed by biological encapsulation of the electrode array. Lastly, we will discuss the relationship between electrical stimulation parameters and impedance measurement, providing a foundation for understanding how impedance can serve as a proxy for cochlear health and device performance. By exploring these aspects, this chapter aims to provide a comprehensive understanding of the role of impedance in enhancing CI outcomes and addressing ongoing challenges in the field.

5.1.1 What Is Electrical Impedance to Begin With

Electrical impedance is a fundamental concept that describes how a material or system opposes the flow of electrical current. Unlike resistance, which accounts only for energy dissipation, impedance provides a broader picture by also considering energy storage and the timing relationship between voltage and current. This makes impedance a more complete measure of how an electrical signal interacts with a system, capturing both the dissipative and reactive behaviors that shape the signal as it passes through (Callegaro 2012).

In electrical circuits, impedance is defined as the ratio of voltage to current and is measured in ohms (Ω). It is not a fixed value but rather a dynamic property that depends on the components of the system and the characteristics of the electrical signal. Mathematically, impedance (Z) is expressed as:

$$Z = \frac{V}{I} \tag{5.1}$$

This equation indicates that impedance represents how much voltage is required to drive a given amount of current through a system. However, impedance is not just a single value. It is a complex quantity that can be broken into two main components: resistance and reactance (Gross and Roppel 2012).

5.1.2 The Two Components of Impedance: Resistance and Reactance

The resistive component of impedance represents the system's opposition to the flow of current that results in energy loss, typically as heat. This resistance is independent of the nature of the electrical signal and remains constant regardless of how quickly or slowly the current changes (Callegaro 2012). In practical terms, resistance is what "slows down" the flow of electricity through a conductor and dissipates energy into the environment.

The reactive component, on the other hand, represents the system's ability to temporarily store energy and release it back into the circuit. Reactive behavior arises from two different physical phenomena. The first is capacitance, which stores energy in electric fields when charges accumulate on opposite surfaces separated by an insulating material. The second is inductance, which stores energy in magnetic fields generated by changes in current flow. Together, these behaviors describe how a system can temporarily "hold onto" electrical energy, introducing timing shifts between the voltage and current signals. Reactance itself is a combination of inductive and capacitive effects. Inductive reactance increases with faster-changing signals, while capacitive reactance decreases under the same conditions. This interplay means that the total impedance of a system depends on how quickly the signal changes and on the system's inherent properties.

5.1.3 Temporal and Frequency-Dependent Effects of Impedance

As we mentioned, the behavior of impedance changes depending on the timing and nature of the electrical signal. Temporally, reactive components like capacitance and inductance can delay the flow of current. A capacitor resists sudden changes in voltage by storing charge and then releasing it slowly, while an inductor resists sudden changes in current by generating opposing magnetic fields (Glisson 2011; Horowitz et al. 1990; Kalbitz 2023). These effects influence how an electrical signal is shaped as it passes through a system, creating smooth transitions or slight delays in the output (Lee and Kim 2022).

Frequency also plays a significant role in determining impedance. Capacitive reactance decreases as the signal becomes faster, making capacitors act almost like short circuits under rapid changes (higher frequencies) (Mazzeo 2009). In contrast, inductive reactance increases with faster signals, as the magnetic fields generated by inductors become more resistant to change. The frequency dependence of impedance creates unique profiles for how systems respond to signals of different speeds, which can be critical in understanding their behavior in practical applications (Horowitz et al. 1990).

5.2 Impedance in Biological Tissue

Biological systems, such as the electrode-tissue interface (ETI) in cochlear implants, exhibit similar behaviors to electrical circuits (Fig. 5.1a). At the ETI, resistance arises from the natural conductivity of tissues and fluids, which is primarily governed by the flow of ions in the cochlear environment. These ions carry charge through the electrolyte, which opposes the flow of electrical current and contributes to the overall resistive component of impedance (Sijgers et al. 2024). Reactance, on the other hand, is dominated by capacitance at the boundary between the electrode and the surrounding biological tissue. This boundary forms a double-layer capacitance, where free ions in the fluid flow and align in response to charges accumulating on the electrode surface (Barsoukov and Macdonald 2005). During stimulation, positive ions (cations) such as Na^+ and K^+ migrate toward the negatively charged electrode, while negative ions (anions) like Cl^- move toward the positively charged electrode. During cochlear implant stimulation, this ion movement occurs within the perilymph, where Na^+ and K^+ cations are drawn toward the cathode, while Cl^- anions migrate toward the anode, influencing local charge distribution and potentially affecting neural excitation, electrode impedance, and long-term electrochemical interactions within the cochlear environment (Zarowski et al. 2023).

As we mentioned before, impedance of the medium can be divided into components. Although inductive effects could theoretically occur in biological systems, they are negligible due to the absence of coiled structures and the relatively slow changes in electrical signals typically used. As a result, impedance in biological systems can be effectively modeled as a combination of resistance and capacitance effects alone. As we are going to see in the next sections, by modeling biological tissues as an equivalent electrical circuit, we can gain insights into how current interacts with electrodes and surrounding tissues. This approach enables us to predict the performance of devices like CIs, ensuring precise and efficient neural stimulation.

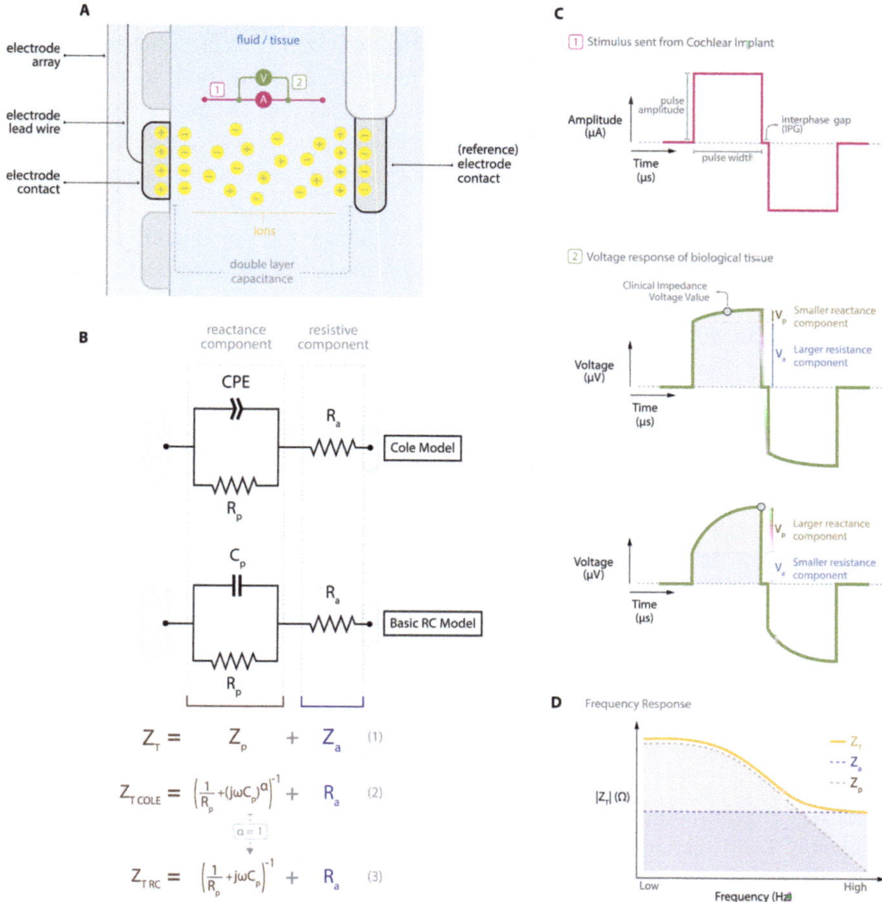

Fig. 5.1 (**a**) Schematic representation of the interaction between the stimulating electrode and the biological medium. A double-layer capacitance forms at the electrode interface, and ions become polarized within the medium between the active and reference electrodes. (**b**) Comparison of the Cole model and the basic RC electrical analog model of the ETI, along with their corresponding impedance equations. (**c**) Illustration of a biphasic current pulse delivered by the cochlear implant and the resulting voltage response of the biological tissue. Examples of the magnitude contributions of resistance and reactance components are depicted. (**d**) Frequency-dependent impedance response of the ETI, showing how the resistive and reactance components of the model contribute to the overall impedance Z_T across frequencies

5.3 Importance of Impedance in Cochlear Implants

The role of impedance in CIs cannot be overstated. Impedance directly affects the efficiency of electrical stimulation, the precision of neural activation, and the overall reliability of the implant (Sijgers et al. 2024). Abnormal impedance values can indicate problems such as broken or short-circuited electrodes, fibrous encapsulation of the electrode array, or changes in the cochlear environment that compromise the effectiveness of stimulation. Additionally, impedance influences the compliance level of the implant, which sets the maximum current that can be delivered based on the device's internal voltage limitations. For example, if impedance is too high, the implant may be unable to deliver sufficient current for effective neural stimulation (Hu et al. 2017).

Monitoring impedance is particularly critical during the intraoperative and postoperative phases of CI use. Intraoperatively, impedance measurements provide a baseline assessment of electrode function, allowing surgeons to detect anomalies early. Postoperatively, routine impedance monitoring enables clinicians to track changes over time, offering valuable information for device programming and patient management. Additionally, impedance can potentially serve as a proxy for cochlear health, reflecting changes in tissue response, fluid dynamics, and electrode positioning that could affect auditory outcomes (Zarowski et al. 2023). The dynamic nature of the cochlear fluid and the continuous electrical stimulation can cause changes in the local tissue environment, requiring careful management of the ETI and CI fitting over time. Understanding the ETI is essential for optimizing CI performance. Advances in electrode design, material science, and measurement techniques aim to enhance the stability of the ETI and minimize its impact in the cochlear inner ear medium.

5.4 Electrical Models of Electrode-Tissue Interface (ETI)

Electrical circuit models are invaluable tools for understanding the ETI in cochlear implants. They provide a simplified representation of the complex interactions between electrical signals and biological tissues, allowing the study of how stimulation pulses behave at the interface. Two widely used models for the ETI are the basic model and the Cole model (Molaee-Ardekani and Donahue 2023) (Fig. 5.1b).

The basic model represents the ETI as a series resistor, also known as access resistance ($Z_a = R_a$), and a polarization impedance (Z_p; Fig. 5.1b, Eq. 5.1). The series resistor R_a accounts for the bulk resistance of the cochlear fluids and tissues, while the polarization impedance models the behavior at the electrode surface. Z_p is typically represented by a parallel configuration of a resistor (R_P, polarization resistance) and a capacitor (C_P, polarization capacitance). This arrangement captures the dual nature of the ETI, where resistance represents energy dissipation, and capacitance reflects energy storage in the double layer at the electrode surface (Hrncirik et al. 2023). This relationship can be mathematically described (Fig. 5.1b, Eq. 5.2), where the total impedance Z_T is given by the combination of the series resistance R_a and the parallel RC network, representing the polarization impedance Z_p. Note that the polarization impedance contains the term "$j\omega$," which represents the frequency-dependent behavior of the system with j being an imaginary unit linked with the phase shift and ω the angular frequency, given $\omega = 2\pi f$, where f is the frequency in hertz.

The Cole model extends the basic model by replacing the ideal capacitor (C_p) with a constant phase element (CPE). The CPE accounts for the nonideal capacitive behavior observed in real electrodes, which arises due to surface roughness, oxide layers, and inhomogeneous reaction rates. By introducing a fractional-order component (α), the Cole model provides a more accurate representation of the ETI's frequency-dependent behavior (Mazzeo 2009) (Fig. 5.1b, Eq. 5.2). This model is particularly useful in spectral analyses, as it can capture the gradual transition between resistive and capacitive dominance across different frequencies.

Both models have been extensively applied in research to analyze the ETI in cochlear implants, helping to optimize electrode design, stimulation protocols, and impedance monitoring techniques (Hu et al. 2017). Each element (or subcomponent) in the electrical circuit model can be understood as representing a specific aspect of the interface's electrical behavior.

5.4.1 Resistive Subcomponents

Access resistance (R_a): This component mainly represents the resistance of the cochlear fluids and tissues, which governs the energy dissipation as current flows through the system. It also includes the resistance of the lead wires, though this contribution is typically minimal and can often be disregarded in practical applications (Schraivogel et al. 2023).

Polarization resistance (R_P): Located in the parallel branch of the circuit, this resistance represents the Faradaic (charge transfer) processes occurring at the electrode surface. These processes involve electrochemical reactions, such as oxidation and reduction, which contribute to charge transfer between the electrode and the surrounding tissue (Schraivogel et al. 2023).

5.4.2 Capacitive Subcomponents

Polarization capacitance (C_p): This capacitance arises from the separation of charges at the electrode-electrolyte boundary (see illustration in Fig. 5.1a). It reflects the interface's ability to store energy temporarily (Schraivogel et al. 2023).

Constant phase element (CPE): Used in the Cole model, the CPE accounts for the imperfect capacitive behavior of real electrodes, where frequency-dependent effects create a fractional-order response (Mazzeo 2009; Zarowski et al. 2023).

Each subcomponent plays a critical role in shaping the impedance profile of the ETI, influencing how electrical stimulation interacts with the cochlea.

5.4.3 Temporal Impedance Modeling

Temporal modeling examines how the medium responds to electrical stimulation over time. This approach is particularly relevant when studying biphasic current pulses, the standard form of stimulation used in cochlear implants (Jiang et al. 2020).

As we know, a cochlear implant delivers biphasic pulses at a fixed current level. If we could measure the current with an ammeter (as illustrated in Fig. 5.1a1), we would observe an exact replica of the pulse delivered by the CI (Fig. 5.1c1). However, measuring the voltage change over time while the biphasic pulse is applied (Fig. 5.1a2) reveals that the resulting waveform does not conserve the same shape (Fig. 5.1c2). This deviation occurs due to the interaction between the electrical stimulus and the surrounding biological medium mentioned earlier (von Rohr 2011).

Initially, a very rapid voltage change is observed, which corresponds to the access resistance of the medium (R_a). According to Ohm's law ($R_a = V_a/I$), a higher resistance component leads to a higher observed voltage. Following this, the measured voltage exhibits an exponential rise while the current is applied. This occurs due to the reactance component of the impedance, which introduces a delay by temporarily storing charge at the electrode surface. The greater the exponential voltage increase, the larger the contribution of the polarization impedance (V_p). Depending on the impedance model used, the extent of this exponential behavior can be associated with C_p or CPE (Hey et al. 2015).

It is important to note, however, that the use of short-duration biphasic pulses primarily probes the medium and high-frequency behavior of the electrode-tissue interface. As a result, it is not possible to fully visualize the low-frequency reactance components, such as slower Faradaic charge transfer processes (Molaee-Ardekani and Donahue 2023). Consequently, models fitted to short-pulse data tend to be valid mainly in the high-frequency range, and extrapolation to lower frequencies must be approached with caution. This can lead to differences in the low-frequency characteristics of the models across studies, depending on experimental conditions and data interpretation (Hu et al. 2017).

5.4.4 Spectral Modeling

Spectral modeling examines how the impedance of the ETI varies across different frequencies. Due to the electrical properties of the interface and the biological medium, impedance is frequency dependent, meaning the way the medium responds to electrical stimulation changes depending on the frequency content of the applied input signal (Sijgers et al. 2024).

A biphasic pulse, like the one used in CIs, is not a single-frequency signal but rather a sum of multiple frequency components, as described by Fourier analysis. For example, consider a biphasic pulse with an amplitude of 100 μA and a pulse width of 20 μs. The fundamental frequency of such a pulse is approximately 25 kHz, with higher harmonics extending into the hundreds of kilohertz (e.g., 50 kHz,

75 kHz, 100 kHz, and beyond). Since impedance varies with frequency, the biological medium will not respond uniformly to all these components (Jiang et al. 2020).

To characterize these frequency-dependent behaviors, electrochemical impedance spectroscopy is often used, measuring impedance across a broad frequency range (Fig. 5.1d) (von Rohr 2011). At lower frequencies, the ETI behaves mostly resistively, governed by access resistance (R_a) and polarization resistance (R_p). However, as frequency increases, capacitive effects from the polarization capacitance (C_p) or the constant-phase element (CPE) play a greater role, leading to a frequency-dependent impedance (Molaee-Ardekani and Donahue 2023). This relationship is reflected in the model equations by the ω term (Fig. 5.1b2–3), where an increase in frequency reduces the overall impedance (Z_T), while lower frequencies result in higher impedance values (Hu et al. 2017).

5.4.5 Voltametric Method

In addition to temporal and spectral modeling techniques, voltammetry could offer a complementary electrochemical approach to assess the electrode-tissue interface (ETI) in cochlear implants. This method involves sweeping the voltage applied to an electrode over a defined range (e.g., −0.8 V to +0.8 V) in a linear fashion over time, meaning the voltage increases at a constant rate (Alimohammadi et al. 2021). The initial slope of the voltage ramp—how quickly the voltage is swept—is called the scan rate, usually expressed in units of volts per second (V/s). When the voltage reaches a preset switching voltage (for example, +0.8 V), the sweep reverses direction, and the voltage is linearly decreased back toward the starting value (e.g., −0.8 V), again at the same scan rate. This complete cycle produces a characteristic current-voltage curve known as a voltammogram, which contains information about capacitive behavior and any redox reactions occurring at the electrode surface (Harris et al. 2023). Redox processes are critically important in biomedical devices because they influence both the efficiency of charge transfer and the long-term stability of the electrode, ultimately impacting the safety, performance, and durability of implanted systems like cochlear implants (Harris et al. 2022).

When compared with other approaches, voltammetry directly probes the electrochemical activity and surface reactions occurring at the electrode-tissue boundary. Peaks or changes in the voltammogram can reveal important processes, such as oxidation and reduction reactions (e.g., oxygen reduction, platinum oxide formation) or adsorption of biological molecules onto the electrode surface. Recent studies (e.g., Harris et al. 2022; Hibbert et al. 2001) have shown that voltammetric signals are highly sensitive to changes in electrode condition, protein adsorption, and the surrounding tissue environment, offering valuable insights into phenomena like fibrotic encapsulation, biofouling, and drug delivery.

5.5 Clinical Impedance in Cochlear Implants

The primary goal of clinical impedance measurement is to assess the status and integrity of the cochlear implant system. This measurement provides critical information about the functionality of the device, identifying potential issues such as open circuits, where an electrode or lead wire is disconnected, or short circuits, where the cables of two or more electrodes are in contact (Alahmadi et al. 2024). Detecting these issues is particularly valuable in cases where the user has experienced head trauma, potentially affecting the implant area or lead wires (Gottfried et al. 2024). Beyond diagnosing hardware integrity, clinical impedance measurement allows the implant to determine the voltage required to achieve the necessary stimulation current for sound perception. Since every cochlear implant has an internal maximum voltage capacity, this information is essential in cases of high impedance, such as a severely ossified cochlea, where stimulation limits may be reached (Sijgers et al. 2024). By visualizing these constraints, clinicians can assess whether the implant is functioning within its safe and effective operating range. Given that this measurement takes only a few seconds, it is a fundamental step during every clinical interaction with a CI user, ensuring that any device-related issues are quickly detected and addressed.

This section focuses on three key aspects of clinical impedance measurement: the characteristics of the biphasic input stimulus and its morphology, the timing of voltage acquisition, and the active-reference electrode configuration used to stimulate and record the response, also known as the stimulation/recording mode. Following this, we will examine how different cochlear implant manufacturers define and implement clinical impedance measurements, highlighting variations in methodology. Finally, we will discuss typical impedance values observed in clinical practice, as well as the defined thresholds used to assess electrode function and overall implant performance.

5.5.1 Stimulation Pulse and Voltage Recording

To measure clinical impedance, a biphasic pulse with a very small, non-audible current is applied to the electrodes, and the resulting voltage at the electrode surfaces is recorded via telemetry. The clinical software then calculates the impedance values by applying Ohm's law, taking the ratio of the measured voltage (in volts) to the known applied current (in amperes) (Di Lella et al. 2020), as shown in Eq. (5.1).

As illustrated in Fig. 5.1c1, the biphasic stimulus pulse consists of three main components: amplitude, width, and interphase gap (IPG). The amplitude and width of the pulse determine the total energy delivered to the tissue, while the IPG represents the period of zero current between the two phases (Gottfried et al. 2024). As we are going to see, each cochlear implant manufacturer defines and regulates this energy using its own conventions and definitions. The energy level is deliberately kept low to ensure that the CI user does not perceive any sound during the measurement (Sijgers et al. 2024).

Once the stimulus pulse has been delivered, the CI system sends a command to measure the voltage at a specific time point and retrieves the value in volts. As depicted in Fig. 5.1c2, voltage measurements are taken at a certain time during the stream of the pulse. The precise timing of this voltage acquisition is critical, as different sampling points can yield significantly different values. This variation occurs because voltage measurements can capture only the resistive component or a combination of the resistive and reactive components of impedance. Thus, the exact moment at which voltage is recorded directly influences the final impedance value. However, it is important to note that this timing does not impact the primary purpose of clinical impedance measurement, which is to assess the integrity of the system and detect any device-related anomalies (Hu et al. 2017).

Once the stimulus pulse morphology and the voltage measurement timing are defined, the impedance value is computed straightforwardly by applying Ohm's law. Since the current is set by the implant and its value is known, dividing the measured voltage by the current provides the final impedance value in ohms (Ramos Miguel et al. 2016).

5.5.2 Stimulation and Recording Coupling Modes

Another crucial factor in impedance measurement is the recording mode, which refers to the specific electrode pair used to measure the voltage. The choice of electrodes has a significant impact on the recorded impedance values, as different configurations involve varying amounts and types of biological tissue in the measurement.

When both electrodes are positioned inside the cochlea, the measured impedance primarily reflects the properties of the perilymph and the internal structures of the inner ear, resulting in a more localized measurement. In contrast, when one electrode is located inside the cochlea and the other is placed beneath the temporalis muscle (between the muscle and the skull), the electrical field extends through a much broader and more dispersed volume of tissue. This configuration includes not only cochlear structures but also surrounding soft tissues, muscle, and bone, leading to a different impedance profile. The choice of recording mode thus directly influences the impedance measurement, as it determines the extent and composition of the biological medium contributing to the recorded voltage.

Usually, the same active electrode used for the stimulation pulse is also used to retrieve and measure the voltage. However, and as we are going to see, some configurations utilize separate circuits, with one set of electrodes for stimulation and another for recording. Here we discuss the main recording/stimulation modes, which are essential for interpreting impedance values in a clinical setting and ensuring accurate assessment of electrode function and implant integrity (Alimohammadi et al. 2021).

Monopolar mode refers to the stimulation of an intracochlear electrode relative to either the implant chassis or an extracochlear electrode (Fig. 5.2a). In this mode, the same electrodes used for stimulation are also used for recording. As previously

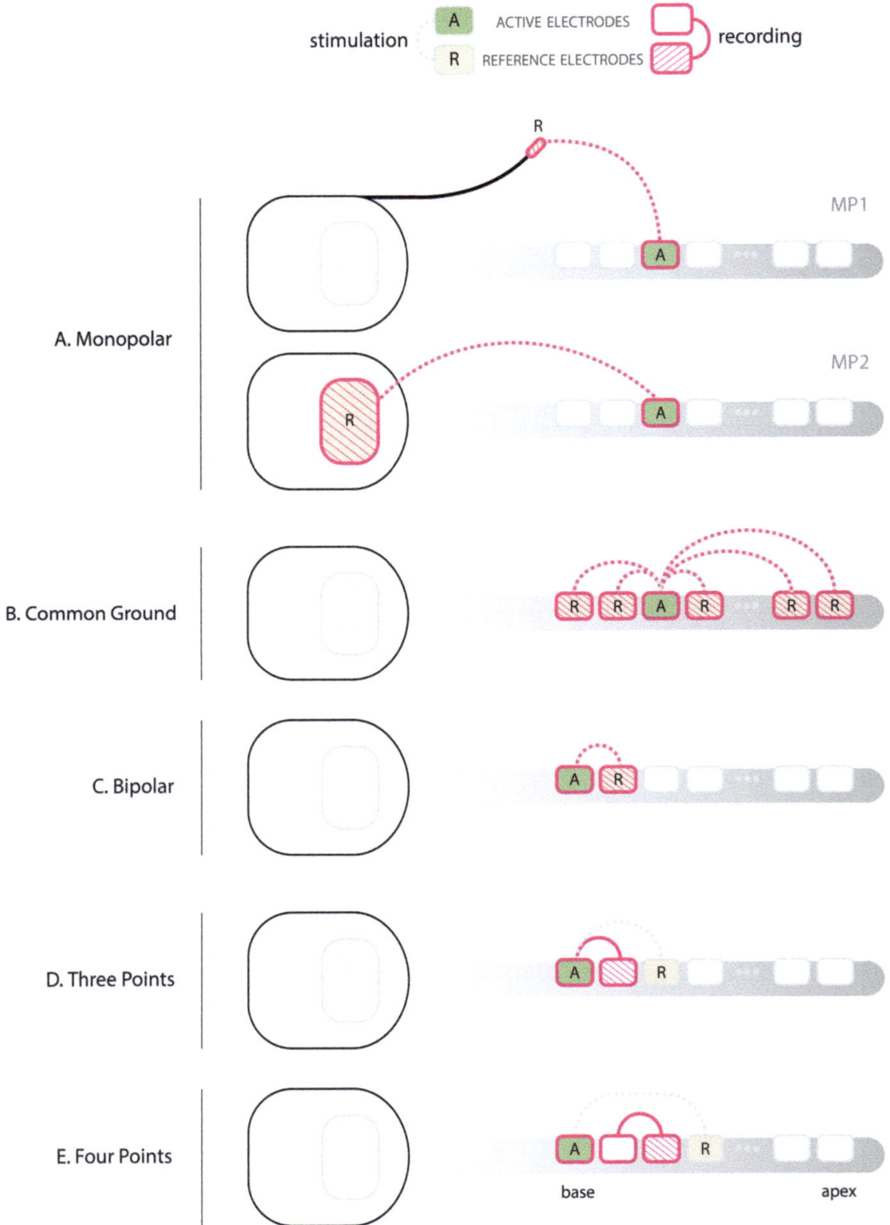

Fig. 5.2 Illustration of stimulation and recording modes for clinical impedance measurement

discussed, the electric field extends over a large volume of tissue, encompassing not only the inner ear but also surrounding structures up to the location of the reference electrode (Hey et al. 2015). A unique configuration found in some cochlear implants includes two monopolar configurations: monopolar 1 (MP1), which uses the implant chassis as the extracochlear return electrode, and monopolar 2 (MP2), which uses a dedicated extracochlear electrode placed under the temporalis muscle. These systems also allow the use of a combined configuration, referred to as MP1+2, where both reference electrodes are short-circuited together to serve as the return path. This combined mode can optimize current spread and electrode field distribution depending on the individual implant anatomy and placement.

Another mode of impedance measurement is common ground mode (Fig. 5.2b). Here, a single intracochlear electrode is set as the active stimulation electrode, while all other intracochlear electrodes are short-circuited together to serve as the reference. Similarly, the obtained voltage to compute the clinical impedance is recorded between these same electrodes. This method provides a more localized impedance measurement compared to monopolar mode but still captures contributions from multiple electrode sites (Molaee-Ardekani and Donahue 2023). Interestingly, this measurement approach is not available in all cochlear implant systems.

The bipolar coupling mode involves only two electrodes: an active electrode and a reference electrode (Fig. 5.2c). Unlike other modes, the bipolar mode confines both stimulation and recording within the inner ear, resulting in a more localized impedance measurement. Since only two adjacent intracochlear electrodes are involved, this configuration is highly sensitive to local tissue properties, providing a more precise assessment of electrode function in relation to nearby neural structures (Sijgers et al. 2024). However, because the current path is shorter and the surrounding medium more resistive, impedance values in bipolar mode are typically lower than those measured in monopolar mode.

The following modes can be seen as adaptations or reconfigurations of the bipolar mode, designed to achieve even more localized impedance measurements while minimizing interference between stimulation and recording. These configurations involve not only a more confined electrical field but also separate circuits for delivering the stimulation pulse and recording the voltage response.

The three-point (3P) mode (Fig. 5.2d) utilizes a single active electrode for both stimulation and recording, but it employs two distinct reference electrodes for each process. Stimulation occurs between the active electrode and the next-to-next electrode, leaving one electrode in between. In contrast, recording is performed between the active electrode and its immediately adjacent electrode, ensuring a localized measurement while maintaining some separation between the stimulation and recording sites (Di Lella et al. 2020).

The four-point (4P) mode (Fig. 5.2e) takes this separation further by employing completely independent circuits for stimulation and recording. Stimulation occurs between the active electrode and a reference electrode that is spaced two positions away, leaving two electrodes in between. Meanwhile, recording is performed between the two electrodes enclosed within the stimulation circuit, further isolating the recording process from potential artifacts introduced by stimulation. This separation effectively reduces electrical noise and minimizes interference between the two processes. Interestingly, recent findings suggest that 4P mode is particularly sensitive to detecting changes in the inner ear medium, such as bleeding and fibrosis, making it a potentially valuable tool for assessing cochlear health (Ramos Miguel et al. 2016).

An important note on the 3P and 4P modes is that they are not yet fully implemented in clinical software and are primarily used for research purposes. While they offer promising insights into electrode function and cochlear health, their clinical application remains limited until further validation and integration into commercial CI systems (Harris et al. 2023).

5.5.3 Voltage Compliance

As previously discussed, cochlear implants contain an internal voltage source that ensures the current source can accurately maintain a fixed amount of current to stimulate the auditory nerve. However, this voltage source has inherent limitations, which also constrain the maximum amount of current the implant can deliver to the user (Joly et al. 2021). According to Ohm's law, voltage and current are directly related by impedance, meaning that the voltage required to achieve a desired stimulation current depends on the impedance of the surrounding medium. This relationship can be expressed as:

$$V_{\text{compliance}} = Z_{\text{clinical}} \times I_{\text{desired}} \tag{5.2}$$

Once clinical impedance is measured, the compliance voltage is determined and can also be displayed in the clinical fitting software, allowing clinicians to monitor the available headroom in cases where high stimulation levels are required, such as in severely ossified cochleae (Saoji et al. 2021). When stimulation levels approach the voltage compliance limit, the implant must operate near its maximum power capacity, which significantly increases battery consumption. As a result, users may experience shorter battery life, requiring more frequent recharging or battery replacements.

Note that cochlear implant manufacturers design electrode arrays with varied contact sizes and surface areas, and some manufacturers incorporate differently sized electrode contacts within the same array. These differences directly influence impedance values, as smaller contacts typically exhibit higher impedance due to their reduced surface area (Dhanasingh et al. 2024). As a result, electrodes with higher impedance are more likely to reach voltage compliance limits faster, whereas larger contacts, with their lower impedance, facilitate greater current delivery within the implant's voltage constraints.

5.5.4 Open and Short Circuits

For a cochlear implant to function properly, electrical current must flow through a complete circuit, with a defined source and return path. Any disruption or unintended alteration in this pathway affects the effectiveness of stimulation and is directly reflected in the impedance measurement. Two critical issues that can arise are open circuits and short circuits, each presenting distinct impedance characteristics and clinical implications.

An open circuit occurs when the electrical path is broken, preventing current from flowing. This results in extremely high impedance values, often in the megohm range, as no conductive pathway exists. Several factors can cause an open circuit, including a disconnected electrode contact, a broken lead wire, or the electrode being surrounded by a nonconductive medium, such as an air bubble or extrusion outside of the cochlea. Clinically, different manufacturers define the threshold for identifying high impedance or potential open circuits based on their specific measurement techniques and device characteristics. It is also important to note that the polarity and measurement mode used during testing can influence the recorded impedance. Since the return path is interrupted, open circuits can be identified across different coupling modes, as none of these configurations can successfully deliver current when a discontinuity is present.

In contrast, a short circuit occurs when two normally independent electrodes or lead wires come into unintended electrical contact, changing the original pathway of the current flow. This is typically due to physical contact between electrodes, such as when the array is kinked or folded back inside the cochlea, or from insulation failure, where a defect allows electrical connection between lead wires. Short circuits are characterized by very low impedance values, generally below 1 kΩ.

However, not all clinical coupling modes can reliably detect short circuits. In monopolar mode, if two intracochlear electrodes are shorted together, current may still find a return path to the extracochlear electrode, meaning the issue may go unnoticed (Fig. 5.3a). Common ground and bipolar modes are more effective for detecting short circuits because they measure the relative impedance between multiple electrodes, making it easier to identify abnormally low values (Fig. 5.3b, c).

A partial short circuit represents an intermediate case, where impedance is lower than expected but not as extreme as a full short circuit (Eitutis et al. 2022). This often results from small defects in insulation, allowing fluid ingress into the electrode carrier or electronic housing, which reduces resistance between electrodes. Unlike full short circuits, partial short circuits may not be flagged by clinical software because their impedance values remain within an acceptable range. However, their progressive nature makes them detectable over time, as impedance gradually decreases as more conductive fluid infiltrates the carrier. The common ground mode is particularly useful for identifying these cases, as it provides a broad view of impedance trends across the array.

Both open and short circuits are critical to the long-term stability and functionality of a cochlear implant. Regular impedance measurements are essential for early detection, allowing clinicians to assess electrode integrity and prevent potential stimulation failures. While clinical software automatically flags extreme impedance values, subtle changes—such as those in partial short circuits—may require longitudinal monitoring to detect slow-developing hardware issues. As shown, the different electrode configurations and coupling modes are crucial for accurate diagnosis and troubleshooting in clinical practice.

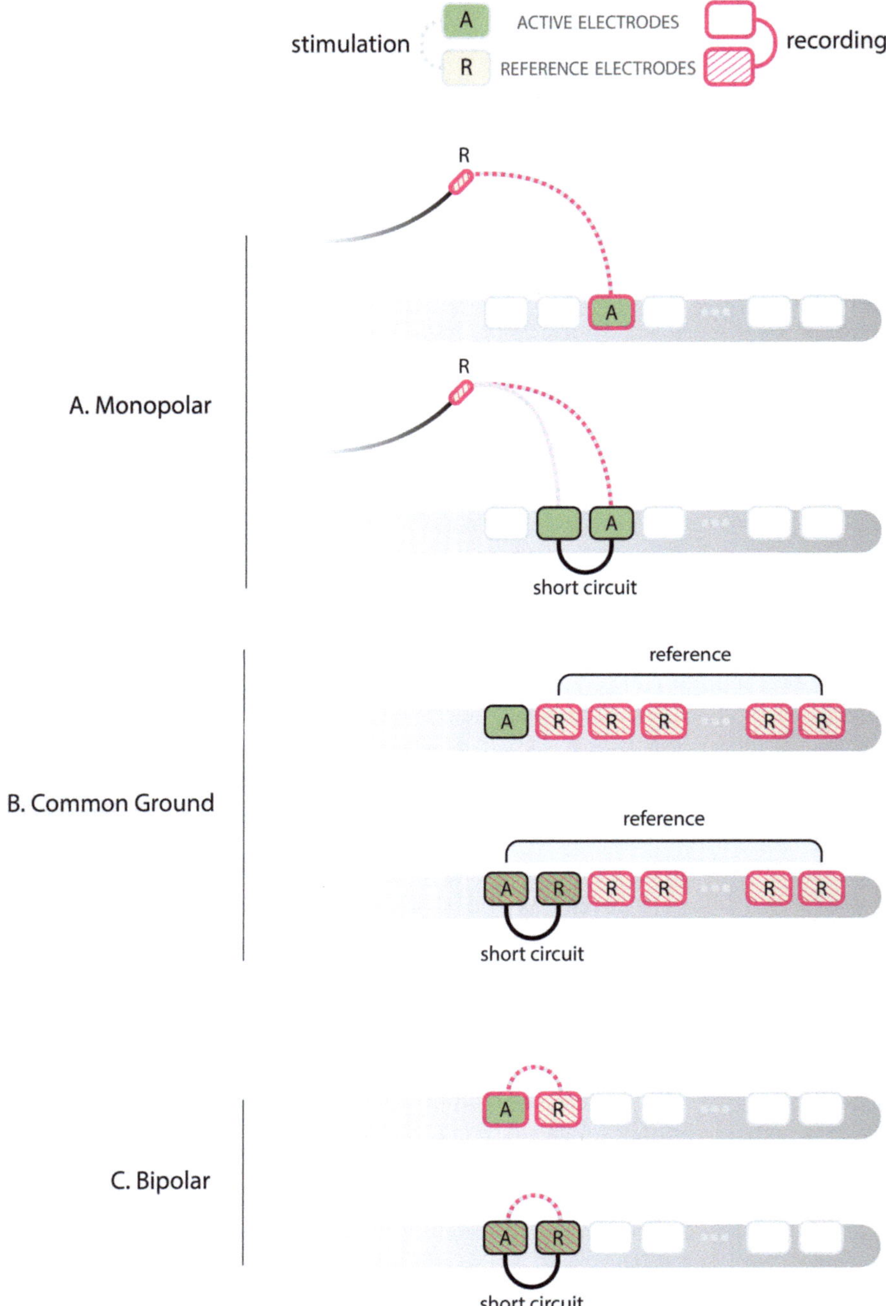

Fig. 5.3 Illustrations of short circuits between electrodes and their relationship with clinical coupling modes

5.5.5 Intraoperative and Postoperative Impedance Measures

Impedance measurements serve as a critical tool for assessing CI function at different stages, from the moment of implantation to long-term device use. These measurements help clinicians evaluate electrode integrity, tissue response, and potential complications that could impact stimulation efficiency.

5.5.5.1 Intraoperative Impedance: Establishing a Baseline

During surgery, intraoperative impedance measurements provide an initial assessment of whether all electrodes are functional and properly positioned within the cochlea. These values set a baseline against which future impedance changes can be compared. Notably, even before removing the device from its sterilized packaging at the operating room, an initial impedance measurement is normally performed to confirm that the CI is functional and does not present any hardware errors. At this stage, it is expected that all electrodes will exhibit high impedance values, as they are in direct contact with air rather than a conductive medium. These open-circuit readings serve as an additional verification step, ensuring that the implant is operating correctly before surgical placement within the cochlea.

At the time of implantation, impedance is generally low, as the electrodes are surrounded by perilymph, a highly conductive fluid (Parreño et al. 2020). However, the surgical environment introduces temporary factors that may affect impedance values. For instance, if air bubbles become trapped around the electrode array, some contacts may initially present open circuits (Oberhoffner et al. 2024). These cases often resolve within the first few days post-surgery as the air is absorbed by surrounding tissues. A more persistent issue can arise if part of the electrode array remains outside the cochlea or in contact with middle ear structures. Initially, these electrodes may exhibit normal impedances due to surgical fluids, but as these fluids dissipate, the electrodes may then present open circuits postoperatively. Imaging techniques can confirm electrode placement when these cases arise. However, if a single electrode consistently presents an open circuit, it is more likely to indicate a true hardware issue rather than a temporary condition, such as a defective electrode contact or a damaged lead wire. Since open circuits prevent current flow, these electrodes will be inactive during stimulation, potentially reducing stimulation coverage and affecting sound perception. If confirmed postoperatively, clinicians may need to disable the affected electrode in the programming software or, in rare cases, consider further diagnostic imaging or reimplantation if multiple failures occur.

5.5.5.2 Postoperative Impedance: Changes over Time

Once the implant is activated, impedance typically increases over the first few weeks before stabilizing (Ausili et al. 2022; Parreño et al. 2020). This increase is attributed to fibrous tissue encapsulation around the electrode array, a natural immune response to the presence of a foreign object. This process, which occurs within 2–3 weeks post-surgery, results in a denser tissue sheath around the electrodes, increasing electrical resistance (Newbold et al. 2004; Tykocinski et al. 2001; Wilk et al. 2016).

After the initial increase, impedances tend to decrease and stabilize within the first few months of device use (Busby et al. 2002; Newbold et al. 2004). This decrease is not due to a reduction in fibrotic tissue itself but rather a restructuring of the tissue matrix around the electrodes, making it less resistive to current flow. The introduction of electrical stimulation also helps to maintain impedance stability over time.

5.5.5.3 Factors Affecting Impedance Beyond the Initial Months

Although impedance typically stabilizes within the first few months, certain factors can cause delayed changes that require clinical attention.

Nonuse of the Implant: If an electrode is left unstimulated for an extended period, its impedance may increase due to a lack of consistent charge transfer (Busby et al. 2002). This can happen if an electrode is turned off in a patient's programming due to perceptual distortions or discomfort. If stimulation is resumed, impedance often returns to previous levels, though map adjustments may be required.

Middle Ear Infections and Inflammation: Transient impedance increases can occur due to middle ear infections, labyrinthitis, or other inflammatory processes affecting the inner ear (Clark et al. 1995; Dixon et al. 2014; Neuburger et al. 2009). Since these conditions can alter the electrical environment, impedance values may fluctuate before returning to normal once the infection or inflammation resolves. In these cases, temporary adjustments to the patient's map may be necessary to maintain sound quality during recovery.

Reimplantation Effects: Patients undergoing reimplantation often exhibit higher impedances with the new electrode array, likely due to increased trauma from the removal and reinsertion process (De Ceulaer et al. 2003). The body's response to the second implant may be more aggressive, resulting in thicker fibrotic encapsulation and higher electrical resistance.

Voltage Compliance and Stimulation Strategy: Impedance can also increase when a patient's stimulation strategy pushes the implant close to its voltage compliance limits. Fast-rate stimulation strategies, which use shorter pulse durations, require higher current amplitudes to maintain adequate loudness perception. If the implant operates near or beyond its voltage compliance limit, asymmetric current pulses may occur, accelerating platinum dissolution at the electrode surface and further increasing impedance (Neuburger et al. 2009). In such cases, widening the pulse width can help lower current demands and restore impedance stability.

Partial Short Circuits: While full short circuits are typically identified early, partial short circuits can emerge gradually over time due to small insulation breaches that allow fluid ingress into the electrode carrier. Unlike full short circuits, partial shorts present as moderately lower impedances rather than an immediate drop (Eitutis et al. 2022; Gärtner et al. 2021). Longitudinal impedance tracking is crucial for detecting these trends, as a steady decline in impedance over time often indicates a developing partial short circuit.

5.5.5.4 Monitoring and Clinical Management

Given the dynamic nature of impedance changes, it is essential to measure impedances at every clinical visit to track trends and detect potential issues early. Significant deviations from the expected impedance trajectory may indicate underlying problems that require further investigation.

In cases of increasing impedance, map levels should be reviewed to ensure the device is operating within its voltage compliance range. If impedance continues to rise in conjunction with other symptoms (e.g., reduced sound quality, distorted perception, or facial nerve stimulation), it may indicate a developing hardware issue, such as an emerging short circuit.

A notable advancement in impedance monitoring is the recent development of mobile applications that allow patients to remotely measure their impedance levels without needing to visit a clinical center (Wasmann et al. 2024). These tools provide real-time impedance data, enabling early detection of potential issues and allowing clinicians to track electrode function remotely. By integrating these technologies with cloud-based systems, impedance trends can be analyzed over time, offering a more personalized and proactive approach to cochlear implant care. Remote monitoring, along with other emerging telehealth solutions, represents a significant step toward individualized medicine in CI users, enhancing accessibility and facilitating early intervention when needed.

For partial short circuits, clinicians should monitor specific electrode patterns. Some CI models, particularly those with differential electrode wiring configurations, may exhibit a zigzag pattern in impedance values, where even- and odd-numbered electrodes are affected differently. When multiple electrodes show signs of progressive impedance reduction, deactivating every other affected electrode can sometimes stabilize performance. However, if a substantial portion of the array is involved, reimplantation may be necessary.

Impedance tracking should always be interpreted alongside other clinical measures, such as map stability, neural response, and patient-reported perceptual changes. Combining these measures allows clinicians to make data-driven programming adjustments that optimize implant performance and longevity.

5.6 Clinical Impedances by Companies

As introduced earlier, the general method for clinical impedance measurement is depicted in Fig. 5.1c1–2. However, the specific parameters used, such as the stimulus pulse characteristics, voltage recording timing, and interpretation thresholds, vary significantly across manufacturers. No universal standard has been established, leading to differences in typical impedance values observed intra- and postoperatively, and in the criteria for identifying open or short circuits. After laying the foundation with general impedance concepts, we now review how the major cochlear implant manufacturers implement impedance measurements clinically.

5.6.1 Advanced Bionics

In Advanced Bionics (AB) devices, impedance is measured exclusively using a monopolar coupling mode. The stimulation consists of a biphasic pulse with a phase duration of approximately 18 μs with no IPG. The amplitude current of the pulse is 32 μA.

As a clinical reference, the charge unit C applied as defined for AB is ~7 CU, which is defined as:

$$C = \frac{\mathrm{I}}{\mathrm{I}_{\mathrm{max}}} \times \frac{\mathrm{T}}{\mathrm{T}_{\mathrm{max}}} \times s_{\mathrm{AB}} = \frac{32}{2040} \times \frac{18}{229} \times 6000 = 7.4\ \mathrm{CU},$$

where I and I_{max} are the applied and maximum allowable current amplitudes in μA, T and T_{max} are the applied and maximum allowable pulse duration in μs, and s_{AB} is the AB scaling factor. The voltage is recorded approximately 4 μs after the onset of the cathodic phase of biphasic pulse. Having the voltage measured close to the pulse onset time, it is important to note that this is measuring mostly the access resistance (R_a) when looking at the electrode-tissue model.

According to the most recent guidelines (Target CI Fitting Guide, 2024[1]), impedances are classified as valid (1.0–30.0 kΩ), open circuit (>30.0 kΩ), short circuit (0.5–1.0 kΩ), and invalid (<0.5 kΩ). This classification is color coded in the software to facilitate its interpretation. Advanced Bionics automatically measures impedances upon processor connection, color-coding the results for immediate feedback.

5.6.2 Cochlear Ltd.

Cochlear devices perform impedance measurements across various coupling modes, including monopolar (MP1, MP2, MP1+2), bipolar and common ground configurations. Stimulation is delivered using a biphasic pulse with a phase duration of 25 μs and an IPG of 7 μs, at a current amplitude of approximately 74 μA. Cochlear's conversion between programming current levels (CL) and microamperes (μA) follows this transformation:

$$I(\mu\mathrm{A}) = 17.5 \cdot 100^{\left(\frac{\mathrm{CL}}{255}\right)}$$

where CL ranges between 0 and 255. We can then compute the clinical CL value applied to measure their default impedance:

[1] https://www.advancedbionics.com/content/dam/advancedbionics/ifus/th/029-M813-02_Rev%20B_Manual%20Target%20CI%20FDA%20EN-US.pdf

$$\mathrm{CL} = 255 \cdot \log_{100}\left(\frac{I(\mu\mathrm{A})}{17.5}\right) = 255 \cdot \log_{100}\left(\frac{74}{17.5}\right) \approx 80\ \mathrm{CL}$$

The voltage is recorded at the end of the first phase of the biphasic pulse (25 μs), known as the trailing edge of the pulse. Interestingly, this measurement time captures the voltage contributions from both the access resistance and the early polarization effects. Cochlear's default coupling configuration for clinical fittings is MP1+2, which combines two extracochlear electrodes (the case electrode and the dedicated plate electrode; see Fig. 5.1a) to ensure grounding redundancy. This allows for continued operation even if one return path fails.

Cochlear defines valid impedance ranges from 1 to 33 kΩ; values above 33 kΩ are flagged as open circuits, while values below 1 kΩ are flagged as short circuits. These thresholds are automatically enforced in the Custom Sound software, which provides visual feedback in both numerical and graphical formats.

5.6.3 MED-EL

MED-EL cochlear implants perform impedance testing using monopolar stimulation, referred to as impedance and field telemetry (IFT) in their Maestro fitting software, with the implant chassis serving as the reference electrode. The standard stimulating pulse has a phase duration of 27 μs and an IPG of 2.1 μs, delivered at a default current amplitude of 300 μA, which corresponds to a stimulation charge level of approximately 8 QU (where 1 QU equals 1 nanocoulomb) based on MED-EL's internal clinical definition:

$$\mathrm{QU} = (I \times T)/1000 = (300\ \mu\mathrm{A} \times 27\ \mu\mathrm{s})/1000 \approx 8$$

The timing of the voltage measurement—at the end of the second phase—allows capturing access resistance as well as the capacitive effect of the electrode-tissue interface.

MED-EL defines typical impedance values as those between 3.3 and 10 kΩ. The Maestro software automatically color-codes and highlights impedance results, indicating normal, shorted, or open channels. Notably, depending on the impedance outcomes, the fitting software suggests actions (e.g., turn off certain channels).

5.6.4 Electrical Field Potential

Clinical impedance measurements provide essential insights into the integrity of the electrode array and the surrounding tissue environment. However, they provide only a simplified, single-point assessment of electrode function. To gain a more comprehensive understanding of current spread within the cochlea and potential malfunctions such as short circuits, a more spatially detailed method is required. This is where Electrical Field Potentials (EPFs) come into play. Instead of recording

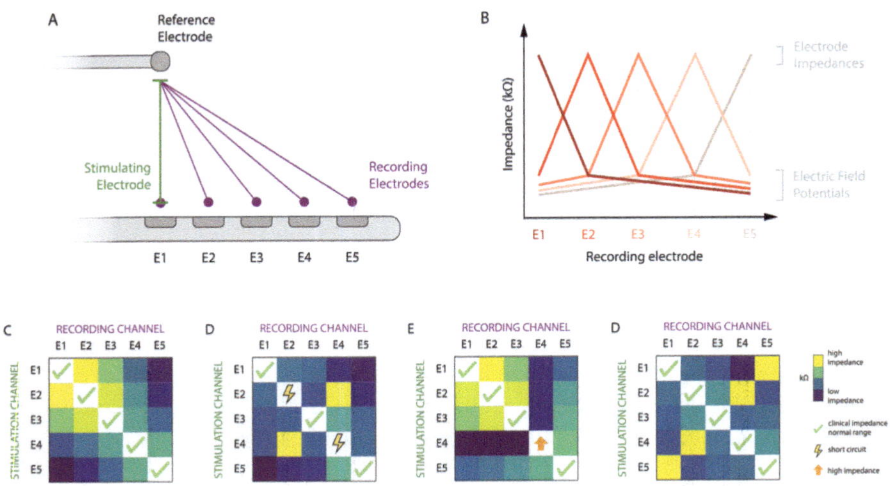

Fig. 5.4 (**a**) Illustration of Electrical Field Potentials measurement, where E1 is the stimulating electrode and voltages are recorded with all the intracochlear electrodes. (**b**) Typical impedance distribution, with peaks indicating electrode impedance and tails showing the spread of the electric field. (**c**–**f**) Examples of clinical patterns: (**c**) normal matrix; (**d**) short circuit between electrodes; (**e**) high-impedance channel suggestive of an open circuit; and (**f**) a tip foldover pattern, where reversed gradients reveal displaced apical electrodes

voltage only at the stimulating contact, voltages are measured at all intracochlear electrodes relative to a monopolar extracochlear reference (Fig. 5.4a). This process generates a voltage matrix, which is then converted to impedance values. The resulting data represent the voltage distribution across the cochlear array in response to stimulation from a single intracochlear electrode (Fig. 5.4b).

EPFs provide a complementary perspective on the electrical properties of the cochlear environment, offering valuable insights beyond what traditional impedance measurements can reveal. By generating a voltage matrix across the electrode array, clinicians can assess the spatial distribution of current and detect subtle anomalies in electrode function and positioning. For example, in a normally functioning electrode array, the voltage profile typically decays monotonically with increasing distance from the stimulation site (Fig. 5.4c). Deviations from this expected pattern can detect issues such as short circuits, where two electrodes exhibit nearly identical voltage responses (Fig. 5.4d), or high-impedance channels, an indication of open circuits or compromised contacts (Fig. 5.4e). Notably, this method has demonstrated particular utility in detecting electrode tip foldovers intraoperatively (Arambula et al. 2025; Ayas et al. 2024). In such cases, the voltage pattern exhibits a non-monotonic reversal, creating a "cross" pattern in the heatmap, indicating that apical contacts are abnormally close to more basal sites due to the array folding back on itself (Fig. 5.4d). This has been validated against intraoperative imaging and shown to prompt immediate surgical correction, thereby preventing postoperative complications (Klabbers et al. 2021). In newer implementations, similar heatmap-based approaches enable clinicians to rapidly and noninvasively detect abnormal positioning, short circuits, or extracochlear contacts. This makes EFPs a powerful tool for both intraoperative and routine postoperative monitoring.

Cochlear implant manufacturers have developed their own implementations of EFPs, each under a distinct name but based on the same core principle: delivering a low-level biphasic pulse in monopolar mode and recording the resulting voltage at each intracochlear electrode. Advanced Bionics refers to its approach as Electrical Field Imaging (EFI), integrated into both clinical and research tools. Cochlear Ltd. implements a similar technique called Transimpedance Matrix (TIM), also applied in both research and investigational clinical settings. Meanwhile, MED-EL presents their Impedance Field Telemetry (IFT) in a heat map through their fitting software.

5.7 Towards Improved Health: Cochlear Health Metrics

Advances in impedance measurement techniques are creating new opportunities to use cochlear implants not only for stimulation but also as diagnostic tools. By leveraging impedance telemetry, it may become possible to monitor cochlear health in real time, optimize device fitting, and detect biological or mechanical issues before they affect auditory outcomes. In this section, we highlight emerging approaches that could significantly improve patient care, ranging from daily impedance monitoring to novel electrochemical and predictive models.

5.7.1 Daily Impedance Monitoring

The measurement of intracochlear electrode impedance is a quick (1–2 min) and safe procedure, because it involves the use of sub- or near-threshold level stimulation (Wolfe and Schafer 2025). During the postoperative period, impedance measurement is routinely performed to guide CI-programming, detect device failures, and identify electrode contacts extrusion. While electrode impedance measurement provides essential information about the CI and cochlear status, it is only performed in the patient's clinical appointments. Consequently, little is known about the behavior of this parameter between visits (Hughes et al. 2001; Shaul et al. 2019). As a result, most impedance variations are diagnosed later in time, when it is often too late to, for example, start pharmacological treatment if necessary. More frequent monitoring of CI impedance with available methods is not feasible, but currently the use of telemedicine can be used to improve clinical practice by performing constant monitoring of electrode impedance values (Parreño et al. 2020).

Telemedicine has made its way into cochlear implant clinics over the past decade, driven by advances in connectivity. This progress has been supported by the development of remote-access applications and new telecommunication systems (McElveen et al. 2010; Rodríguez et al. 2010). Audiologists began performing remote fitting and monitoring of implanted patients, allowing medical care to be delivered while patients remain at home without significant differences compared to standard programming sessions (Eikelboom et al. 2014; Goehring et al. 2012; Hughes et al. 2018; Kuzovkov et al. 2014; McElveen et al. 2010; Rodríguez et al. 2010; Schepers et al. 2019; Wesarg et al. 2010). This data provides valuable insights into the post-surgical healing process and its effects on the electrode-tissue interface. Detecting atypical impedance variations also highlights the potential for early intervention in response to inflammatory or pathological changes. This approach represents a significant step toward enhancing patient autonomy, reducing the need for clinical visits, and providing comprehensive longitudinal data. With advancements in automated and continuous measurement techniques, future developments could include real-time monitoring systems capable of triggering alerts for clinicians, thereby further optimizing the management and outcomes of cochlear implant users.

5.7.2 Novel Measurement Approaches

Beyond standard impedance telemetry, which is primarily used to confirm electrode integrity and detect short or open circuits, new research is expanding the clinical utility of electrical impedance data. As discussed before, one promising approach is voltammetry, which applies a controlled voltage sweep to capture redox reactions and subtle changes at the electrode surface. This method can detect early signs of degradation, such as platinum dissolution or protein adsorption, which may precede functional failure (Harris et al. 2022; Hibbert et al. 2001). Electrochemical impedance spectroscopy is another technique under investigation, offering a frequency-dependent analysis of impedance to disentangle bulk tissue effects from those at the electrode interface (Mc Carthy et al. 2022). Weltin et al. (2022) have also shown the potential of cochlear implant electrodes as in vivo electrochemical oxygen sensors, further broadening the diagnostic scope of implanted devices.

Another important advancement involves using impedance data to estimate electrode positioning within the cochlea. Researchers have demonstrated that intraoperative impedance measurements can be analyzed through predictive modeling to estimate the insertion depth of basal electrodes with submillimeter accuracy, providing a radiation-free alternative to imaging (Schraivogel et al. 2025). Gottfried et al. (2024) reported the clinical feasibility of real-time depth tracking during implantation, enhancing intraoperative decision-making. Complementing these efforts, the EFPs generate spatial maps of voltage across the array to detect faults like short circuits and tip foldovers (Klabbers et al. 2021; Vanpoucke et al. 2004). Studies using three-point and four-point impedance methods have also demonstrated increased sensitivity to local intracochlear changes (Razmovski et al. 2022; Sijgers et al. 2024).

Finally, high-resolution impedance data is driving the development of time-based predictive models that monitor electrode stability and anticipate clinical issues. Alohali et al. (2023) used machine learning to identify abnormal postimplantation impedance trajectories up to 1 year after the day of surgery, while Schraivogel et al. (2023) linked near-field impedance components to poorer residual hearing, highlighting the importance of dissecting interface versus bulk properties. In a related study, Lei et al. (2021) combined EFP data and neural networks to model patient-specific cochlear anatomy, offering a personalized view of current flow and electrode positioning. Collectively, these novel measurement approaches are transforming impedance telemetry from a simple integrity check into a powerful tool for diagnostics, surgical guidance, and patient-specific CI programming.

5.7.3 Impedance as a Proxy for Cochlear Health

Impedance measurements in cochlear implants can serve as an indirect indicator of cochlear health by reflecting the electrode-tissue interface. Following cochlear implantation, the immune system initiates an inflammatory response that results in protein adsorption onto the electrode surface, formation of fibrotic tissue, and eventual stabilization of the foreign body reaction (Nadol et al. 2014; Noonan et al. 2020; Seyyedi and Nadol 2014). Impedance subcomponents, such as polarization resistance and capacitance, offer further insights into the biological processes at play. Their variations over time highlight the extent of tissue changes surrounding the electrodes and can provide early indications of pathological responses. Mathematical modeling of impedance dynamics can effectively serve to capture these temporal changes, facilitating a deeper understanding of the cochlear healing process (Ausili et al. 2022). With the advent of next-generation electrode designs incorporating self-eluting dexamethasone, impedance measurements play a critical role in objectively tracking intracochlear changes over time. This approach enables rapid, precise comparisons between treatment strategies, offering valuable insights into tissue response, drug efficacy, and long-term cochlear health across patient populations.

All the different impedance-related methods hold great potential for obtaining information about the health of the implanted cochlea, contributing to the improvement of long-term auditory preservation, and optimizing neural electrical stimulation. Enhanced stimulation efficiency not only contributes to better auditory discrimination but also reduces energy consumption, supporting advancements toward fully implantable cochlear implant systems. Furthermore, protecting the inner ear' delicate structure and its residual hearing opens new horizons for auditory restoration through gene or biological therapies, opening new ways for innovative treatment and monitoring strategies that go beyond traditional implant functionality.

Despite its promise, leveraging impedance telemetry for reliable monitoring of cochlear health still faces several technical challenges. One key limitation lies in the short-duration stimulation pulses and low sampling rates typically used in clinical settings, which can hinder the accurate estimation of model parameters over time. As shown by Molaee-Ardekani and Donahue (2023), the robustness of parameter extraction is sensitive to noise, temporal variability, and fitting algorithm stability. These factors complicate longitudinal tracking of cochlear status, particularly when relying on complex biophysical models. Overcoming these challenges will require optimized stimulation paradigms, more precise acquisition protocols, and, potentially, machine learning techniques to ensure consistent and clinically meaningful interpretations.

Acknowledgments The authors acknowledge the use of Microsoft Copilot for proofreading and enhancing the readability of the chapter. All suggested changes were carefully reviewed and edited. The authors assume full responsibility for the content of the final publication.

References

Alahmadi A, Abdelsamad Y, Yousef M, Almuhawas F, Hafez A, Alzhrani F, Hagr A (2024) Cochlear implantation: long-term effect of early activation on electrode impedance. J Clin Med 13:3299. https://doi.org/10.3390/jcm13113299

Alimohammadi S, Kiani MA, Imani M, Rafii-Tabar H, Sasanpour P (2021) A proposed implantable voltammetric carbon fiber–based microsensor for corticosteroid monitoring by cochlear implants. Microchim Acta 188:357. https://doi.org/10.1007/s00604-021-04994-9

Alohali YA, Fayed MS, Abdelsamad Y, Almuhawas F, Alahmadi A, Mesallam T, Hagr A (2023) Machine learning and cochlear implantation: predicting the post-operative electrode impedances. Electronics 12:2720. https://doi.org/10.3390/electronics12122720

Arambula AM, Piper R, Murray G, Mowry SE, Rivas A, Semaan M, Manzoor NF, Killeen DE, Kelsall DC, Lupo JE, Biever A, Stidham KR, Albinus R, Morgan S, Kaplan BJ, Presley R, Menendez S, Tejani VD (2025) Sensitivity and specificity of intraoperative TransImpedance matrix recordings compared with X-ray imaging in detecting perimodiolar cochlear implant tip foldovers: a multicenter study. Otol Neurotol 46:627–633. https://doi.org/10.1097/MAO.0000000000004502

Ausili SA, Parreño M, Fernandez F, Boccio CM, Di Lella FA (2022) Biological response dynamics to cochlear implantation: modeling and modulating the electrode–tissue interface. Ear Hear 43: 1687–1697. https://doi.org/10.1097/AUD.0000000000001236

Ayas M, Muzaffar J, Borsetto D, Eitutis S, Phillips V, Tam YC, Salorio-Corbetto M, Bance ML (2024) A scoping review on the clinical effectiveness of Trans-Impedance Matrix (TIM) measurements in detecting extracochlear electrodes and tip fold overs in Cochlear Ltd devices. PLoS One 19:e0299597. https://doi.org/10.1371/journal.pone.0299597

Barsoukov E, Macdonald JR (eds) (2005) Impedance spectroscopy: theory, experiment, and applications, 1st edn. Wiley. https://doi.org/10.1002/0471716243

Busby P, Plant K, Whitford L (2002) Electrode impedance in adults and children using the Nucleus 24 cochlear implant system. Cochlear Implants Int 3:87–103. https://doi.org/10.1179/cim.2002.3.2.87

Callegaro L (2012) Electrical impedance, 0th edn. CRC Press. https://doi.org/10.1201/b13069

Clark G (ed) (2003) Cochlear implants: fundamentals and applications. Springer, New York. https://doi.org/10.1007/b97263

Clark GM, Shute SA, Shepherd RK, Carter TD (1995) Cochlear implantation: osteoneogenesis, electrode-tissue impedance, and residual hearing. Ann Otol Rhinol Laryngol Suppl 166:40–42

De Ceulaer G, Johnson S, Yperman M, Daemers K, Offeciers FE, O'Donoghue GM, Govaerts PJ (2003) Long-term evaluation of the effect of intracochlear steroid deposition on electrode impedance in cochlear implant patients. Otol Neurotol 24:769–774. https://doi.org/10.1097/00129492-200309000-00014

Dhanasingh A, Nielsen SB, Beal F, Schilp S, Hessler R, Jolly C, Hochmair I (2024) Cochlear implant electrode design for safe and effective treatment. Front Neurol 15:1348439. https://doi.org/10.3389/fneur.2024.1348439

Di Lella FA, Parreño M, Fernandez F, Boccio CM, Ausili SA (2020) Measuring the electrical status of the bionic ear. Re-thinking the impedance in cochlear implants. Front Bioeng Biotechnol 8: 568690. https://doi.org/10.3389/fbioe.2020.568690

Dixon JF, Shinn JB, Adkins M, Hardin BD, Bush ML (2014) Middle ear disease and cochlear implant function: a case study. Hear Balance Commun 12:155–158. https://doi.org/10.3109/21695717.2014.918757

Eikelboom RH, Jayakody DM, Swanepoel DW, Chang S, Atlas MD (2014) Validation of remote mapping of cochlear implants. J Telemed Telecare 20:171–177. https://doi.org/10.1177/1357633X14529234

Eitutis ST, Tam YC, Roberts I, Swords C, Tysome JR, Donnelly NP, Axon PR, Bance ML (2022) Detecting and managing partial shorts in cochlear implants: a validation of scalp surface potential testing. Clin Otolaryngol 47:641–649. https://doi.org/10.1111/coa.13953

Gärtner L, Büchner A, Illg A, Lenarz T (2021) Hidden electrode failure in a cochlear implant user. Laryngoscope 131. https://doi.org/10.1002/lary.29023

Glisson TH (2011) Introduction to circuit analysis and design. Springer Netherlands, Dordrecht. https://doi.org/10.1007/978-90-481-9443-8

Goehring JL, Hughes ML, Baudhuin JL, Valente DL, McCreery RW, Diaz GR, Sanford T, Harpster R (2012) The effect of technology and testing environment on speech perception using telehealth with cochlear implant recipients. J Speech Lang Hear Res 55:1373–1386. https://doi.org/10.1044/1092-4388(2012/11-0358)

Gottfried TM, Galeazzi P, Föger A, Dejaco D, Tröger A, Fischer N, Innerhofer V, Di Trapani F, Weiss N, Seebacher J, Dierker A, Schmutzhard J (2024) Evaluation of an impedance-based method to monitor the insertion of the electrode array during cochlear implantation. Eur Arch Otorrinolaringol 281:4121–4131. https://doi.org/10.1007/s00405-024-08584-2

Gross CA, Roppel TA (2012) Fundamentals of electrical engineering, 0th edn. CRC Press. https://doi.org/10.1201/b11786

Harris AR, Newbold C, Stathopoulos D, Carter P, Cowan R, Wallace GG (2022) Comparison of the in vitro and in vivo electrochemical performance of bionic electrodes. Micromachines 13:103. https://doi.org/10.3390/mi13010103

Harris AR, Grayden DB, John SE (2023) Electrochemistry in a two- or three-electrode configuration to understand monopolar or bipolar configurations of platinum bionic implants. Micromachines 14:722. https://doi.org/10.3390/mi14040722

Hey M, Böhnke B, Dillier N, Hoppe U, Eskilsson G, Löwgren K, Cullington H, Mauch H, Müller-Deile J (2015) The intra-cochlear impedance-matrix (IIM) test for the nucleus® cochlear implant. Biomed Eng Biomed Tech 60. https://doi.org/10.1515/bmt-2014-0058

Hibbert DB, Weitzner K, Carter P (2001) Voltammetry of platinum in artificial perilymph solution. J Electrochem Soc 148:E1. https://doi.org/10.1149/1.1344543

Horowitz P, Hill W, Rollefson RJ (1990) The art of electronics. Am J Phys 58:702–703. https://doi.org/10.1119/1.16385

Hrncirik F, Roberts I, Sevgili I, Swords C, Bance M (2023) Models of cochlea used in cochlear implant research: a review. Ann Biomed Eng 51:1390–1407. https://doi.org/10.1007/s10439-023-03192-3

Hu H-C, Chen JK-C, Tsai C-M, Chen H-Y, Tung T-H, Li LP-H (2017) Evolution of impedance field telemetry after one day of activation in cochlear implant recipients. PLoS One 12: e0173367. https://doi.org/10.1371/journal.pone.0173367

Hughes ML, Vander Werff KR, Brown CJ, Abbas PJ, Kelsay DMR, Teagle HFB, Lowder MW (2001) A longitudinal study of electrode impedance, the electrically evoked compound action potential, and behavioral measures in nucleus 24 cochlear implant users. Ear Hear 22:471–486. https://doi.org/10.1097/00003446-200112000-00004

Hughes ML, Sevier JD, Choi S (2018) Techniques for remotely programming children with cochlear implants using pediatric audiological methods via telepractice. Am J Audiol 27:385–390. https://doi.org/10.1044/2018_AJA-IMIA3-18-0002

Jiang C, De Rijk SR, Malliaras GG, Bance ML (2020) Electrochemical impedance spectroscopy of human cochleas for modeling cochlear implant electrical stimulus spread. APL Mater 8:091102. https://doi.org/10.1063/5.0012514

Joly C-A, Reynard P, Hermann R, Seldran F, Gallego S, Idriss S, Thai-Van H (2021) Intra-cochlear current spread correlates with speech perception in experienced adult cochlear implant users. J Clin Med 10:5819. https://doi.org/10.3390/jcm10245819

Kalbitz R (2023) Impedance spectra of different capacitor technologies. J Adv Dielectr 13:2341010. https://doi.org/10.1142/S2010135X23410102

Klabbers TM, Huinck WJ, Heutink F, Verbist BM, Mylanus EAM (2021) Transimpedance matrix (TIM) measurement for the detection of intraoperative electrode tip foldover using the slim modiolar electrode: a proof of concept study. Otol Neurotol 42:e124–e129. https://doi.org/10.1097/MAO.0000000000002875

Kuzovkov V, Yanov Y, Levin S, Bovo R, Rosignoli M, Eskilsson G, Willbas S (2014) Remote programming of MED-EL cochlear implants: users' and professionals' evaluation of the remote programming experience. Acta Otolaryngol (Stockh) 134:709–716. https://doi.org/10.3109/00016489.2014.892212

Lee J, Kim J (2022) Theoretical and empirical verification of electrical impedance matching method for high-power transducers. Electronics 11:194. https://doi.org/10.3390/electronics11020194

Lei IM, Jiang C, Lei CL, De Rijk SR, Tam YC, Swords C, Sutcliffe MPF, Malliaras GG, Bance M, Huang YYS (2021) 3D printed biomimetic cochleae and machine learning co-modelling provides clinical informatics for cochlear implant patients. Nat Commun 12:6260. https://doi.org/10.1038/s41467-021-26491-6

Mazzeo BA (2009) Parasitic capacitance influence of potential-sensing electrodes on four-electrode liquid impedance measurements. J Appl Phys 105:094106. https://doi.org/10.1063/1.3124365

Mc Carthy K, Gullapalli H, Ryan KM, Kennedy T (2022) Electrochemical impedance correlation analysis for the estimation of Li-ion battery state of charge, state of health and internal temperature. J Energy Storage 50:104608. https://doi.org/10.1016/j.est.2022.104608

McElveen JT, Blackburn EL, Green JD, McLear PW, Thimsen DJ, Wilson BS (2010) Remote programming of cochlear implants: a telecommunications model. Otol Neurotol 31:1035–1040. https://doi.org/10.1097/MAO.0b013e3181d35d87

Messersmith JJ, Entwisle L, Warren S, Scott M (2019) Clinical practice guidelines: cochlear implants. J Am Acad Audiol 30:827–844. https://doi.org/10.3766/jaaa.19088

Molaee-Ardekani B, Donahue MJ (2023) Investigating the electrode-electrolyte interface modelling in cochlear implants. Biomed Phys Eng Express 9:055024. https://doi.org/10.1088/2057-1976/aceafb

Nadol JB, O'Malley JT, Burgess BJ, Galler D (2014) Cellular immunologic responses to cochlear implantation in the human. Hear Res 318:11–17. https://doi.org/10.1016/j.heares.2014.09.007

Neuburger J, Lenarz T, Lesinski-Schiedat A, Büchner A (2009) Spontaneous increases in impedance following cochlear implantation: suspected causes and management. Int J Audiol 48:233–239. https://doi.org/10.1080/14992020802600808

Newbold C, Richardson R, Huang CQ, Milojevic D, Cowan R, Shepherd R (2004) An *in vitro* model for investigating impedance changes with cell growth and electrical stimulation: implications for cochlear implants. J Neural Eng 1:218–227. https://doi.org/10.1088/1741-2560/1/4/005

Noonan KY, Lopez IA, Ishiyama G, Ishiyama A (2020) Immune response of macrophage population to cochlear implantation: cochlea immune cells. Otol Neurotol 41:1288–1295. https://doi.org/10.1097/MAO.0000000000002764

Oberhoffner T, Mlynski R, Schraven S, Brademann G, Dierker A, Spitzer P, Hey M (2024) Effects of intraoperative cochlear implant electrode conditioning on impedances and electrically evoked compound action potentials. IEEE Trans Biomed Eng 71:650–659. https://doi.org/10.1109/TBME.2023.3313198

Parreño M, Di Lella FA, Fernandez F, Boccio CM, Ausili SA (2020) Toward self-measures in cochlear implants: daily and "homemade" impedance assessment. Front Digit Health 2:582562. https://doi.org/10.3389/fdgth.2020.582562

Ramos Miguel A, Ramos Macías A, Artiles JV, Perez Zaballos MT (2016) The effect of reference electrode position in cochlear implants. J Int Adv Otol 11:222–228. https://doi.org/10.5152/iao.2015.1746

Razmovski T, Bester C, Collins A, O'Leary SJ (2022) Four-point impedance changes in the early post-operative period after cochlear implantation. Otol Neurotol 43:e730–e737. https://doi.org/10.1097/MAO.0000000000003592

Rodríguez C, Ramos A, Falcon JC, Martínez-Beneyto P, Gault A, Boyle P (2010) Use of telemedicine in the remote programming of cochlear implants. Cochlear Implants Int 11:461–464. https://doi.org/10.1179/146701010X12671177204624

Saoji AA, Adkins WJ, Olund AP, Nelson-Bakkum ER, Koka K (2021) Effect of exceeding compliance voltage on speech perception in cochlear implants. Hear Res 400:108112. https://doi.org/10.1016/j.heares.2020.108112

Schepers K, Steinhoff H-J, Ebenhoch H, Böck K, Bauer K, Rupprecht L, Möltner A, Morettini S, Hagen R (2019) Remote programming of cochlear implants in users of all ages. Acta Otolaryngol (Stockh) 139:251–257. https://doi.org/10.1080/00016489.2018.1554264

Schraivogel S, Aebischer P, Weder S, Caversaccio M, Wimmer W (2023) Cochlear implant electrode impedance subcomponents as biomarker for residual hearing. Front Neurol 14:1183116. https://doi.org/10.3389/fneur.2023.1183116

Schraivogel S, Weder S, Mantokoudis G, Caversaccio M, Wimmer W (2025) Predictive models for radiation-free localization of cochlear implants' most basal electrode using impedance telemetry. IEEE Trans Biomed Eng 72:1453–1464. https://doi.org/10.1109/TBME.2024.3509527

Seyyedi M, Nadol JB (2014) Intracochlear inflammatory response to cochlear implant electrodes in humans. Otol Neurotol 35:1545–1551. https://doi.org/10.1097/MAO.0000000000000540

Shannon RV (1992) A model of safe levels for electrical stimulation. IEEE Trans Biomed Eng 39:424–426. https://doi.org/10.1109/10.126616

Shaul C, Bester CW, Weder S, Choi J, Eastwood H, Padmavathi KV, Collins A, O'Leary SJ (2019) Electrical impedance as a biomarker for inner ear pathology following lateral wall and peri-modiolar cochlear implantation. Otol Neurotol 40:e518–e526. https://doi.org/10.1097/MAO.0000000000002227

Sijgers L, Geys M, Geissler G, Boyle P, Huber A, Pfiffner F (2024) Electrical bioimpedance-based monitoring of intracochlear tissue changes after cochlear implantation. Sensors 24:7570. https://doi.org/10.3390/s24237570

Tykocinski M, Duan Y, Tabor B, Cowan RS (2001) Chronic electrical stimulation of the auditory nerve using high surface area (HiQ) platinum electrodes. Hear Res 159:53–68. https://doi.org/10.1016/S0378-5955(01)00320-3

Vanpoucke F, Zarowski A, Casselman J, Frijns J, Peeters S (2004) The facial nerve canal: an important cochlear conduction path revealed by clarion electrical field imaging. Otol Neurotol 25:282–289. https://doi.org/10.1097/00129492-200405000-00014

von Rohr R (2011) Implant impedance telemetry measurements and model calculations to estimate modiolar currents. Dissertation, University of Zurich

Wasmann J-WA, Huinck WJ, Lanting CP (2024) Remote cochlear implant assessments: validity and stability in self-administered smartphone-based testing. Ear Hear 45:239–249. https://doi.org/10.1097/AUD.0000000000001422

Weltin A, Kieninger J, Urban GA, Buchholz S, Arndt S, Rosskothen-Kuhl N (2022) Standard cochlear implants as electrochemical sensors: intracochlear oxygen measurements in vivo. Biosens Bioelectron 199:113859. https://doi.org/10.1016/j.bios.2021.113859

Wesarg T, Wasowski A, Skarzynski H, Ramos A, Falcon Gonzalez JC, Kyriafinis G, Junge F, Novakovich A, Mauch H, Laszig R (2010) Remote fitting in nucleus cochlear implant recipients. Acta Otolaryngol (Stockh) 130:1379–1388. https://doi.org/10.3109/00016489.2010.492480

Wilk M, Hessler R, Mugridge K, Jolly C, Fehr M, Lenarz T, Scheper V (2016) Impedance changes and fibrous tissue growth after cochlear implantation are correlated and can be reduced using a dexamethasone eluting electrode. PLoS One 11:e0147552. https://doi.org/10.1371/journal.pone.0147552

Wilson BS (2008) Cochlear implants: current designs and future possibilities. J Rehabil Res Dev 45:695–730. https://doi.org/10.1682/JRRD.2007.10.0173

Wolfe J, Schafer EC (2025) Programming cochlear implants, Core clinical concepts in audiology series, 3rd edn. Plural Publishing, Inc, San Diego

Zarowski A, Leblans M, Molisz A, Sismono F, van Dinther J, Offeciers F (2023) Calculation of the faradaic impedance of the electrode–tissue interface improves prediction of behavioral T/C levels in cochlear implant patients. J Int Adv Otol 19:368–375. https://doi.org/10.5152/iao.2023.23695

Electrocochleography (eCochG): Origin and Applications in Cochlear Implant Surgery

6

Pierre Stahl

Abstract

Electrocochleography (eCochG) is a neurophysiological technique used to assess cochlear and auditory nerve function by recording electrical potentials generated in response to sound stimuli. Since its inception in the mid-twentieth century, eCochG has evolved significantly in both methodology and clinical applications. Initially, extratympanic eCochG employed electrodes placed in the ear canal to measure cochlear potentials noninvasively, although this approach offered lower signal clarity. The development of transtympanic eCochG improved signal quality by positioning electrodes near the promontory, aiding in the diagnosis of conditions such as Ménière's disease. However, its invasive nature limited widespread clinical adoption. With advancements in cochlear implants, intracochlear eCochG has emerged as a powerful tool for real-time monitoring of cochlear function. Early research in animal models paved the way for integrating eCochG into cochlear implants, enabling direct measurement of summating and action potentials. This approach helps optimize electrode placement, assess residual hearing, and improve implant outcomes. The historical transition from extratympanic to intracochlear eCochG reflects a shift toward more precise and clinically valuable techniques. Today, intracochlear eCochG continues to advance, enhancing auditory diagnostics and hearing preservation. As technology progresses, it holds promise for further refining cochlear implant programming and auditory rehabilitation.

Keywords

Electrocochleography · Cochlear potentials · Inner ear diagnostics action potential · Summating potential

P. Stahl (✉)
Oticon Medical, ITSA Medical, Vallauris, France
e-mail: pisl@oticonmedical.com

B. Molaee-Ardekani, M. Segovia-Martinez (eds.), *Advancing Cochlear Implants*,
https://doi.org/10.1007/978-3-032-15719-5_6

6.1 Introduction

6.1.1 History of Electrocochleography

Investigations into cochlear electrical potentials began with seminal animal experiments by Wever and Bray (1930), in which an electrode was placed on the auditory nerve to record cochlear responses to acoustic stimuli. Remarkably, these recorded potentials mirrored the waveform of the emitted sound. Subsequent research by Adrian (1931) identified the cochlea as the origin of these signals, leading to the term “cochlear microphonics” (CM). This discovery spurred further exploration into cochlear potentials (Davis et al. 1934; Tasaki et al. 1954).

By the mid-1930s, the first human cochlear electrical potentials were successfully recorded (Fromm et al. 1935), marking a significant milestone in auditory physiology. Initially, cochlear microphonics were captured using a clinical technique known as the cochleogram, which involved placing a transtympanic needle electrode on the promontory under general anesthesia (Lempert et al. 1950). Although effective, this invasive method was later eclipsed by the development of the noninvasive auditory brainstem response (ABR) technique.

Between 1950 and 1960, the term “cochleogram” evolved into electrocochleography (eCochG), encompassing the broader physiological events associated with cochlear potentials. The first compound action potential (CAP) was documented in 1962, along with observations on the impact of Ménière’s disease on CAP responses (Ruben et al. 1962).

From 1960 to 1980, various clinical applications for cochlear electrical potentials were proposed. eCochG was suggested as an objective method for assessing auditory thresholds in children who were unresponsive to behavioral testing (Portman and Aran 1971). Concurrently, it was used to detect Ménière’s disease (Eggermont 1979). However, the emergence of noninvasive techniques such as ABR, which offered comparable clinical utility, led to a decline in the use of eCochG, relegating it primarily to academic and research settings. Moreover, the diagnostic value of eCochG for Ménière’s disease remained a topic of debate within the otolaryngology community (Nguyen et al. 2010).

Despite these challenges, research on eCochG continued, exploring its potential in diagnosing vestibular schwannomas (Tanaka et al. 1999) and, more recently, auditory neuropathy spectrum disorders (Gibson 2017).

After 2000, eCochG found a novel application in the context of cochlear implants (CIs). During CI surgeries, eCochG has been utilized to provide real-time monitoring of cochlear function during the insertion of the electrode array into the scala tympani (Kumar and Peepal 2012; Campbell et al. 2016; Giardina et al. 2018; Adunka et al. 2016; Haumann et al. 2019). The invasive nature of this measurement is considered acceptable during surgery, as CI implantation does not require additional surgical placement of the eCochG recording probe. The visibility of the cochlear promontory during the procedure allows for optimal probe positioning, significantly enhancing the signal-to-noise ratio compared to traditional transtympanic needle electrode placements.

6.1.2 Electrocochleography in CI

The variability in outcomes following cochlear implantation is notably high. While some patients achieve near-normal speech recognition scores, others perform as low as 20%, even in quiet listening conditions (McDermott 2004). This variability extends to other perceptual domains, including loudness perception (Nelson et al. 1996), pitch sensation (Kong and Carlyon 2010), and temporal resolution (Stahl et al. 2016). One contributing factor to this interindividual variability may be cochlear trauma incurred during the implantation procedure (Hoskison et al. 2017).

To mitigate insertion-related trauma, extensive research has focused on electrode array design (Bento et al. 2016) and surgical techniques (Jiam et al. 2016). Given that eCochG can capture cochlear potentials in response to acoustic stimuli, Campbell et al. (2010) employed eCochG in gerbils during electrode insertion. Using an endoscope for direct visualization, they demonstrated that cochlear trauma—later confirmed through histological analysis—was associated with an irreversible reduction in the cochlear microphonic component of the eCochG response.

Subsequent human feasibility studies confirmed the utility of real-time cochlear feedback during implantation. Advances in recording techniques enabled eCochG measurements from the stapes (Harris et al. 2011) and the round window (Choudhury et al. 2012), facilitating faster and safer probe placement.

Until the late 2010s, eCochG recordings were obtained exclusively using external diagnostic equipment, making the quality of the response highly dependent on the type and precise placement of the recording electrode during surgery. However, technological advancements have since enabled eCochG recordings via the CI electrode array itself. In this approach, an acoustic burst is delivered through a small speaker, and the most apical electrode of the implant records the cochlear response (Calloway et al. 2014).

Currently, two main eCochG recording techniques are available:

- Extracochlear (EC) technique: A recording probe is placed outside the cochlea (e.g., on the round window or promontory).
- Intracochlear (IC) technique: A recording probe is placed within the cochlea, typically using one of the CI electrodes.

Both techniques are now employed to monitor cochlear health in real time during CI surgery, offering valuable feedback to optimize outcomes and preserve residual hearing.

6.1.3 Eligibility for eCochG and Expected Benefits

Electrocochleography involves acoustic stimulation of the cochlea, which requires the presence of sufficient residual hearing in the patient. Typically, audiometric thresholds should not exceed 80–85 dB HL for frequencies below 500 Hz to ensure

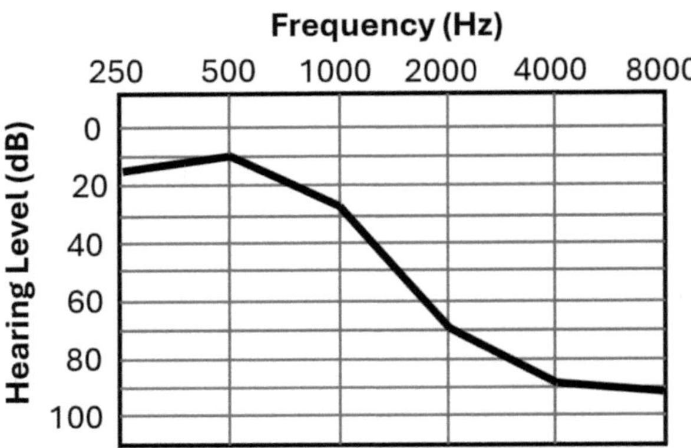

Fig. 6.1 Drawing representing a typical ski slope audiogram, showing residual hearing in the low frequencies (250–750 Hz) characteristic to a CI candidate with residual hearing

reliable recordings. As a result, eCochG is feasible only in patients with measurable residual hearing. Figure 6.1 illustrates typical audiograms of patients eligible for eCochG recordings.

6.1.4 Preoperative eCochG

Given the critical role of functional auditory nerve fibers in transmitting information from a CI system, Formeister et al. (2015) hypothesized that preimplantation eCochG could serve as a predictor of postoperative hearing outcomes.

6.1.5 Perioperative eCochG

The use of eCochG during CI surgery enables real-time assessment of potential trauma to the inner cochlea during electrode array insertion. The primary goal is to estimate postoperative residual hearing, which may correlate with the patient's speech perception performance using the CI. Although perioperative eCochG has shown promise in predicting residual hearing, results remain variable (Radeloff et al. 2012; Dalbert et al. 2015; Campbell et al. 2014). Moreover, the relationship between residual acoustic hearing and electrically stimulated hearing outcomes remains inconclusive.

To improve predictive accuracy, several studies have aggregated eCochG components into a total response metric, termed "eCochG-TR" (McClellan et al. 2014). This approach demonstrated that eCochG-TR accounted for 40% of the variance in consonant-nucleus-consonant (CNC) word scores. In a larger cohort of 77 children, Formeister et al. (2015) found that eCochG-TR explained 50% of the variance in word scores. More recently, Walia et al. (2022) combined eCochG-TR with the Montreal Cognitive Assessment (MoCA) in a multivariate model, explaining 60% of the variance in speech perception in noise, with eCochG-TR alone accounting for 34%. For comparison, two of the most widely recognized predictors of CI outcomes—duration of deafness and age at implantation—each explain only about 20% of the variance (Lazard et al. 2012; Blamey et al. 2012).

6.1.6 Postoperative eCochG

More recently, Attias et al. (2020) proposed the use of eCochG postoperatively to estimate residual hearing after CI placement. Their study reported strong correlations between eCochG responses and behavioral hearing thresholds in children. However, cochlear microphonic signals were rarely measurable in their cohort, with only 29 ears out of 88 children showing detectable CM responses.

6.2 Neuropsychological Components of eCochG

Electrocochleography typically involves the use of loud acoustic stimuli, such as clicks or tone bursts, usually presented at intensities above 80 dB HL. Stimulating frequencies generally range from 250 to 750 Hz, depending on the specific study design. Tone bursts can alternate in polarity, producing condensation (C) and rarefaction (R) responses (Dalbert et al. 2018). These measurements are always repeated and averaged to enhance signal reliability. The resulting response comprises multiple components that appear based on the input stimulus and post-processing methods (for clicks and tone bursts): C alone, R alone, C–R (to enhance neural components), or C + R (to emphasize hair cell responses) (Barnes et al. 2020).

The four components are the summating potential (SP), compound action potential (CAP), auditory nerve neurophonic (ANN), and cochlear microphonics (CM). The presence and clarity of each component depend on the stimulus type and the method used to process the condensation and rarefaction responses. Tone bursts are preferentially used to monitor CI surgery, while clicks are commonly employed for diagnosing hearing pathologies such as Ménière's disease or auditory nerve neuropathy.

6.2.1 Summating Potentials (SP)

SP is a direct current (DC) shift primarily generated by the hair cells in the cochlea (Pappa et al. 2019). It occurs during the stimulus and reflects the sustained receptor potential. SP is typically obtained in response to clicks (e.g., Santarelli et al. 2021) or tone bursts (e.g., Jwair et al. 2023). It is measured as the baseline shift from the pre-stimulus period to the sustained portion (i.e., plateau) of the waveform during the stimulus.

The SP response is thought to predominantly originate from the inner hair cells but may represent a mixed response from all hair cells. Its exact source remains unclear, as it can change polarity (i.e., positive or negative voltage) depending on the frequency and level of the stimulus (Dallos et al. 1972) and the position of the recording electrode.

The SP represents the sum of multiple nonlinearities associated with the transduction process in the cochlea (Tasaki et al. 1954; Davis 1968; Dallos et al. 1972). Clinically, the magnitude of the SP is significant for detecting Ménière's disease (Hornibrook 2017), often associated with endolymphatic hydrops, which is characterized by enlarged SPs (Koyuncu et al. 1994).

6.2.2 Compound Action Potentials (CAP)

CAP is generated by the synchronous firing of auditory nerve fibers, similar to wave I of an auditory brainstem response (ABR). CAPs are obtained from clicks (e.g., Santarelli et al. 2021) or tone bursts (e.g., Jwair et al. 2023) and calculated from the negative peak occurring approximately 2 milliseconds after stimulus onset. CAPs are clearly visible when eCochGs are evoked by broadband clicks, appearing as a voltage-negative peak around 2 milliseconds post-stimulus onset (see CAP in Fig. 6.2).

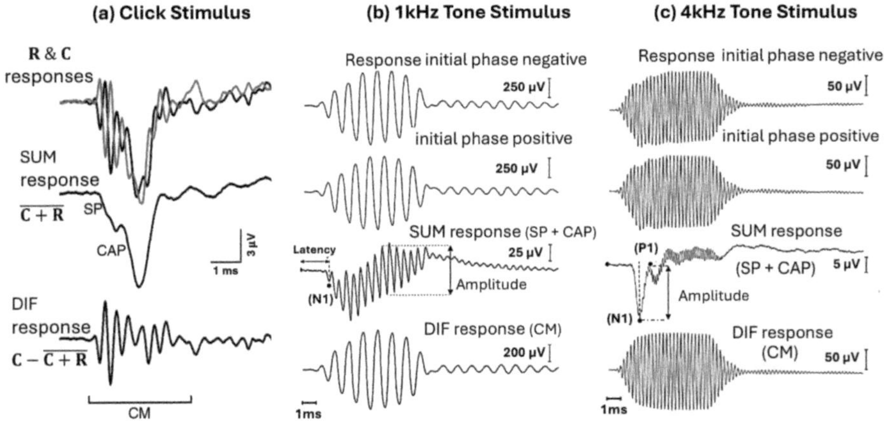

Fig. 6.2 Electrocochleography (eCochG) waveforms vary depending on the type of acoustic stimulus used. (**a**) Click stimulus. (Adapted from Santarelli et al. (2021) under CC BY license). (**b**) 1 kHz tone burst, and (**c**) 4 kHz tone burst. (Adapted from Jwair et al. (2023) under CC BY license). In the top panels, responses to condensation (*C*) and rarefaction (*R*) phases of the stimulus are shown. Any summation function of the *R* and *C* responses (e.g., $\overline{C+R}$ or $C + R$) enhances the SP and CAP components—this is referred to as the SUM response. Conversely, subtracting functions (e.g., $C-\overline{C+R}$ or $C - R$) enhances the CM component—this is referred to as the DIF response

Endolymphatic hydrops, characteristic of Ménière's disease, is often accompanied by an enlarged CAP. The ratio of SP to CAP is a good predictor for this pathology. Additionally, as shown in ABR studies (Bramhall 2021), wave I is a marker of cochlear synaptopathy.

6.2.3 Auditory Nerve Neurophonic (ANN)

ANN is defined as the early neural response, primarily reflecting phase-locked activity of auditory nerve fibers. It is elicited exclusively by tone burst stimuli and is calculated from the peak-to-peak (P2P) amplitude of the averaged sum of condensation and rarefaction responses. ANNs arise from the addition of the alternating current (AC) components of both stimulus polarities (i.e., R and C), resulting in a response frequency that is twice the stimulus frequency (Barnes et al. 2020; Choudhury et al. 2012). A significant decrease or absence of the ANN indicates auditory neuropathy.

6.2.4 Cochlear Microphonics (CM)

CMs responses are typically elicited using tone burst stimuli and are calculated from the peak-to-peak amplitude of the difference between condensation and rarefaction responses. The CM response originates almost exclusively from the outer hair cells following stimulation. CMs can be observed from both clicks and tone burst stimulations. Represented as a sine wave matching the frequency and duration of the stimulus, CMs provide real-time feedback of the hair cells' responses. They reflect the resulting AC component created by the difference between condensation and rarefaction stimuli.

6.3 Measurement and Analysis Techniques for CI Surgery Monitoring

6.3.1 Cochlear Microphonics: The Best eCochG Component for Intraoperative Monitoring

In CI patients, eCochG recordings often do not exhibit clear SP or CAP components, and the ANN typically does not provide useful real-time feedback. In contrast, CMs have proven to be the most sensitive indicators of hair cell activity (Giardina et al. 2019) and are now widely used for real-time monitoring of cochlear function during CI surgery. In many scientific studies investigating eCochG in the context of CI, as well as this chapter, the terms "eCochG" and "CM" may be used interchangeably.

During CI surgery, CMs are usually recorded in response to single-frequency tone bursts. The origin of the eCochG signal can help identify the location of trauma caused by electrode array insertion (Barnes et al. 2020). However, depending on the

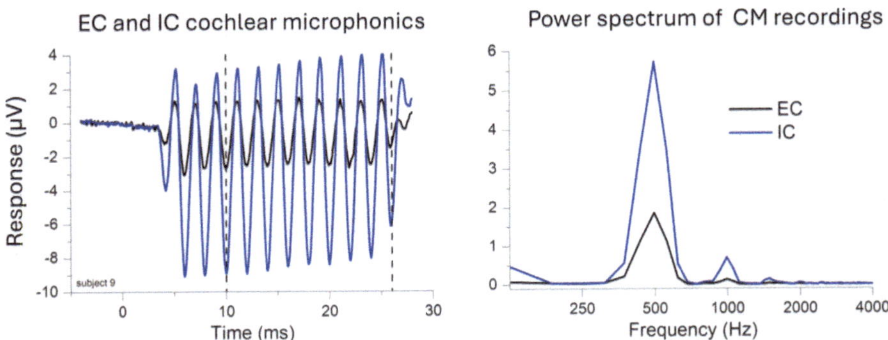

Fig. 6.3 Example of cochlear microphonic responses recorded from intracochlear (IC, shown in blue) and extracochlear (EC, shown in black) eCochG recordings. (Figure adapted from Calloway et al. (2014) with permission from Wolters Kluwer Health, Inc.)

recording technique, eCochGs are the result of the combination of hair cells corresponding to the tonotopic place of the stimulated frequency (i.e., targeted hair cells) and hair cells located at more basal regions.

When using an EC eCochG probe, basal hair cells contribute more to the signal since it is positioned on the cochlear promontory. In contrast, an IC eCochG probe captures a more complex signal, originating both from basal cells and from hair cells at the stimulus frequency. As the electrode array is inserted deeper into the cochlea, the amplitude and phase of the CM exhibit substantial variations. Additionally, as depicted in Fig. 6.3, because the IC probe is positioned closer to the cells, the amplitude of the eCochG signal tends to be significantly greater when measured intracochlearly than when measured using an EC probe (Calloway et al. 2014). Dalbert et al. (2021) correlated EC and IC simultaneously during the CI insertion of ten patients. They found that IC is always larger, with an average amplitude difference of 14 dB.

6.3.2 Temporal Peak-to-Peak Versus Spectral Peak Measurement

The peak-to-peak (P2P) CM response measures the amplitude of the entire waveform by calculating the difference between the maximum positive and minimum negative peaks in the time domain (see Fig. 6.4). This comprehensive measurement offers a holistic view of the response magnitude, crucial for clinical assessments of cochlear health and function. P2P is particularly effective in diagnosing conditions like Ménière's disease, capable of detecting changes in cochlear function over time (Campbell et al. 2014). It also aids in evaluating cochlear health in cases of suspected endolymphatic hydrops, where significant variations in P2P amplitude have been observed (Dalbert et al. 2018).

The spectral peak (SPk) CM response specifically measures the amplitude of the response at the frequency of the stimulus, offering insights into the signal intensity at

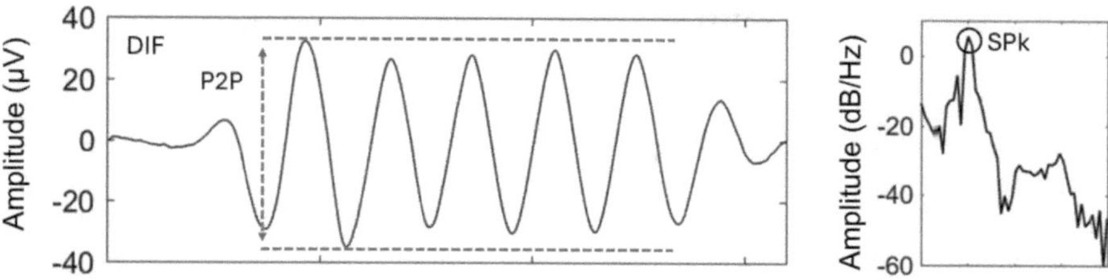

Fig. 6.4 Example of P2P response obtained from the CM waveform and SPk response obtained from its spectrum. The stimulus is a 0.5-kHz tone burst at 100 dB HL. Power spectral density function on the right is expressed as dB relative to 1 μV. (Figure adapted from Bester et al. (2017), under CC BY license)

that precise frequency component. This capability allows detailed analysis of cochlear responses at specific frequencies, critical for research into frequency-specific cochlear physiology. For instance, Fontenot et al. (2017) used spectral peak analysis as a key method for distinguishing between two overlapping components of eCochG signal: the CM and the ANN. The authors developed a model-based approach, validated by both animal and clinical data, which leverages the frequency-domain characteristics of eCochG signals—particularly focusing on the spectral peaks at the stimulus frequency and its harmonics. By analyzing the amplitude and phase of these spectral peaks, they were able to isolate the CM component, which reflects hair cell activity, from the ANN, which represents neural phase-locking. This spectral analysis enabled more accurate interpretation of eCochG data, especially in contexts where both components are present and difficult to separate using traditional time-domain methods. In another study, Giardina et al. (2018) utilized SPk CM responses to study cochlear function in patients with auditory neuropathy spectrum disorder, demonstrating its utility in understanding impairments in auditory nerve function.

Both peak-to-peak (P2P) and spectral peak (SPk) CM response measurements are commonly used to assess cochlear trauma during CI surgery. Several studies have shown results from both techniques (e.g., Bester et al. 2017; Calloway et al. 2014; Walia et al. 2022; Haumann et al. 2019), consistently showing that P2P measurements yield larger amplitude responses than SPk measurements. Our unpublished simulation data support these findings, indicating that P2P amplitudes are approximately twice as large as SPk amplitudes. This discrepancy can be attributed to the characteristics of FFT-based spectral analysis, where energy from non-pure-tone signals is distributed across multiple adjacent frequency bins. As a result, relying solely on the maximum spectral peak may underestimate the true amplitude, particularly when more sophisticated frequency-based estimation methods are not employed. In other words, the less pure the signal, the lower the amplitude observed in SPk measurements.

6.3.3 Stimulation Setup

The stimulation setup for eCochG remains consistent regardless of whether the technique is extracochlear (EC) or intracochlear (IC). It involves a speaker connected to approximately 1.5 m of silicone tubing, which transmits sound through an ear foam placed in the ipsilateral ear canal. A calibration microphone is typically positioned near the speaker to measure the sound level delivered to the ear canal. Experimental setups, such as the one shown in Fig. 6.5 (on a cadaver), may include a second microphone to measure sound levels within the middle ear cavity. In some cases, the ear canal is filled with a perilymph-like fluid, such as 0.9% NaCl solution, to simulate physiological conditions.

As previously mentioned in Sect. 6.2, the stimulation protocol used for perioperative eCochG monitoring during CI implantation employs tone bursts alternating in polarity (i.e., the C and R stimuli). To ensure smooth transitions and avoid sudden bursts of stimulation, the stimulus amplitude is gradually increased or decreased at the onset and offset. The slope and duration of these ramps may vary. Typically, the stimulation amplitude is set around 100 dB HL, tailored to the patient's residual hearing level, although it can range from 80 dB HL (in rare cases) to 110 dB HL if the system permits generating very high-level tones. The frequency is usually set at 500 Hz but can vary from 250 to 2000 Hz depending on the application and the capabilities of the speaker.

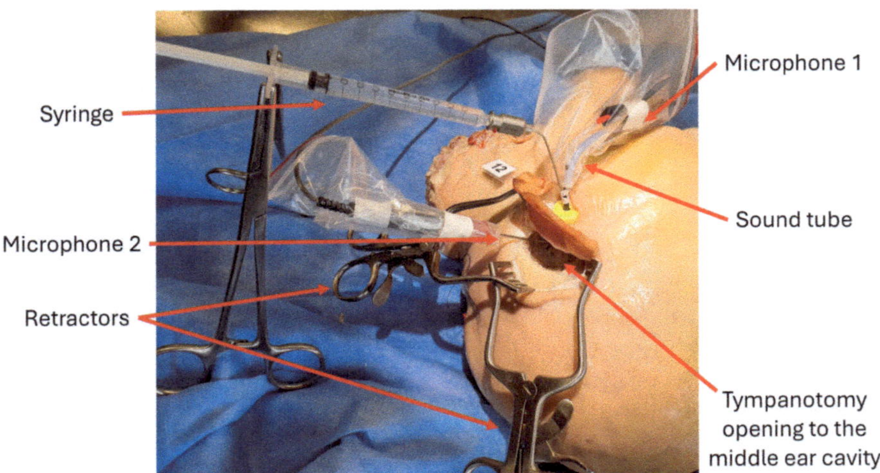

Fig. 6.5 Ex vivo eCochG experimental setup. Standard middle ear access with posterior tympanotomy has been performed in a whole-head specimen (here with a wound retractor positioned away from the ear canal). Microphone 1 and microphone 2 are used to measure sound pressure levels (SPLs) in the ear canal and the middle ear cavity, respectively. The syringe is used to fill the ear canal with NaCl 0.9%. (Figure adapted from Schuerch et al. (2022), with permission from Springer Nature)

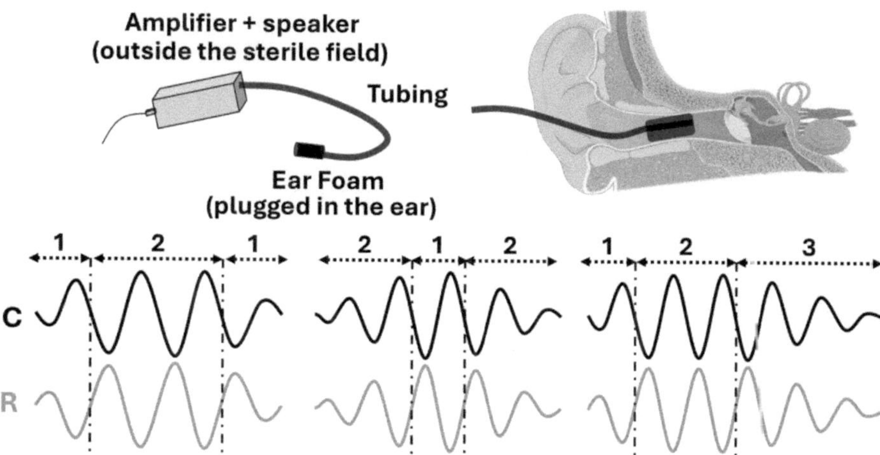

Fig. 6.6 (Top) Typical clinical setup and electrode placement for perioperative eCochG recording. (Bottom) Example of condensation (*C*) and rarefaction (*R*) acoustic tone burst stimuli used for eCochG and CM measurements. Dashed vertical lines indicate the boundaries of the onset ramp, plateau, and offset ramp segments of the stimuli. The number of cycles in each segment is shown above the illustrations. Onset and offset ramps are applied to the stimulus envelope to avoid click-like stimulation artifacts. (The anatomical illustration of the ear is adapted from Gordon Betts et al. (2022) under CC BY license)

Lowering the frequency below 250 Hz could potentially expand the pool of eligible patients by accessing more residual hearing. However, many speakers struggle to produce such low frequencies at high levels without distortion. Conversely, recording eCochGs for frequencies higher than 1000 Hz is generally impractical due to the rapid decrease in residual hearing sensitivity with increasing frequency, making even a 110 dB HL stimulus insufficient to elicit a cochlear potential. Therefore, 500 Hz is considered a practical compromise—high enough to avoid distortions yet still within the residual hearing range for most CI candidates.

Figure 6.6 illustrates a few different stimulus waveforms, each numbered to indicate cycles during the ramping, flat, and falling phases of stimulation. Commonly used formats in literature include the 1-2-1 and 2-1-2 patterns, chosen for their efficiency in facilitating faster recordings. The duration of stimulation is typically constrained by the recording system's capabilities. For instance, some systems limit the maximum applicable recording window to 30 ms, which sets an upper bound on stimulation duration and number of cycles that can be supported in a ramp-flat-fall stimulation paradigm.

6.3.3.1 Effect of Stimulus Frequency on eCochG

The impact of stimulus frequency on eCochG amplitude responses has been extensively studied in the literature. Generally, higher stimulation frequencies (>1000 Hz) tend to elicit lower responses (Campbell et al. 2014, 2016, 2017; Dalbert et al. 2018; Fontenot et al. 2017; Giardina et al. 2018). However, results for frequencies below 1000 Hz vary with the setup. Haumann et al. (2019) investigated eCochG responses in human subjects across different surgical stages—preoperative (EC), perioperative (EC), and postoperative (both EC and IC). Among the 250 Hz, 500 Hz, and 1000 Hz stimulus frequencies used, the 500 Hz and 1000 Hz tones produced the clearest responses with minimal artifacts. The study found that intraoperative EC eCochG thresholds—defined as the minimum sound level required to elicit a measurable eCochG response—correlated strongly with preoperative audiometric thresholds at 1000 Hz. For IC recordings, stronger correlations were observed at 250 Hz and 500 Hz.

Walia et al. (2022) reported that responses to a 250 Hz tone were larger than those to a 500 Hz tone during the electrode insertion phase of cochlear implantation. The study also demonstrated that the correlation between the largest eCochG drop using a 250 Hz stimulus and the shift in low-frequency pure tone audiometric average (measured at 125, 250, and 500 Hz) was stronger than with the 500 Hz stimulus ($r = 0.58$ vs. $r = 0.31$). Figure 6.7 illustrates the evolution of eCochG

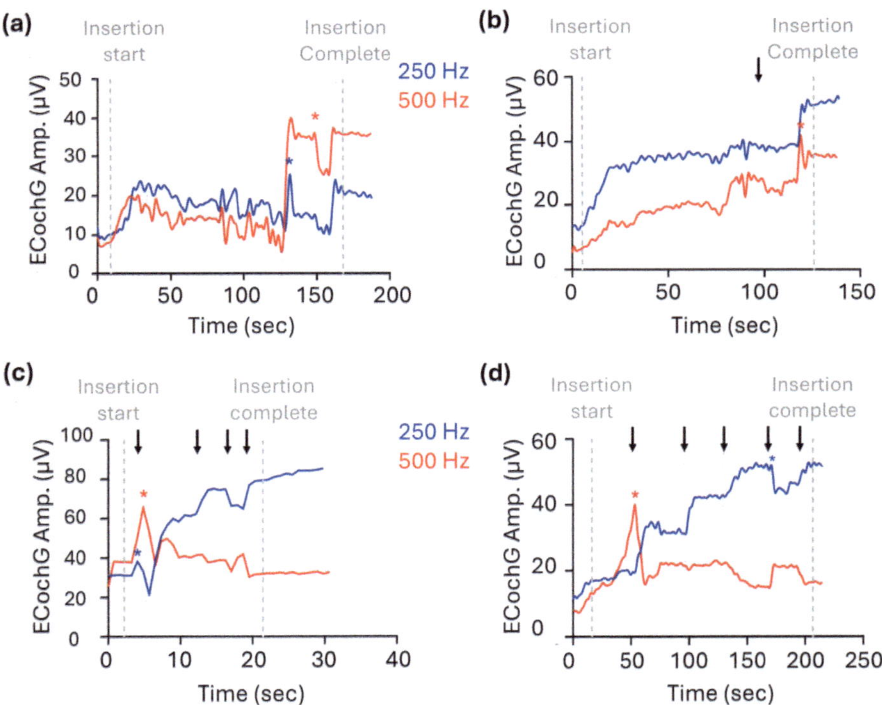

Fig. 6.7 Patient examples of eCochG response to multifrequency eCochG using 250 and 500 Hz tone burst. (**a**–**d**) Show four different insertion tracks. (**a**) No asynchrony between 250 and 500 Hz response. (**b**) Overall rise in response with one dyssynchronous point (marked with black arrow) where there is stability in 250 Hz response and drop in 500 Hz response. (**c**, **d**) Multiple dyssynchronous points (black arrows) noted between 250 and 500 Hz response with overall rise in 250 Hz response and drop in 500 Hz response. CM drops >30% are shown as asterisks. (Figure adapted from Walia et al. (2022), under the CC BY license)

measurements—specifically the CM—over time for four patients, highlighting asynchronous eCochG drops observed with 250 Hz and 500 Hz stimulations. Ultimately, the study concludes that multifrequency eCochG can help distinguish between cochlear trauma and the advancement of the apical electrode beyond the 500 Hz tonotopic region. While the 250 Hz eCochG provides a stronger signal and demonstrates a higher correlation with postoperative low-frequency hearing thresholds, the 500 Hz eCochG reaches its peak earlier due to the cochlea's tonotopic organization. This earlier peak makes it particularly useful for monitoring the depth of electrode insertion relative to specific frequency regions.

6.3.3.2 Effect of Stimulus Shape on eCochG

As mentioned earlier, the envelope of the stimulus can be defined by the number of cycles in the onset/offset ramp and the plateau. A longer signal plateau may facilitate easier signal detection, either because there is less constraint on defining the time window that selects a steady response part or because a longer signal offers more resolution in the frequency domain. A literature review reveals that a variety of stimulus shapes have been used (Table 6.1).

One of the main challenges in continuous (i.e., ongoing) eCochG monitoring is the reliable detection of CM responses within the shortest possible time. Using short-duration stimuli can help reduce overall recording time; however, stimuli with a longer plateau phase extend the effective stimulation period (i.e., longer window), potentially improving the consistency and reliability of CM detection. This can be investigated by examining the impact of various stimulus shapes on the convergence and stabilization of the CM response.

We conducted a simulation (unpublished data) to illustrate the impact of stimulus shape on CM responses. Various stimulation configurations—including 1-2-1, 1-3-1, 2-1-2, 2-2-2, and 2-9-2—were used to provide different stimulation durations and duty cycles (i.e., the duration of the flat phase in the signal). Figure 6.8 presents the simulation results after 2000 repetitions of a 500 Hz stimulus. The stimuli were delivered at a rate of 60 Hz, except for the 2-9-2 configuration, which was reduced to 30 Hz due to its longer duration. Among all configurations, the 2-9-2 yielded the clearest result, showing a distinct peak at the 500 Hz stimulation frequency.

The evolution of P2P and SPk values over recording time (i.e., number of repetitions) reveals a progression from fastest to slowest stabilization across the following stimulation configurations: 2-9-2, 1-3-1, 1-2-1, 2-2-2, and 2-1-2 (Fig. 6.9). SPk converges more rapidly than P2P, reaching within 0.3 dB of its final value after approximately 1 s. In contrast, P2P convergence to a similar threshold is only achieved after about 10 s—and only for the four most effective stimulation configurations.

Table 6.1 Literature review showing a sample of eCochG investigations and differences in the stimulus shape (i.e., rise/fall and plateau times)

Publication	Stimulus frequency [Hz]	Rise/fall time	Plateau	Repetition rate [Hz]
Choudhury et al. (2012)	250, 500, 700	1–4 ms (Blackman)	28 ms	23.3
Choudhury et al. (2012)	1000, 2000, 4000	1–4 ms (Blackman)	10 ms	29.3
Campbell et al. (2014)	500, 1000, 1500	1 ms (Linear)	4 ms	20
Campbell et al. (2016)	250, 500, 1000, 2000, 4000	1 ms (Linear)	4–10 ms	20
Harris et al. (2017)	500	5 ms	40 ms	5
Dalbert et al. (2018)	250, 500, 750, 1000	2 cycles (Blackman)	4, 10, 14, 20 cycles	–
Giardina et al. (2018)	250, 500, 750, 1000, 2000, 4000	–	16, 14 ms for 250 and 500 Hz	–
Fontenot et al. (2019)	250, 500, 750, 1000, 2000, 4000	1 cycle or 1 ms (Blackman)	@ 250–2000: 32 ms @ 4000: 10.66 ms	@ 250–2000: 17.3 @ 4000: 29.3
Haumann et al. (2019)	Mostly 500 and 1000	–	15 ms	19.1

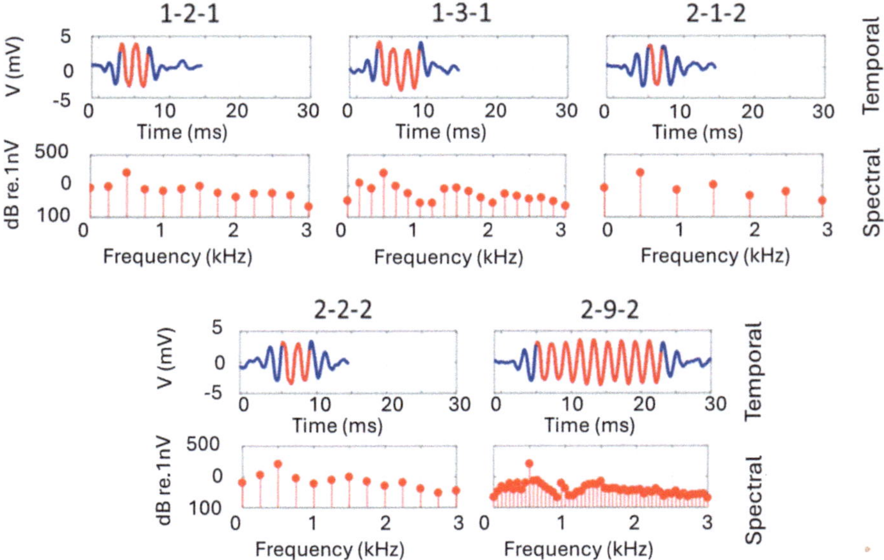

Fig. 6.8 CM responses in time and frequency domain to stimuli with various ramp-plateau-ramp configurations as indicated in the titles. The top row shows the complete recordings in time domain with red color highlighting the plateau part that is used to extract the single-sided amplitude spectrum shown in the bottom row

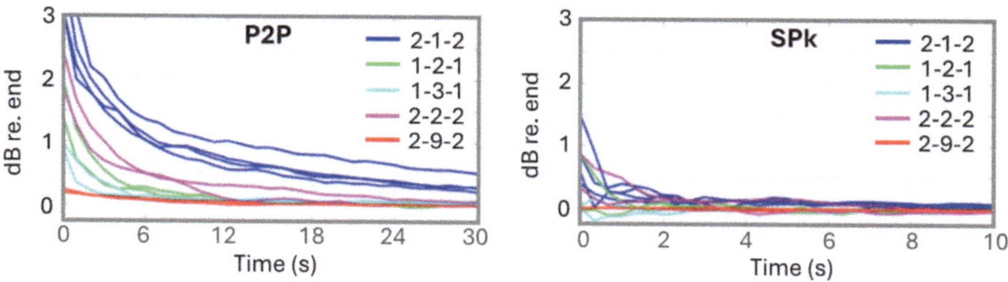

Fig. 6.9 Convergence of P2P and SPk values to their final true values. Each line represents one round of simulation for a given color-coded configuration

6.3.4 Extracochlear eCochG Recording

6.3.4.1 Artifact Removal from Low-Amplitude Electrical Signals

EC eCochG requires recording very small electrical signals from contact electrodes. Therefore, applying automatic artifact rejection methods is necessary to distinguish noise from biological signals. Any recording system equipped with at least an active electrode (+), reference electrode (−), and ground (GND) scalp electrode automatically rejects minor biological artifacts generated by the body.

The active electrode (also called the non-inverting electrode) detects the electrical activity generated by the auditory nerve and is the main electrode that picks up the electrical response. The reference electrode (called the inverting electrode) serves as a point of comparison for the active electrode. The ground electrode provides a stable electrical baseline. The typical artifact rejection involves a differential amplification of the signal. It amplifies the voltage difference between the active and reference electrode and reduces noise common to both electrodes so that the resulted signal is as free as possible from electrical noise or artifact. More advanced setup uses a four-electrode setup, the fourth electrode being a second reference.

The placement of the reference electrode depends on the eCochG recording probe chosen. The configuration of active and ground electrodes can vary based on measured electrode impedance and achieved signal-to-noise ratio. Figure 6.10 provides examples of electrode setups for EC eCochG recording. The left and middle setups are adapted from Carpi and Migliorini (2009), while the right setup is recommended in the Interacoustics® Eclipse Instruction for Use.

6.3.4.2 Recording Electrode Placement

Depending on the clinical or academic context, various types of electrodes are used for recording eCochG signals. Four commonly used types include the tiptrode, TM-trode (Simpson et al. 2020), transtympanic (TT) electrode (Haapaniemi et al. 2000), and small-surface cotton wick, cup, or ball electrode (see Fig. 6.11).

While the signal amplification required varies by electrode type, strong amplification is generally necessary, typically ranging from 5000 to 1,000,000. eCochG peak-to-peak (P2P) amplitudes usually fall between 500 and 2000 nanovolts (nV).

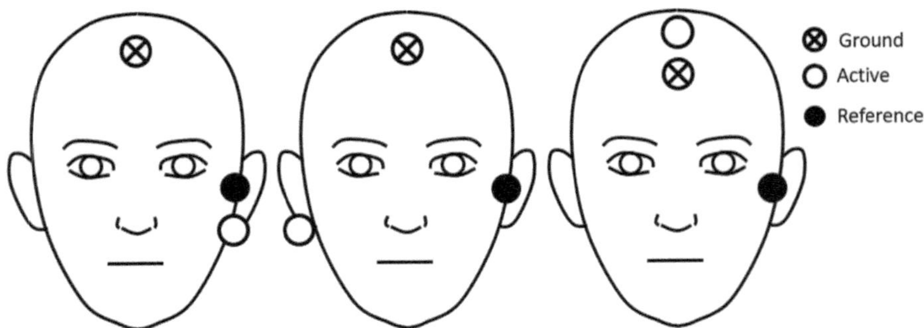

Fig. 6.10 Possible montages using three scalp electrodes for EC eCochG recording. (The left and middle configurations are taken from Carpi and Migliorini (2009), while the right configuration is recommended in the Interacoustics® Eclipse Instruction for Use)

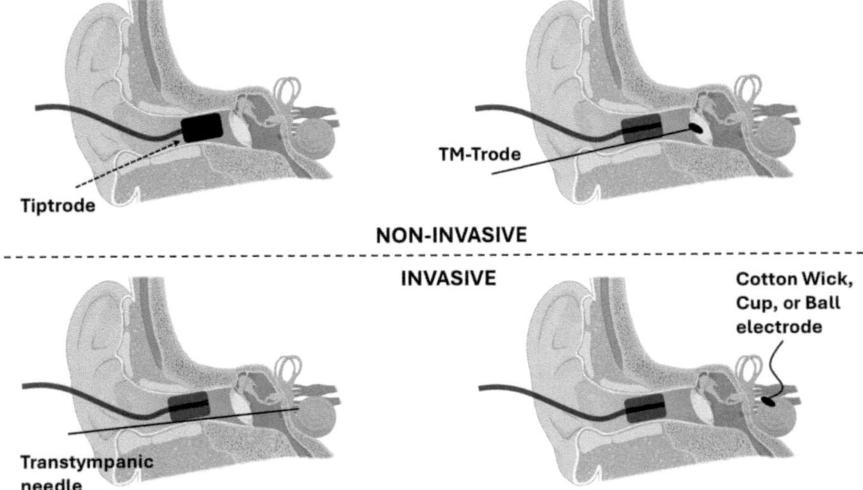

Fig. 6.11 Schematic representation of electrode placements: tiptrode (top left), TM-trode (top right), transtympanic (TT) needle (bottom left), and small-surface cotton wick, cup, or ball electrode (bottom right). (The anatomical illustration of the ear is adapted from Gordon Betts et al. (2022) under CC BY license)

The placement of these electrodes relative to the cochlea differs significantly, and most are designed to be disposable:

- Tiptrode: A minimally invasive and widely used probe. It consists of a conductive metallic (gold) coating wrapped around a foam earplug. It can be moistened with saline to improve conductivity. The tiptrode is considered one of the safest and least invasive options for eCochG recordings.

- TM-trode (or tymptrodes): This is a cup-shaped electrode (with variable shapes and surface areas) placed directly on the tympanic membrane (TM). While still noninvasive, its placement typically requires medical training or supervision.
- Transtympanic (TT) electrode: A sterile needle electrode that penetrates the eardrum to reach the cochlear promontory. This method is invasive and must be performed by a trained physician, often under anesthesia.
- Cotton wick, cup, or ball electrode: A small-surface sterile electrode placed directly on the cochlear promontory after tympanometry. Placement requires an ENT surgeon and is typically done under general anesthesia.

Because eCochGs originate from the cochlea, the proximity of the recording probe to it significantly impacts signal quality. While the amplitude of TM-trode measurements can be up to an order of magnitude larger than tiptrode measurements, transtympanic amplitudes can be far more than an order of magnitude larger than those on the eardrum. Small-surface cotton wick, cup, or ball electrodes may generate even larger-amplitude signals. It is thus suggested that extratympanic measurements are best classified as far-field and transtympanic measurements as near-field (Pienkowski et al. 2018). Recording far-field eCochG with a tiptrode in patients with severe to profound hearing loss may be nearly impossible. Depending on the circumstances, large transtympanic electrodes may be needed to obtain the strongest extracochlear eCochG signals.

6.3.5 Intracochlear eCochG Recording

In intracochlear (IC) eCochG, the cochlear implant (CI) itself serves as the recording system. In many respects, IC eCochG recording setups are simpler compared to EC eCochGs. An intracochlear electrode, typically the most apical one which may cause trauma during array insertion, is used as the probe. It is connected to an external reference electrode, and the CI circuitry is switched to record electrical potentials. This is similar to the mode used to record electrically evoked compound action potentials (eCAPs) with the CI. The recording time is set to a much longer period (~12–20 ms for eCochG rather than ~4–6 ms for eCAP), so the CI processor needs enough memory to record such long-period signal with good enough sample rate.

Unlike EC eCochGs, IC eCochG does not require artifact removal methodologies akin to external, small electrical potential recording techniques. Because the probe is closest to the origin of the electrical potentials, the amplitude of the response is higher, typically ranging from approximately 2000 to 20,000 nV peak-to-peak (P2P).

6.4 Monitoring Electrode Array Insertion During Surgery: Normal Versus Traumatic Insertion

6.4.1 Extracochlear eCochG Monitoring

6.4.1.1 Behavior of eCochGs with Insertion Depth

Since EC eCochGs are accompanied by a fixed probe placement situated outside the cochlea, the eCochG recordings (1) mostly originate from basal hair cell responses; (2) do not depend on the presence of a CI electrode, meaning that CM can be recorded on a non-operated cochlea (with no cochlear opening and no presence of an electrode array) to provide a useful baseline for identifying response drops; and (3) are only affected by the traumatic presence of the CI electrode array (Fig. 6.12).

6.4.1.2 Advantages of EC eCochGs

There are four primary advantages associated with acquiring eCochG using an EC setup, being availability of the material, baseline measurement, stability of the measurement, and compatibility with other types of objective measures:

- Numerous commercially systems specialized in acquiring very small electrical signals are available. Some also offer embedded stimulation, meaning only one compact setup is required for all functions. Such systems typically offer extensive stimulation and recording capabilities.
  - Stimulation: No triggering needed, configurable waveforms
  - Recording: Filtering, windowing, automatic stop feature, and extended recording capabilities
- Before conducting EC eCochG monitoring, a baseline measurement can be taken without the presence of the electrode array to facilitate easy detection of eCochG responses. This baseline can be used as a reference to detect any response drop.
- CM amplitude is generally stable during electrode insertion (unless trauma occurs) because the recording probe is not moving into the cochlea, as it would be in an IC setup.

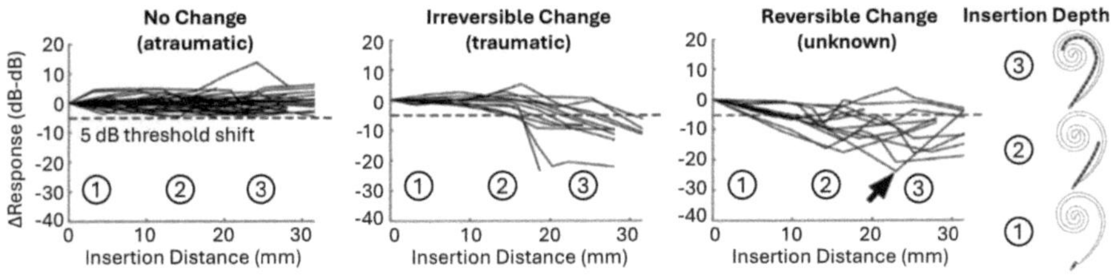

Fig. 6.12 Example of EC eCochG recorded at three different insertion depths noted 1, 2, and 3. Giardina et al. (2018) sorted eCochGs into three categories as compared to the baseline: those with no change (left), supposedly from atraumatic insertion, those with an irreversible change (middle) from a traumatic insertion, and those with a reversible change (right). (Figure adapted with modification from Giardina et al. (2018), with permission from Wolters Kluwer Health, Inc.)

- The CI can operate concurrently with other types of objective measures. For instance, simple impedance measurement of an Oticon Medical CI system in distributed all-polar (DAP) configuration (Stahl et al. 2022) could be used to determine the number of electrodes inserted.

6.4.2 Intracochlear eCochG Monitoring

6.4.2.1 Behavior of eCochGs with Insertion Depth

Intracochlear eCochG recordings involve a moving probe, which influences CM responses in several key ways:

- No response is detected until the CI electrode array is sufficiently inserted into the cochlea, meaning that a baseline response from a healthy cochlea cannot be established at the start of the insertion.
- As the electrode array advances toward the cochlear apex, CM amplitude increases, due to the probe approaching the region of the targeted hair cells, enhancing signal strength.
- The CM is a composite response, initially dominated by basal hair cells. As the probe progresses deeper, the contribution from basal hair cells diminishes, while the response from apical (targeted) hair cells becomes more prominent.
- In case of a traumatic insertion, the CM drops as shown in Fig. 6.13. The eCochG can be monitored and analyzed continuously using automatic methods for the early and robust detection of trauma (Wijewickrema et al. 2022).

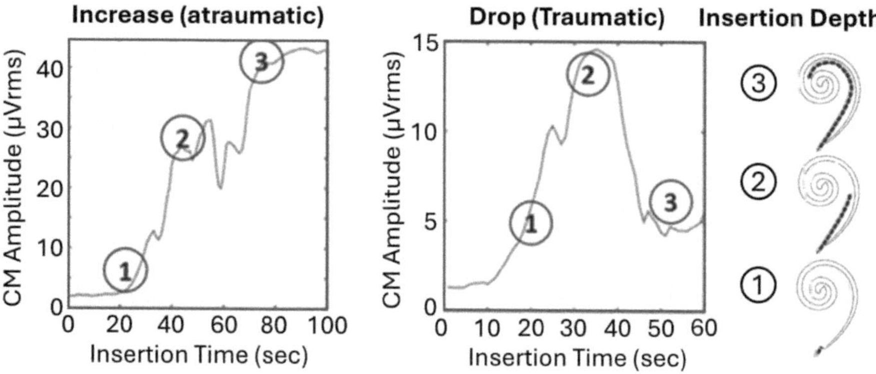

Fig. 6.13 Example of IC eCochG recorded at three different insertion depths noted 1, 2, and 3. Campbell et al. (2016) showed two possible behaviors: those with pseudo-steady increase to full insertion, sometimes reaching a plateau (left) supposedly from a traumatic insertion, and those with a major drop of response (right) from a traumatic insertion. (Figure adapted with modification from Campbell et al. (2016), with permission from Wolters Kluwer Health, Inc.)

6.4.2.2 Advantages of IC eCochGs

There are three primary advantages associated with acquiring eCochG using an IC setup: easiness of electrode placement, eCochG response strength, and postoperative noninvasiveness.

The cochlear implant is used as the recording system; no extended surgery time is needed from the placement of additional EEG recording electrodes such as in EC eCochGs.

- The eCochG quality is also not dependent on a multiple recording electrodes placement.
- The resulting CM amplitude response is generally larger than in EC eCochGs (Campbell et al. 2016), although more complex to interpret.
- Postoperative noninvasive eCochGs can easily be made to evaluate residual hearing.

6.5 Advanced eCochG Methods

6.5.1 Robotic Assisted

Current clinical trials involving robotic surgery are focusing on minimally invasive cochlear implantation techniques. eCochG may be effectively integrated with robot-assisted procedures, particularly when using intra-cochlear recording methods: First, eCochG signals typically exhibit a delay of up to 1 s between the hair cell-evoked potential and its display on the monitoring system. To avoid traumatic insertions, the electrode must be inserted as smoothly as possible, with rapid responses to any sudden drops in eCochG amplitude—something that may be too demanding for manual control. Second, since intra-cochlear eCochG uses the most apical electrode of the cochlear implant as a moving probe, signal drops are more predictable when the insertion is controlled, constant, and smooth by a robot. Pilot studies have already begun to demonstrate promising results in this area (Gawęcki et al. 2022).

6.5.2 Multi-frequency eCochGs

IC eCochGs can capture electrical potentials coming from the responding hair cells. From previous works (Polak et al. 2022; Barnes et al. 2020), a study used different stimulation frequencies in the range of the patient's residual hearing to evaluate any trauma cause in the cochlea, at different insertion depths (Skarżyński et al. 2022). This case report showed promising results.

6.5.3 Bone-Anchored Stimulated eCochG

Up to date, the only available stimulation is coming from a speaker from which the sound is conveyed to the eardrum with a silicone tube. This comes with some inconvenience during a CI surgery. The ear receiving the sound is ipsilateral to the CI, so high-level watch shall be made on the sterility level during the patient's preparation. Moreover, the earplug shall be inserted with caution because the ear is sealed during the patient's preparation phase. Any misplacement/displacement of the earplug will result in the loss of eCochG monitoring. If care is not taken during placement of the ear foam, complications during the surgery may happen. Small quantities of blood may flood inside the ear canal and obstruct the stimulation tube.

The acoustical stimulation may be replaced with bone vibrations that could be placed anywhere on the skull, further away from the surgical area (Harder et al. 1983; Saoji et al. 2022). Stahl et al. (2021) introduced an eCochG acquisition setup utilizing bone vibration. An animal experiment demonstrated the feasibility of this concept. In the study, bone vibrator stimulation was combined with IC eCochG recordings using a prototype of the Neuro Zti implant (Oticon Medical, Smørum, Denmark) in a guinea pig. For comparison, conventional acoustically induced eCochG recordings were also obtained from the same animal.

Figure 6.14 illustrates the basic setup for both bone vibration and acoustic stimulation. Acoustic stimulation is controlled by a personal computer (PC), and

Acoustic-induced eCochG setup

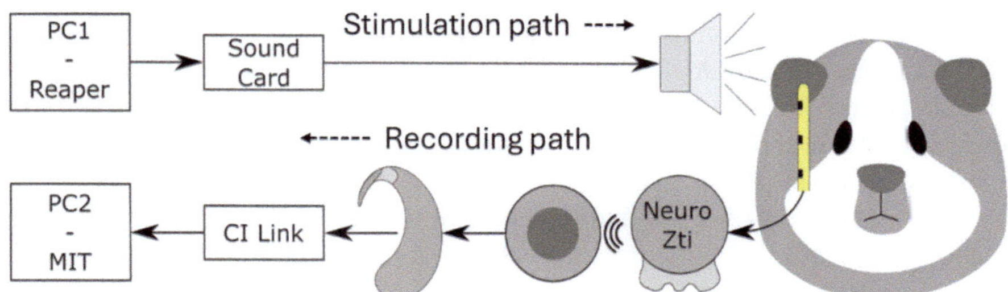

Bone-conduction-induced eCochG setup

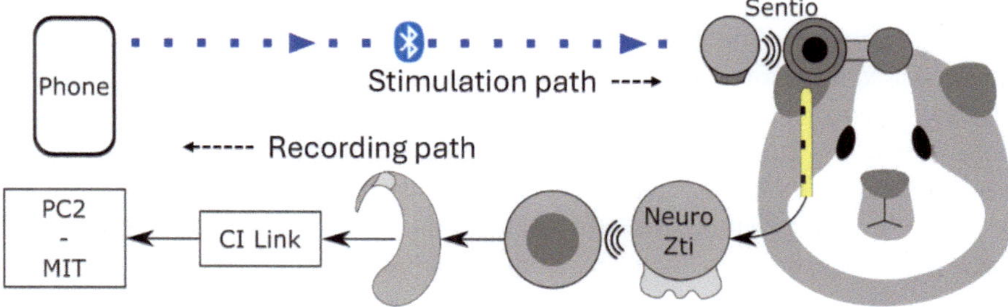

Fig. 6.14 Acoustic stimulation and bone vibration stimulation setups. The bone vibrator is sealed on a socket to allow vibrations to be conducted on the largest surface possible

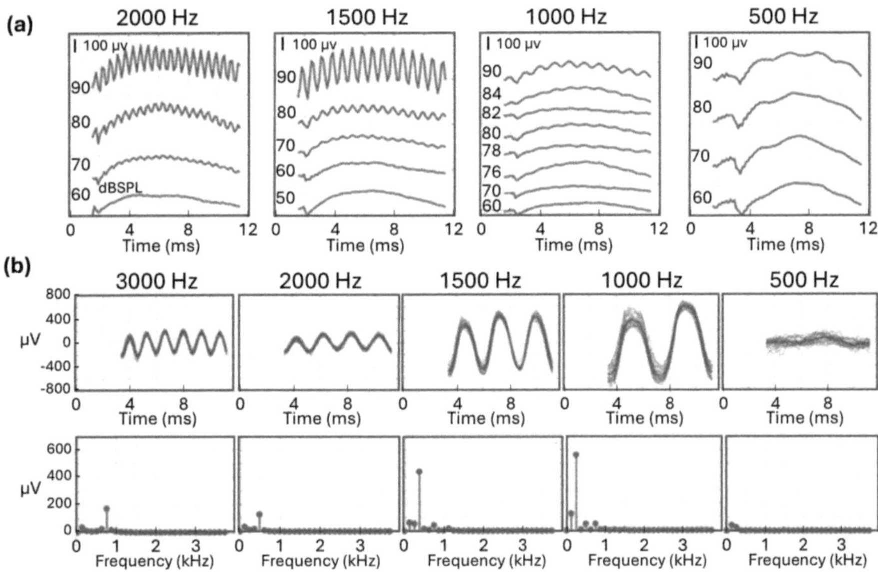

Fig. 6.15 (**a**) Baseline IC eCochG recordings in a guinea pig using four different acoustic input frequencies (500 Hz, 1000 Hz, 1500 Hz, and 2000 Hz) at varying intensity levels. (**b**) IC eCochG recordings from the same animal following bone conduction stimulation. The equivalent dB SPL delivered to the ear is unknown. Notably, the recorded eCochG response frequency was unexpectedly one-quarter of the stimulus frequency

continuous acoustic stimulation is sent by the Reaper software via a sound card (StarTech.com—ICUSBAUDIO2D). This stimulation is transmitted through a modified Neuro 2 case incorporating a CS45 input connection to a speaker (Oticon speaker Unit Power 3L, with a maximum output of 132 dB SPL) snugly fitted in the animal's ear canal. Bone-anchored stimulation is controlled with preloaded tone bursts from a smartphone, with information sent via Bluetooth to the Sentio™ bone anchor system. eCochG recordings are obtained from the most apical electrode of a prototype animal electrode array containing six 0.3 mm diameter electrodes. Retro feedback information is sent from the CI to the speech processor and analyzed using custom Matlab codes.

The results of the pilot study (see Fig. 6.15) indicate the feasibility of recording eCochG from bone vibrations, with a lower level of noise observed compared to acoustical stimulations. However, further investigations are necessary due to challenges encountered, such as difficulty in managing the stimulation level of bone vibrations. This difficulty arises because bone-anchored systems incorporate equivalency for vibration strengths to hearing levels in humans, which may not apply to guinea pigs with cemented sockets. Additionally, a consistent finding is that the IC eCochG responses are approximately one-quarter of the stimulus frequency.

6.6 Existing Tools and Software

6.6.1 EC eCochG Equipment

Any system capable of recording very small potentials can be used. In academic research, high-quality recording hardware is typically employed, such as equipment from Tucker-Davis Technologies (TDT, Alachua, FL, USA). These systems must be integrated to enable stimulus-triggered, amplified recordings across multiple channels. A typical setup might include a workstation (TDT-SW4), signal processor (TDT-RZ6), amplifier (TDT-RA4PA), and insert earphones (ER-3C, Etymotic Research, Elk Grove, IL, USA), all operated via TDT's Synapse software.

For medical diagnostics, more user-friendly systems are available for eCochG recordings. These typically consist of a recording unit with built-in amplifiers and a speaker for stimulus delivery. Note that disposable components are usually provided separately.

Several systems offer versatile options for recording eCochG (and EEG) potentials. For example, the Eclipse system (Interacoustics® A/S, Middelfart, Denmark) uses EPxx software[1] with dedicated eCochG programming. Similarly, the Neuro Audio system (Neuromed) operates with Neuro-Audio software. Other systems, such as the Elite (Echodia) with Elios software, prioritize portability. Note that user interfaces may offer many customizations for stimulation and may be complex for use without training.

6.6.2 IC eCochG Equipment

Intracochlear eCochG recordings are entirely dependent on the solutions provided by CI manufacturers (see Fig. 6.16). This section presents some currently available options for IC eCochG recordings, offered by cochlear implant manufacturers Advanced Bionics LLC, Cochlear Ltd., and MED-EL GmbH.

Advanced Bionics has introduced the AIM™ System, which stands for Active Insertion Monitoring. It is a commercially available, fully integrated system designed for clinical use of IC eCochG to monitor CI electrode array insertion. The AIM™ System enables stimulation monitoring and provides a real-time display of post-processed eCochG data from Advanced Bionics CI devices. In this system, a sophisticated tablet delivers both electrical and acoustic stimulation while also processing and visualizing the acquired eCochG signals.

In the Cochlear Ltd. solution, the eCochG software is installed on a computer. Stimulation data—both electrical and acoustic—is transmitted from the computer to the programming pod. A sound processor (e.g., CP920) with an acoustic component

[1] https://www.interacoustics.com/images/files/manuals/D-0120572-H_2025_01_Additional_Information_-_Eclipse_OAE_Suite.pdf

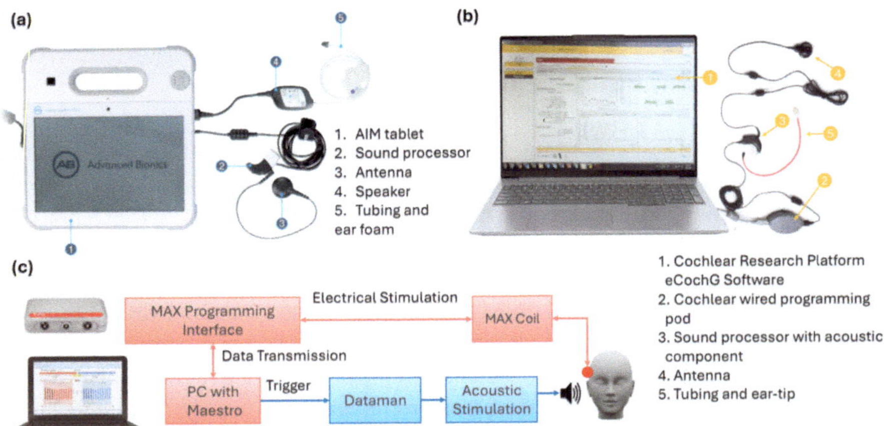

Fig. 6.16 Various setups for obtaining real-time IC eCochG signals across three different CI brands. (**a**) The AIM™ System by Advanced Bionics. (Image © 2025, provided courtesy of Advanced Bionics LLC). (**b**) The Cochlear Research Platform eCochG software. (Image © Cochlear Limited, provided courtesy of Cochlear Limited). (**c**) A setup compatible with MED-EL implants, consisting of the Maestro software, MAX programming interface, and an external acoustic stimulation system such as the Dataman system. (Image adapted with modification from Skarżyński et al. (2022), under CC BY license)

(e.g., EAC200) then delivers the electrical and acoustic stimulation to the patient via an antenna and an ear-tip, respectively.

In the MED-EL system, the Maestro software communicates with the MAX programming interface (MED-EL GmbH, Austria), which transmits electrical stimulation to the implant via the MAX coil. Maestro also provides a trigger signal to synchronize acoustic stimulation generated by an external device (e.g., the Dataman system by Dataman Programmers Ltd., Dorset, UK). This trigger ensures that both electrical and acoustic stimuli are delivered synchronously to the patient.

Acknowledgments The author would like to thank Dr. Jean-Marc Edeline and Dr. Elie Partouche of the NeuroPSI Institute at Paris-Saclay University for providing the components necessary for the acquisition of animal data.

The author acknowledges the use of Microsoft Copilot for proofreading and enhancing the readability of the chapter. All suggested changes were carefully reviewed and edited. The author assumes full responsibility for the content of the final publication.

References

Adrian E (1931) The microphonic action of the cochlea: an interpretation of Wever and Bray' experiments. In: Report of a discussion on audition. Physical Society of London, London, pp 5–9

Adunka OF, Giardina CK, Formeister EJ, Choudhury B, Buchman CA, Fitzpatrick DC (2016) Round window electrocochleography before and after cochlear implant electrode insertion. Laryngoscope 126:1193–1200. https://doi.org/10.1002/lary.25602

Aran JM (1971) The electro-cochleogram. Recent results in children and in some pathological cases. Arch Klin Exp Ohren Nasen Kehlkopfheilkd 198(1):128–143

Attias J, Ulanovski D, Hilly O, Greenstein T, Sokolov M, HabibAllah S, Mormer H, Raveh E (2020) Postoperative intracochlear electrocochleography in pediatric cochlear implant recipients: association to audiometric thresholds and auditory performance. Ear Hear 41(5): 1135–1143. https://doi.org/10.1097/AUD.0000000000000833

Barnes JH, Yin LX, Saoji AA, Carlson ML (2020) Electrocochleography in cochlear implantation: development, applications, and future directions. World J Otorhinolaryngol Head Neck Surg 7(2):94–100. https://doi.org/10.1016/j.wjorl.2020.04.006. PMID: 33997718; PMCID: PMC8103527

Bento RF, Danieli F, Magalhães AT, Gnansia D, Hoen M (2016) Residual hearing preservation with the Evo® cochlear implant electrode array: preliminary results. Int Arch Otorhinolaryngol 20(4):353–358. https://doi.org/10.1055/s-0036-1572530

Bester CW, Campbell L, Dragovic A, Collins A, O'Leary SJ (2017) Characterizing electrocochleography in cochlear implant recipients with residual low-frequency hearing. Front Neurosci 23(11):141. https://doi.org/10.3389/fnins.2017.00141

Blamey P, Artieres F, Başkent D et al (2012) Factors affecting auditory performance of postlinguistically deaf adults using cochlear implants: an update with 2251 patients. Audiol Neurotol 18(1):36–47. https://doi.org/10.1159/000343189

Bramhall NF (2021) Use of the auditory brainstem response for assessment of cochlear synaptopathy in humans. J Acoust Soc Am 150(6):4440. https://doi.org/10.1121/10.0007484

Calloway NH, Fitzpatrick DC, Campbell AP et al (2014) Intracochlear electrocochleography during cochlear implantation. Otol Neurotol 35:1451–1457. https://doi.org/10.1097/MAO.0000000000000451

Campbell AP, Suberman TA, Buchman CA, Fitzpatrick DC, Adunka OF (2010) Correlation of early auditory potentials and intracochlear electrode insertion properties: an animal model featuring near real-time monitoring. Otol Neurotol 31(9):1391–1398. https://doi.org/10.1097/MAO.0b013e3181f6c899. PMID: 20856155; PMCID: PMC3051407

Campbell L, Kaicer A, Briggs R, O'Leary S (2014) Cochlear response telemetry: intracochlear electrocochleography via cochlear implant neural response telemetry. Oto Neurotol 36:399–405. https://doi.org/10.1097/MAO.0000000000000678

Campbell L, Kaicer A, Sly D, Iseli D, Wei B, Briggs R, O'Leary S (2016) Intraoperative real-time cochlear implant telemetry predicts hearing preservation in cochlear implantation. Otol Neurotol 37:332–338. https://doi.org/10.1097/MAO.0000000000000972

Campbell L, Bester C, Iseli C, Sly D, Dragovic A, Gummer AW, O'Leary S (2017) Electrophysiological evidence of the basilar-membrane travelling wave and frequency place coding of sound in cochlear implant recipients. Audiol Neurotol 22(3):180–189. https://doi.org/10.1159/000478692

Carpi F, Migliorini S (2009) Non-invasive wet electrocochleography. IEEE Trans Biomed Eng 56: 2744–2747. https://doi.org/10.1109/TBME.2009.2026178

Choudhury B, Fitzpatrick DC, Buchman CA (2012) Intraoperative round window recordings to acoustic stimuli from cochlear implant patients. Otol Neurotol 33:1507–1515. https://doi.org/10.1097/MAO.0b013e31826dbc80

Dalbert A, Sim JH, Gerig R, Pfiffner F, Roosli C, Huber A (2015) Correlation of electrophysiological properties and hearing preservation in cochlear implant patients. Otol Neurotol 36:1172–1180. https://doi.org/10.1097/MAO.0000000000000768

Dalbert A, Pfiffner F, Hoesli M, Koka K, Veraguth D, Roosli C, Huber A (2018) Assessment of cochlear function during cochlear implantation by extra- and intracochlear electrocochleography. Front Neurosci 26(12):18. https://doi.org/10.3389/fnins.2018.00018

Dalbert A, Sijgers L, Grosse J, Veraguth D, Roosli C, Huber A, Pfiffner F (2021) Simultaneous intra- and extracochlear electrocochleography during electrode insertion. Ear Hear 42(2): 414–424. https://doi.org/10.1097/AUD.0000000000000935

Dallos P, Schoeny ZG, Cheatham MA (1972) Cochlear summating potentials descriptive aspects. Acta Otolaryngol Suppl 302:1–46

Davis H (1968) Mechanisms of the inner ear. Ann Otol Rhinol Laryngol 77:644–656. https://doi.org/10.1177/000348946807700405

Davis H, Derbyshire M, Lurie MH, Saul LJ (1934) The electric response of the cochlea. Am J Phys 102(2):311–332

Eggermont JJ (1979) Summating potentials in Meniere's disease. Arch Otorhinolaryngol 222:63–75. https://doi.org/10.1007/BF00456340

Fontenot TE, Giardina CK, Fitzpatrick DC (2017) A model-based approach for separating the cochlear microphonic from the auditory nerve neurophonic in the ongoing response using electrocochleography. Front Neurosci 11:Article 592. https://doi.org/10.3389/fnins.2017.00592

Fontenot TE, Giardina CK, Dillon M et al (2019) Residual cochlear function in adults and children receiving cochlear implants: correlations with speech perception outcomes. Ear Hear 40(3): 577–591. https://doi.org/10.1097/AUD.0000000000000630

Formeister EJ, McClellan JH, Merwin WH, Iseli CE, Calloway NH, Teagle HFB, Buchman CA, Adunka OF, Fitzpatrick DC (2015) Intraoperative round window electrocochleography and speech perception outcomes in pediatric cochlear implant recipients. Ear Hear 36(2):249–260. https://doi.org/10.1097/AUD.0000000000000106

Fromm B, Nylen CO, & Zotterman Y (1935) Studies in the mechanism of the Wever and Bray effect. Acta oto-laryngologica, 22(3):477–486. https://doi.org/10.3109/00016483509118125

Gawęcki W, Balcerowiak A, Podlawska P, Borowska P, Gibasiewicz R, Szyfter W, Wierzbicka M (2022) Robot-assisted electrode insertion in cochlear implantation controlled by intraoperative electrocochleography—a pilot study. J Clin Med 11(23):7045. https://doi.org/10.3390/jcm11237045

Giardina CK, Khan TE, Pulver SH et al (2018) Response changes during insertion of a cochlear implant using extracochlear electrocochleography. Ear Hear 39(6):1146–1156. https://doi.org/10.1097/AUD.0000000000000571

Giardina CK, Brown KD, Adunka OF (2019) Intracochlear electrocochleography: response patterns during cochlear implantation and hearing preservation. Ear Hear 40:833–848. https://doi.org/10.1097/AUD.0000000000000659

Gibson WP (2017) The clinical uses of electrocochleogrphy. Front Neurosci 11:274. https://doi.org/10.3389/fnins.2017.00274

Gordon Betts J, Desaix P, Johnson E et al (2022) Anatomy_and_physiology 2e. OpenStax™. Digital version ISBN-13: 978-1-951693-42-8. https://openstax.org/details/books/anatomy-and-physiology-2e

Haapaniemi EL, Johansson R, Seppo Karjalainen J (2000) Transtympanic versus tympanic membrane electrocochleography in examining cochleovestibular disorders. Acta Otolaryngol 120(545):127–129. https://doi.org/10.1080/000164800454189

Harder H, Arlinger S, Kylen P (1983) Electrocochleography with bone-conducted stimulation: a comparative study of different methods of stimulation. Acta Otolaryngol 95(1–4):35–45. https://doi.org/10.3109/00016488309130913

Harris R, Cruise A, Gibson W, Bate K, Sanli H (2011) Preliminary results and technique for electrophysiological intra-operative monitoring of residual hearing during cochlear implantation. Cochlear Implants Int 12:209–215. https://doi.org/10.1179/146701011X12950038111657

Harris MS, Riggs WJ, Koka K, Litvak LM, Malhotra P, Moberly AC, O'Connell BP, Holder J, Di Lella FA, Boccio CM, Wanna GB, Labadie RF, Adunka OF (2017) Real-Time Intracochlear Electrocochleography Obtained Directly Through a Cochlear Implant. Otol Neurotol. 38(6): e107–e113. https://doi.org/10.1097/MAO.0000000000001425

Haumann S, Imsiecke M, Bauernfeind G, Büchner A, Helmstaedter V, Lenarz T, Salcher RB (2019) Monitoring of the inner ear function during and after cochlear implant insertion using electrocochleography. Trends Hear 23:1–18. https://doi.org/10.1177/2331216519833567

Hornibrook J (2017) Tone burst electrocochleography for the diagnosis of clinically certain Meniere's disease. Front Neurosci 11:301. https://doi.org/10.3389/fnins.2017.00301

Hoskison E, Mitchell S, Coulson C (2017) Systematic review: radiological and histological evidence of cochlear implant insertion trauma in adult patients. Cochlear Implants Int 18(4): 192–197. https://doi.org/10.1080/14670100.2017.1330735

Jiam NT, Jiradejvong P, Pearl MS, Limb CJ (2016) The effect of round window vs cochleostomy surgical approaches on cochlear implant electrode position: a flat-panel computed tomography study. JAMA Otolaryngol Head Neck Surg 142(9):873–880. https://doi.org/10.1001/jamaoto.2016.1512

Jwair S, Ramekers D, Thomeer Hans GXM, Versnel H (2023) Acute effects of cochleostomy and electrode-array insertion on compound action potentials in normal-hearing guinea pigs. Front Neurosci 17. https://doi.org/10.3389/fnins.2023.978230

Kong Y-Y, Carlyon RP (2010) Temporal pitch perception at high rates in cochlear implants. J Acoust Soc Am 125:1649–1657. https://doi.org/10.1121/1.3372713

Koyuncu M, Mason SM, Saunders MW (1994) Electrocochleography in endolymphatic hydrops using tone-pip and click stimuli. Clin Otolaryngol Allied Sci 19(1):73–78. https://doi.org/10.1111/j.1365-2273.1994.tb01152.x

Kumar P, Peepal P (2012) Electrocochleography in individuals with Meniere's disease. Indian J Otol 18(2):72–75

Lazard DS, Vincent C, Venail F et al (2012) Pre-, per- and postoperative factors affecting performance of postlinguistically deaf adults using cochlear implants: a new conceptual model over time. PLoS One 7(11):e48739. https://doi.org/10.1371/journal.pone.0048739

Lempert J, Meltzer PE, Wever EG, Lawrence M (1950) The cochleogram and its clinical application. Arch Otolaryngol 51:307–311

McClellan JH, Formeister EJ, Merwin WH 3rd, Dillon MT, Calloway N, Iseli C, Buchman CA, Fitzpatrick DC, Adunka OF (2014) Round window electrocochleography and speech perception outcomes in adult cochlear implant subjects: comparison with audiometric and biographical information. Otol Neurotol 35(9):245–252. https://doi.org/10.1097/MAO.0000000000000557

McDermott HJ (2004) Music perception with cochlear implants: a review. Trend Amplif 8(2): 49–82. https://doi.org/10.1177/108471380400800203

Nelson DA, Schmitz JL, Donaldson GS, Viemeister NF, Javel E (1996) Intensity discrimination as a function of stimulus level with electric stimulation. J Acoust Soc Am 100:2393–2414. https://doi.org/10.1121/1.417949

Nguyen LT, Harris JP, Nguyen QT (2010) Clinical utility of electrocochleography in the diagnosis and management of Meniere's disease: AOS and ANS membership survey data. Otol Neurotol 31:455–459. https://doi.org/10.1097/MAO.0b013e3181d2779c

Pappa AK, Hutson KA, Scott WC, Wilson JD, Fox KE, Masood MM, Giardina CK, Pulver SH, Grana GD, Askew C, Fitzpatrick DC (2019) Hair cell and neural contributions to the cochlear summating potential. J Neurophysiol 121(6):2163–2180. https://doi.org/10.1152/jn.00006.2019

Pienkowski M, Adunka OF, Lichtenhan JT (2018) Editorial: new advances in electrocochleography for clinical and basic investigation. Front Neurosci 8(12):310. https://doi.org/10.3389/fnins.2018.00310. PMID: 29867322; PMCID: PMC5951982

Polak M, Lorens A, Walkowiak A, Furmanek M, Skarzynski PH, Skarzynski H (2022) In vivo basilar membrane time delays in humans. Brain Sci 12(3):400. https://doi.org/10.3390/brainsci12030400

Radeloff A, Shehata-Dieler W, Scherzed A (2012) Intraoperative monitoring using cochlear microphonics in cochlear implant patients with residual hearing. Otol Neurotol 33:348–354. https://doi.org/10.1097/MAO.0b013e318248ea86

Ruben RJ, Lieberman AT, Bordley JE (1962) Some observations on cochlear potentials and nerve action potentials in children. Laryngoscope 72:545–554. https://doi.org/10.1288/00005537-196205000-00001

Santarelli R, Scimemi P, La Morgia C, Cama E, del Castillo I, Carelli V (2021) Electrocochleography in auditory neuropathy related to mutations in the OTOF or OPA1 gene. Audiol Res 11:639–652. https://doi.org/10.3390/audiolres11040059

Saoji AA, Graham MK, Adkins WJ, Nassiri AM, Neff BA, Carlson ML, Driscoll CLW (2022) Relationship between intraoperative electrocochleography responses and immediate postoperative bone conduction thresholds in cochlear implantation. Otol Neurotol 43(8):e880–e887. https://doi.org/10.1097/MAO.0000000000003620

Schuerch K, Waser M, Mantokoudis G et al (2022) Increasing the reliability of real-time electrocochleography during cochlear implantation: a standardized guideline. Eur Arch Otorrinolaringol 279:4655–4665. https://doi.org/10.1007/s00405-021-07204-7

Simpson MJ, Jennings SG, Margolis RH (2020) Techniques for obtaining high-quality recordings in electrocochleography. Front Syst Neurosci 15(14):18. https://doi.org/10.3389/fnsys.2020.00018. PMID: 32351368; PMCID: PMC7176302

Skarżyński PH, Loren A, Walkowiak A, Polak M, Skarżyński H (2022) Multi-frequency intraoperative monitoring of hearing preservation during cochlear implantation. Life 12:636. https://doi.org/10.3390/life12050636

Stahl P, Macherey O, Meunier S, Roman S (2016) Rate discrimination at low pulse rate in normal-hearing and cochlear implantees: influence of intracochlear stimulation site. J Acoust Soc Am 139:1578. https://doi.org/10.1121/1.4944564

Stahl P, Frater A, Segovia Martinez M, Gnansia D, Molaee B, Demarcy T (2021) Method for detecting a plurality of health conditions of a cochlea. United State patent application No. US 2021/0138236 A1

Stahl P, Dang K, Vandersteen C, Guevara N, Clerc M, Gnansia D (2022) Current distribution of distributed all-polar cochlear implant stimulation mode measured in-situ. PLoS One 17(10): e0275961. https://doi.org/10.1371/journal.pone.0275961

Tanaka F, Tsukasaki N, Nakao Y, Shigeno K, Kobayashi T (1999) Electrocochleographic evaluation of hearing loss in acoustic neuromas. Am J Otolaryngol 20:479–483

Tasaki I, Davis H, Eldredge D (1954) Exploration of cochlear potentials in guinea pig with a microelectrode. J Acoust Soc Am 26:765–773

Walia A, Shew MA, Kallogjeri D, Wick CC, Durakovic N, Lefler SM, Ortmann AJ, Herzog JA, Buchman CA (2022) Electrocochleography and cognition are important predictors of speech perception outcomes in noise for cochlear implant recipients. Sci Rep 12(1):3083. https://doi.org/10.1038/s41598-022-07175-7

Wever EG, & Bray CW (1930) The nature of acoustic response: The relation between sound frequency and frequency of impulses in the auditory nerve. Journal of experimental psychology, 13(5):373. https://doi.org/10.1037/h0075820

Wijewickrema S, Bester C, Gerard JM, Collins A, O'Leary S (2022) Automatic analysis of cochlear response using electrocochleography signals during cochlear implant surgery. PLoS One 17(7): e0269187. https://doi.org/10.1371/journal.pone.0269187

Electrically Evoked Compound Action Potentials (ECAPs) 7

Charlotte Garcia

Abstract

The electrically evoked compound action potential (ECAP) is a synchronous response of the auditory nerve to stimulation with a current pulse from a cochlear implant electrode. The cochlear implant platform offers a unique opportunity to record neural activity using hardware very close to the nerves, as the electrodes of the device used to stimulate and evoke an auditory percept can also be used to record the neural response. ECAP measures are used clinically, especially in cases where it is not possible to assess auditory perception behaviorally. They can provide insight into patient-specific patterns of neural responsiveness and the interface between the electrodes and the auditory nerves more generally. The ability of the ECAP to predict for biophysical or behavioral indicators of auditory perception is, however, limited by its recording techniques and the fact that it only captures the response at the very periphery of the auditory system. Nonetheless, it is the most widely used objective measure of neural activity in CI healthcare, especially as it is simple to measure clinically. This chapter covers methods for recording ECAPs, interpretation of the measurement waveform, relationships to other methods of assessing CI perception, and extended techniques for modeling the electrode-neuron interface.

Keywords

Evoked potential · Objective measure · Reverse telemetry · Neural health · Electrode-neuron interface · Spread of excitation

C. Garcia (✉)
Cambridge Hearing Group, MRC Cognition & Brain Sciences Unit, University of Cambridge, Cambridge, UK
e-mail: charlotte.garcia@mrc-cbu.cam.ac.uk

B. Molaee-Ardekani, M. Segovia-Martinez (eds.), *Advancing Cochlear Implants*,
https://doi.org/10.1007/978-3-032-15719-5_7

7.1 Introduction

Electrically evoked compound action potentials (ECAPs) have increasingly become the most widely used objective measures of neural activity in cochlear implants (CIs) in a clinical setting since the initial commercial availability of reverse telemetry systems with the Nucleus CI24M from Cochlear Ltd. in 1998. They are reflections of the synchronized peripheral neural response to electrical stimulation with the implant electrodes and correspond to the first wave of electrically evoked auditory brainstem response (EABR) recordings. EABRs are synchronous physiological responses from the auditory nerve to various structures in the brainstem and are characterized by five waves. EABRs have the advantage over ECAPs that they provide slightly more central neural recordings than ECAPs; the fifth wave of the EABR (Wave V) represents the response from the lateral lemniscus and inferior colliculus, whereas the ECAPs reflect only the most peripheral activity of the auditory nerve (Hall 2006). However, ECAPs have the advantage of being simpler measures to acquire than EABRs as they do not require placement of extra electrodes (Brown et al. 1994), nor do they require the CI user to keep their head movements to a minimum to avoid affecting the measurements with myogenic activity (Abbas et al. 1999). This is because the recordings leverage electrodes for stimulation, recording, and ground that are entirely internal to the CI user's head and are unaffected by neural activity related to things like head movement. They leverage only the hardware already existing within a CI for stimulation, and it is possible to record them using the clinical software available from each of the three major cochlear implant manufacturers: Advanced Bionics, Cochlear, and MED-EL. It is likely that the ECAP is the most widely used objective measure of neural responsiveness to a CI because of the ease and speed with which clinical personnel may record them.

This chapter first describes ECAP morphology and techniques for extracting the neural responses from associated stimulation artifacts before providing an overview of their relationships with various physiological and perceptual measures of auditory function with a cochlear implant. It then introduces a number of extended methods that exploit attributes of ECAP recording paradigms to model the electrode-neuron interface, especially spread of electrical excitation to surrounding tissues.

7.2 Recording and Interpreting ECAPs

7.2.1 Morphology and Neurological Representation

ECAPs are measured from the extra- and intra-cochlear electrodes of a cochlear implant and—as previously stated—represent the synchronous physiological response of the population of auditory nerve fibers excited by an electrical stimulus (Brown et al. 1990; Charlet de Sauvage et al. 1983). The standard stimulus is a monopolar, symmetric, biphasic current pulse presented on one electrode in the intra-cochlear electrode array. Their latencies are quite short in comparison to the most robust and primary wave of interest in EABR signals (Wave V), starting with a

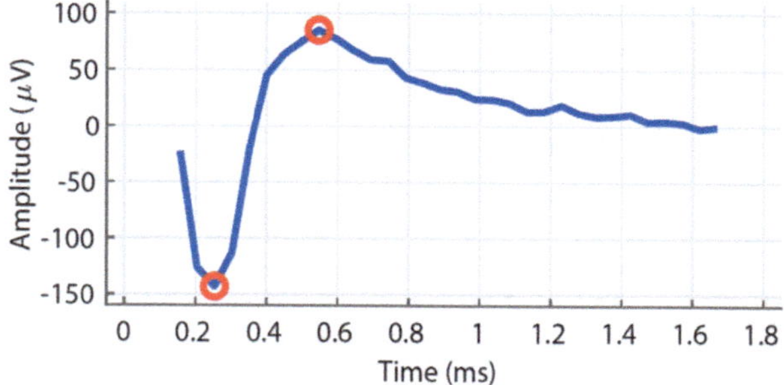

Fig. 7.1 An example of an electrically evoked compound action potential (ECAP) from a cochlear CI user recorded at the MRC Cognition & Brain Sciences Unit at the University of Cambridge. The waveform's morphology (the solid blue line) is characterized by a negative peak (N1) just after 0.2 ms and a subsequent positive peak (P2) just before 0.6 ms, indicated here by the red circles. *ms* milliseconds, μV microvolts

negative N1 peak at 0.2–0.4 milliseconds (ms) after the electrical stimulus followed by a positive P2 peak, as seen in Fig. 7.1 (Abbas et al. 1999). They are also several orders of magnitude larger than typical EABR peak-to-peak amplitudes, generally between tens of microvolts (μV) and ≈1 millivolt (mV) depending on the CI patient's loudness comfort levels (Hughes 2013).

It is important to note that ECAPs represent what has been referred to as the *whole-nerve* potential (Abbas et al. 1999). The morphology of the ECAP signal, characterized by a sharp negative peak followed by a wider positive one before returning to its resting potential, is similar to what would be expected from a signal-nerve cation potential recording when it is firing: the negative peak indicates the depolarization wherein positively charged particles cross the membrane of the neuron through an open channel causing the neuron to "fire" or "spike," and the positive peak indicates a period of hyperpolarization before the nerve returns to its resting potential (Purves et al. 2012). The ECAP we observe is representative of not just one nerve but the sum of each of the potentials of the whole *population* of nerves that respond to an electrical stimulus with a CI. When responding to an electrical stimulus individual nerves may not spike at the same time, creating jitter in the timing of their individual action potential peaks. Because we only record the *sum* and not the individual nerve potentials, we do not directly capture this jitter, but instead the recorded signal is dominated by spiking activity that is synchronized with each other and may be temporally broadened by any asynchronous activity. For this reason, we describe ECAPs as measuring the *synchronized* peripheral neural responsiveness to electrical stimulation through a cochlear implant. The "peripheral" portion of this sentence refers to the fact that we only capture the activity of the auditory nerve itself, without any contribution of additional neural structures further up in the auditory pathway such as from the brainstem.

Since the introduction of the commercial reverse telemetry system, there have been a number of studies looking at the ECAP measurements both in humans and animals to evaluate what parameters affect the signal. Most animal studies that investigate ECAPs include implantation of short, experimental cochlear implants (these generally have between one and eight electrodes) and have allowed the comparison of ECAP metrics to things such as spiral ganglion nerve (SGN) density calculated via harvesting and imaging of the neural tissue following objective and/or behavioral experiments (Prado-Guitierrez et al. 2006; Ramekers et al. 2014). However, there are differences between humans and animal models, such as the effective stimulating polarity; in most animal studies, the negative or cathodic phase of the stimulating biphasic current pulse excites the SGN tissue more effectively than the positive or anodic phase, whereas this is the reverse in humans (Macherey et al. 2008; Undurraga et al. 2010). One possible reason for this comes from the manner of deafening: animal studies often employ acute deafening and evaluation of responses subsequently to implanting cochlear implant arrays, whereas humans often receive their implants after prolonged periods of different types of hearing loss. The differences in the process of deafening could affect different neural structures differently, and there is evidence that different phases of stimulation preferentially stimulate different portions of the auditory nerve (i.e., the peripheral processes vs. the cochlear nucleus) (Brochier et al. 2021a). Besides the differences in residual acoustic hearing processes and/or the health of the auditory nerve between individual human CI patients, there is evidence that SGN health declines after even short periods of hearing loss, possibly partially due simply to the lack of signal input to the relevant neural pathways and resulting brain plasticity (Lewis and Steel 2010; Steel 2000). There is mixed evidence for this, however, as Macherey and Cazals found no effect of duration of deafness on polarity sensitivity in Guinea pigs (Macherey and Cazals 2016). There is also evidence that neural anatomy and electrode-modiolus distances (distances between an electrode and their target neurons calculated from computerized tomography or CT scans) influence the polarity effect (Kalkman et al. 2022; Rattay et al. 2001).

7.2.2 Artifact Cancellation Techniques

One of the obstacles to recording ECAPs is that there is a significant electrical artifact that occurs as the amplifier switches on prior to recording. This is both larger in magnitude than the subsequent neural response and overlapping in latency; so therefore it must be detangled from the recorded waveform in order to extract the (neural) signal of interest. There are multiple artifact cancellation techniques—also sometimes referred to as artifact reduction methods—that take advantage of various properties of the stimulated neuronal population.

7.2.2.1 The Alternating Polarity Method

Perhaps the most commonly employed method is the alternating polarity artifact cancellation technique. This method takes advantage of the electrical properties of the artifact itself and is available for recording ECAPs in all three of the major cochlear implant manufacturer's clinical software. When the polarity of a stimulating pulse reverses, so does the polarity of the stimulus artifact. However,

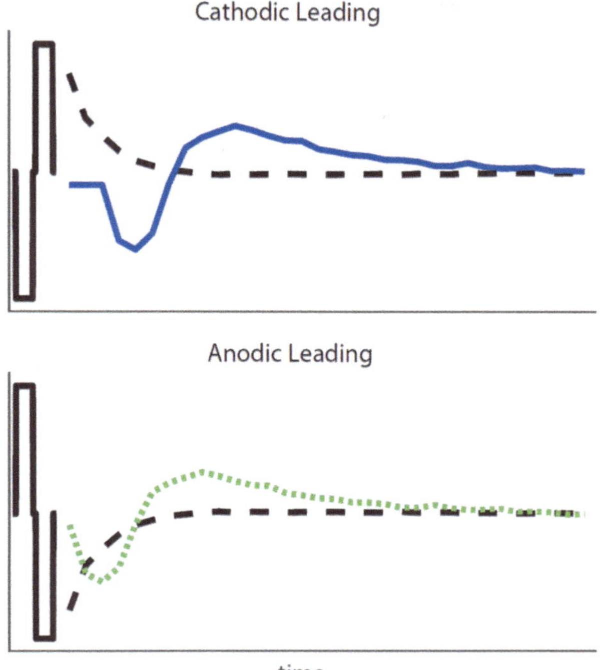

Fig. 7.2 The two primary frames of the ECAP involved in the alternating polarity artifact cancellation technique. The solid black lines on the left indicate the biphasic stimulating current pulses (the y-axis in this case is measured in units of current), and the dashed black lines indicate the electrical stimulus artifact (in units of voltage) as a result of the stimulating current pulses. *Top:* The solid blue line indicates the neural response to the cathodic-leading stimulation pulse (in units of voltage) that is orders of magnitude smaller than the overlaid electrical artifact. *Bottom:* The dotted green line similarly indicates the neural response to the anodic-leading stimulation pulse. Note that this represents a case where the effective phase has an anodic polarity (as in humans), and this is why the neural response to the cathodic-leading pulse is delayed in time compared to the neural response to the anodic-leading pulse

the neural response's polarity does not. Therefore, two biphasic current pulses can be presented with alternating leading polarity, the responses recorded, and the two responses averaged together to cancel out the artifact. This is depicted graphically in Fig. 7.2.

A disadvantage to this method, however, is that since there is a primarily effective phase in a biphasic current pulse (Macherey et al. 2008; Undurraga et al. 2010), the responses to the two pulses will have differences in latency for cathodic-leading- vs. anodic-leading pulses (Miller et al. 1998), temporally separated by the phase duration (PD) plus the interphase gap (IPG). This means that the shape of the measured and averaged waveform will have a reduced peak-to-peak amplitude compared to either the true cathodic-leading or anodic-leading case alone, since it is smoothed over time (depicted graphically in Fig. 7.3). However, one characteristic of the alternating polarity method of removing the stimulation artifact is that it does

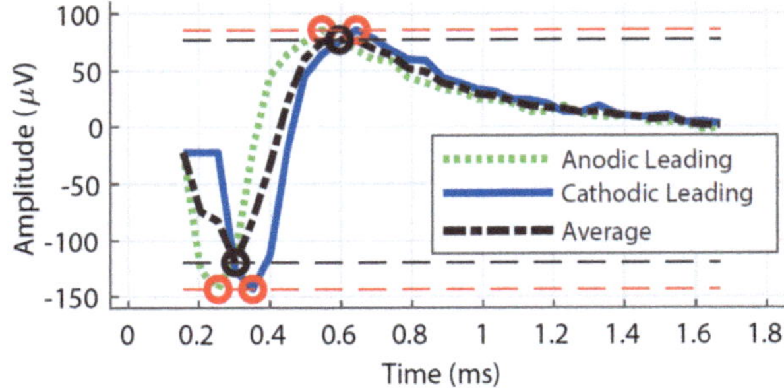

Fig. 7.3 Graphical representation of the latency effect of recording ECAPs using the alternating polarity artifact cancellation technique. The dotted green and solid blue lines represent the neural responses to anodic-leading and cathodic-leading biphasic stimuli, respectively, unobscured by the stimulus artifact, and assuming a phase duration + inter-phase gap (PD + IPG) length of approximately 100 μs. The dash-dotted black line represents the average of the two waveforms at the different latencies. The resultant ECAP amplitude for the averaged waveform (the height between black horizontal dashed lines) is calculated at approximately 196 μV, whereas the true amplitude (the height between the red horizontal dashed lines) is approximately 228 μV. This underestimation effect is simulated here by taking a recorded ECAP waveform and plotting the same waveform again with a latency of 100 μs; the nature of the stimulus artifacts does not allow extraction of the neural response waveforms with different latencies to be extracted from their artifacts separately. *ms* milliseconds, *μs* microseconds, *μV* microvolts

not rely on refractory properties of the neural population. This is the case for the forward masking method described next.

7.2.2.2 The Forward Masking Method

The second most commonly used forward masking artifact cancellation technique takes advantage of the refractory properties of the neural population in order to remove the stimulation artifact and extract the neural response. There are four frames involved in this paradigm, first described by Charlet de Sauvage and later Brown (Brown et al. 1990; Charlet de Sauvage et al. 1983), with different stimulating conditions: (A) a probe, (B) a masker followed by the probe, (C) the masker alone, and (D) non-stimulation. The paradigm requires that the time interval between the masker pulse and the probe pulse (referred to as the masker-probe interval—MPI) be sufficiently short such that the neural population that would normally be excited by the probe pulse is still in a fully refractory state in the masker + probe condition (B). The signal from condition A should result in the stimulation artifact from the probe and the neural response to the probe. In the case where the masker has successfully masked the neural response of the probe, then condition B should result in the stimulation artifact from the masker, the stimulation artifact from the probe, the neural response to the masker, but crucially *not* the neural response to the probe. Condition C will result in the stimulation artifact from the masker and the neural response to the masker, and condition D will only contain the system signature, or any electrical artifact that results any time a recording is taken (this would also be present in each of the other three conditions). The neural response to the probe can be

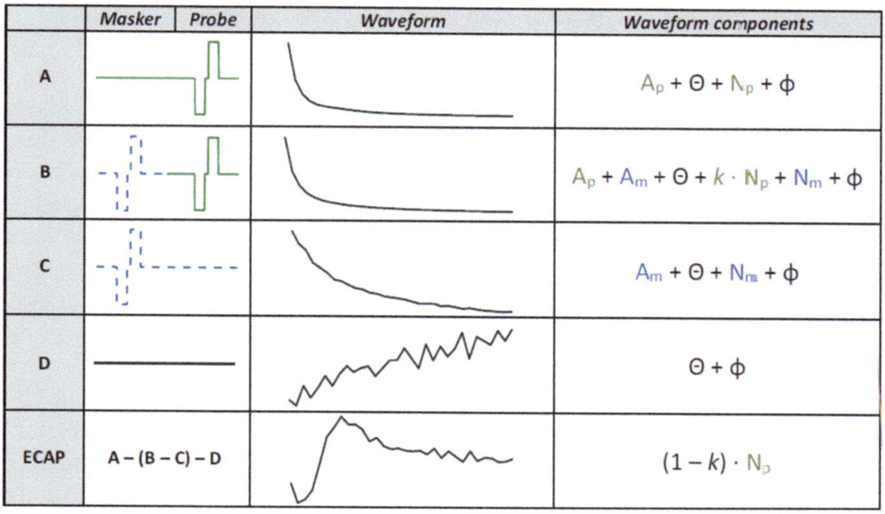

	Masker	Probe	Waveform	Waveform components
A				$A_p + \Theta + N_p + \phi$
B				$A_p + A_m + \Theta + k \cdot N_p + N_m + \phi$
C				$A_m + \Theta + N_m + \phi$
D				$\Theta + \phi$
ECAP	A – (B – C) – D			$(1 - k) \cdot N_p$

Fig. 7.4 Schematics and components for the forward masking artifact reduction technique. The maskers (dashed blue lines) and probes (solid green lines) are presented in biphasic current units, and the waveforms (solid black lines) were measured (with the probe and masker both on electrode 3) in μV (y-axis) over a period of 2 ms (x-axis). The data in the "Waveform" column was recorded from a research participant and user of a cochlear device at the MRC Cognition & Brain Sciences Unit at the University of Cambridge. A_p is the stimulus artifact as a result of the probe, A_m is the same for the masker, N_p is the neural response to the probe, N_m is the same for the masker, θ is the amplifier switch-on artifact, ϕ is the baseline neural activity, and k is the proportion of the probe's neural response that is *not* masked by the masker. For the maskers and probes presented on the same electrode, the assumption is that the entire neural response to the probe is masked. Therefore, $k = 0$ and the resulting ECAP waveform is N_p. Any partial masking occurs in extraction of only a proportion of N_p. (This figure was originally published open-access by Garcia et al. (2021))

extracted from the artifacts and the neural response to the masker by combining the different conditions using formula in Eq. 7.1. This is depicted both graphically and mathematically in Fig. 7.4.

$$\text{ECAP} = A - (B - C) - D \tag{7.1}$$

This paradigm has the advantage over the alternating polarity method that it does not underestimate the ECAP amplitude due to temporal smoothing, but does take longer to record since it requires four frames instead of two or three (the alternating polarity method requires the cathodic-leading frame, the anodic-leading frame, and sometimes also a system signature like the D frame in the forward masking method). The final extracted neural signal has a higher noise floor than the alternating polarity method with the same signal-to-noise ratio (SNR) of each frame, because the mathematics for extracting the neural response requires addition of four frames as opposed to two. Adding four frames together as in the forward masking method increases the noise floor by a factor of 2, whereas averaging two frames together as in the alternating polarity method only increases the noise floor by a factor of $\sqrt{2}$ (Stronks et al. 2019; Undurraga et al. 2012).

The method also relies on the refractory properties of the stimulated neural population, assuming that the MPI is sufficiently short such that the probe pulse is presented within the *absolute* refractory period of the masker pulse. Of the three major cochlear implant manufacturers, only Cochlear Ltd.'s clinical software includes an option to record ECAPs using this method, with a default MPI setting of 400 μs. While it has been assumed that this is sufficiently short, there is some evidence that some neurons have absolute refractory periods even shorter than this, suggesting that it may be advantageous to shorten the standard MPI to 350 or even 300 μs to ensure complete masking of the probe and improve the efficiency of the artifact cancellation technique (Skidmore et al. 2024).

7.2.2.3 A Modified Forward Masking Method

There is another, similar artifact cancellation technique presented as a modified version of the forward masking method (Miller et al. 2000). In this scenario, the probe-alone condition (A) is not presented, but instead there are two conditions with both the probe and the masker and different masker-probe intervals (MPIs), as displayed graphically in Fig. 7.5. In one condition (E), it is assumed that the masker effectively masks the probe, meaning that the MPI is short enough such that the probe is presented when the stimulated neural population is in a fully refractory state. In the other condition (B), the MPI is stretched such that the probe excites the neural population when it is no longer in a refractory state. Two masker-alone conditions (C and C') are also presented and subtracted from the first two conditions to cancel out the artifacts and the neural responses to the masker. One advantage of this method is that it can be used to investigate the refractory properties of a population of neurons directly, by adjusting the length of the interval between masker and probe in the B condition and observing the MPI at which the amplitude of the extractable neural response to the probe starts to decrease. This technique is not a standard method available in CI clinical software but can be deployed in various research platforms.

7.2.2.4 The Template Subtraction Method

The fourth and final artifact cancellation technique described here is the artifact template subtraction method. It requires two stimuli, the first of which is required to be sufficiently subthreshold that it will not excite a neural response. The resulting waveform is therefore a template of the stimulation artifact when the amplifier switches on. A waveform is then collected from a supra-threshold stimulus. The artifact from this second condition is then scaled up by the same amount as the stimulus was between the subthreshold and supra-threshold conditions and subtracted from the waveform recorded in the supra-threshold condition. This effectively cancels out the stimulus artifact in some circumstances (Miller et al. 2000), but a limitation of this approach is that it assumes the effect of the amplifier on the size of the stimulus artifact is linear, which it often is not.

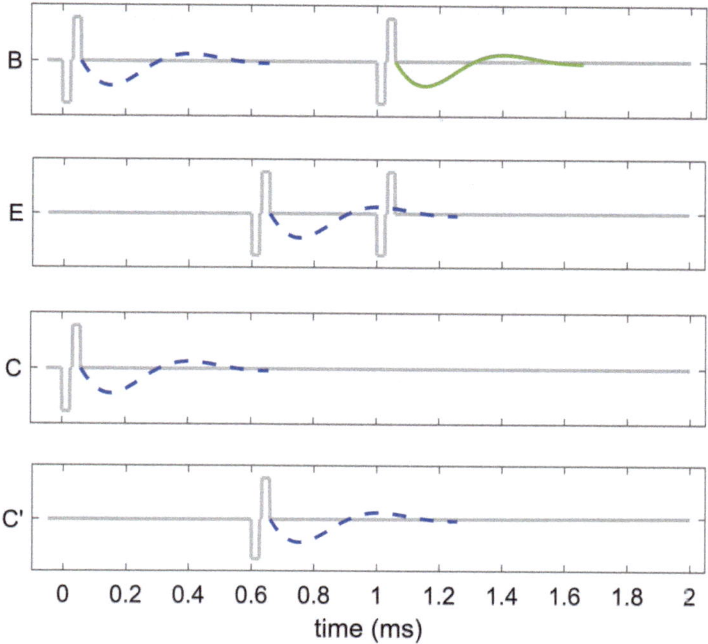

Fig. 7.5 Graphical depiction of the stimuli and waveform responses of the modified forward masking artifact cancellation technique as described by Miller et al. (2000). The gray square pulses represent the stimulating pulses, the blue dashed line represents the neural response to the masker pulse, and the solid green line represents the neural response to the probe pulse (without the electrical stimulation artifacts). In the modified method, two different masker-probe intervals (MPIs) are used in frames *B* and *E* in order to extract the neural response using the equation ECAP = (*B*–*C*)–(*E*–*C'*). *ms* milliseconds

7.2.3 Stimulation and Recording Parameters

There are various stimulus and recording parameters that can affect ECAP waveforms and therefore the metrics that are calculated from them. In most manufacturers, the electrical stimuli used to elicit ECAPs are cathodic-leading, symmetric, biphasic, charge-balanced current pulses as shown on the left of Fig. 7.2 and in the first column of Fig. 7.4. One exception is that the Oticon Medical device uses anodic-leading pulses. While using some research platforms, it is possible to stimulate using asymmetric pulses in CIs, it is important (especially in humans) to maintain equal positive and negative charge stimulation per electrode to avoid damage due to electrochemical reactions (Brummer and Turner 1977). Some parameters that can be adjusted using clinically available software include the phase duration (PD), the interphase gap (IPG), the level (in μA or current units (CU)), the amplifier gain, the recording delay (the delay after stimulation before recording is initiated), and the recording electrode used (with the constraint that it should not be the same as the stimulating electrode to avoid amplifier saturation and large electrical artifacts that cannot be removed with the artifact cancellation techniques described above).

Some adjustable parameters are specific to the forward masking artifact cancellation technique, such as the inter-pulse interval (IPI) between the masker and the probe pulses, or whether the probe and masker are placed on the same electrode. An initial investigation of the effects of varying a number of these parameters on the ECAP waveform was executed shortly after the commercial platform was launched in the Nucleus CI24M (Abbas et al. 1999; Charlet de Sauvage et al. 1983). For instance, increasing either the level or the PD of the stimulating current pulse tends to increase both the ECAP amplitude and the perceived loudness of the stimuli. However, these two parameters cannot be used entirely interchangeably to adjust the loudness of the stimuli: if the total amount of applied charge is equal, longer pulses of lower current level often produce less neural excitation than shorter pulses of higher current level, a phenomenon that has led scientists to refer to the neural pathways involved in the perception of loudness as a "leaky integrator" (Abbas and Miller 2004; McKay 2021; Miller et al. 2001; Pfingst et al. 1991).

7.2.4 Basic ECAP-Derived Metrics

There are a couple of metrics that can be calculated from an ECAP waveform. The primary measurement is the ECAP amplitude: the voltage difference between the positive "P2" peak and the negative "N1" peak of the waveform. For the ECAP waveform displayed in Fig. 7.1, the P2 peak is $\approx$ 85 μV, and the N1 peak is ≈ -145 μV. This results in an amplitude of P2–N1 $\approx$ 230 μV. Another metric of interest may be the latency of the N1 or P2 peaks that in our example in Fig. 7.1 are $\approx$0.25 ms and 0.55 ms, respectively. The majority of ECAP metrics are calculated using the peak-to-peak amplitude.

One of these amplitude-derived metrics is the ECAP threshold level, i.e., the lowest stimulation current level at which an ECAP amplitude above 0 μV can be observed. This level can be affected by other characteristics of the pulse, such as its duration, and is often calculated by measuring ECAPs at a few above-threshold current levels and fitting a regression line to these data to estimate the 0-crossing (as depicted in Fig. 7.6c), the highest stimulation current level at which a 0 μV ECAP amplitude would be elicited (Stronks et al. 2019).

Another of these amplitude-derived metrics is known as the input/output (I/O) function or amplitude growth function (AGF), where stimulus pulse levels are increased in steps from threshold, and the output amplitudes are regressed against the input pulse levels to calculate a slope. The literature varies in its methods for calculating this metric. Linear regression is generally used, but the units with which the calculation are done vary. Sometimes the input values are measured directly in pulse current μA (Schvartz-Leyzac and Pfingst 2018), and sometimes in current level measured in dB (Hughes et al. 2018). The inputs are then related directly to output voltage (generally expressed in either μV or mV), as shown in Fig. 7.6. However, these curves do not always appear linear when plotting input current in units of μA to output ECAP amplitude in units of μV, and different methods have been used to select which data points are used for regression. For example, some researchers

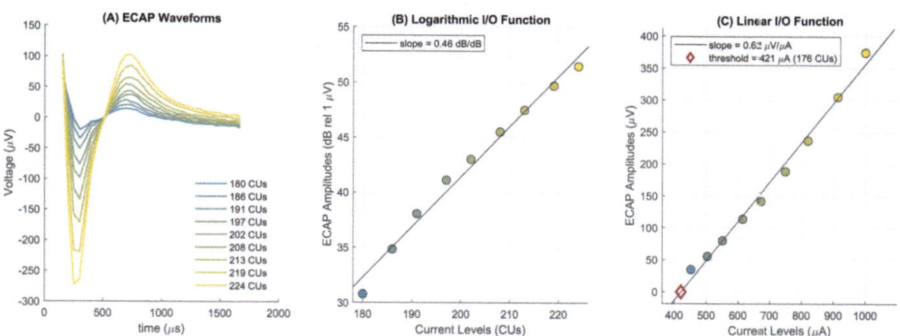

Fig. 7.6 An example of an amplitude growth function (AGF) or input/output (I/O) function. (**a**) The ECAP waveforms as a result of 9 different stimulation levels from 180 to 244 CUs. (**b**) An AGF calculated on a logarithmic scale and its associated slope calculated using linear regression. (**c**) An AGF calculated on a linear scale and its associated slope, from which an ECAP threshold of 176 CUs is calculated. (These data are a Cochlear Ltd. CI user at the MRC Cognition & Brain Sciences Unit at the University of Cambridge). *μs* microseconds, *μV* microvolts, *CUs* current units, *dB* decibels

exclude all data points after which the output voltage decreases with increasing input current, therefore truncating non-normative AGFs (Schvartz-Leyzac and Pfingst 2016), whereas some fit a Boltzmann sigmoid and calculate a linear slope centered on the current level required to achieve 50% voltage between threshold and maximum comfort levels (MCL) (Ramekers et al. 2014). The latter sigmoidal method has been used in animal models where it is possible to stimulate at a wide range of current levels, but may be less applicable in human studies where it is not always possible to stimulate at high enough current levels to observe the top of the sigmoid curve without risking uncomfortable over-stimulation in the patient.

An argument can also be made that the logarithm of the output ECAP amplitude, generally taken in units of voltage, should be taken in order to calculate the AGF slope metric using linear regression, as it characterizes dynamic range in the evaluated electrode, and loudness models suggest that loudness perception occurs on a logarithmic scale in cochlear implant users, not on a linear one (McKay et al. 2003). However, even a logarithm of the AGF slope does not successfully control for the effects of non-neural factors in modeling neural health using ECAP measures. It has been suggested that a better approach to control for these non-neural factors is to calculate the dB offset between the linear portions of the ECAP AGFs for two stimuli differing only in IPG, referred to as the IPG offset effect (Brochier et al. 2021b) (Fig. 7.7).

When using the forward masking method for artifact cancellation, there are also various approaches to estimating spread of neural excitation using ECAPs when placing the probe and masker pulses on different electrodes. This is discussed in further detail in Sect. 7.4.

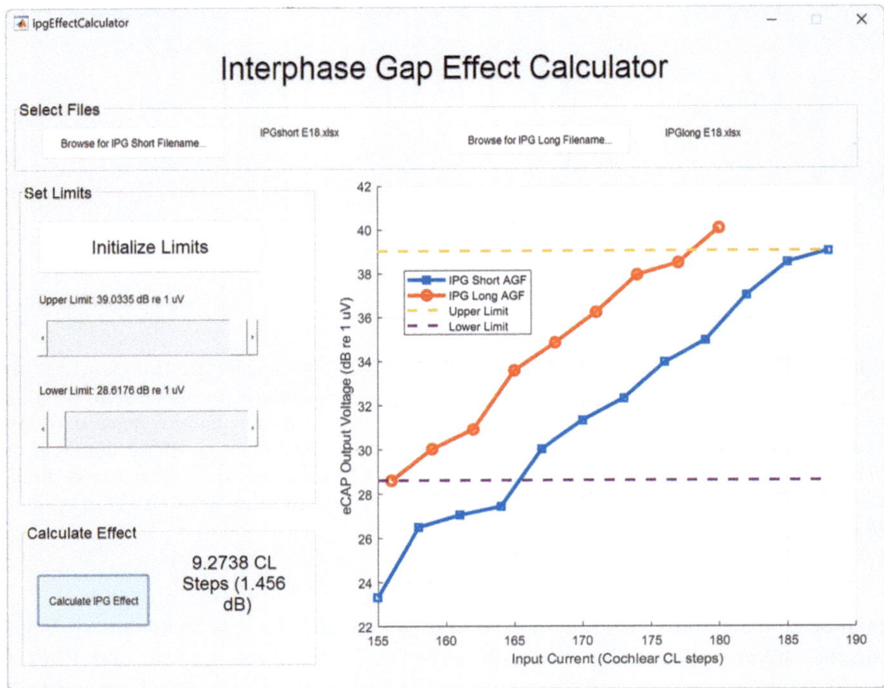

Fig. 7.7 A graphical user interface developed and provided as a piece of companion software with Brochier et al.'s paper demonstrating that calculating the IPG effect on the AGF slope successfully isolates non-neural factors when predicting for neural health within the cochlea (Brochier et al. 2021b). The program calculates the horizontal gap between the long-IPG (orange) and short-IPG (blue) AGF curves. (These data are a Cochlear Ltd. CI user at the MRC Cognition & Brain Sciences Unit at the University of Cambridge). *IPG* inter-phase gap, *CLs* current level, *dB* decibels

7.3 Predicting Neural and Perceptual Aspects of CI Stimulation Using ECAPs

In addition to stimulating and recording parameters, there are also factors that affect the ECAP measurements that give it potential diagnostic properties. The key aim of the ECAP is to provide an objective measure that has predictive power for explaining a patient's performance with their CI with a view to guiding treatment. As such, a number of studies have been done to assess the relationship between ECAP metrics and speech performance tests, as well as with biophysical measures either directly characterizing neural health or other perceptual tests thought to indirectly estimate it. The relevance of ECAP-derived metrics to CI performance is varied and is a topic of ongoing investigation in the research community.

7.3.1 Relationship to Speech Tests

One well-known study looked at relationships between ECAP measurements and speech in users of hybrid Cochlear Ltd. devices (the Nucleus Hybrid M and Hybrid RE, generally used with patients that have good residual hearing and can use a combination of acoustic and electrical stimulation with these devices). While they did find significant correlations to both Consonant-Nucleus-Consonant (CNC) and Bamford-Kowal-Bench Speech-in-Noise (BKB-SIN) scores and the slope of the AGF in the hybrid users, this was sensitive to device type (Kim et al. 2010). They only found these correlations in the newer Hybrid RE device and suggested that this may be due to the higher noise floor in the older Hybrid M device.

Another study looked at comparing the spread of excitation (measured both by changing the position of the recording electrode and the masker electrode along the array, described in more detail in Sect. 7.4) to speech perception measured with a Dutch speech test similar to the CNC words consisting of phonetically balanced monosyllabic (Consonant-Vowel-Consonant or CVC) words (van der Beek et al. 2012). This study found no significant correlation between the speech scores and the excitation profiles. Using a metric called a Channel Separation Index (CSI) (Hughes 2008), another study found significant correlations between this index and CNC scores using the activation of different CI programming MAPs designed to introduce variation in channel interactions (Scheperle and Abbas 2015). The researchers were surprised by the strength of the correlation between the CSI and the vowel recognition task (CNC scores). It is possible that its strength in contrast to the findings of van der Beek et al. (2012) might be due to the usage of multiple electrodes at a time in the test stimulus, as well as the way in which the channel separation index is calculated. The CSI is calculated by comparing ECAP amplitudes between electrodes, whereas van der Beek et al. used AGFs calculated from three electrodes and subsequently averaged them. Most previous studies only stimulate using one electrode at a time, which may not be the best model for clinical use where all activated electrodes are employed at once to transmit auditory information. One note they do make, however, is that the CSI outlined by Hughes (2008) is limited in that it is only calculated for pairs of electrodes and does not examine higher-order interactions that are likely present in clinical use.

There have also been a few recent studies that found significant correlations between ECAP measures and speech measures. For example, a study of bilateral CI users found correlations between the IPG effect on the AGF slope and both consonant recognition test results and City University of New York (CUNY) sentence recognition, controlling for individual cognitive differences by evaluating differences between ears for each metric (Schvartz-Leyzac and Pfingst 2018). Note that the "IPG effect on the AGF slope" refers to the change observed in the slope of the AGF as a result of changing the IPG. One of the advantages of bilateral CI studies is that they allow control of differences between subjects such as cognitive ability that might affect speech perception scores, as they calculate differences between ears within subjects. Interestingly, Schvartz-Leyzac and Pfingst also looked at ECAP amplitude sensitivity to IPG changes, and both ECAP peak amplitude and

AGF slopes for each of the two IPG levels individually and found no significant correlations between these metrics and any of the speech perception tests (Schvartz-Leyzac and Pfingst 2018). This lies in contrast to results of two other studies that did find significant correlations between ECAP peak amplitude and speech tests: one found a significant correlation between consonant recognition and mean ECAP amplitudes averaged across electrodes (DeVries et al. 2016), and another found significant correlation between ECAP amplitude from apical electrodes and CNC scores in hybrid users, both in situations with electrical and acoustic stimulation and with electrical stimulation alone (Kim et al. 2017). A further study included a number of different ECAP-derived metrics in a multiple regression model and found that the speed of recovery from adaptation (estimated using stimulation methods similar to the modified forward masking technique described in Sect. 7.2.2.3) best predicted speech scores, and their multidimensional model predicted speech metrics better in noisy scenarios compared to quiet ones (Skidmore et al. 2023).

While the studies described above demonstrate some significant correlations found between ECAP measures and speech perception, there have also been several studies done which fail to find significant correlations, contributing to a lack of consistency in the literature and slowing progress toward fully objective fitting of CIs (Berg et al. 2024; de Vos et al. 2018; McKay et al. 2013). One of the arguments for the lack of a clear, consistent correlation between ECAP measures and speech performance is that ECAP measures are only representative of peripheral auditory neural activity, whereas central auditory processing as well as higher-order cognition presumably additionally play a role in a patient's speech perception when using their cochlear implant(s). Another is that ECAPs need to be recorded with slower stimulation rates than CI processing strategies in order to extract neural responses. The default pulse rate in the Cochlear Ltd. platform is 80 pulses per second (pps) (and sometimes rates as slow as 15–20 pps are necessary to extract neural responses i.e., in CI patients with Cochlear Nerve Deficiency (He et al. 2018)), whereas coding strategies generally stimulate at ≈500–2000 pps (Lamping et al. 2020; Wouters et al. 2015). Confounded with this effect of differences in pulse rate is the increased stimulation current level required to achieve the same perceptual loudness when using lower pulse rates, further separating the ECAP and clinical CI processing strategy stimulation paradigms.

7.3.2 Relationship to Neural Health

Among the peripheral characteristics that one would expect to affect a CI user's ability to hear speech is the neural health of the spiral ganglion cell population the CI electrodes stimulate. Significant correlations have been found in animal studies between ECAP measures and SGN density. A study with implanted guinea pigs found that SGN density correlated positively with the duration of deafness (this was controlled and varied intentionally between subject groups) and negatively with ECAP thresholds (Prado-Guitierrez et al. 2006). They also found that the current change required to evoke the same ECAP amplitude when increasing the IPG was higher in subjects with lower SGN density. Another study with a similar design (also with guinea pigs) confirmed the relationship between the IPG effect on the AGF slope and SGN density and additionally found that ECAP amplitudes themselves correlated significantly with SGN density (Ramekers et al. 2014). A further study showed that a metric coined the "Failure Index" that takes the ratio of the input stimulation current and the resultant ECAP amplitude at the upper saturation point of an AGF is sensitive to the presence and location of induced lesions in guinea pigs (Konerding et al. 2025). Although this Failure Index is easily calculable from retrospective clinical data, it was found to be less isolated from non-neural factors of the electrode-neuron interface than another ECAP-derived method called the "panoramic ECAP method," described in Sect. 7.4.2.1 (Garcia et al. 2025). There are certain perceptual measures other than ECAPs that have also been found to correlate with SGN density. One of these is multiple pulse integration (MPI) (Pfingst et al. 2011). MPI has been found to correlate with speech perception in cochlear implant users (Zhou and Pfingst 2014), and while it is only a measurement of peripheral auditory processing, this further suggests the importance of neural health in speech perception.

There are a number of other factors that have been shown to significantly correlate with changes in ECAP metrics. For example, the effect of changing the IPG has been shown to correlate with ECAP thresholds (Hughes et al. 2018) and loudness perception (McKay and Henshall 2003). A non-adjustable factor that has been shown to correlate with ECAP peak amplitude and AGF slope is the impedance of the recording electrode used. However, the IPG effect on these ECAP metrics did not show this correlation, which the authors suggest means the IPG effect is likely better at isolating neural factors from electrode ones (Schvartz-Leyzac and Pfingst 2016). As mentioned earlier, the polarity of a pulse also affects ECAP measures, where the more effective polarity in exciting neural activity is the cathodic phase in animals, but the anodic phase in humans (Macherey et al. 2008; Undurraga et al. 2010). There is other evidence that polarity-sensitivity effects and IPG effects on ECAP measurements (both metrics that have been proposed as correlates to neural health) are likely governed by different mechanisms (Hughes et al. 2018). However, this work was done with symmetric pulses, so they were looking at the interaction of polarities as opposed to difference in sensitivity to pulse polarity. While there are studies that show no significant relationship between the slope of the AGF and the distance between electrodes and stimulated neural populations (Eisen and Franck 2004; Van Weert et al. 2005), significant correlations have been found between this electrode-modiolar distance (EMD) and clinically programmed T and C levels (Davis et al. 2016), as well as channel interaction functions (CIF) computed from forward masking ECAP amplitude responses (DeVries et al. 2016). All these items should be considered when attempting to correlate ECAP measures to various physiological characteristics such as neural health. However, although many of them are considered proxies to estimate neural health when it cannot be measured directly, there is evidence that they do not all correlate with each other, suggesting that they may represent different aspects of neural health (Brochier et al. 2021a).

While many ECAP studies measure responses on a single electrode, some groups such as Kim et al. (2010) have gathered and analyzed ECAP data at multiple electrodes in an attempt to correlate with speech recognition tests. However, even if multiple electrodes are considered, many of the studies only examine three or four electrodes along the array to represent apical, middle, and basal areas of the cochlea (Undurraga et al. 2010). When multiple electrodes have been considered, it has been observed that there are variations in ECAP measurements between electrodes even when using the same or normalized recording parameters (DeVries et al. 2016; Kim et al. 2010, 2017), indicating that recording electrode impedance, electrode-modiolar distance, SGN density, and/or possibly other factors that are believed to have an impact on ECAP measurements also vary along the length of the electrode array. One of the explanations by Kim et al. for the lack of a significant correlation found with ECAP measurements and speech recognition tests in the standard devices, when there was a significant correlation in the Hybrid devices, was that while only three electrodes were evaluated in the short Hybrid electrode array, many more electrodes were considered for the standard CI correlations, among which more variation was present (Kim et al. 2010). These measures were averaged along the length of the array before correlating with speech metrics. It was suggested that the increase in variability in the data that arises along the length of the longer, standard electrode array interferes with finding significant correlations to speech metrics. However, the differences in ECAP metrics along the array surely arise from *somewhere* and may be better accounted for in more complex models that capture variation throughout the length of the array, instead of taking an average response along it.

7.3.3 Automated Fitting

ECAPs have been used in automatic fitting techniques such as AutoNRT™ in the Cochlear Ltd. devices and are available in all three major manufacturers' clinical software. These approaches are most often used in pediatric recipients of cochlear implants who are not able to provide reliable indicators of perceived loudness behaviorally during a post-surgical clinical appointment. In these cases, ECAPs are often performed intraoperatively and in a sweeping fashion from below threshold to above the current level beyond which an ECAP is observable. These data are collected for multiple electrodes and are then used to determine the ECAP threshold, and this can be used as a proxy for the behavioral threshold at which clinicians can set CI software settings for individual patients (Botros et al. 2007; Gärtner et al. 2022). One of the key advantages of this approach includes measurements from multiple electrodes within the cochlea of an individual patient, whereas many of the studies described earlier investigating the relationship between ECAPs and either biophysical or behavioral indicators of auditory perception have leveraged single-electrode measures.

Although these systems are used clinically and are useful especially in clinical situations where there is no better option since participants may not be able to respond to behavioral assessments, there are those who have questioned whether ECAPs are sufficient for estimating perception of loudness due to their inconsistent correlation with behavioral measures (McKay et al. 2013; McKay and Smale 2017), and many have suggested that objective measures that represent more central (i.e., cortical) auditory perception would be better tools for objective fitting as they demonstrate much stronger correlations to behavioral indications of loudness perception (Mao et al. 2019; Visram et al. 2015). Multiple attempts have been made to record these more central objective measurements using the same telemetry system used for ECAPs in order to increase clinical translatability (Attias et al. 2022; Bell-Souder et al. 2024; Beynon et al. 2021; Mc Laughlin et al. 2012; Somers et al. 2021), but generally these measurements require additional external electrodes to be places as in an electroencephalography (EEG) system.

7.4 Advanced and Multielectrode Characterization of the Electrode-Neuron Interface Using ECAPs

One ECAP metric only briefly referred to in Sects. 7.2 and 7.3 that involves ECAP measurements along the whole electrode array is the spread of excitation (SOE) curve (sometimes also referred to as the spatial masking (SM) (Hughes and Stille 2010) or spatial forward masking (SFM) curve (Biesheuvel et al. 2022)). There are a few metrics that can be calculated from ECAPs that represent channel interaction: the phenomenon where the number of distinct channels delivered to the auditory nerve is reduced compared to those presented at the electrodes because neighboring electrodes stimulate overlapping areas of neural tissue. It is understood that a reduced number of individual channels can limit speech perception (Goehring et al. 2020, 2021).

7.4.1 Spread of Excitation Curves

SOE curves are widely used measures for estimating channel interaction that consist of ECAPs recorded using the standard forward masking artifact cancellation technique and were first presented by Cohen et al. (2003). They involve placing the probe stimulus on one electrode and recording a range of ECAPs with the masker placed in turn on *every* electrode along the length of the electrode array. This means that while the recording parameters for the A and D frames of the forward masking frames of the ECAP, as shown in Fig. 7.4, remain the same for all ECAPs recorded, the parameters for frames B and C will change every time the masker is placed on a new electrode. An example of an SOE curve from a CI user with a cochlear device is displayed in Fig. 7.8. They are used to estimate channel interaction by exploiting the refractory properties of the neurons involved when recording ECAPs using the forward-masking method. As the masker moves away from the probe, the measured ECAP amplitude will decrease due to the reduction in overlap between the excited neural populations of the probe and the masker (Abbas et al. 2004).

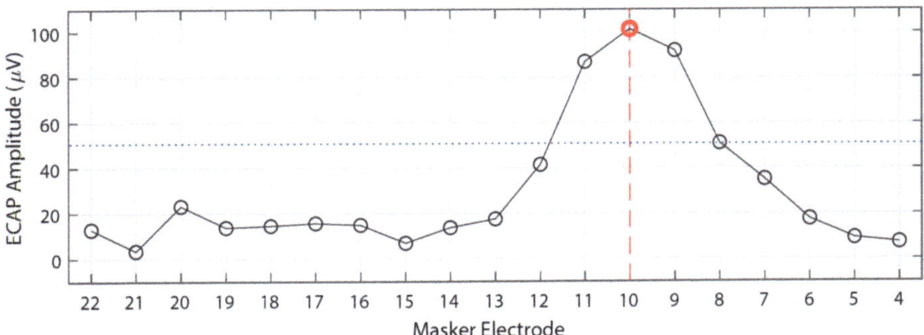

Fig. 7.8 An example of spread of excitation (SOE) curve where the probe is located on electrode 10 and an ECAP is recorded with the masker on every electrode from 4 to 22. The black circles indicate the ECAP amplitudes. The thick red circle is displayed for the ECAP amplitude measured when the masker and the probe are both located on electrode 10. This co-location point is also indicated by the red dashed line when the masker is presented on electrode 10. The horizontal blue dotted line is displayed to indicate a 50% reduction in ECAP amplitude from the peak (50.6 μV). Using the method presented by Cohen et al. to determine spread of excitation, the spread of excitation from electrode 10 (on which the probe is situated) is estimated at ≈ ±2 electrodes. (These data are a Cochlear Ltd. CI user at the MRC Cognition & Brain Sciences Unit at the University of Cambridge). *μV* microvolts

Recall that the subtraction method of the forward masking artifact cancellation technique requires that the neural population excited by the probe in the probe-alone frame (A) is in a fully refractory state in response to the masker pulse when the probe occurs in the masker + probe frame (B). If the masker does not mask the entire neural population excited by the probe, this reduces the effectiveness of the forward masking subtraction method in extracting the neural response to the probe by a function of the overlap between the neural populations excited by both the masker and the probe. The extractable portion of the neural response to the probe (and therefore the recorded ECAP) will be the neural response of the probe that is effectively masked by the masker. This is represented mathematically by k in Fig. 7.4 and displayed graphically in Fig. 7.9. This k value will equal 0 if the probe and the masker are located on the same electrode and therefore excite the same neural populations (note that this requires that the stimulating current level of the masker be greater than or equal to that of the probe).

The fact that the reduction in efficiency of the forward masking subtraction method in extracting the neural response to the probe stimulus is a function of the overlapping areas of neural tissue that are stimulated by two electrodes (the ones that the masker and the probe are placed on) means that SOE curves can be used to estimate the degree of overlap in stimulated neural populations between electrodes. Cohen et al. described a method for doing this by calculating the width of the SOE curves that counted the number of electrodes away from the probe that the masker must be in order for the ECAP amplitude to drop by 50% compared to the ECAP amplitude measured when the probe and the masker are on the same electrode. A few things should be noted about parameters that affect this measurement. First, the

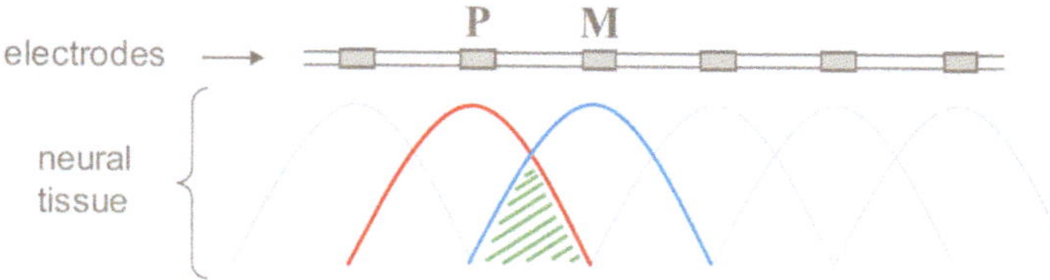

Fig. 7.9 Graphic of the overlap in stimulated neural tissue from the scenario when the probe and the masker are not located on the same electrode. The green shaded area represents the overlapping area of stimulated neural tissue between the masker and probe electrodes and therefore represents the neural tissue for the extractable neural response using the forward masking artifact cancellation technique. Since its area is smaller than the area of neural tissue stimulated by the probe (represented by the red curve), it is expected that the measured ECAP amplitude would be lower than if both the maskers were presented on the same electrode as the probe

degree of overlap is dependent on stimulation level: SOE curves are wider if the stimulation current level is higher (Abbas et al. 2004). This makes sense intuitively, as an increase in stimulus level would be expected to reach a wider area within the cochlea and therefore may reach and recruit more neurons also recruited by nearby electrodes. Additionally, there has not been consistency in the % reduction used to calculate the widths, with studies calculating widths between 50 and 75% of the peak ECAP amplitude (Abbas et al. 2004; Hughes and Abbas 2006; van der Beek et al. 2012).

7.4.1.1 Spatial Spread Curves

While varying the location of the masker electrode along the array while recording ECAPs gives an indication of the overlap of neural populations given stimulation from different electrodes, changing the location of the recording electrode when the masker and probe are co-located likely represents the same neural population (Kashio et al. 2016), but gives a measure of the current spread throughout the cochlea as well as the conduction of the voltage generated by the ECAP response along the fluid-filled cochlea (Hughes 2013). Because it involves changing the recording electrode location as opposed to the masker, this metric is often referred to as spatial spread (as opposed to spatial masking) and can be evaluated using multiple artifact cancellation techniques. Spatial spread has been shown to result in wider excitation curves than SOE curves typically show (Hughes and Stille 2010; van der Beek et al. 2012). One study compared the forward masking and alternating polarity artifact cancellation techniques when calculating spatial spread curves and found no significant effect of technique on the shape of the excitation pattern (assessed by calculating its width) (van der Beek et al. 2012) but did not investigate the effect of artifact cancellation technique on ECAP amplitudes directly.

There is also evidence of asymmetry in excitation patterns both in cases of spatial spread and SOE such that the apical side of the pattern is wider (van der Beek et al. 2012). This is consistent with other findings that recording ECAPs using a recording electrode located apically to the stimulating electrode may generate higher amplitudes than when recording the same number of electrodes away, basally (Frijns et al. 2002).

7.4.1.2 Relationship to Speech Scores

While SOEs are peripheral measures that characterize the electrode-neuron interface (ENI), it is expected that they may influence higher-order auditory perception such as pitch or speech perception. Although correlations do not prove causal relationships, the degree to which channels interact as characterized by SOE functions has been shown to correlate with pitch discrimination, when characterized using the channel separation index (CSI) metric mentioned previously (Hughes 2008). As opposed to calculating the width of individual SOE curves, it evaluates the difference in SOE curves between two electrodes by summing up the differences between normalized peak ECAP amplitude for each masker electrode. This metric has also been shown to correlate with speech performance (Scheperle and Abbas 2015). Another metric, the equivalent rectangular bandwidth (ERB) of the area under the ECAP SOE function, has been shown to correlate with electrode-modiolus distances (EMD) measured from CT scans (DeVries et al. 2016). EMDs are generally understood to affect current spread, as the farther the electrodes are from their closest responsive neural population, the higher the level of the current that is required to stimulate them, and therefore the wider the spread of that current is likely to be.

However, correlations between higher-order auditory perception measures and the width of the SOE curves show mixed results. The significant correlations described above are accompanied by a lack of evidence for similar comparisons. Abbas and Hughes failed to find a significant interaction between the width of the SOE curves and pitch-ranking results, as well as performance on speech tests (Hughes and Abbas 2006). In a large-scale study, Berg et al. also failed to find robust relationships between widths of SOE curves and speech-perception scores, although one moderate correlation was found at 6 months with perimodiolar arrays specifically (Berg et al. 2024). DeVries et al. also found no significant correlation between ERBs and speech scores (DeVries et al. 2016). This lack of consistency in correlations between SOE-derived metrics and behavioral speech perception may be partially due to differences in pulse rate and stimulation current level between ECAPs and CI coding strategies as well as other factors (perhaps more central ones) that impact speech perception and are not captured by the SOE curves but also suggest that there may be room for improvement while estimating channel interactions with ECAP data.

7.4.1.3 Oversights

All of the above metrics derived from SOE curves (width, CSI, ERBs) are oversimplifications of neural excitation, because the amplitude of each ECAP waveform is determined by the spread from *both* the probe and the masker electrodes. Essentially, these methods assume uniform (and, in fact, infinitely narrow) spread from each electrode on which the masker is placed, which is not the case (this assumption is not present in spatial spread curves). They make this assumption by interpreting each SOE curve as the spread of excitation from the electrode on which the probe is presented alone, and not as having been influenced by variation in spread of excitation from each electrode the masker is placed on. Consider the case

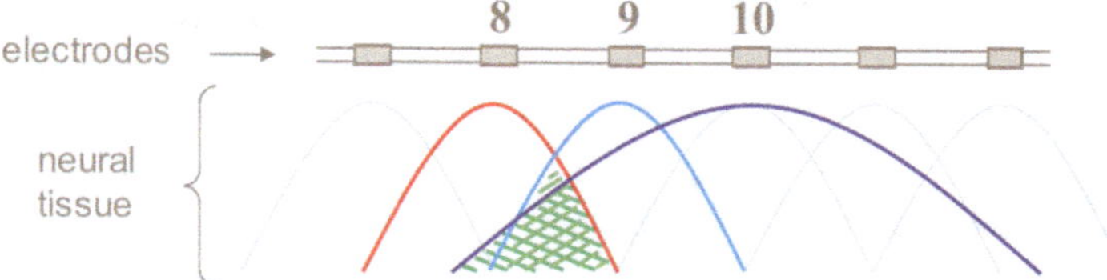

Fig. 7.10 Graphic of the overlapping stimulated neural populations for the scenario where the probe is on electrode 8, and the masker is either on the adjacent electrode 9 (blue curve) or two electrodes array from the probe on electrode 10 (purple curve). Note that the wide area of excitation from electrode 2 away from the probe creates a scenario where the overlapping area of neural excitation between the probe electrode and both masker electrodes is roughly equivalent (represented by the green shading). This would normally result in wider estimation of spread of excitation and would be attributed to the probe electrode if using the Cohen et al. method (Cohen et al. 2003), when in fact it is due to the masker electrode's wide pattern of excitation

where the probe is fixed on electrode 8 and the masker varied between electrodes 1 and 16. The resulting SOE curve may be interpreted as the spread of current from electrode 8 into the neural populations also stimulated by electrodes 1 through 16 and would be expected to show less overlap as the masker moves farther away from electrode 8. However, the measured ECAP amplitudes are also a product of the spread of neural excitation from the electrodes on which the *masker* is presented. For example, the excited neural population could be narrow at electrode 9 and wide at 10, as shown in Fig. 7.10. In recording the SOE for electrode 8, the overlapping areas of neural excitation for the conditions where the masker is on 9 and 10 may produce the same ECAP amplitude and therefore be interpreted as widespread of excitation from electrode 8, when in reality, it is the result of widespread of excitation from the masker on electrode 10.

This is the basic error in standard SOE measures and interpretation: the model assumes incorrectly that the distributions of the masker's excitation patterns are all equal and infinitely narrow. For this reason, it may be beneficial to consider methods that take this additional variation into account when modeling the electrode-neuron interface using ECAPs, especially as they may better predict biophysical and perceptual measures of cochlear implant function.

7.4.2 Deconvolution and Nonlinear Optimization Methods for Characterizing Spread of Excitation

It has been suggested that SOE curves and similar data could better describe the neural activity in response to stimulation with a cochlear implant along the length of a patient's cochlear using more complex modeling than the methods described in Sect. 7.3.3. There are two similar methods that have been proposed that take into consideration the interaction between the excitation patterns of the probe and the masker and therefore avoid the oversimplification of previous SOE curve metrics. These methods use the forward masking artifact cancellation technique to record

ECAPs for all combinations of probe and masker electrodes. This is equivalent to recording all possible SOE curves as opposed to just one or two and interpreting the width of the curve as the spread of excitation from the probe electrode.

7.4.2.1 The Panoramic ECAP Method

The first of these methods described was the first version of the panoramic ECAP method (PECAP) proposed by Cosentino et al. (2016). It leveraged single-level ECAP data from all combinations of masker and probe electrodes collected at current levels that were scaled to the most comfortable level (MCL) across the electrode array and submitted it to a two-stage nonlinear optimization algorithm that modeled the patterns of neural excitation along the length of the cochlea as combinations of all the SOE curves. It was able to identify single exceptions to an ideal neural scenario in individual CI users: either an instance of cross-turn stimulation (this is when a current pulse from one electrode may excite neural populations not only located in the section of the scala that it is situated in but also located across a cochlear turn) or a neural dead region. However, it struggled (by design) to identify more than one instance of a neural dead region or cross-turn stimulation in a single CI user and largely modeled each of the exceptions to the ideal cochlear scenario using one electrode at a time, whereas a more realistic method would be to model all electrodes simultaneously.

A revised version of this method separated the relative contributions to the neural activation patterns of current spread and neural responsiveness along the length of the electrode array, as shown in Fig. 7.11 (Garcia et al. 2021). The neural responsiveness estimate was validated by confirming local reduction in the neural responsiveness estimate where a localized area of reduced neural responsiveness was simulated by presenting two pre-masker pulses on a single electrode throughout ECAP recording (Garcia et al. 2021). This estimate has also been shown to correlate to two additional behavioral measures that are thought to represent aspects of neural health: focused thresholds (Garcia et al. 2021) and a variation metric calculated from the across-electrode changes in the difference between monopolar and focused thresholds, although the latter metric correlated with vowel identification scores, whereas the PECAP-derived metric did not (Peng et al. 2025). PECAP has also been shown to isolate neural from non-neural aspects of the electrode-neuron interface with greater success than other ECAP-derived metrics (Garcia et al. 2025).

The current-spread estimate was validated by identifying localized increases in current spread when multiple electrodes were simulated simultaneously to "blur" current (Garcia et al. 2024) and has also identified apical-specific differences in current spread between different electrode array types (Garcia and Carlyon 2025). Advancements have also been made to this method such that all necessary ECAP data can be recorded in 8 min using research software (whereas using the clinically available software from Cochlear Ltd. requires >40 min) to increase clinical viability (Garcia et al. 2023), and an online platform has been developed to enable users to both collect and analyze PECAP data themselves without requiring specialist programming competencies (Garcia and Lazauskas 2024).

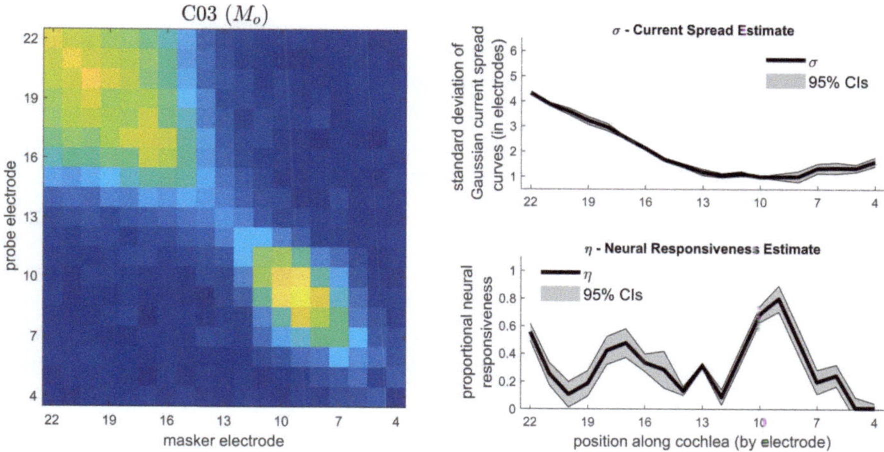

Fig. 7.11 Graphical overview of the PECAP method. *Left:* ECAP amplitudes by masker and probe electrode for participant C03, collected at the MRC Cognition & Brain Sciences Unit at the University of Cambridge. These data constitute the M_0 matrix and the input data to the PECAP algorithm. *Right:* Current spread (top) and neural responsiveness (bottom) estimates by electrode. The solid black lines are the primary outcome and the shaded regions represent 95% confidence intervals based on ECAP recording variation. These data are the two outputs of the revised PECAP method. (Data from Garcia et al. 2021). *CIs* confidence limits

7.4.2.2 The Deconvolution Method

Another broadly similar method is proposed by Biesheuvel et al. (2016, 2022). It also used single-level ECAP recordings from all combinations of masker and probe electrodes, stimulating these intraoperatively at equal current levels across the array (in contrast to the method of Cosentino et al. and Garcia et al. where ECAPs are recorded postoperatively and at most comfortable level). This model similarly estimates neural excitation patterns of each electrode along the array (referred to as excitation density profiles or EDPs) and has the advantage of being fully objective as it does not require behavioral indications of loudness from CI patients, although it does not directly predict for the presence of neural dead regions nor cross-turn stimulation. It treats the SOE curves like a convolution of EDPs of maskers with the EDPs of probes. The EDPs represent the percentage of neurons excited as a function of their position along the length of the cochlea and are characterized either by square, symmetric double-sided exponential, or symmetric Gaussian-shaped curves centered at the stimulating electrode. To indirectly deconvolve the measured SOEs, the EDPs for maskers and probes are convolved with each other and iteratively adjusted until they most closely approximate the SOE curves. The authors found that the exponential and Gaussian EDPs provided better fits to the measured SOE measures than the square functions and were able to achieve a good fit with root mean squared errors (RMSEs) between measured and predicted SOE curves averaging 0.042 mV across participants using Gaussian EDPs. They also found that even with symmetric EDPs, they were able to model observed asymmetry in the SOE curves.

The primary differences between this method and the panoramic ECAP method include the ECAP measurement procedure (equal current vs. equal loudness), the fact that multiple curve shapes are investigated (PECAP uses only Gaussian curves), and that it does not separate the relative contributions of current spread and neural responsiveness to the EDPs. Neither method accounts for potential variation in ECAP amplitudes due to recording electrode variation.

7.4.2.3 Clinical Application Potential

Approaches such as the PECAP and Deconvolution methods may serve to better characterize the electrode-neuron interface and ultimately predict perception with cochlear implants in an objective way. Tools such as these that characterize individual cochlear-implant user's unique patterns of neural activation can provide important information for optimizing efficacy and improving speech perception (Long et al. 2014; Pfingst et al. 2015). While many models have been developed to attempt to project highly variable speech perception outcomes in CI users (Firszt et al. 2004; Friesen et al. 2001) using a myriad of behavioral and objective measurements (Carlyon and Goehring 2021; Kim et al. 2010, 2017; Lazard et al. 2012; Scheperle and Abbas 2015; Schvartz-Leyzac and Pfingst 2018), they have varying degrees of accuracy and often use summary metrics for CI users that do not provide detailed estimates of a patient's unique ENI that Long et al. (2014) and Pfingst et al. (2015) have both indicated may be important. Patient-specific information about electric hearing with a CI may assist clinicians in personalizing CI programming in order to optimize speech perception and may ultimately lead to greater success in speech performance in patients who would otherwise have struggled to use their implants.

7.5 Summary

The electrically evoked compound action potential (ECAP) measures the peripheral, synchronous response of the auditory nerve to stimulation with a cochlear implant. This chapter has provided an overview of the ECAP, including methods for measurement, interpretation of the recorded waveforms, and relationships to other biophysical markers and behavioral assessments of auditory perception with a CI. While there are a number of extended techniques that leverage multiple ECAPs to build as complete a model of the electrode-neuron interface as possible with the measure, their ability to predict auditory perception in a CI is limited compared to some other objective measurements of more cortical responses. However, until the higher-order objective measurements of responses to electrical stimulation with a CI are more user-friendly and clinically viable, ECAPs provide an easy, efficient tool to support and inform CI fitting sessions as a first objective indicator of perception.

Acknowledgments and Declarations The author declares that no generative artificial intelligence (AI) was used in the creation of this chapter.

This chapter was developed as an extension of the introductory chapter of a dissertation written by the author for her Ph.D. thesis (Garcia 2022). It has been updated and extended to reflect subsequent research for the purposes of this chapter.

References

Abbas PJ, Miller CA (2004) Biophysics and physiology. In: Zeng F-G, Popper AN, Fay RR (eds) Cochlear implants: auditory prostheses and electric hearing. Springer, pp 149–212. https://doi.org/10.1007/978-0-387-22585-2_5

Abbas PJ, Brown CJ, Shallop JK, Firszt JB, Hughes ML, Hong SH, Staller SJ (1999) Summary of results using the nucleus CI24M implant to record the electrically evoked compound action potential. Ear Hear 20(1):45–59. https://doi.org/10.1097/00003446-199902000-00005

Abbas PJ, Hughes ML, Brown CJ, Miller CA, South H (2004) Channel interaction in cochlear implant users evaluated using the electrically evoked compound action potential. Audiol Neurotol 9(4):203–213. https://doi.org/10.1159/000078390

Attias J, HabibAllah S, Aditya Tarigoppula VS, Glick H, Chen C, Kanthaiah K, Litvak L (2022) Cortical auditory evoked potentials recorded directly through the cochlear implant in cochlear implant recipients: a feasibility study. Ear Hear 43(5):1426. https://doi.org/10.1097/AUD.0000000000001212

Bell-Souder D, Chen C, Spahr A, Sharma A (2024) Validation of direct recording of electrically evoked cortical auditory evoked potentials through a cochlear implant system. Sci Rep 14(1): 28366. https://doi.org/10.1038/s41598-024-79528-3

Berg KA, DeFreese AJ, Sisler-Dinwiddie AL, Labadie RF, Tawfik KO, Gifford RH (2024) Clinical applications for spread of excitation functions obtained via electrically evoked compound action potentials (eCAP). Otol Neurotol 45(7):790. https://doi.org/10.1097/MAO.0000000000004242

Beynon AJ, Luijten BM, Mylanus EAM (2021) Intracorporeal cortical telemetry as a step to automatic closed-loop EEG-based CI fitting: a proof of concept. Audiol Res 11(4):691–705. https://doi.org/10.3390/audiolres11040062

Biesheuvel JD, Briaire JJ, Frijns JHM (2016) A novel algorithm to derive spread of excitation based on deconvolution. Ear Hear 37(5):572. https://doi.org/10.1097/AUD.0000000000000296

Biesheuvel JD, Briaire JJ, Kalkman RK, Frijns JHM (2022) The effect of stimulus level on excitation patterns of individual electrode contacts in cochlear implants. Hear Res 420: 108490. https://doi.org/10.1016/j.heares.2022.108490

Botros A, van Dijk B, Killian M (2007) AutoNR: an automated system that measures ECAP thresholds with the nucleus freedom cochlear implant via machine intelligence. Artif Intell Med 40(1):15–28. https://doi.org/10.1016/j.artmed.2006.06.003

Brochier T, Guérit F, Deeks JM, Garcia C, Bance M, Carlyon RP (2021a) Evaluating and comparing behavioural and electrophysiological estimates of neural health in cochlear implant users. J Assoc Res Otolaryngol 22(1):67–80. https://doi.org/10.1007/s10162-020-00773-0

Brochier T, McKay CM, Carlyon RP (2021b) Interpreting the effect of stimulus parameters on the electrically evoked compound action potential and on neural health estimates. J Assoc Res Otolaryngol 22(1):81–94. https://doi.org/10.1007/s10162-020-00774-z

Brown CJ, Abbas PJ, Gantz B (1990) Electrically evoked whole-nerve action potentials: data from human cochlear implant users. J Acoust Soc Am 88(3):1385–1391. https://doi.org/10.1121/1.399716

Brown CJ, Abbas PJ, Fryauf-Bertschy H, Kelsay D, Gantz BJ (1994) Intraoperative and postoperative electrically evoked auditory brain stem responses in nucleus cochlear implant users: implications for the fitting process. Ear Hear 15(2):168–176. https://doi.org/10.1097/00003446-199404000-00006

Brummer SB, Turner MJ (1977) Electrochemical considerations for safe electrical stimulation of the nervous system with platinum electrodes. IEEE Trans Biomed Eng 24(1):59–63. https://doi.org/10.1109/TBME.1977.326218

Carlyon RP, Goehring T (2021) Cochlear implant research and development in the twenty-first century: a critical update. J Assoc Res Otolaryngol 22(5):481–508. https://doi.org/10.1007/s10162-021-00811-5

Charlet de Sauvage R, Cazals Y, Erre JP, Aran JM (1983) Acoustically derived auditory nerve action potential evoked by electrical stimulation: an estimation of the waveform of single unit contribution. J Acoust Soc Am 73(2):616–627. https://doi.org/10.1121/1.388872

Cohen LT, Richardson LM, Saunders E, Cowan RSC (2003) Spatial spread of neural excitation in cochlear implant recipients: comparison of improved ECAP method and psychophysical forward masking. Hear Res 179(1):72–87. https://doi.org/10.1016/S0378-5955(03)00096-0

Cosentino S, Gaudrain E, Deeks JM, Carlyon RP (2016) Multistage nonlinear optimization to recover neural activation patterns from evoked compound action potentials of cochlear implant users. IEEE Trans Biomed Eng 63(4):833–840. https://doi.org/10.1109/TBME.2015.2476373

Davis TJ, Zhang D, Gifford RH, Dawant BM, Labadie RF, Noble JH (2016) Relationship between electrode-to-modiolus distance and current levels for adults with cochlear implants. Otol Neurotol 37(1):31. https://doi.org/10.1097/MAO.0000000000000896

de Vos JJ, Biesheuvel JD, Briaire JJ, Boot PS, van Gendt MJ, Dekkers OM, Fiocco M, Frijns JHM (2018) Use of electrically evoked compound action potentials for cochlear implant fitting: a systematic review. Ear Hear 39(3):401. https://doi.org/10.1097/AUD.0000000000000495

DeVries L, Scheperle R, Bierer JA (2016) Assessing the electrode-neuron interface with the electrically evoked compound action potential, electrode position, and behavioral thresholds. J Assoc Res Otolaryngol 17(3):237–252. https://doi.org/10.1007/s10162-016-0557-9

Eisen MD, Franck KH (2004) Electrically evoked compound action potential amplitude growth functions and HiResolution programming levels in pediatric CII implant subjects. Ear Hear 25(6):528–538

Firszt JB, Holden LK, Skinner MW, Tobey EA, Peterson A, Gaggl W, Runge-Samuelson CL, Wackym PA (2004) Recognition of speech presented at soft to loud levels by adult cochlear implant recipients of three cochlear implant systems. Ear Hear 25(4):375–387. https://doi.org/10.1097/01.aud.0000134552.22205.ee

Friesen LM, Shannon RV, Baskent D, Wang X (2001) Speech recognition in noise as a function of the number of spectral channels: comparison of acoustic hearing and cochlear implants. J Acoust Soc Am 110(2):1150–1163. https://doi.org/10.1121/1.1381538

Frijns JHM, Briaire JJ, de Laat JAPM, Grote JJ (2002) Initial evaluation of the Clarion CII cochlear implant: speech perception and neural response imaging. Ear Hear 23(3):184–197. https://doi.org/10.1097/00003446-200206000-00003

Garcia C (2022) The panoramic ECAP method: estimating patient-specific patterns of current spread and neural health in cochlear-implant users. PhD thesis, University of Cambridge. https://doi.org/10.17863/CAM.89113

Garcia C, Carlyon RP (2025) Assessing array-type differences in cochlear implant users using the panoramic ECAP method. Ear Hear 46(5):1355–1368. https://doi.org/10.1097/AUD.0000000000001673

Garcia C, Lazauskas T (2024) Panoramic ECAP (Version 1) [Computer software]. https://panoramic-ecap.mrc-cbu.cam.ac.uk/

Garcia C, Goehring T, Cosentino S, Turner RE, Deeks JM, Brochier T, Rughooputh T, Bance M, Carlyon RP (2021) The panoramic ECAP method: estimating patient-specific patterns of current spread and neural health in cochlear implant users. J Assoc Res Otolaryngol 22(5):567–589. https://doi.org/10.1007/s10162-021-00795-2

Garcia C, Deeks JM, Goehring T, Borsetto D, Bance M, Carlyon RP (2023) SpeedCAP: an efficient method for estimating neural activation patterns using electrically evoked compound action-potentials in cochlear implant users. Ear Hear 44(3):627. https://doi.org/10.1097/AUD.0000000000001305

Garcia C, Morse-Fortier C, Guérit F, Hislop S, Goehring T, Carlyon RP, Arenberg JG (2024) Investigating the effect of blurring and focusing current in cochlear implant users with the panoramic ECAP method. J Assoc Res Otolaryngol 25:591–609. https://doi.org/10.1007/s10162-024-00966-x

Garcia C, Sismono F, Goehring T, Guérit F, Arzounian D, Carlyon RP (2025) A comparison of electrophysiological measures for characterizing the cochlear-implant electrode-neuron interface. JASA Express Lett 5(8):082001. https://doi.org/10.1121/10.0038746

Gärtner L, Bräcker T, Kals M, Penninger RT, Billinger-Finke M, Lenarz T, Büchner A (2022) ARTFit—a quick and reliable tool for performing initial fittings in users of MED-EL cochlear implants. Life 12(2):269. https://doi.org/10.3390/life12020269

Goehring T, Arenberg JG, Carlyon RP (2020) Using spectral blurring to assess effects of channel interaction on speech-in-noise perception with cochlear implants. J Assoc Res Otolaryngol 21(4):353–371. https://doi.org/10.1007/s10162-020-00758-z

Goehring T, Archer-Boyd AW, Arenberg JG, Carlyon RP (2021) The effect of increased channel interaction on speech perception with cochlear implants. Sci Rep 11(1):Article 1. https://doi.org/10.1038/s41598-021-89932-8

Hall JW (2006) New handbook of auditory evoked responses, 1st edn. Pearson

He S, Shahsavarani BS, McFayden TC, Wang H, Gill KE, Xu L, Chao X, Luo J, Wang R, He N (2018) Responsiveness of the electrically stimulated cochlear nerve in children with cochlear nerve deficiency. Ear Hear 39(2):238. https://doi.org/10.1097/AUD.0000000000000467

Hughes ML (2008) A re-evaluation of the relation between physiological channel interaction and electrode pitch ranking in cochlear implants. J Acoust Soc Am 124(5):2711–2714. https://doi.org/10.1121/1.2990710

Hughes ML (2013) Objective measures in cochlear implants, 1st edn. Plural Publishing

Hughes ML, Abbas PJ (2006) The relation between electrophysiologic channel interaction and electrode pitch ranking in cochlear implant recipients. J Acoust Soc Am 119(3):1527–1537. https://doi.org/10.1121/1.2163273

Hughes ML, Stille LJ (2010) Effect of stimulus and recording parameters on spatial spread of excitation and masking patterns obtained with the electrically evoked compound action potential in cochlear implants. Ear Hear 31(5):679–692. https://doi.org/10.1097/AUD.0b013e3181e1d19e

Hughes ML, Choi S, Glickman E (2018) What can stimulus polarity and interphase gap tell us about auditory nerve function in cochlear-implant recipients? Hear Res 359:50–63. https://doi.org/10.1016/j.heares.2017.12.015

Kalkman RK, Briaire JJ, Dekker DMT, Frijns JHM (2022) The relation between polarity sensitivity and neural degeneration in a computational model of cochlear implant stimulation. Hear Res 415:108413. https://doi.org/10.1016/j.heares.2021.108413

Kashio A, Tejani VD, Scheperle RA, Brown CJ, Abbas PJ (2016) Exploring the source of neural responses of different latencies obtained from different recording electrodes in cochlear implant users. Audiol Neurootol 21(3):141–149. https://doi.org/10.1159/000444739

Kim J-R, Abbas PJ, Brown CJ, Etler CP, O'Brien S, Kim L-S (2010) The relationship between electrically evoked compound action potential and speech perception: a study in cochlear implant users with short electrode array. Otol Neurotol 31(7):1041. https://doi.org/10.1097/MAO.0b013e3181ec1d92

Kim J-R, Tejani VD, Abbas PJ, Brown CJ (2017) Intracochlear recordings of acoustically and electrically evoked potentials in nucleus hybrid L24 cochlear implant users and their relationship to speech perception. Front Neurosci 11:216. https://doi.org/10.3389/fnins.2017.00216

Konerding W, Arenberg J, Sznabel D, Kral A, Baumhoff P (2025) An electrically evoked compound action potential marker for local spiral ganglion neuron degeneration: the failure index. J Neurosci 45(7):e0954242024. https://doi.org/10.1523/JNEUROSCI.0954-24.2024

Lamping W, Goehring T, Marozeau J, Carlyon RP (2020) The effect of a coding strategy that removes temporally masked pulses on speech perception by cochlear implant users. Hear Res 391:107969. https://doi.org/10.1016/j.heares.2020.107969

Lazard DS, Vincent C, Venail F, Van de Heyning P, Truy E, Sterkers O, Skarzynski PH, Skarzynski H, Schauwers K, O'Leary S, Mawman D, Maat B, Kleine-Punte A, Huber AM, Green K, Govaerts PJ, Fraysse B, Dowell R, Dillier N et al (2012) Pre-, per- and postoperative factors affecting performance of postlinguistically deaf adults using cochlear implants: a new conceptual model over time. PLoS One 7(11):e48739. https://doi.org/10.1371/journal.pone.0048739

Lewis MA, Steel KP (2010) MicroRNAs in mouse development and disease. Semin Cell Dev Biol 21(7):774–780. https://doi.org/10.1016/j.semcdb.2010.02.004

Long CJ, Holden TA, McClelland GH, Parkinson WS, Shelton C, Kelsall DC, Smith ZM (2014) Examining the electro-neural interface of cochlear implant users using psychophysics, CT scans, and speech understanding. J Assoc Res Otolaryngol 15(2):293–304. https://doi.org/10.1007/s10162-013-0437-5

Macherey O, Cazals Y (2016) Effects of pulse shape and polarity on sensitivity to cochlear implant stimulation: a chronic study in guinea pigs. Adv Exp Med Biol 894:133–142. https://doi.org/10.1007/978-3-319-25474-6_15

Macherey O, Carlyon RP, van Wieringen A, Deeks JM, Wouters J (2008) Higher sensitivity of human auditory nerve fibers to positive electrical currents. J Assoc Res Otolaryngol 9(2):241–251. https://doi.org/10.1007/s10162-008-0112-4

Mao D, Innes-Brown H, Petoe MA, Wong YT, McKay CM (2019) Fully objective hearing threshold estimation in cochlear implant users using phase-locking value growth functions. Hear Res 377:24–33. https://doi.org/10.1016/j.heares.2019.02.013

Mc Laughlin M, Lu T, Dimitrijevic A, Zeng F-G (2012) Towards a closed-loop cochlear implant system: application of embedded monitoring of peripheral and central neural activity. IEEE Trans Neural Syst Rehabil Eng 20(4):443–454. https://doi.org/10.1109/TNSRE.2012.2186982

McKay CM (2021) Applications of phenomenological loudness models to cochlear implants. Front Psychol 11:611517. https://doi.org/10.3389/fpsyg.2020.611517

McKay CM, Henshall KR (2003) The perceptual effects of interphase gap duration in cochlear implant stimulation. Hear Res 181(1):94–99. https://doi.org/10.1016/S0378-5955(03)00177-1

McKay CM, Smale N (2017) The relation between ECAP measurements and the effect of rate on behavioral thresholds in cochlear implant users. Hear Res 346:62–70. https://doi.org/10.1016/j.heares.2017.02.009

McKay CM, Henshall KR, Farrell RJ, McDermott HJ (2003) A practical method of predicting the loudness of complex electrical stimuli. J Acoust Soc Am 113(4):2054–2063. https://doi.org/10.1121/1.1558378

McKay CM, Chandan K, Akhoun I, Siciliano C, Kluk K (2013) Can ECAP measures be used for totally objective programming of cochlear implants? J Assoc Res Otolaryngol 14(6):879–890. https://doi.org/10.1007/s10162-013-0417-9

Miller CA, Abbas PJ, Rubinstein JT, Robinson BK, Matsuoka AJ, Woodworth G (1998) Electrically evoked compound action potentials of guinea pig and cat: responses to monopolar, monophasic stimulation. Hear Res 119(1–2):142–154. https://doi.org/10.1016/s0378-5955(98)00046-x

Miller CA, Abbas PJ, Brown CJ (2000) An improved method of reducing stimulus artifact in the electrically evoked whole-nerve potential. Ear Hear 21(4):280

Miller CA, Robinson BK, Rubinstein JT, Abbas PJ, Runge-Samuelson CL (2001) Auditory nerve responses to monophasic and biphasic electric stimuli. Hear Res 151(1–2):79–94. https://doi.org/10.1016/s0300-2977(00)00082-6

Peng T, Garcia C, Haneman M, Shader MJ, Carlyon RP, McKay CM (2025) Comparing patient-specific variations in intra-cochlear neural health estimated using psychophysical thresholds and panoramic electrically evoked compound action potentials (PECAPs). J Assoc Res Otolaryngol 26(1):77–91. https://doi.org/10.1007/s10162-024-00972-z

Pfingst BE, DeHaan DR, Holloway LA (1991) Stimulus features affecting psychophysical detection thresholds for electrical stimulation of the cochlea. I: phase duration and stimulus duration. J Acoust Soc Am 90(4 Pt 1):1857–1866. https://doi.org/10.1121/1.401665

Pfingst BE, Colesa DJ, Hembrador S, Kang SY, Middlebrooks JC, Raphael Y, Su GL (2011) Detection of pulse trains in the electrically stimulated cochlea: effects of cochlear health. J Acoust Soc Am 130(6):3954–3968. https://doi.org/10.1121/1.3651820

Pfingst BE, Zhou N, Colesa DJ, Watts MM, Strahl SB, Garadat SN, Schvartz-Leyzac KC, Budenz CL, Raphael Y, Zwolan TA (2015) Importance of cochlear health for implant function. Hear Res 322:77–88. https://doi.org/10.1016/j.heares.2014.09.009

Prado-Guitierrez P, Fewster LM, Heasman JM, McKay CM, Shepherd RK (2006) Effect of interphase gap and pulse duration on electrically evoked potentials is correlated with auditory nerve survival. Hear Res 215(1–2):47–55. https://doi.org/10.1016/j.heares.2006.03.006

Purves D, Augustine GJ, Fitzpatrick D, Hall WC, LaMantia A-S, White LE (2012) Electrical signals of nerve cells. In: Neuroscience, 5th edn. Sinauer Associates Inc., pp 25–40

Ramekers D, Versnel H, Strahl SB, Smeets EM, Klis SFL, Grolman W (2014) Auditory-nerve responses to varied inter-phase gap and phase duration of the electric pulse stimulus as predictors for neuronal degeneration. J Assoc Res Otolaryngol 15(2):187–202. https://doi.org/10.1007/s10162-013-0440-x

Rattay F, Lutter P, Felix H (2001) A model of the electrically excited human cochlear neuron: I. Contribution of neural substructures to the generation and propagation of spikes. Hear Res 153(1):43–63. https://doi.org/10.1016/S0378-5955(00)00256-2

Scheperle RA, Abbas PJ (2015) Relationships among peripheral and central electrophysiological measures of spatial and spectral selectivity and speech perception in cochlear implant users. Ear Hear 36(4):441. https://doi.org/10.1097/AUD.0000000000000144

Schvartz-Leyzac KC, Pfingst BE (2016) Across-site patterns of electrically evoked compound action potential amplitude-growth functions in multichannel cochlear implant recipients and the effects of the interphase gap. Hear Res 341:50–65. https://doi.org/10.1016/j.heares.2016.08.002

Schvartz-Leyzac KC, Pfingst BE (2018) Assessing the relationship between the electrically evoked compound action potential and speech recognition abilities in bilateral cochlear implant recipients. Ear Hear 39(2):344–358. https://doi.org/10.1097/AUD.0000000000000490

Skidmore J, Oleson JJ, Yuan Y, He S (2023) The relationship between cochlear implant speech perception outcomes and electrophysiological measures of the electrically evoked compound action potential. Ear Hear 44(6):1485. https://doi.org/10.1097/AUD.0000000000001389

Skidmore J, Yuan Y, He S (2024) A new method for removing artifacts from recordings of the electrically evoked compound action potential: single-pulse stimulation. medRxiv, p 2024.01.17.24301435. https://doi.org/10.1101/2024.01.17.24301435

Somers B, Long CJ, Francart T (2021) EEG-based diagnostics of the auditory system using cochlear implant electrodes as sensors. Sci Rep 11(1):5383. https://doi.org/10.1038/s41598-021-84829-y

Steel KP (2000) New interventions in hearing impairment. BMJ 320(7235):622–625. https://doi.org/10.1136/bmj.320.7235.622

Stronks HC, Biesheuvel JD, de Vos JJ, Boot MS, Briaire JJ, Frijns JHM (2019) Test/retest variability of the eCAP threshold in advanced bionics cochlear implant users. Ear Hear 40(6): 1457–1466. https://doi.org/10.1097/AUD.0000000000000721

Undurraga JA, van Wieringen A, Carlyon RP, Macherey O, Wouters J (2010) Polarity effects on neural responses of the electrically stimulated auditory nerve at different cochlear sites. Hear Res 269(1):146–161. https://doi.org/10.1016/j.heares.2010.06.017

Undurraga JA, Carlyon RP, Wouters J, van Wieringen A (2012) Evaluating the noise in electrically evoked compound action potential measurements in cochlear implants. IEEE Trans Biomed Eng 59(7):1912–1923. https://doi.org/10.1109/TBME.2012.2194292

van der Beek FB, Briaire JJ, Frijns JHM (2012) Effects of parameter manipulations on spread of excitation measured with electrically-evoked compound action potentials. Int J Audiol 51(6): 465–474. https://doi.org/10.3109/14992027.2011.653446

Van Weert S, Stokroos RJ, Rikers MMJG, Van Dijk P (2005) Effect of peri-modiolar cochlear implant positioning on auditory nerve responses: a neural response telemetry study. Acta Otolaryngol 125(7):725–731. https://doi.org/10.1080/00016480510028492

Visram AS, Innes-Brown H, El-Deredy W, McKay CM (2015) Cortical auditory evoked potentials as an objective measure of behavioral thresholds in cochlear implant users. Hear Res 327:35–42. https://doi.org/10.1016/j.heares.2015.04.012

Wouters J, McDermott HJ, Francart T (2015) Sound coding in cochlear implants: from electric pulses to hearing. IEEE Signal Process Mag 32(2):67–80. IEEE Signal Processing Magazine. https://doi.org/10.1109/MSP.2014.2371671

Zhou N, Pfingst BE (2014) Relationship between multipulse integration and speech recognition with cochlear implants. J Acoust Soc Am 136(3):1257–1268. https://doi.org/10.1121/1.4890640

Advanced Performance Measurement Techniques in Cochlear Implant Patients: EEG, eABR

8

Olivier Valentin, Don Luong Nguyen, and Alexandre Lehmann

Abstract

This chapter provides a comprehensive overview of the use of eABR and EEG as two important electrophysiological measures in cochlear implant (CI) users. It outlines the motivations, methodologies, and applications of these measures, distinguishing between clinical and research aspects. The chapter is divided into two main parts: eABR and EEG. In the eABR part, we begin with the clinical applications of eABR during surgery and the programming of CI devices. We explain best practices for eABR acquisition and interpretation and discuss

O. Valentin (✉)
School of Communication Sciences and Disorders, Faculty of Health, Dalhousie University, Halifax, NS, Canada

Laboratory for Brain, Music and Sound Research & Centre for Research on Brain, Language or Music, Montreal, QC, Canada
e-mail: olivier.valentin@dal.ca

D. L. Nguyen
Laboratory for Brain, Music and Sound Research & Centre for Research on Brain, Language or Music, Montreal, QC, Canada

Department of Speech-Language Pathology and Audiology, McGill University Health Centre, Montreal, QC, Canada

Research Institute of the McGill University Health Centre, Montreal, QC, Canada
e-mail: don.nguyen@mail.mcgill.ca

A. Lehmann
Laboratory for Brain, Music and Sound Research & Centre for Research on Brain, Language or Music, Montreal, QC, Canada

Research Institute of the McGill University Health Centre, Montreal, QC, Canada

Department of Otolaryngology, Head and Neck Surgery, Faculty of Medicine and Health Sciences, McGill University, Montreal, QC, Canada
e-mail: alexandre.lehmann@mcgill.ca

B. Molaee-Ardekani, M. Segovia-Martinez (eds.), *Advancing Cochlear Implants*,
https://doi.org/10.1007/978-3-032-15719-5_8

cutting-edge research on eABR, focusing on its use in understanding auditory system maturation and detecting cochlear abnormalities. The EEG part explores the application of EEG to assess cognitive and sensory functions in CI patients. We explain how EEG can provide objective markers for processes such as speech and music cognition, neural plasticity, sensory integration, and listening effort. Additionally, we address the challenges of recording EEG signals in the presence of CI artifacts and strategies for minimizing these artifacts.

Keywords

eABR · EEG · Speech and music cognition · Neural plasticity · Sensory integration · Listening effort · CI artifacts · Signal processing

8.1 Introduction

In this chapter, we focus on electrophysiological objective measures originating from the brain, specifically the electrically evoked auditory brainstem response (eABR) and electroencephalography (EEG). These techniques are essential in both clinical and research settings for assessing auditory function and cognitive processes in cochlear implant (CI) users. Recording brain activity requires specialized equipment, including electrodes, amplifiers, and filters. Electrodes are placed on the scalp to record the activity of groups of neurons near the electrode. The recorded responses can result from specific events, such as auditory stimuli, or from spontaneous brain activity. One primary advantage of these techniques is their high temporal resolution, allowing for the recording of neural events on the order of milliseconds. However, they are limited by poor spatial resolution, making it challenging to pinpoint the precise location of neural activity. These electrophysiological techniques are particularly valuable in clinical settings as they provide objective information about the patient's auditory system and the performance of the cochlear implant. This is especially important for patients who are unable to communicate reliably, such as young children or individuals with certain disabilities. By offering objective measures, clinicians can better assess and adjust CI parameters to optimize auditory outcomes. Clinically, eABR provides crucial information about the integrity of the brainstem's auditory pathways. It is similar to air-conducted auditory brainstem response (ABR) but uses electrical stimulation via the cochlear implant. eABR threshold values are closely correlated with threshold current levels (T-levels), providing a reliable measure for setting initial stimulation levels.

Beyond its clinical utility, eABR has significant research applications. It can assess the maturation of the auditory system, with studies indicating that eABR morphology obtained during surgery correlates with postoperative speech outcomes. This suggests that eABR could predict future auditory performance with a CI. Additionally, eABR can help identify cochlear malformations, nerve deficiencies, and auditory neuropathies, making it a versatile tool for both diagnosis and prognosis. Section 8.2 provides an in-depth exploration of eABR, starting with its introduction and historical context. It then discusses its clinical applications for verifying electrode insertion during surgery and estimating minimum electrical thresholds during programming, covers best practices for acquisition and interpretation, and explores its cutting-edge research applications.

Section 8.3 is dedicated to the EEG in CI. EEG provides information about brain activity at a higher level than the brainstem. It can be recorded passively or in response to specific stimuli, such as auditory signals. When recording cortical activity time-locked to events, this method is called evoked potentials. Evoked potentials allow researchers to investigate more complex cognitive functions by analyzing how the brain processes specific sensory inputs. This method enables researchers to study speech and music cognition among CI users and to establish correlations between CI users' speech perception scores and specific patterns of brain activity, aiming to understand how different brain regions contribute to auditory processing. For example, EEG can investigate differences in neural processing of emotional perception between CI users and normal-hearing individuals (Sect. 8.3.1).

EEG can also play a role in assessing brain plasticity (Sect. 8.3.2). This is particularly relevant in the context of cochlear implantation, as the brain undergoes significant changes in response to hearing deprivation and restoration. EEG can assess these changes both before and after implantation, providing valuable insights into how the brain adapts to the implant. Understanding brain plasticity is crucial for explaining why some individuals achieve better auditory outcomes with a CI than others. Moreover, preimplantation EEG recordings might eventually help predict individual variability in CI outcomes, guiding personalized rehabilitation strategies.

Multisensory integration is crucial for complex cognitive tasks involved in everyday activities, such as focusing on a conversation in a noisy environment. While CIs are considered the most successful neural prosthesis, the signal transmitted to the auditory nerve remains degraded compared to natural hearing. Consequently, the ability to process and integrate information from different senses, such as combining voice and lip movement, is impaired in CI users. EEG can provide a deeper understanding of how CI users integrate auditory information with other senses, such as visual and tactile inputs (Sect. 8.3.3), which could ultimately lead to the development of more effective rehabilitation strategies that enhance overall communication outcomes and quality of life for CI users.

Listening effort, the mental exertion required to attend to and understand an auditory message, can be particularly taxing for hearing-impaired individuals, including CI users. Section 8.3.4 explores how EEG studies can quantify the cognitive load associated with listening tasks, providing insights into the effort CI users need to expend to achieve understanding.

Recording EEG in CI users presents unique challenges due to potential artifacts introduced by the CI device. For example, the electrical activity generated by the CI's internal electrode array during stimulation can produce artifacts in EEG recordings, complicating the interpretation of brain signals. To minimize CI artifacts, researchers and clinicians employ various strategies during experimental design and signal processing. Section 8.3.5 reviews the main strategies to optimize signal-to-noise ratio, along with tips for achieving clean EEG recordings.

In summary, this chapter examines the applications of eABR and EEG in clinical and research settings among cochlear implant recipients. The discussion encompasses the assessment of auditory and cognitive functions, such as speech and music cognition, neural plasticity, and sensory integration. Furthermore, the chapter addresses methodological challenges associated with EEG signal recording in cochlear implant users, focusing on improving cochlear implant programming and rehabilitation methodologies, and providing recommendations for optimizing signal acquisition quality.

8.2 eABR: From Clinical Practice to Cutting-Edge Research

The eABR is essentially the ABR but stimulated electrically. It has proven to be an important objective measurement that can be used for adjusting current levels. However, in day-to-day clinic, this measurement is often overlooked since it has been overshadowed by simpler objective measurements such as the evoked compound action potential (eCAP). Nonetheless, the eABR is still one of the only ways of assessing function beyond the cochlea and should continue to be considered in future research. The following section will provide additional information about clinical applications of eABR, best practices for eABR acquisition, and latest advances with eABR.

8.2.1 Introduction to ABR and eABR

The auditory brainstem response (ABR) is a well-established technique used to record neural activity along the auditory pathway between the cochlea and the brainstem. This neurophysiological technique is used to evaluate the integrity and function of different neurological structures located in this pathway. The ABR waveform is obtained by recording electrical activity triggered by the presentation of auditory stimuli. It is recorded using scalp electrodes, while the subject remains very still or is sleeping. The ABR waveform has five distinctive peaks that are associated with different generators that are located along the auditory pathway. This objective measurement technique has proven to be very valuable, and it provides the gold standard for the diagnosis of hearing loss (HL) in newborn children. It is also used to obtain signs of brainstem lesions, and nerve damage associated with vestibular schwannomas (Young et al. 2024).

The cochlear implant is a neural prothesis that can partially restore hearing in individuals with severe to profound HL. It is composed of an external processor and the implanted electrode array. The external processor houses the microphones and the components responsible for sound processing. The electrode array that is implanted in the cochlea will substitute the hair cells and stimulate the auditory nerve via electrical impulses. For success with the cochlear implant, it is crucial to have adequate function of the auditory nerve and brainstem. Unfortunately, the traditional ABR is not compatible with most cochlear implant candidate since their HL has progressed to a severity that prevents acquisition of ABR responses. Therefore, it is not possible to assess the auditory nerve and brainstem using traditional ABR with implanted patients.

Starr and Brackmann (1979) successfully recorded an ABR in humans using electrical impulses from a cochlear implant rather than auditory stimuli. This response would eventually be termed the eABR (electrical auditory brainstem response). The eABR shares many similarities with the traditional ABR. The main difference with the ABR is the delivery of the stimuli is done electrically rather than acoustically. In most cases, the electrical stimuli are delivered using the cochlear implant. The ability to acquire this objective measure not only allows for the evaluation of key structures important for hearing but has also proven important for cochlear implant programming. In the next section, we will discuss the various clinical applications related to the eABR.

8.2.2 Clinical Applications

The eABR is a valuable tool that allows clinicians to assess the auditory pathway beyond the cochlea. It has many applications similar to those of the ABR. Interestingly, while the ABR is often used for diagnostic or monitoring, the eABR is primarily utilized to guide the adjustment of current levels. In this section, we will concentrate on the main clinical applications of the eABR.

After recovering from surgery, all patients will go through a few appointments for programming the cochlear implant. The main objective is to provide access to as many sounds as possible. This will set the building blocks for the patient to learn and understand speech again. Throughout the process, many iterations of a "map" will be created. A "map" is a collection of parameters such as current levels, stimulation rate, and coding strategies. These maps will shape the sound experience of the implanted individual. Every version will be fine-tuned and hopefully provide clearer and more natural sounds. Programming a cochlear implant involves adjusting, among other parameters, current levels for individual electrodes along the electrode array. For a given electrode, an upper current limit and a lower current limit can be adjusted.

In order to determine the upper limit, patients are asked to judge the loudness of different stimulations on the electrode array. The audiologist's objective is to adjust the current levels to reach a certain loudness rating, generally loud but comfortable. Reliable patient feedback is necessary to determine the appropriate current levels for each electrode. For certain populations, such as children or individuals with cognitive difficulties, obtaining a reliable response can be difficult. Even with a pleasant and cooperative patient, reporting loudness can be difficult when they have not heard for decades. To compliment patient feedback, clinicians can use the eSRT to guide their programming.

The eSRT leverages a routine test, the stapedial muscle reflex, to estimate upper current levels (Pérez-Rodríguez et al. 2023; Holder et al. 2024). This measurement only requires a tympanometer (in addition to the CI programming software) that is usually available in most audiology clinics. Instead of presenting acoustic stimuli to trigger the stapedial reflex, the eSRT is measured when the muscle reflex is triggered by delivering electrical impulses via the cochlear implant. Usually, the reflex is measured in the contralateral ear from the cochlear implant. The goal is to find the lowest current level that elicits the stapedial reflex. The measurement can be repeated on different electrodes to obtain an upper limit value for different electrodes. All CI manufacturers define their upper limit differently. There is a correction factor applied depending on the manufacturer of the cochlear implant. In summary, the eSRT is a test that can be performed with minimal equipment. It is well correlated with the upper current levels. However, to measure a reliable reflex, the patient must have normal middle ear function. Additionally, some individuals will not display this reflex despite normal middle ear function.

Similar to eSRT, the eABR provides objective information that can be used as a guide when programming a cochlear implant. Unlike the eSRT, the eABR is loosely associated to the lower current limit (T-level) when programming a cochlear implant (Chao et al. 2023). Determining T-level can be a tedious task for certain patients. Patients are asked to detect very soft sounds multiple times. This exercise is repeated for multiple electrodes. This task can be very difficult early on during cochlear implant programming. Patients may have not heard for a very long time, or their tinnitus can be overwhelming making soft sounds hard to detect. Furthermore, some patients are not even able to report this information back to the audiologist (e.g., very young children). In the absence of reliable patient feedback, the eABR is a valuable measurement that provides some information for setting the T-levels. The eABR threshold will usually fall between the upper and lower current limits. The eABR threshold is usually closer to the lower limit than the eCAP threshold, another objective measure that can aid with programming (Ries et al. 2024). The eCAP has one major advantage over the eABR. It does not require any additional equipment other than the equipment used to program a cochlear implant, making this measure particularly interesting in a clinical setting.

The eABR can also be used intraoperatively to assess the integrity and function of the cochlear nerve. This technique can be useful during acoustic neuromas resection and help determine whether a cochlear implant is still an interesting option following surgery (Alzhrani et al. 2024). The idea is similar to monitoring with the standard ABR. Again, instead of using an acoustic stimulus to elicit the ABR, an electrical stimulation is delivered via the cochlear implant. Morphology and latency of the brainstem response is assessed with varying presentation parameters.

Postoperatively, the eABR could be used to confirm auditory function or continue monitoring cochlear nerve function (Nada et al. 2022). For example, the eABR can be useful when implanting an individual with auditory neuropathy spectrum disorder (ANSD). ANSD occurs when the signal from the cochlea is not properly transmitted through the auditory nerve, often because of desynchrony. With cochlear implantation, the hope is to provide enough current to reestablish this synchronized firing from the auditory nerve. With the eABR, a clinician could verify if the brainstem response is intact following implantation. This would suggest success at overcoming the desynchrony from ANSD (Jeon et al. 2013).

8.2.3 Best Practices for eABR Acquisition

As with other electrophysiological measurements, the quality of recording can depend on patient setup, equipment, and recording and presentation parameters. Optimal conditions and setup for recording eABRs are fairly similar to those for recording a traditional ABR. Below are some practical considerations regarding patient preparation, electrode montage, necessary equipment, and suggested recording parameters.

8.2.3.1 Patient Preparation

Patient preparation is key for obtaining clean eABR recordings. The brainstem response is a relatively weak electrical signal that can be easily dominated by muscle contractions or movement. For this reason, the eABR is acquired, while the patient is under anesthesia. This prevents excessive motion and allows for the cleanest recordings. Interestingly, it has been shown that anesthesia or sedation does not influence the eABR response. This protocol is certainly used for young children but can be avoided in adults who can remain relaxed and calm during the recording. Patients are either lying down or in a seated and reclined position. Patients who are able to sleep during the procedure are encouraged to do so. To maximize patient cooperation, it is important to clearly explain the procedure before starting any manipulations. It is important to not forget that patients undergoing eABR are hard of hearing and will greatly benefit from their hearing devices during the procedure explanations.

8.2.3.2 Electrode Montage

The recording can be performed with a standard 3 electrode montage for ABR. The inverting electrode (−) is positioned on the contralateral ear to the cochlear implant. The noninverting electrode (+) is positioned on the high forehead. Finally, the ground electrode is placed on the lower forehead. The inverting electrode is placed contralaterally to the cochlear implant to minimize electrical interference when the impulse is delivered. Skin preparation is similar to the standard ABR. Scrubbing is required before placing the electrodes to ensure adequate impendence levels. Ideally, the impedance should be similar between electrodes and under 5 kΩ. Finally, ensure that the electrodes are well attached to the skin by using conductive paste, gel, or adhesive electrodes.

8.2.3.3 Recording Equipment

The hardware used to record an ABR can also be used to record the eABR. However, the stimuli will not be delivered through the ABR equipment. Rather, stimulation will be initiated through the cochlear implant programming software. This is a crucial difference since it will require an additional trigger cable to synchronize the timing of stimuli onset with the recording software. It is best to verify with the manufacturer for the appropriate trigger cable. It is also important to verify if your current ABR system can accommodate a trigger cable.

8.2.3.4 Recording Parameters

Exact parameters will differ depending on the various cochlear implant manufacturers and the software used for acquisition. It is best to verify the manufacturer's recommendation. In general, parameters on the recording software need to be changed to indicate that the stimulation will be initiated by an external apparatus. Filters are generally set between 50 and 5000 Hz. The clinician can choose through which electrode the current will be delivered. The stimulus rate is also determined via the cochlear implant programming software. Depending on the objective of the recording, the current level will be varied throughout the recording.

8.2.3.5 eABR Characteristics

The eABR response is similar to the ABR response. It is characterized by five waves, where wave III and V are the most prominent. Wave I is often obscured by the artifact that occurs before 2 ms caused by the implant itself. Latencies are shorter for eABR when comparing with ABR. This is due to the direct stimulation of the cochlea through the implant. There is no delay caused by the sound wave travelling through the structures of the ear as seen with the acoustic stimuli. Generally, there are no shifts in latency with change in intensity. The observed response is also usually larger in amplitude than the ABR response.

8.2.4 Cutting-Edge Research on eABR: Latest Advances and Insights

As mentioned in the previous sections, the eABR is an involved measurement. Optimal conditions, sometimes requiring anesthesia, is required for adequate recordings. The information is occasionally useful for programing a cochlear implant but is often not required. For these reasons, it is not surprising that the eABR is not a routine practice in cochlear implant clinics. In fact, the eABR is only used in very specific situations or entirely omitted. Despite the limitations associated with this measurement, the eABR still holds valuable research potential. This technique is one of the only methods of peering at the auditory pathways beyond the cochlea. In the following, we describe the latest knowledge surrounding the use of eABRs and how these may eventually translate into future applications.

8.2.4.1 eABR as a Maturation Marker for the Auditory System

As the external processor is activated for the first time, an entirely new input is delivered to the auditory nerve. The ear must learn how to understand from this novel signal. At first, when the programming is adequate, most sounds should be audible at a conversational level. However, clarity is often not immediately present.

With time and practice, these foreign sounds become words and words become sentences. In other words, there is some form of change happening along the auditory pathway.

It has been suggested the eABR can be used as a marker of maturation of the auditory system. Similar to the standard ABR, the waveform associated with the ABR changes in latency and morphology as the auditory system matures. This characteristic of the ABR waveform is maintained in the eABR. It has been shown that the eABR peak latencies decrease with cochlear implant use. This effect was independent of age of implantation and duration of deafness (Gordon et al. 2006). Furthermore, greater change that occurred between 0.5 and 1 year was observed for the latency of wave V and the inter-latency difference between waves III and V, suggesting more change in the rostral brainstem. Interestingly, the magnitude of change is independent of the age of implantation. The ability to tract maturation of the auditory pathways can be particularly valuable when assessing young children. Contrary to adults, it is much more difficult to track the progress of very young children with speech scores. With an objective method that tracks maturation of the auditory system, one could potentially use the eABR to determine whether the brainstem response of an implanted patient is progressing according to an appropriate time course.

The eABR response has also been investigated in sequential bilateral CI recipients. Here, researchers questioned whether electrical stimulation of the second ear will result in different eABR responses overtime. The researchers measured eABR response in both implanted ear at different time points up to 6 months after implantation. They found that the latencies were longer in the newly implanted ear and were longer even 6 months after implantation. This suggests that the newly implanted ear is still maturing even half a year after implantation (Deniz et al. 2021).

Researchers looked at the difference in eABR response between prelingual and postlingual deafened adults (Lammers et al. 2015). They noted that wave V, the most robust component of the eABR response, was significantly delayed in prelingual hearing-impaired patients. These results suggest that the auditory system in prelingually deaf adults shows residual delays from an impairment of maturation during infancy. This is particularly interesting because their results not only show a delay with the electrophysiological results but also behavioral results. They found that shorter wave V latencies appeared to be a predictor for better speech perception.

8.2.4.2 Predicting Cochlear Implant Performance Outcomes

One of the most difficult tasks related to the CI process is to predict performance outcomes. Many patients wait too long before starting the process for a CI. Most patients that are implanted could have benefited from an implant years before they commit to the process. In fact, it is estimated that only about 5% of the people who could benefit from a CI get a CI in the right time. Among the many different factors that contribute to this abysmal number, difficulty to predict outcomes is a main factor leading to delayed implantation. Committing to the CI is seen as a point of no return. One cannot simply trial a CI. Currently, it is difficult for clinicians to be able to quantify the expected benefit of a CI. The inability to promise significant benefits is certainly contributing to the hesitation related to cochlear implantation. The current factors that are currently examined during candidacy are only able to explain about 25% of the variability in outcomes postimplantation. The eABR has been suggested as an additional metric that can be used to predict postoperative speech outcomes (Lammers et al. 2015; Jafari et al. 2024).

Researchers recorded eABR in unilaterally implanted children and correlated these results with various outcome measures (Abohelal et al. 2024). They found that the amplitude growth function of the eABR as well as the maximum amplitude correlated with speech in quiet, speech in noise, and phonological assessment scores. Similar findings were reported by Wang et al. (2015). These predictive qualities of the eABR are truly valuable when assessing young children. Before a certain age, it is not possible to obtain results with speech tests. Besides subjective parent reports, it is difficult to track progress. A tool that could confirm proper signal transmission and that is correlated with future performance in speech tests would help tremendously in counseling and confirming adequate programming.

8.3 Application of EEG to Assess Cognitive and Sensory Functions

Decoding the intricate neural processes of CI users presents both challenges and opportunities. EEG offers an ideal solution for exploring sensory and cognitive functions beyond the brainstem in CI users. Its noninvasive and CI-compatible nature allows researchers to track neural activity across different time points, offering valuable insights into brain maturation and functional progress postimplantation.

In this section, we will distinguish three related cortical measures—continuous EEG, event-related potentials (ERPs), and electrically evoked auditory steady-state responses (EASSRs)—and show how each is used in CI research as well as how their signal processing pipelines differ (and overlap) with those for electrically evoked brainstem responses (eABRs):

- Continuous EEG: Records ongoing scalp potentials without stimulus-locked repetition. In CI users it reveals spontaneous cortical rhythms and state changes (e.g., attention, arousal).
- Event-Related Potentials (ERPs): Extract time-locked responses by averaging 100–200 trials of EEG segments synchronized to stimulus onset.
- Electrically Evoked Auditory Steady-State Responses (EASSRs): Capture phase-locked cortical responses to periodic stimuli by averaging 200–500 trials in the frequency domain.

Researchers employ both continuous EEG (to probe dynamic, nonrepetitive cortical activity) and ERP/EASSR paradigms (to isolate stimulus-locked components). Signal processing methods have been designed for both raw EEG and averaged signals: ICA and artifact rejection are applied before averaging to clean each trial, while filtering, spectral analysis, and residual denoising may also be applied after averaging to the composite ERP or EASSR waveforms.

By contrast, eABR reflects brainstem activity and therefore uses much higher-frequency filters (100–3000 Hz) and hundreds to thousands of repetitions (>500 trials) to extract waves I–V. Although both eABR and cortical EEG measures rely on trial averaging to boost weak, stimulus-evoked signals, their distinct neural origins demand different bandpass settings, artifact-rejection strategies, and averaging approaches.

The following subsections delve into specific applications of EEG in the context of CI users. The "Speech and Music Cognition" section examines how EEG markers such as mismatch negativity (MMN; an ERP component) illuminate the neural complexities of auditory and emotional processing. The "Neural Plasticity" subsection explores how EEG can track intramodal and cross-modal plastic changes, shedding light on the factors influencing CI success. The "Sensory Integration" subsection unpacks the neural mechanisms underpinning the integration of multiple sensory inputs, leveraging methods such as steady-state evoked potentials (SSEPs; frequency-domain averaging). In "Listening Effort," we explore the role of EEG, to measure cognitive load in challenging listening scenarios. Finally, "Recording Considerations and Signal Processing" offers practical strategies for minimizing artifacts and ensuring the accuracy of EEG data, paving the way for more effective clinical and research applications.

8.3.1 Speech and Music Cognition

Speech and music cognition are complex processes that play pivotal roles in communication, social interaction, and cultural engagement. For CI users, these processes present challenges due to the technological limitations of CIs, which provide limited spectral and temporal resolution. Speech cognition, essential for daily communication, relies on recognizing subtle phonetic cues, emotional prosody, and linguistic patterns that are often difficult for CI users to discern, especially in noisy environments. Music cognition, on the other hand, involves perceiving intricate elements such as pitch, melody, rhythm, and timbre, which current CI technology struggles to replicate. This dual challenge affects CI users' ability to fully engage in social contexts and experience the emotional and aesthetic aspects of sound. Understanding how CI users process speech and music and ways to measure neurophysiological markers associated with these processes is critical for developing tailored interventions and advancing CI technology.

8.3.1.1 Speech Cognition

CIs were engineered primarily to restore speech comprehension by bypassing damaged hair cells and directly stimulating auditory nerve fibers. Despite their innovative design, CIs inherently offer limited spectral and temporal resolution, which impairs the user's ability to distinguish finer phonetic details, particularly in complex auditory environments. These constraints hinder essential aspects of speech processing, from prosody recognition to understanding emotion in voices. For example, children using CIs often have difficulty identifying emotional prosody—critical for effective social communication—as they miss important acoustic cues for emotion, such as pitch and intonation (Nakata et al. 2012).

Speech perception in CI users varies significantly based on factors such as the age at which deafness occurred and the timing of implantation. Early implantation (before age 4) is associated with improved long-term outcomes in speech and language development, underscoring the importance of early auditory exposure (Dunn et al. 2014). EEG has proven essential in studying these auditory processes in CI users, particularly through markers such as mismatch negativity (MMN). MMN reflects the brain's automatic response to changes in sound and serves as a reliable indicator of speech discrimination abilities in CI users (Kraus et al. 1993). CI users with stronger MMN responses tend to have better speech recognition abilities, making MMN a valuable tool for assessing speech processing in both clinical and research settings (Turgeon et al. 2014; Prévost and Lehmann 2018). Other EEG markers, such as cortical alpha oscillations, have been linked to effortful decoding of speech. Increased alpha power in parietal regions reflects the additional cognitive resources CI users allocate to understand speech in noisy environments (Paul et al. 2021). A study used EEG to compare emotional prosody recognition in CI users with two different processing strategies: ACE and MP3000 (Agrawal et al. 2013). Through event-related potentials (ERPs), the study found that CI users with the MP3000 strategy showed greater accuracy in recognizing prosodic information, particularly for happy emotions, with a significantly higher P200 amplitude than ACE users. Gamma power analysis further revealed early and late gamma activities in MP3000 users, suggesting enhanced attention and cognitive control. The temporal response function (TRF) is an EEG tool that tracks how brain activity aligns with the envelope of natural, continuous speech, providing a realistic view of speech perception in everyday listening situations. Notably, TRF has been effectively used with CI users, helping to reveal that the brain can still follow speech rhythms despite hearing challenges, as shown in Xiu et al. (2022). CI users often rely heavily on visual cues, such as lip movements, to compensate for auditory limitations. EEG studies of multisensory integration reveal that CI users demonstrate heightened visual cortex activation during audiovisual speech tasks, indicating a unique compensatory mechanism that improves speech comprehension, especially in noisy settings (Layer et al. 2022a).

In conclusion, EEG offers objective markers, which could be used to guide the development of tailored auditory training programs. Such programs could be optimized to address individual speech processing needs and listening challenges in CI users, fostering improvements in both quiet and noisy environments. Future research may focus on enhancing CI functionality for multisensory integration, allowing users to leverage visual cues more effectively in real-world situations. Improving CI users' speech comprehension can also be achieved by advancing CI technology to provide finer spectral and temporal resolution, as well as emerging "machine hearing" algorithms, which would enhance the user's ability to process complex auditory signals.

8.3.1.2 Music Cognition

Music cognition has two key aspects: music perception and music appraisal. Music perception is challenging for CI users due to the limited ability of CIs to transmit fine-grained spectral information essential for pitch, timbre, and melody recognition. Unlike speech, which requires more basic spectral resolution, music involves complex auditory features that current CI technology cannot fully replicate (Limb and Roy 2014). These limitations significantly impact CI users' experiences with music, affecting both the emotional and aesthetic enjoyment derived from musical exposure (Mc Dermott 2004). Regarding music appraisal, while some normal-hearing (NH) listeners may perceive music accurately without fully enjoying it, many CI users, despite perceptual limitations like poor pitch discrimination, still feel powerful emotions in response to music, dance to it, and appreciate it. This appraisal supports social engagement and enhances quality of life (Looi et al. 2012).

ERPs are valuable tools for examining neural responses to musical features. A synthesis of ERP research indicates that cochlear implantation significantly impacts the perception of musical elements, including complex rhythm and duration. CI users display neural activation patterns that differ notably from those of individuals with normal hearing (Sharp et al. 2018).

The role of neuroplasticity is apparent in music perception, as repeated musical exposure can lead to improvements in neural encoding, which may enhance CI users' auditory experience and music enjoyment. The MMN has also proven valuable in music research. A study found that CI users show diminished cortical encoding of timbre changes compared to normal-hearing (NH) listeners, with MMN present in all NH listeners but only about half of CI users. For CI users who exhibited MMNs, these responses were reduced in amplitude and duration, in line with their challenges in distinguishing melodies and harmonics (Zhang et al. 2013). Another study used MMN to show that adolescent CI users exhibit neural responses to changes in timbre, intensity, and rhythm, indicating some residual music processing ability. However, CI users showed weaker MMN responses for pitch changes, aligning with their lower behavioral performance in pitch discrimination compared to normal-hearing peers (Petersen et al. 2015). Lehmann and Paquette (2015) propose that EEG could be used to explore how CI users process musical emotions through non-pitch cues like rhythm and tempo, given their difficulties with pitch perception. This approach may reveal compensatory neural mechanisms in music perception. Using alpha band EEG asymmetry, Vecchiato et al. (2012) found that bilateral CI users demonstrated brain activation patterns related to music pleasantness similar to healthy children, while unilateral CI users showed different patterns. Marsella et al. (2014) showed that CI users exhibited atypical alpha asymmetry in response to musical stimuli, suggesting an impaired ability to distinguish pleasant music from dissonant sounds, which may reflect challenges in affective engagement with music.

In conclusion, while music perception remains challenging for CI users, EEG provides insights that could drive improvements in CI technology and optimizes training interventions aimed at enhancing music perception. These advancements could significantly improve CI users' quality of life, allowing them to experience music with greater richness and clarity.

8.3.2 Neural Plasticity

Neural plasticity—the brain's remarkable ability to reorganize itself—is critical for adaptation following cochlear implantation. After receiving a CI, the brain must adjust to interpreting the device's electrical signals, requiring substantial reorganization, particularly within the auditory pathways. Adaptive plasticity refers to beneficial neural changes that enhance CI users' ability to process auditory stimuli. For instance, early-implanted users often exhibit stronger neural encoding in response to speech stimuli, which is associated with more successful auditory outcomes (Glennon et al. 2020). This form of plasticity is particularly prominent in younger individuals who receive CIs early in life, often during critical developmental windows when the brain is most receptive to reorganization. Research has shown that early-implanted children develop auditory processing capabilities closer to those of normal-hearing peers, reflecting a high degree of plasticity that supports effective adaptation to the implant's input (Sharma et al. 2015). The causes of variability in CI outcomes remain unclear. Clinical factors, such as age at deafness onset, duration of deafness, age at implantation, and length of CI use, significantly affect speech perception outcomes (Blamey et al. 2015). Brain plasticity postimplantation also plays a key role, accounting for over 30% of auditory performance improvement (Lazard et al. 2012). While adaptive plasticity can facilitate improved speech and sound processing, prolonged auditory deprivation may lead to cross-modal plasticity, where nonauditory brain regions—such as the visual and somatosensory cortices—take over areas traditionally dedicated to auditory processing. The extent to which cross-modal plasticity can be adaptive or maladaptive remains debated, with differences observed between prelingually and postlingually deafened patients (Heimler et al. 2014; Glennon et al. 2020).

Cross-modal plasticity, as observed in functional MRI (fMRI) and positron emission tomography (PET), is characterized by increased visual cortex activation during auditory tasks, especially in individuals with prolonged deafness (Heimler et al. 2014). This plasticity, however, can be maladaptive for postlingually deaf CI users who often develop a dependency on visual cues like lip reading, which may complicate their ability to prioritize auditory input postimplantation (Lazard et al. 2023). A functional MRI study suggests that after deafness, increased coupling between the occipital lobe and the temporal lobe may predict poor CI outcomes (Lazard and Giraud 2017). While imaging techniques can explore neuroplastic changes, most are not CI compatible. MRI, though commonly used, is limited by CI-related artifacts, and PET, though feasible, involves invasive tracers. EEG, however, offers a noninvasive, implant-compatible alternative. EEG provides valuable insights into the neural mechanisms underlying plasticity in CI users, offering both spatial and temporal resolution that enables researchers to track changes in brain activity over time.

EEG studies have showed evidence that cross-modal activation negatively correlated with outcomes in postlingually deafened cochlear implant users. It was shown that poor CI performers often exhibit broader, anteriorly distributed high P2 amplitudes, indicating extensive cross-modal reorganization, whereas good performers display higher P2 amplitudes in visual occipital regions, suggesting better audiovisual coupling crucial for speech discrimination (Doucet et al. 2006). Additionally, EEG studies have identified smaller P100 amplitudes and reduced visual cortex activation in CI users, with P100-related activation in the right auditory cortex inversely related to speech recognition, implying that incomplete reversal of cross-modal reorganization may limit CI efficacy (Sandmann et al. 2012). Increased visual cortex activity in CI users during visual tasks was reported using high-density EEG to record the visual mismatch negativity (vMMN) potential and sLORETA for cortical source estimation (Intartaglia et al. 2022a). These findings suggest that intramodal plasticity may serve as a compensatory mechanism for auditory deficits in this population. Taken together, this supports the use of EEG as an objective, implant compatible, and cost-effective method be used to track longitudinal cortical plasticity from pre- to postimplantation rehabilitation. Another, more recent and promising method, functional near-infrared spectroscopy (fNIRS), offers improved spatial resolution without electrical artifacts, making it ideal for investigating cortical plasticity in hearing-impaired individuals (Basura et al. 2018).

To conclude, understanding neural plasticity in CI users is essential for developing more effective individualized rehabilitation strategies. Early intervention is critical, as timely CI implantation can harness adaptive plasticity during sensitive developmental stages, enhancing not only auditory outcomes but also supporting broader cognitive development, integrating language, attention, and memory networks with auditory pathways. EEG-based monitoring could allow real-time tracking of plasticity and enable tailored rehabilitation plans based on individual neural responses. Future research should continue to explore how targeted training and CI enhancements can foster adaptive plasticity, improving auditory outcomes for CI users across different age groups. Additionally, investigating the role of structured training programs—such as music or multisensory training—in promoting adaptive neural plasticity could further optimize CI outcomes (Valentin et al. 2024a). Ultimately, a deeper understanding of the factors influencing plasticity in CI users will lead to innovations that improve quality of life and communication abilities for individuals with HL.

8.3.3 Sensory Integration

Sensory integration, the process by which the brain combines information from multiple senses, is essential for effective communication and interaction with the environment. The auditory and visual systems play complementary roles in facilitating speech understanding, improving perceptual accuracy, and enriching social interactions. This integration supports complex tasks such as focusing on a conversation in a noisy environment by synchronizing a speaker's voice with their lip movements or interpreting emotions conveyed through facial expressions and vocal tones. When hearing is impaired, the degradation of auditory input leads to increased reliance on visual cues to enhance speech comprehension and social communication. Following hearing restoration after a period of deafness, dynamic adaptive neuroplastic processes enable the brain to reorganize and reestablish sensory integration, which can significantly influence outcomes for individuals with CIs.

Multisensory integration is commonly studied using behavioral experiments and neuroimaging techniques. Research has shown that sensory integration is altered in CI users, reflecting the complex interplay between sensory deficits and neural adaptation. Behavioral studies have revealed significant differences in sensory reliance between NH individuals and CI users. NH individuals typically prioritize auditory inputs when processing conflicting auditory and visual stimuli, leveraging the auditory system's superior temporal resolution. Tremblay et al. (2010) demonstrated, using the McGurk effect, that proficient CI users exhibit audiovisual fusion abilities comparable to NH individuals. While proficient users rely on auditory cues, nonproficient users depend more on visual cues when faced with conflicting multisensory inputs. This highlights the role of auditory proficiency in shaping sensory reliance and integration strategies. Valentin et al. (2024b, 2025) observed that although CI users achieve auditory synchronization abilities comparable to NH individuals, postimplantation neural reorganization diminishes their preimplant visual rhythm synchronization advantage. This finding underscores the impact of sensory recovery on multimodal integration and suggests that sensory restoration may shift the balance between modalities. Layer et al. (2022b) further emphasized individual variability, reporting that unilateral and bilateral CI users exhibit distinct audiovisual speech processing patterns influenced by the degree of hearing loss and auditory experience.

EEG research has provided critical insights into how CI users integrate sensory information and the neural mechanisms underlying these processes. Doucet et al. (2006) used EEG to assess visual evoked potentials (P2 component) and their relationship to cross-modal reorganization in CI users. Poor performers exhibited anteriorly distributed high P2 amplitudes, while good performers showed higher P2 amplitudes in occipital regions. This differential neural reorganization reflects the variability in how CI users adapt to sensory deficits and suggests that successful integration involves targeted reorganization of specific cortical areas. Stropahl and Debener (2017) investigated auditory cross-modal reorganization and its relationship to audiovisual (AV) integration in CI users using EEG. Source localization of the N170 visual evoked potential in response to faces revealed significantly stronger activation in CI users' right auditory cortex than NH controls. Additionally, cross-modal activation correlated positively with AV integration strength, as measured by the McGurk illusion. These findings suggest that cross-modal recruitment of the auditory cortex supports enhanced AV speech processing, illustrating the brain's adaptive capacity in response to sensory deprivation.

Hu et al. (2021) examined sensory processing deficits in CI children using EEG. Diminished mismatch negativity (MMN) responses to tone contrasts in complex speech contexts and the absence of late discriminative negativity (LDN) highlighted challenges in integrating and processing auditory information in CI users. These findings point to deficits in both low-level sensory encoding and higher-level cognitive integration, which may contribute to difficulties in complex listening environments. Cardon and Sharma (2019) utilized high-density EEG to investigate somatosensory-auditory integration in children with CIs. Vibrotactile stimulation evoked typical somatosensory responses in both CI users and NH children, but CI users uniquely exhibited activation in auditory cortical areas, including the superior temporal gyrus, indicating cross-modal recruitment. Earlier latencies in somatosensory evoked potentials (CSEP), particularly the P50 component, were observed in CI users compared to NH controls. Notably, stronger cross-modal activation was linked to poorer speech perception in noise, suggesting that while somatosensory-auditory integration may serve as an adaptive mechanism, it is insufficient to fully compensate for deficits.

Schierholz et al. (2017) conducted an EEG study to compare auditory and audiovisual processing in CI users and NH individuals, as well as patients with auditory brainstem implants (ABI) and auditory midbrain implants (AMI). ERP analyses revealed that CI users exhibited N1 amplitudes and latencies closer to NH listeners, particularly in audiovisual tasks, indicating effective cortical processing. Both CI users and NH listeners showed significant benefits in audiovisual conditions compared to auditory-only ones, reflecting robust multisensory integration capabilities in CI users. This study highlights the unique processing advantages afforded by the combination of auditory and visual inputs, even in the context of hearing loss and sensory restoration.

Collectively, these findings underscore the intricate interplay between hearing deprivation, sensory integration, and neuroplasticity in CI users. Clinicians and researchers should consider individual variability in sensory reliance, neural adaptation, and cross-modal reorganization when designing personalized interventions for CI users. These insights have broad implications for improving auditory performance, enhancing quality of life, and advancing the development of CI technologies.

8.3.4 Listening Effort

Hearing ability is typically assessed through pure-tone audiometry, which evaluates absolute sensitivity across a range of frequencies important for speech perception. However, these audiometric measures do not reflect the difficulty listeners experience, even in instance of correct response (Mackersie and Cones 2011; Wingfield et al. 2005). Pure-tone audiograms and speech perception tests in quiet do not always accurately represent the extent of HL (Humes 2013). A common complaint among individuals with HL is understanding speech in noisy settings (Vermiglio et al. 2012). These difficulties not captured by standard audiometric tests are called "hidden hearing loss" (HHL).

Among seniors, reduced speech perception in complex or noisy situations involves factors not predictable by the audiogram alone (Martin and Jerger 2005).

Aging is linked to slower processing speeds, increasing the cognitive effort required to follow conversations (Mc Coy et al. 2005; Tun et al. 2009). Research indicates older listeners may perform as well as younger individuals on certain tasks, but their brain activation patterns differ significantly, likely reflecting compensatory cognitive mechanisms (Wingfield and Grossman 2006; Getzmann and Falkenstein 2011).

As the population ages, there is increasing concern about the rise in HHL (Amieva et al. 2015). This issue is particularly significant given the growing body of evidence linking age-related HL to cognitive decline and the onset of dementia (Deal et al. 2017; Meister et al. 2015; Lin et al. 2011). Recent studies have shown that every 10 decibels in HL increases dementia risk by 20% (Lin et al. 2013). Another study found that individuals with age-related hearing loss (ARHL) have a 24% higher risk of incident cognitive impairment compared to those with normal hearing (NH) (Livingston et al. 2017). A recent article by the Lancet Commission reported that age-related HL accounts for 9% of the risk of developing dementia (Nachtegaal et al. 2009).

Listening effort (LE) has drawn significant interest among audiologists and clinicians, particularly because the performance of individuals with hearing difficulties does not always align with the standard audiogram. For example, individuals with HL often demonstrate elevated LE (Bronkhorst 2015) even when their HL is mild (Mc Coy et al. 2005) or when they achieve correct word recognition (Tun et al. 2009). Unfortunately, the relationship between psychophysical task performance and LE is not straightforward: Two listeners might achieve the same intelligibility score while exerting different levels of effort, with pupil dilation reflecting subjective difficulty despite equal intelligibility (Mattys et al. 2012). This discrepancy may arise because speech perception relies not only on auditory processing but also on cognitive and linguistic abilities (Ohlenforst et al. 2017; Rönnberg et al. 2013). Consequently, individuals with moderate HL may achieve comparable intelligibility scores to those with normal hearing, but only by exerting more effort and relying on compensatory cognitive strategies (Pichora-Fuller et al. 1995; Granholm et al. 1996; Koeritzer et al. 2018). These compensatory mechanisms, however, can lead to diminished performance in other cognitive or physical tasks, or impaired memory of heard words (Beatty and Wagoner 1978; Mc Coy et al. 2005).

Increased LE is linked to several negative and undesirable outcomes, such as longer recovery times after work (Danermark and Gellerstedt 2004), higher rates of sick leave (Järvelin et al. 1997), early retirement (Ringdahl and Grimby 2000), unemployment (Bernarding et al. 2013), and withdrawal from social activities (Kramer et al. 1997). Despite the significant health and social impacts associated with elevated LE, a reliable and objective measure of LE has yet to be established (Koelewijn et al. 2012).

LE has been primarily measured via subjective questionnaires, but there has been a growing interest in identifying reliable and objective measures of LE.

Pupillometry is increasingly recognized as a valuable tool for measuring LE, as it provides a continuous signal that reflects processing load with sufficient temporal resolution to capture effort changes within a single sentence (Zekveld et al. 2010;

Winn et al. 2015). Pupil responses have been shown to be sensitive to various factors, including intelligibility levels (Koelewijn et al. 2012), the degree of spectral degradation of the signal (Zekveld et al. 2014), masking conditions (Kahneman and Beatty 1966; Piquado et al. 2010), syntactic complexity and sentence length (Hillyard et al. 1973), memory load (Zekveld et al. 2014), and selective attention (Mc Garrigle et al. 2014). In speech perception tasks, pupil size increases as performance levels decrease, indicating greater LE at lower intelligibility thresholds (Kahneman and Beatty 1966; Başar et al. 2001; Piquado et al. 2010; Koelewijn et al. 2012). Despite this sensitivity, the link between pupil dilation and LE has only recently been established (Klimesch 2012; Zekveld et al. 2014; Nguyen et al. 2024).

Changes in brain electrical oscillations measured using electroencephalography (EEG) are another physiological measure that has shown systematic changes associated with various cognitive processes (Karrasch et al. 2004; Pesonen et al. 2006). For example, increased activity in the alpha frequency band (8–12 Hz) is observed in response to greater cognitive demand during working memory tasks with various types of stimuli, including syllables and single words (Obleser et al. 2012; Becker et al. 2013). During speech perception tasks in challenging listening environments (i.e., noise), alpha band activity increases in response to acoustic degradation and heightened task difficulty (Mc Mahon et al. 2016; Miles et al. 2017). However, simultaneous investigation of pupil dilation and alpha power has been rare, with only a couple of studies addressing this in NH listeners (Steel et al. 2015; Ohlenforst et al. 2018).

Unfortunately, not all the proposed measures of LE correlate with one another, e.g., pupillometry measures and EEG alpha power (Haro et al. 2022). Changes in the pupillary response are said to reflect increased processing load (Neagu et al. 2023), but the relationship between EEG alpha power, pupil dilation, self-reported LE, and auditory intelligibility remains unclear. Additionally, the available research investigating these variables in CI populations remains limited (Steel et al. 2015).

8.3.5 EEG Recordings Considerations and Signal Processing

8.3.5.1 CI-Induced EEG Artifacts

CI devices comprise two main components: an external sound processor and an internal implant with electrode array inserted in the scala tympani of the cochlea. The sound processor, powered by batteries, transmits both energy and stimulus parameters, such as stimulation mode, current amplitude, phase width, phase shape, and polarity, to the electrode array through radiofrequency (RF) transmission (Zeng et al. 2008). This RF transmission, along with the electrical activity generated during stimulation of the CI electrodes, has been identified as a primary source of artifacts in EEG recordings (Debener et al. 2008; Hofmann and Wouters 2010; Martin 2007). These artifacts can manifest as electrical noise, often overlapping in frequency and timing with the neural signals of interest, complicating the interpretation of EEG data.

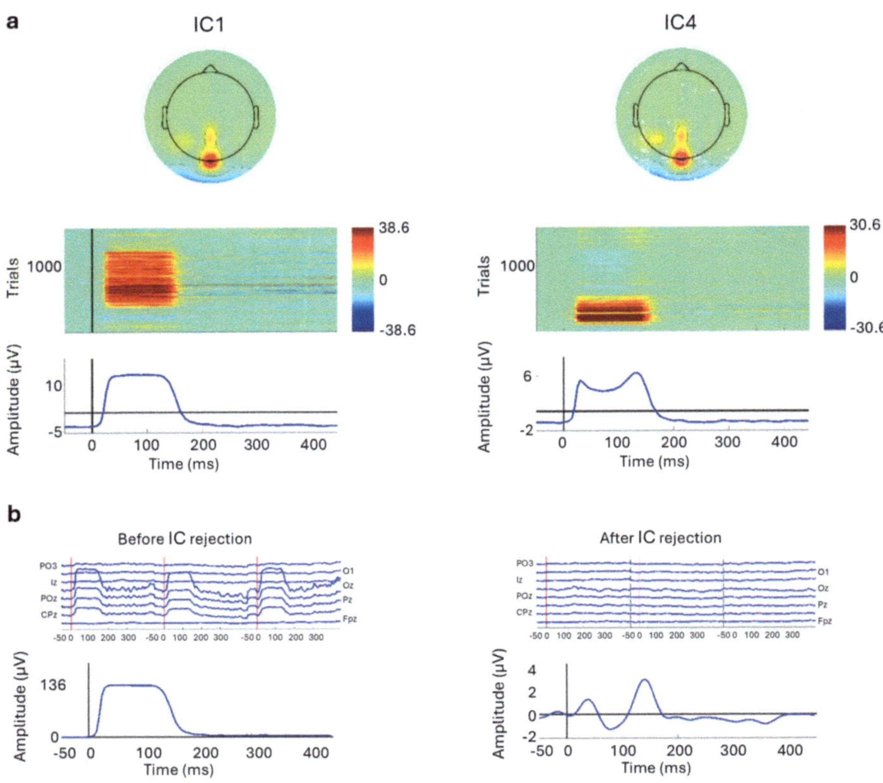

Fig. 8.1 Illustrative example of successful CI artifact rejection via independent component analysis (ICA) decomposition from one CI participant's data recorded during an auditory task. (**a**) Before ICA. (**b**) After ICA. (Reproduced from Intartaglia et al. (2022b) with permission from Elsevier)

Another potential source of EEG signal contamination is the wireless connectivity of CIs. Wireless technologies, such as Bluetooth, are increasingly prevalent among CI users, generating electromagnetic fields that can degrade EEG signal quality. However, to date, there is limited research specifically investigating the artifacts introduced by CI wireless connectivity. When conducting EEG recordings, it is recommended to disable these wireless features to minimize additional signal contamination.

The characteristics of CI-induced artifacts can vary significantly depending on the task parameters, including amplitude, duration, and shape. These variations are especially problematic in tasks designed to elicit phase-locked or time-locked brain activities, such as ERPs, because artifacts can overlap with auditory evoked potentials (AEPs), making it challenging to dissociate the two (see example in Fig. 8.1). This issue is particularly critical when studying auditory processing since CI-induced artifacts may distort or obscure the genuine neural responses, potentially leading to inaccurate conclusions. In tasks involving nonauditory stimuli, this problem can be mitigated by simply deactivating the CI.

8.3.5.2 Recording Considerations

Minimizing CI artifacts during EEG recordings requires careful planning of experimental design and consideration of various factors that influence artifact generation. One effective approach is to adjust stimulation strategies; for instance, alternating stimulus polarity can help reduce artifact contamination by canceling out the non-neural components of the EEG signal (Hofmann and Wouters 2010; Klop et al. 2004; Miller et al. 2000). Additionally, using low-rate pulse trains or specific pulse configurations can mitigate the overlap between CI artifacts and brain responses, helping to preserve the integrity of recorded neural signals.

When designing auditory EEG experiments, it is important to carefully select the type of stimulation, which can be acoustic (e.g., stimuli are presented via speakers), direct streaming (e.g., auditory stimuli are directly delivered to the audio processor using a cable to bypass the CI's external microphone), or direct electric stimulation (e.g., auditory signals are delivered directly to the inner part of the CI device, bypassing the audio processor). Each of these has different implications for artifact generation. Acoustic stimulation in free field using speakers reduces contamination from the acoustic delivering system but can be prone to artifacts from the CI's conversion of analog signals into electrical pulses (Alemi and Lehmann 2019; BinKhamis et al. 2019). In comparison, direct electric stimulation allows finer control over stimulus parameters, compared to acoustic or direct streaming stimulations: When using speakers delivered stimuli, or those delivered via audio streaming, the acoustic wave is then converted by the implant processor into an electrodogram, which describes how each stimulation electrode will deliver electric pulses to the auditory nerve over time. Since the electrodogram is dependent on the processor and the coding strategy of the implant, it is impossible to control the actual shape, length, and amplitude of the electrical pulses delivered to stimulation electrodes of the implant. This limitation can be overcome by using a direct electric stimulation as the computer-CI interface allows the experimenter to bypass the processor and finely control all the stimulus parameters, although it requires specialized equipment and expertise (Litovsky et al. 2017).

The complexity, amplitude, and length of the auditory stimuli play crucial roles in determining the characteristics of CI artifacts. For example, higher stimulus amplitudes tend to produce larger artifacts (Wagner et al. 2018), while complex sounds (e.g., speech syllables) can lead to more diffused artifacts as they engage multiple electrodes in the CI system. Conversely, simpler sounds, like pure sine waves, may result in more localized and distinct artifact patterns (Li et al. 2010; Wagner et al. 2018).

One way to further reduce artifacts is to use low-rate pulse trains or introduce short silent gaps (typically a few milliseconds) within the stimulation sequence, a technique known as "stimulation gaps," which minimizes overlap between CI-induced artifacts and the neural signals of interest (Somers et al. 2018). Another effective method is to alternate the polarity of pulses across trials: CI artifacts are sensitive to stimulus polarity (e.g., their polarity is reversed when the stimulus polarity is reversed), whereas the neural response's polarity is not. Therefore, alternating polarities helps cancel out artifacts, while retaining genuine neural responses (Hofmann and Wouters 2010; Klop et al. 2004).

The choice of EEG equipment and electrode setup is crucial for minimizing CI artifacts. High-density EEG systems can provide detailed spatial information but require careful consideration of electrode placement to avoid physical interference with the CI components, such as the sound processor and coil. Active electrodes, which amplify signals directly at the electrode site, are often recommended to improve the signal-to-noise ratio, especially in setups where electrode impedance is above 2 kΩ (Laszlo et al. 2014). However, passive electrodes may perform better if impedance remains below this threshold.

EEG systems with high sampling rates (>100 kHz) are beneficial for minimizing artifact distortion, as they allow more accurate representation and easier removal of rapid variations caused by CI-induced pulses (Gransier et al. 2016; Mc Laughlin et al. 2013). Additionally, using an amplifier with a large dynamic range can help prevent clipping and distortion, which is crucial for preserving the fidelity of recorded neural signals.

Finally, ensuring that participants remain calm and relaxed during the recording session can significantly improve the signal-to-noise ratio. Muscle tension, eye movements, and other physiological factors can introduce additional noise, making it harder to isolate neural responses. Encouraging a comfortable and quiet recording environment is essential for reducing both external and internal sources of artifacts.

8.3.5.3 Signal Processing

Reducing CI artifacts from EEG recordings involves a combination of preprocessing methods and offline signal processing techniques. These strategies aim to minimize the influence of CI-induced artifacts and ensure the preservation of true neural activity. Below, we outline several effective approaches used in CI EEG studies.

One of the most common preprocessing steps is re-referencing, which can significantly impact the quality of EEG recordings. Optimal referencing helps maximize the signal-to-noise ratio by amplifying neural signals of interest and minimizing noise, including CI artifacts. For unilateral CI users, placing the reference electrode contralaterally along the isopotential line is recommended, while for bilateral CI users, midline references (e.g., nose tip) can be effective (Bauer et al. 2006; Gilley et al. 2008). Although rejecting electrodes that capture excessive artifacts might seem tempting, it is often more beneficial to retain as many electrodes as possible, especially when using methods like independent component analysis (ICA) that require ample data to disentangle artifacts from neural responses.

High-pass and low-pass filtering can effectively reduce frequency components associated with CI artifacts. Low-pass filters, for example, have been used to attenuate high-frequency artifacts originating from stimulation pulses, typically setting a cutoff around 35 Hz for cortical auditory evoked potentials (CAEPs) (Mc Laughlin et al. 2013). However, higher frequencies may be necessary for other responses, such as ABRs, which can go up to 3000 Hz (BinKhamis et al. 2019). Care should be taken to avoid excessive filtering that might inadvertently remove neural signals of interest.

Template subtraction is another preprocessing method that involves creating a template of the artifact based on repeated measurements and then subtracting it from the contaminated signal. However, this method is initially developed to address direct current (DC) artifacts in transient CAEPs (Mc Laughlin et al. 2013), and it is not always successful, as artifacts and neural responses can overlap, complicating the removal process.

ICA has emerged as a powerful tool for isolating CI artifacts. By decomposing the EEG signal into independent components, ICA can separate artifacts from genuine brain activity, provided the components meet certain criteria (e.g., time-locked to the stimulus and consistent across trials). Although this method can be time-consuming and requires manual inspection, it has been successfully used in several studies to recover clean neural responses (Debener et al. 2008; Zhang et al. 2010, 2011; Liang et al. 2018). Variants of ICA, like SOBI (second-order blind identification), have shown promise, especially when combined with frequency filtering to improve artifact removal (Paul et al. 2020).

Empirical mode decomposition (EMD) is an adaptive method that breaks down the signal into intrinsic mode functions (IMFs), which can then be filtered to isolate and remove CI artifacts (Huang et al. 1998). The ensemble empirical mode decomposition (EEMD), a noise-assisted data analysis method that consists of adding white noise to the signal during the sifting process to force different portions of the signal to have a comparable scale to collate into one IMF, has been shown to improve artifact separation and has been used in combination with ICA to further enhance artifact removal (Martínez-Camacho and Castaneda-Villa 2018). Although promising, these methods require more research for widespread adoption.

Similarly to ERPs, EASSR recordings are contaminated with CI artifacts from stimulation and power-up pulses, which distort response properties such as shape, amplitude, and phase. Some significant work has been made in addressing the challenges associated with improving EASSR signals in CI users. EASSR CI artifacts can be removed using three steps: (1) averaging sweeps with alternating pulse polarities to eliminate slow-decay components and reduce peak artifact amplitudes, (2) compensating for distortions introduced by the recording setup using an inverse filter constructed from the filter coefficients of the recording system, and (3) removing the remaining time-limited artifacts by interpolating the neural responses during the artifact duration (Hofmann and Wouters 2010, 2012). Additionally, an automatic EASSR CI artifact-rejection technique based on ICA has been proved effective (Deprez et al. 2014).

By combining preprocessing methods and offline signal processing techniques (filtering, linear interpolation, template subtraction, ICA, or more advanced approaches like EMD), it is possible to significantly improve the quality of EEG recordings in CI users by minimizing the influence of CI-induced artifacts. While some methods are more established, emerging techniques like EEMD and SOBI offer promising avenues for future research.

Acknowledgements The authors declare that no Generative AI was used in the creation of this manuscript.

References

Abohelal M, El Mahallawi T, El-Gharib A, Serag S, Nada N (2024) Electric Auditory Brainstem Response (EABR) as a predictor of speech perception outcome in cochlear implanted children. Egypt J Ear Nose Throat Allied Sci 25:1–10

Agrawal D, Thorne JD, Viola FC, Timm L, Debener S, Buchner A, Dengler R, Wittfoth M (2013) Electrophysiological responses to emotional prosody perception in cochlear implant users. NeuroImage Clin 2:229–238. https://doi.org/10.1016/j.nicl.2013.01.001

Alemi R, Lehmann A (2019) Middle latency responses to optimized chirps in adult cochlear implant users. J Am Acad Audiol 30(5):396–405. https://doi.org/10.3766/jaaa.18014

Alzhrani F, Aljazeeri I, Abdelsamad Y et al (2024) International consensus statements on intraoperative testing for cochlear implantation surgery. Ear Hear 45(6):1418. https://doi.org/10.1097/AUD.0000000000001526

Amieva H, Ouvrard C, Giulioli C, Meillon C, Rullier L, Dartigues JF (2015) Self-reported hearing loss, hearing aids, and cognitive decline in elderly adults: a 25-year study. J Am Geriatr Soc 63(10):2099–2104. https://doi.org/10.1111/jgs.13649

Başar E, Başar-Eroglu C, Karakaş S, Schürmann M (2001) Gamma, alpha, delta, and theta oscillations govern cognitive processes. Int J Psychophysiol 39(2–3):241–248. https://doi.org/10.1016/s0167-8760(00)00145-8

Basura GJ, Koehler SD, Shore SE (2018) Human central auditory plasticity: a review of functional changes, clinical applications, and limitations. J Am Acad Audiol 29(6):466–479

Bauer PW, Sharma A, Martin K, Dorman M (2006) Central auditory development in children with bilateral cochlear implants. Arch Otolaryngol Head Neck Surg 132(10):1133–1136. https://doi.org/10.1001/archotol.132.10.1133

Beatty J, Wagoner BL (1978) Pupillometric signs of brain activation vary with the level of cognitive processing. Science 199(4334):1216–1218. https://doi.org/10.1126/science.628837

Becker R, Pefkou M, Michel CM, Hervais-Adelman AG (2013) Left temporal alpha-band activity reflects single-word intelligibility. Front Syst Neurosci 7:121. https://doi.org/10.3389/fnsys.2013.00121

Bernarding C, Strauss DJ, Hannemann R, Seidler H, Corona-Strauss FI (2013) Neural correlates of listening effort-related factors: influence of age and hearing impairment. Brain Res Bull 91:21–30. https://doi.org/10.1016/j.brainresbull.2012.11.005

BinKhamis G, Perugia E, O'Driscoll M, Kluk K (2019) Speech-ABRs in cochlear implant recipients: feasibility study. Int J Audiol 58(10):678–684. https://doi.org/10.1080/14992027.2019.1619100

Blamey PJ, Maat B, Başkent D, et al (2015) A Retrospective Multicenter Study Comparing Speech Perception Outcomes for Bilateral Implantation and Bimodal Rehabilitation. Ear & Hearing 36:408–416. https://doi.org/10.1097/aud.0000000000000150

Bronkhorst AW (2015) The cocktail-party problem revisited: early processing and selection of multi-talker speech. Atten Percept Psychophysiol 77(5):1465–1487. https://doi.org/10.3758/s13414-015-0882-9

Cardon G, Sharma A (2019) Somatosensory cross-modal reorganization in children with cochlear implants. Front Neurosci 13:469. Published 2019 Jun 26. https://doi.org/10.3389/fnins.2019.00469

Chao X, Wang R, Luo J, Wang H, Fan Z, Xu L (2023) Relationship between electrically evoked compound action potential thresholds and behavioral T-levels in implanted children with cochlear nerve deficiency. Sci Rep 13(1):4309. https://doi.org/10.1038/s41598-023-31411-3

Danermark B, Gellerstedt LC (2004) Psychosocial work environment, hearing impairment, and health. Int J Audiol 43(7):383–389. https://doi.org/10.1080/14992020400050049

Deal JA, Betz J, Yaffe K, Harris T, Purchase-Helzner E, Satterfield S et al (2017) Hearing impairment and incident dementia and cognitive decline in older adults: the health ABC study. J Gerontol A 72(5):703–709. https://doi.org/10.1093/gerona/glw069

Debener S, Hine J, Bleeck S, Eyles J (2008) Source localization of auditory evoked potentials after cochlear implantation. Psychophysiology 45(1):20–24. https://doi.org/10.1111/j.1469-8986.2007.00610.x

Deniz B, Kara E, Polat Z, Deniz R, Ataş A (2021) Changes in electrically evoked auditory brainstem responses in children with sequential bilateral cochlear implants. Int J Pediatr Otorhinolaryngol 141:110555. https://doi.org/10.1016/j.ijporl.2020.110555

Deprez H, Hofmann M, van Wieringen A, Wouters J, Moonen M (2014) Cochlear implant artifact rejection in electrically evoked auditory steady state responses. In: Signal processing conference (EUSIPCO), 2014 proceedings of the 22nd European, 1995–1999

Doucet ME, Bergeron F, Lassonde M, Ferron P, Lepore F (2006) Crossmodal reorganization and speech perception in cochlear implant users. Brain 129:3376–3383. https://doi.org/10.1093/brain/awl264

Dunn CC, Walker EA, Oleson J et al (2014) Longitudinal speech perception and language performance in pediatric cochlear implant users: the effect of age at implantation. Ear Hear 35(2):148–160. https://doi.org/10.1097/AUD.0b013e3182a4a8f0

Getzmann S, Falkenstein M (2011) Understanding of spoken language under challenging listening conditions in younger and older listeners: a combined behavioral and electrophysiological study. Brain Res 1415:8–22. https://doi.org/10.1016/j.brainres.2011.08.001

Gilley PM, Sharma A, Dorman MF (2008) Cortical reorganization in children with cochlear implants. Brain Res 1239:56–65. https://doi.org/10.1016/j.brainres.2008.08.026

Glennon M, Prell C, Sobczyk-Kojiro K et al (2020) Auditory cortical plasticity in cochlear implant users: an update and future directions. Front Hum Neurosci 14:93

Gordon KA, Papsin BC, Harrison RV (2006) An evoked potential study of the developmental time course of the auditory nerve and brainstem in children using cochlear implants. Audiol Neurootol 11(1):7–23. https://doi.org/10.1159/000088851

Granholm E, Asarnow RF, Sarkin AJ, Dykes KL (1996) Pupillary responses index cognitive resource limitations. Psychophysiology 33(4):457–461. https://doi.org/10.1111/j.1469-8986.1996.tb01071.x

Gransier R, Deprez H, Hofmann M, Moonen M, van Wieringen A, Wouters J (2016) Auditory steady-state responses in cochlear implant users: effect of modulation frequency and stimulation artifacts. Hear Res 335:149–160. https://doi.org/10.1016/j.heares.2016.03.006

Haro S, Rao HM, Quatieri TF, Smalt CJ (2022) EEG alpha and pupil diameter reflect endogenous auditory attention switching and listening effort. Eur J Neurosci 55(5):1262–1277. https://doi.org/10.1111/ejn.15616

Heimler B, Weisz N, Collignon O (2014) Revisiting the adaptive and maladaptive effects of crossmodal plasticity. Neuroscience 283:44–63. https://doi.org/10.1016/j.neuroscience.2014.08.003

Hillyard SA, Hink RF, Schwent VL, Picton TW (1973) Electrical signs of selective attention in the human brain. Science 182(4108):177–180. https://doi.org/10.1126/science.182.4108.177

Hofmann M, Wouters J (2010) Electrically evoked auditory steady state responses in cochlear implant users. J Assoc Res Otolaryngol 11(2):267–282. https://doi.org/10.1007/s10162-009-0201-z

Hofmann M, Wouters J (2012) Improved electrically evoked auditory steady-state response thresholds in humans. J Assoc Res Otolaryngol 13(4):573–589. https://doi.org/10.1007/s10162-012-0321-8

Holder JT, Hoffman J, Williams H, Gifford RH (2024) The relationship between eSRTs and upper stimulation levels in a large cohort of adult cochlear implant recipients. Otol Neurotol 45(10): e756–e762. https://doi.org/10.1097/MAO.0000000000004329

Hu Z, Sun JQ, Guan RR, Chen L, Sun JW, Guo XT (2021) Deficient sensory and cognitive processing in children with cochlear implants: an event-related potential study. Hear Res 408: 108295. https://doi.org/10.1016/j.heares.2021.108295

Huang NE, Zheng S, Long SR, Wu MC, Shih HH, Quanan Z, Nai-Chyuan Y, Chao TC, Liu HH (1998) The empirical mode decomposition and the Hilbert spectrum for nonlinear and non-stationary time series analysis. Proc R Soc Lond A 454:903–995. https://doi.org/10.1098/rspa.1998.0193

Humes LE (2013) Understanding the speech-understanding problems of older adults. Am J Audiol 22(2):303–305. https://doi.org/10.1044/1059-0889(2013/12-0066)

Intartaglia B, Prud'homme M-A, Foster NEV, Zeitouni AG, Lehmann A (2022a) Visual cortex plasticity in cochlear implant users revealed in a visual motion detection task. Clin Neurophysiol 137:11–24. https://doi.org/10.1016/j.clinph.2022.02.005

Intartaglia B, Zeitnouni AG, Lehmann A (2022b) Recording EEG in cochlear implant users: guidelines for experimental design and data analysis for optimizing signal quality and minimizing artifacts. J Neurosci Methods 375:109592. https://doi.org/10.1016/j.jneumeth.2022.109592

Jafari Z, Fitzpatrick EM, Schramm DR, Rouillon I, Koravand A (2024) Prognostic value of electrophysiological and MRI findings for pediatric cochlear implant outcomes: a systematic review. Am J Audiol 33(3):1023–1040. https://doi.org/10.1044/2024_AJA-23-00272

Järvelin MR, Mäki-Torkko E, Sorri MJ, Rantakallio PT (1997) Effect of hearing impairment on educational outcomes and employment up to the age of 25 years in northern Finland. Br J Audiol 31(3):165–175. https://doi.org/10.3109/03005364000000019

Jeon JH, Bae MR, Song MH, Noh SH, Choi KH, Choi JY (2013) Relationship between electrically evoked auditory brainstem response and auditory performance after cochlear implant in patients with auditory neuropathy spectrum disorder. Otol Neurotol 34(7):1261–1266. https://doi.org/10.1097/MAO.0b013e318291c632

Kahneman D, Beatty J (1966) Pupil diameter and load on memory. Science 154(3756):1583–1585. https://doi.org/10.1126/science.154.3756.1583

Karrasch M, Laine M, Rapinoja P, Krause CM (2004) Effects of normal aging on event-related desynchronization/synchronization during a memory task in humans. Neurosci Lett 366(1):18–23. https://doi.org/10.1016/j.neulet.2004.05.010

Klimesch W (2012) Alpha-band oscillations, attention, and controlled access to stored information. Trends Cogn Sci 16(12):606–617. https://doi.org/10.1016/j.tics.2012.10.007

Klop WMC, Hartlooper A, Briare JJ, Frijns JHM (2004) A new method for dealing with the stimulus artefact in electrically evoked compound action potential measurements. Acta Otolaryngol 124(2):137–143. https://doi.org/10.1080/00016480310016901

Koelewijn T, Zekveld AA, Festen JM, Rönnberg J, Kramer SE (2012) Processing load induced by informational masking is related to linguistic abilities. Int J Otolaryngol 2012:865731. https://doi.org/10.1155/2012/865731

Koeritzer MA, Rogers CS, Van Engen KJ, Peelle JE (2018) The impact of age, background noise, semantic ambiguity, and hearing loss on recognition memory for spoken sentences. J Speech Lang Hear Res 61(3):740–751. https://doi.org/10.1044/2017_JSLHR-H-17-0077

Kramer SE, Kapteyn TS, Festen JM, Kuik DJ (1997) Assessing aspects of auditory handicap by means of pupil dilatation. Audiology 36(3):155–164. https://doi.org/10.3109/00206099709071969

Kraus N, Micco AG, Koch DB et al (1993) The mismatch negativity cortical evoked potential elicited by speech in cochlear-implant users. Hear Res 65:118–124. https://doi.org/10.1016/0378-5955(93)90206-g

Lammers MJW, van Eijl RHM, van Zanten GA, Versnel H, Grolman W (2015) Delayed auditory brainstem responses in prelingually deaf and late-implanted cochlear implant users. J Assoc Res Otolaryngol 16(5):669–678. https://doi.org/10.1007/s10162-015-0532-x

Laszlo S, Ruiz-Blondet M, Khalifian N, Chu F, Jin Z (2014) A direct comparison of active and passive amplification electrodes in the same amplifier system. J Neurosci Methods 235:298–307. https://doi.org/10.1016/j.jneumeth.2014.05.012

Layer N, Weglage A, Müller V et al (2022a) The timecourse of multisensory speech processing in unilaterally stimulated cochlear implant users revealed by ERPs. NeuroImage Clin 34:102982. https://doi.org/10.1016/j.nicl.2022.102982

Layer N, Weglage A, Müller V et al (2022b) Electrophysiological differences and similarities in audiovisual speech processing in CI users with unilateral and bilateral hearing loss. Curr Res Neurobiol 3:100059. https://doi.org/10.1016/j.crneur.2022.100059

Lazard DS, Giraud A-L, Gnansia D, et al (2012) Understanding the deafened brain: Implications for cochlear implant rehabilitation. European Annals of Otorhinolaryngology, Head and Neck Diseases 129:98–103. https://doi.org/10.1016/j.anorl.2011.06.001

Lazard DS, Giraud A-L (2017) Faster phonological processing and right occipito-temporal lateralization in deaf patients during word recognition after cochlear implantation. Cereb Cortex 27(6): 2785–2795

Lazard DS, Doelling KB, Arnal LH (2023) Plasticity After Hearing Rehabilitation in the Aging Brain. Trends in Hearing 27:23312165231156412. https://doi.org/10.1177/23312165231156412

Lehmann A, Paquette S (2015) Cross-domain processing of musical and vocal emotions: EEG evidence. Front Psychol 6:123. https://doi.org/10.3389/fnins.2015.00343

Li X, Nie K, Karp F, Tremblay KL, Rubinstein JT (2010) Characteristics of stimulus artifacts in EEG recordings induced by electrical stimulation of cochlear implants. In: 2010 3rd international conference on biomedical engineering and informatics, 2, pp 799–803. https://doi.org/10.1109/BMEI.2010.5639942

Liang C, Houston LM, Samy RN, Abedelrehim LMI, Zhang F (2018) Cortical processing of frequency changes reflected by the acoustic change complex in adult cochlear implant users. Audiol Neurootol 23(3):152–164. https://doi.org/10.1159/000492170

Limb CJ, Roy AT (2014) Technological, biological, and acoustical constraints to music perception in cochlear implant users. Hear Res 308:13–26. https://doi.org/10.1016/j.heares.2013.04.009

Lin FR et al (2011) Hearing loss and incident dementia. Arch Neurol 68:214–220. https://doi.org/10.1001/archneurol.2010.362

Lin FR, Yaffe K, Xia J, Xue Q, Harris TB, Purchase-Helzner E et al (2013) Hearing loss and cognitive decline in older adults. JAMA Intern Med 173:293–299. https://doi.org/10.1001/jamainternmed.2013.1868

Litovsky RY, Goupell MJ, Kan A, Landsberger DM (2017) Use of research interfaces for psychophysical studies with cochlear-implant users. Trends Hear 21. https://doi.org/10.1177/2331216517736464

Livingston G, Sommerlad A, Orgeta V, Costafreda SG, Huntley J, Ames D et al (2017) Dementia prevention, intervention, and care. Lancet 390(10113):2673–2734. https://doi.org/10.1016/S0140-6736(17)31363-6

Looi V, Gfeller K, Driscoll V (2012) Music appreciation and training for cochlear implant recipients: a review. Semin Hear 33(4):307–321. https://doi.org/10.1055/s-0032-1329222

Mackersie CL, Cones H (2011) Subjective and psychophysiological indexes of listening effort in a competing-talker task. J Am Acad Audiol 22(2):113–122. https://doi.org/10.3766/jaaa.22.2.6

Marsella P, Scorpecci A, Vecchiato G, Maglione AG, Colosimo A, Babiloni F (2014) Neuroelectrical imaging investigation of cortical activity during listening to music in prelingually deaf children with cochlear implants. Int J Pediatr Otorhinolaryngol 78(5): 737–743. https://doi.org/10.1016/j.ijporl.2014.01.030

Martin BA (2007) Can the acoustic change complex be recorded in an individual with a cochlear implant? Separating neural responses from cochlear implant artifact. J Am Acad Audiol 18(2): 126–140. https://doi.org/10.3766/jaaa.18.2.5

Martin JS, Jerger JF (2005) Some effects of aging on central auditory processing. J Rehabil Res Dev 42(4 Suppl 2):25–44. https://doi.org/10.1682/jrrd.2004.12.0164

Martínez-Camacho MA, Castañeda-Villa N (2018) Cochlear implant artifact reduction on one channel mismatch negativity recordings based on ensemble empirical mode decomposition and independent component analysis. In: 2018 40th annual international conference of the IEEE Engineering in Medicine and Biology Society (EMBC), pp 6018–6021. https://doi.org/10.1109/EMBC.2018.8513632

Mattys SL, Davis MH, Bradlow AR, Scott SK (2012) Speech recognition in adverse conditions: a review. Lang Cogn Process 27(7–8):953–978

Mc Coy SL, Tun PA, Cox LC, Colangelo M, Stewart RA, Wingfield A (2005) Hearing loss and perceptual effort: downstream effects on older adults' memory for speech. Q J Exp Psychol A 58(1):22–33. https://doi.org/10.1080/02724980443000151

Mc Dermott HJ (2004) Music perception with cochlear implants: a review. Trends Amplif 8(2): 49–82. https://doi.org/10.1177/108471380400800203

Mc Garrigle R, Munro KJ, Dawes P, Stewart AJ, Moore DR, Barry JG, Amitay S (2014) Listening effort and fatigue: what exactly are we measuring? A British Society of Audiology Cognition in Hearing Special Interest Group 'white paper'. Int J Audiol 53(7):433–440. https://doi.org/10.3109/14992027.2014.890296

Mc Laughlin M, Lopez Valdes A, Reilly RB, Zeng F-G (2013) Cochlear implant artifact attenuation in late auditory evoked potentials: a single channel approach. Hear Res 302:84–95. https://doi.org/10.1016/j.heares.2013.05.006

Mc Mahon CM, Boisvert I, de Lissa P, Granger L, Ibrahim R, Lo CY et al (2016) Monitoring alpha oscillations and pupil dilation across a performance-intensity function. Front Psychol 7:745. https://doi.org/10.3389/fpsyg.2016.00745

Meister H, Rählmann S, Walger M, Margolf-Hackl S, Kießling J (2015) Hearing aid fitting in older persons with hearing impairment: the influence of cognitive function, age, and hearing loss on hearing aid benefit. Clin Interv Aging 10:435–443. https://doi.org/10.2147/CIA.S77096

Miles K, Mc Mahon C, Boisvert I, Ibrahim R, de Lissa P, Graham P, Lyxell B (2017) Objective assessment of listening effort: coregistration of pupillometry and EEG. Trends Hear 21: 2331216517706396. https://doi.org/10.1177/2331216517706396

Miller CA, Abbas PJ, Brown CJ (2000) An improved method of reducing stimulus artifact in the electrically evoked whole-nerve potential. Ear Hear 21(4):280–290. https://doi.org/10.1097/00003446-200008000-00003

Nachtegaal J, Kuik DJ, Anema JR, Goverts ST, Festen JM, Kramer SE (2009) Hearing status, need for recovery after work, and psychosocial work characteristics: results from an internet-based national survey on hearing. Int J Audiol 48(10):684–691. https://doi.org/10.1080/14992020902962421

Nada N, Kolkaila E, Schendzielorz P, El Mahallawi T (2022) Electrically evoked auditory brainstem response in cochlear implantation: what you need to know. Egypt J Otolaryngol 38(1):67. https://doi.org/10.1186/s43163-022-00259-1

Nakata T, Trehub SE, Kanda Y (2012) Effect of cochlear implants on children's perception and production of speech prosody. J Acoust Soc Am 131(2):1307–1314. https://doi.org/10.1121/1.3672697

Neagu MB, Kressner AA, Relaño-Iborra H, Bækgaard P, Dau T, Wendt D (2023) Investigating the reliability of pupillometry as a measure of individualized listening effort. Trends Hear 27: 23312165231153288. https://doi.org/10.1177/23312165231153288

Nguyen DL, Valentin O, Lehmann A, Prévost F (2024) A multimodal investigation of listening effort in single-sided deafness. Am J Audiol 33(4):1341–1349. https://doi.org/10.1044/2024_AJA-24-00073

Obleser J, Wöstmann M, Hellbernd N, Wilsch A, Maess B (2012) Adverse listening conditions and memory load drive a common α oscillatory network. J Neurosci 32(36):12376–12383. https://doi.org/10.1523/JNEUROSCI.4908-11.2012

Ohlenforst B, Zekveld AA, Jansma EP, Wang Y, Naylor G, Lorens A et al (2017) Effects of hearing impairment and hearing aid amplification on listening effort: a systematic review. Ear Hear 38(3):267–281. https://doi.org/10.1097/AUD.0000000000000396

Ohlenforst B, Wendt D, Kramer SE, Naylor G, Zekveld AA, Lunner T (2018) Impact of SNR, masker type, and noise reduction processing on sentence recognition performance and listening effort as indicated by the pupil dilation response. Hear Res 365:90–99. https://doi.org/10.1016/j.heares.2018.05.003

Paul BT, Uzelac M, Chan E et al (2020) Poor early cortical differentiation of speech predicts perceptual difficulties of severely hearing-impaired listeners in multi-talker environments. Sci Rep 10:6141. https://doi.org/10.1038/s41598-020-63103-7

Paul BT, Chen JM, Dimitrijevic A (2021) Cortical alpha oscillations in cochlear implant users reflect subjective listening effort during speech-in-noise perception. PLoS One 16(7):e0254162. https://doi.org/10.1371/journal.pone.0254162

Pérez-Rodríguez MÁ, Torres-Gómez SF, Conde-Pacheco ME, Romero-Díaz A (2023) Electrically evoked stapedial reflex threshold and behavioral method usefulness in the cochlear implant programming in pediatric population. Int J Pediatr Otorhinolaryngol 166:111473. https://doi.org/10.1016/j.ijporl.2023.111473

Pesonen M, Björnberg CH, Hämäläinen H, Krause CM (2006) Brain oscillatory 1-30 Hz EEG ERD/ERS responses during the different stages of an auditory memory search task. Neurosci Lett 399(1–2):45–50. https://doi.org/10.1016/j.neulet.2006.01.053

Petersen B, Weed E, Sandmann P, Brattico E, Hansen M, Sørensen SD, Vuust P (2015) Brain responses to musical feature changes in adolescent cochlear implant users. Front Hum Neurosci 9:7. https://doi.org/10.3389/fnhum.2015.00007

Pichora-Fuller MK, Schneider BA, Daneman M (1995) How young and old adults listen to and remember speech in noise. J Acoust Soc Am 97(1):593–608. https://doi.org/10.1121/1.412282

Piquado T, Isaacowitz D, Wingfield A (2010) Pupillometry as a measure of cognitive effort in younger and older adults. Psychophysiology 47(3):560–569. https://doi.org/10.1111/j.1469-8986.2009.00947.x

Prévost F, Lehmann A (2018) Saliency of vowel features in neural responses of cochlear implant users. Clin EEG Neurosci 49(6):388–397. https://doi.org/10.1177/1550059418770051

Ries M, Kelava I, Ajduk J, Košec A, Žaja R, Trotić R (2024) Correlation be-tween neural response telemetry measurements and fitting levels. Int J Pediatr Otorhinolaryngol 182:112001. https://doi.org/10.1016/j.ijporl.2024.112001

Ringdahl A, Grimby A (2000) Severe-profound hearing impairment and health-related quality of life among post-lingual deafened Swedish adults. Scand Audiol 29(4):266–275. https://doi.org/10.1080/010503900750022907

Rönnberg J, Lunner T, Zekveld A, Sörqvist P, Danielsson H, Lyxell B et al (2013) The ease of language understanding (ELU) model: theoretical, empirical, and clinical advances. Front Syst Neurosci 7:31. https://doi.org/10.3389/fnsys.2013.00031

Sandmann P, Dillier N, Eichele T, Meyer M, Kegel A, Pascual-Marqui RD, Marcar VL, Jäncke L, Debener S (2012) Visual activation of auditory cortex reflects maladaptive plasticity in cochlear implant users. Brain J Neurol 135(Pt 2):555–568. https://doi.org/10.1093/brain/awr329

Schierholz I, Finke M, Kral A et al (2017) Auditory and audio-visual processing in patients with cochlear, auditory brainstem, and auditory midbrain implants: an EEG study. Hum Brain Mapp 38(4):2206–2225. https://doi.org/10.1002/hbm.23515

Sharp K, Delcenserie A, Champoux F (2018) Auditory event-related potentials associated with musical pitch processing in cochlear implant users. Front Neurosci 12:103. https://doi.org/10.3389/fnins.2018.00538

Somers B, Verschueren E, Francart T (2018) Neural tracking of the speech envelope in cochlear implant users. J Neural Eng 16(1):016003. https://doi.org/10.1088/1741-2552/aae6b9

Starr A, Brackmann DE (1979) Brain stem potentials evoked by electrical stimulation of the cochlea in human subjects. Ann Otol Rhinol Laryngol 88(4 Pt 1):550–556. https://doi.org/10.1177/000348947908800419

Steel MM, Papsin BC, Gordon KA (2015) Binaural fusion and listening effort in children who use bilateral cochlear implants: a psychoacoustic and pupillometric study. PLoS One 10(2): e0117611. https://doi.org/10.1371/journal.pone.0117611

Stropahl M, Debener S (2017) Auditory cross-modal reorganization in cochlear implant users indicates audio-visual integration. Neuroimage Clin 16:514–523. Published 2017 Sep 4. https://doi.org/10.1016/j.nicl.2017.09.001

Sharma A, Campbell J, Cardon G (2015) Developmental and cross-modal plasticity in deafness: Evidence from the P1 and N1 event related potentials in cochlear implanted children. International Journal of Psychophysiology 95:135–144. https://doi.org/10.1016/j.ijpsycho.2014.04.007

Tremblay C, Champoux F, Lepore F, Théoret H (2010) Audiovisual fusion and cochlear implant proficiency. Restor Neurol Neurosci 28(2):283–291. https://doi.org/10.3233/RNN-2010-0498

Tun PA, Mc Coy S, Wingfield A (2009) Aging, hearing acuity, and the attentional costs of effortful listening. Psychol Aging 24(3):761–766. https://doi.org/10.1037/a0014802

Turgeon C, Lazzouni L, Lepore F et al (2014) An objective auditory measure to assess speech recognition in adult cochlear implant users. Clin Neurophysiol 125:827–835. https://doi.org/10.1016/j.clinph.2013.09.035

Valentin O, Lehmann A, Nguyen D, Paquette S (2024a) Integrating emotion perception in rehabilitation programs for cochlear implant users: a call for a more comprehensive approach. J Speech Lang Hear Res 67(5):1635–1642. https://doi.org/10.1044/2024_JSLHR-23-00660

Valentin O, Foster N, Intartaglia B, Prud'homme M-A, Schönwiesner M, Nozaradan S, Lehmann A (2024b) Multimodal sensorimotor investigation of audio-visual integration in cochlear implant users. Can Acoust 52(1) https://jcaa.caa-aca.ca/index.php/jcaa/article/view/4165

Valentin O, Foster NEV, Intartaglia B, Prud'homme MA, Schönwiesner M, Nozaradan S, Lehmann A (2025) Intact sensorimotor rhythm abilities but altered audiovisual integration in cochlear implant users. PLoS One 20(4):e0320815. https://doi.org/10.1371/journal.pone.0320815

Vecchiato G, Maglione AG, Scorpecci A, Malerba P, Marsella P, Di Francesco G, Vitiello S, Colosimo A, Babiloni F (2012) EEG frontal asymmetry related to pleasantness of music perception in healthy children and cochlear implanted users. In: Conference proceedings: annual international conference of the IEEE Engineering in Medicine and Biology Society. https://doi.org/10.1109/EMBC.2012.6347026

Vermiglio AJ, Soli SD, Freed DJ, Fisher LM (2012) The relationship between high-frequency pure-tone hearing loss, hearing in noise test (HINT) thresholds, and the articulation index. J Am Acad Audiol 23(10):779–788. https://doi.org/10.3766/jaaa.23.10.4

Wagner L, Maurits N, Maat B, Baskent D, Wagner AE (2018) The cochlear implant EEG artifact recorded from an artificial brain for complex acoustic stimuli. IEEE Trans Neural Syst Rehabil Eng 26(2):392–399. https://doi.org/10.1109/TNSRE.2018.2789780

Wang Y, Pan T, Deshpande SB, Ma F (2015) The relationship between EABR and auditory performance and speech intelligibility outcomes in pediatric cochlear implant recipients. Am J Audiol 24(2):226–234. https://doi.org/10.1044/2015_AJA-14-0023

Wingfield A, Grossman M (2006) Language and the aging brain: patterns of neural compensation revealed by functional brain imaging. J Neurophysiol 96(6):2830–2839. https://doi.org/10.1152/jn.00628.2006

Wingfield A, Tun PA, Mc Coy SL (2005) Hearing loss in older adulthood: what it is and how it interacts with cognitive performance. Curr Dir Psychol Sci 14:144–148

Winn MB, Edwards JR, Litovsky RY (2015) The impact of auditory spectral resolution on listening effort revealed by pupil dilation. Ear Hear 36(4):e153–e165. https://doi.org/10.1097/AUD.0000000000000145

Xiu B, Paul BT, Chen JM et al (2022) Neural responses to naturalistic audiovisual speech are related to listening demand in cochlear implant users. Front Hum Neurosci 16:1043499. https://doi.org/10.3389/fnhum.2022.1043499

Young A, Cornejo J, Spinner A (2024) Auditory brainstem response. In: StatPearls. StatPearls Publishing. http://www.ncbi.nlm.nih.gov/books/NBK564321/

Zekveld AA, Kramer SE, Festen JM (2010) Pupil response as an indication of effortful listening: the influence of sentence intelligibility. Ear Hear 31(4):480–490. https://doi.org/10.1097/AUD.0b013e3181d4f251

Zekveld AA, Heslenfeld DJ, Johnsrude IS, Versfeld NJ, Kramer SE (2014) The eye as a window to the listening brain: neural correlates of pupil size as a measure of cognitive listening load. NeuroImage 101:76–86. https://doi.org/10.1016/j.neuroimage.2014.06.069

Zeng F-G, Rebscher S, Harrison WV, Sun X, Feng H (2008) Cochlear implants: system design, integration and evaluation. IEEE Rev Biomed Eng 1:115–142. https://doi.org/10.1109/RBME.2008.2008250

Zhang F, Anderson J, Samy R, Houston L (2010) The adaptive pattern of the late auditory evoked potential elicited by repeated stimuli in cochlear implant users. Int J Audiol 49(4):277–285. https://doi.org/10.3109/14992020903321759

Zhang F, Hammer T, Banks HL, Benson C, Xiang J, Fu QJ (2011) Mismatch negativity and adaptation measures of the late auditory evoked potential in cochlear implant users. Hear Res 275(1–2):17–29. https://doi.org/10.1016/j.heares.2010.11.007

Zhang F, Benson C, Cahn SJ (2013) Cortical encoding of timbre changes in cochlear implant users. J Am Acad Audiol 24(1):46–58. https://doi.org/10.3766/jaaa.24.1.6

9 Plasticity and Perceptual Learning in Cochlear Implants

John J. Galvin, III, Sean Lang, and Qian-Jie Fu

Abstract

As has often been said, it is the brain that hears, not the ear. Indeed, brain plasticity accounts for much of the success of the cochlear implant (CI). In postlingually deaf CI users, the coarse spectro-temporal patterns provided by the CI are often sufficient to activate central auditory speech templates developed during previous acoustic hearing (AH). Prelingually deaf CI users are able to develop these central auditory templates with the CI alone. Current CI candidacy allows for increasing amounts of AH in the implanted and/or non-implanted ears which can greatly improve CI users' access to pitch information, important for challenging listening conditions such as prosody perception, music perception, talker identification, and segregation of competing speech. Again, brain plasticity plays a strong role in combining acoustic and electric stimulation patterns. In all these scenarios, CI users will "passively" adapt to these new stimulation patterns during the first 6–12 months of implant use. However, passive adaptation is often incomplete. For CI users with available AH, the more familiar acoustic stimulation patterns may limit adaptation to new electric stimulation patterns. "Active" auditory training has been shown to significantly improve CI users' speech and music perception, even with extensive experience with their device. Given the limited access to clinical auditory rehabilitation, computer-based auditory train-

J. J. Galvin, III (✉) · S. Lang
House Institute Foundation, Los Angeles, CA, USA
e-mail: jgalvin@hifla.org; slang@hifla.org

Q.-J. Fu
House Institute Foundation, Los Angeles, CA, USA

David Geffen School of Medicine, University of California Los Angeles, Los Angeles, CA, USA
e-mail: qfu@mednet.ucla.edu

B. Molaee-Ardekani, M. Segovia-Martinez (eds.), *Advancing Cochlear Implants*,
https://doi.org/10.1007/978-3-032-15719-5_9

ing has benefitted many CI users. As CI technology continues to improve and AH plays a more substantial role in CI users, auditory training remains an important component of successful CI outcomes, especially when challenging listening conditions are considered.

Keywords

Cochlear implant · Plasticity · Adaptation · Auditory training

9.1 Introduction

Over the last 50 or more years, there have been great advances in cochlear implants (CIs), from technological, surgical, and/or clinical perspectives. Despite the many advances, there remains great variability in CI outcomes. Some of this variability may be due to extrinsic factors (e.g., outcome measures, clinical/surgical expertise, etc.) or more likely due to intrinsic factors (e.g., age at implantation, duration of deafness, etiology of deafness, device experience, etc.). CI outcomes are largely limited by the transmission, reception, and perception of the acoustic input patterns (Fig. 9.1).

9.1.1 Transmission of Acoustic Cues

Transmission of acoustic cues involves the speech processor (e.g., the microphone, signal processing, "pre-processing" of the acoustic input, etc.). CI research and development has largely focused on improving the transmission of acoustic cues.

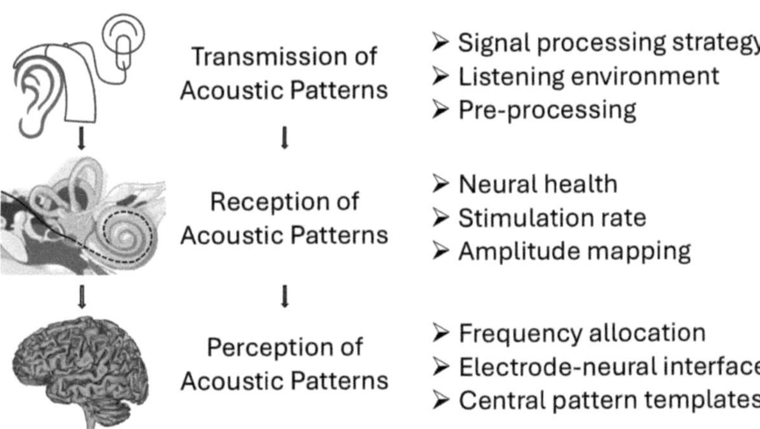

Fig. 9.1 Factors that limit CI outcomes include the Transmission, Reception, and Perception of acoustic patterns. (Some elements of the figure were sourced from https://openclipart.org/ licensed under CC0)

However, the transmission of acoustic patterns interacts with the listening environment, with generally good transmission in quiet environments but worsening as background noise increases (Fetterman and Domico 2002; Fu and Nogaki 2005). Transmission may also depend on the type of information; speech patterns are generally well transmitted by the CI, but pitch cues are not. With increased processing power and cloud computing, preprocessing the acoustic input can help to denoise the signal of interest and/or optimize the input for individual CI users. CI signal processing strategies seem to continuously evolve but have not shown great gains for challenging listening conditions (Vandali et al. 2000; Plant et al. 2007; Firszt et al. 2009; Donaldson et al. 2011). For example, attempts to provide fine-structure (FS) cues in the CI signal processing (e.g., MED-EL's FS strategies) have not shown substantial improvements in pitch perception. This suggests that the bottleneck may be elsewhere, such as the electrode-neural interface.

9.1.2 Reception of Acoustic Cues

One limitation to the reception of acoustic patterns with the CI is neural health. Duration of deafness is a major limiting factor to CI outcomes (Loizou 1999; Green et al. 2005). Poor neural health may limit reception of patterns with electric stimulation. Poor neural health may also limit the effectiveness of high stimulation rates that allow better sampling of the acoustic input (Drennan and Pfingst 2006). High stimulation rates can also increase the electric dynamic range (DR), primarily by reducing threshold stimulation levels, theoretically improving the intensity resolution (Kreft et al. 2004; Galvin and Fu 2005). However, the extended DR and better temporal sampling with high stimulation rates have not been shown to be consistently advantageous (Galvin and Fu 2005; Spahr et al. 2007). For example, high stimulation rates have been associated with poorer perception of amplitude modulation (Pfingst et al. 2007; Middlebrooks 2008; Galvin and Fu 2009) and interaural time differences (ITDs; Van Hoesel et al. 2009).

In CIs, the acoustic input DR must be mapped onto a more limited electric DR. The approach to amplitude mapping varies across CI devices. Some systems map a wide acoustic range, mostly tied to microphone sensitivity, onto a narrow electric DR, which can result in loss of detail and center clipping of the acoustic input—though well-designed automatic gain control (AGC) may mitigate this issue. Alternatively, some devices map the full acoustic input DR onto a broader electric DR, spanning from the minimum stimulation level to the comfortable loudness level. This can result in electric stimulation falling below perceptual thresholds, making parts of the input signal inaudible. In both cases, low-level acoustic noise may be effectively mapped to low stimulation levels, which can improve the signal-to-noise ratio (SNR). However, most CI systems apply some degree of compression in the acoustic-to-electric amplitude mapping. While this helps fit the signal into the limited electric DR, it can reduce the electric SNR compared to the original acoustic SNR. Excessive compression may further degrade perceptual SNR, making it harder for users to distinguish signal from noise.

9.1.3 Perception

Whatever the limitations are to transmission and reception of acoustic patterns, the brain perceives these patterns and must adapt to the representation with electric stimulation. This adaptation will be different for prelingual and postlingual CI users, and for adults and children. For postlingual adults, adaptation to electric stimulation patterns will occur in relation to central pattern templates developed with previous acoustic hearing (AH) experience. Postlingual adult CI users must adapt to electric hearing, as well as to the new stimulation pattern, which is spectrally degraded and likely frequency-shifted and/or -warped, relative to central pattern templates. The limited number of inserted electrodes (12–22) necessarily degrades the spectral resolution of the electric patterns. The electrode-neural interface may further limit the spectral resolution. Due to current spread, spread of excitation, and uneven neural health, the functional spectral resolution is further degraded, with CI users able to access 6–8 spectral channels, regardless of the number of implanted electrodes (Fu and Nogaki 2005). In typical CI fitting, a large acoustic input frequency range (to maximize the information delivered to the CI) is mapped onto a limited electrode extent that is not fully inserted into the cochlea. As such, most clinical frequency allocations result in some degree of tonotopic mismatch.

Postlingual CI users must adapt to both degraded and distorted spectral patterns with electric hearing. Prelingual pediatric CI users develop with electric stimulation, with no previous central pattern templates. Still, spectral degradation and distortion may limit development of these patterns. For example, the limited spectral resolution may be insufficient to develop a "normal" formant vowel space (Harnsberger et al. 2001). However, academic and social development has been shown to be similar between normal-hearing and early-implanted children with CIs, attesting to the strong plasticity during the critical period of childhood development (Nicholas and Geers 2003; Sarant et al. 2018). Some aspects of speech perception (e.g., voice pitch) continue to develop into adolescence (e.g., Buss et al. 2017). While there are relatively few adult prelingual CI users, outcomes are much poorer than those for pediatric prelingual CI users (Wooi Teoh et al. 2004; Santarelli et al. 2008). The prolonged period of auditory deprivation and presumably poor neural health likely limits the efficacy of cochlear implantation in terms of speech perception. Still, the CI can provide extremely beneficial sound awareness in prelingual adults, and some prelingual adult CI users have demonstrated speech performance that is comparable to that of postlingual adult CI users (Wooi Teoh et al. 2004), suggesting great plasticity well beyond the critical period.

Both development in children and aging in adults will also affect perception with CIs. As noted above, early implantation will put pediatric CI users on the same developmental trajectory as their normal-hearing peers. Cognitive decline affects many older adults, regardless of hearing status. However, hearing loss can accelerate and/or deepen cognitive decline (Lin et al. 2013; Uchida et al. 2019). Interventions such as hearing aids and CIs can improve communication, social interaction, and quality of life, all of which can help to slow cognitive decline (Lin et al. 2012, 2023; Yeo et al. 2023). In CIs, perception of key acoustic cues has been shown to decline with age. For example, perception of temporal cues (e.g., amplitude modulation) has been shown to worsen with advanced age in CI users and may contribute to declines in speech performance (Xie et al. 2019; Shader et al. 2020). This suggests that developmental and aging effects on brain function can also limit CI outcomes across the lifespan.

9.1.4 Plasticity and Training in CIs

Brain plasticity accounts for much of the success of the CI. While CI recipients can passively adapt to electric stimulation during the first 6–12 months of implant use, passive adaptation is often incomplete (Sagi et al. 2010). Active auditory training has been shown to benefit CI users with years of experience with their device. Current CI candidacy allows for increasing amounts of AH in the implanted and/or non-implanted ears which can greatly improve CI users' access to pitch information, important for challenging listening conditions such as prosody perception, music perception, talker identification, and segregation of competing speech. However, adaptation to the CI may be limited by the degree of AH, with greater adaptation when residual AH is restricted and less adaptation when AH is unrestricted. The degree of CI adaptation may limit integration of acoustic and electric hearing as well as integration of sequentially implanted bilateral CIs. The delay in CI signal processing may also result in poor fusion across ears, especially when combining acoustic and electric hearing (e.g., precedence effects; Brown et al. 2015; Litovsky et al. 1999). Active training may improve adaptation to the CI and therefore benefit acoustic-electric integration. In turn, training may be needed for more challenging outcome measures (e.g., prosody perception, music perception, segregation of competing speech, etc.).

9.2 Passive Adaptation in CI Users

Postlingual adult CI users must adapt to the degraded spectral patterns and tonotopic mismatch between the acoustic input frequency and electrode place of stimulation. Tonotopic mismatch occurs due to compression of the wide acoustic input frequency range onto the limited extent of the electrode array. The degree of tonotopic mismatch can vary among CI recipients according to the length of the electrode array, insertion depth, and patterns of neural health. If tonotopic mismatch is not too severe, CI users can "passively" adapt via everyday listening with the device (i.e., without structured feedback). Longitudinal studies show that the greatest gains with such passive adaptation occur during the first 3 months post-implantation (Waltzman et al. 1986; Spivak and Waltzman 1990; George et al. 1995; Gray et al. 1995; Loeb and Kessler 1995), with continued improvement observed over longer periods (Tyler et al. 1997b), but with limited gains after the first 6–12 months of device use (e.g., Dorman and Loizou 1997; Pelizzone et al. 1999). If the tonotopic mismatch and/or

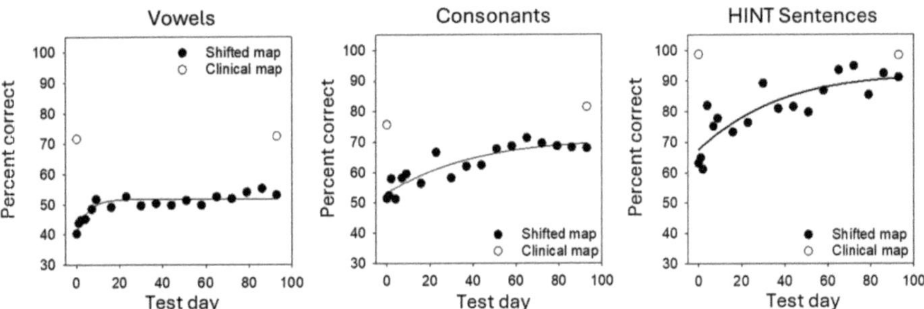

Fig. 9.2 Performance with a 1-octave upshifted frequency allocation as a function of test day (black circles). The white circles show performance with the clinical frequency allocation before and after training with the upshifted allocation

changes in signal processing are sufficiently severe, the time course of passive adaptation may be much longer, and CI users may never fully adapt.

During the initial period of CI use, there is strong adaptation to the clinical map, which may influence adaptation to changes in the acoustic-to-electric frequency allocation. For example, Fu et al. (2002) conducted a 3-month longitudinal study in CI users where the input frequency allocation was downshifted by ~1 octave, resulting in a perceptual upshift in speech. Baseline performance was acutely measured using participants' clinical maps immediately before and after the study period. Figure 9.2 shows mean vowel, consonant, and sentence recognition scores across CI users as a function of test day with the upshifted map. The acute deficit in performance with the shifted map was largest for vowels, followed by consonants, and sentences. Asymptotic performance with the shifted map was observed for vowels by day 58 (mean improvement = 9.5 percentage points), but post-adaptation performance remained 19.3 points poorer than with the clinical map (incomplete adaptation). A similar pattern was observed for consonants, although the acute deficit was smaller than for vowels. The mean acute deficit with the shifted map was largest for sentences (35.6 points), but adaptation was largest for sentences, where asymptotic performance was only 3.5 points poorer than that with the clinical map. Fu et al. (2002) also suggested that CI users can passively develop new speech templates in response to changes in frequency allocation while retaining previously developed templates. When adapting to tonotopic mismatch (e.g., as with initial CI activation), it may be advantageous to gradually introduce the mismatch to allow CI users to first adjust to electric stimulation and then gradually adapt to the mismatch (Svirsky et al. 2004; Fu and Galvin 2007). This may be especially appropriate for CI users with usable AH in the implanted or non-implanted ear. In these patients, the frequency allocation could be gradually expanded toward the edge of AH. This would also be appropriate as the extent of AH diminishes over time, requiring adjustment to the CI frequency allocation.

The above studies suggest that postlingual adult CI users can passively adapt to novel electrical stimulation patterns if the patterns are sufficiently similar to central pattern templates developed with previous AH. However, the degree of similarity between new electric patterns and central templates may limit passive adaptation. In a CI simulation study with normal-hearing listeners, Li and Fu (2007) repeatedly tested vowel recognition with moderately or severely upshifted speech over a 5-day study period. Response boxes were either lexically labeled (e.g., heed, had, hawed, etc.) or non-lexically labeled (e.g., 1, 2, 3). With moderately shifted speech, vowel recognition improved with both the lexical and non-lexical labels, suggesting some degree of automatic learning. With severely shifted speech, vowel recognition improved with the non-lexical labels, but not with the lexical labels. The upshifted speech often sounded like "heed" or "hayed," and listeners selected the response box that best matched the speech pattern. With the non-lexical labels, greater automatic learning occurred, as listeners were not bound by the lexical labels. The data suggest that adaptation to severe tonotopic mismatch or other difficult listening conditions may require meaningful structured feedback.

For early-implanted prelingual children, adaptation to the CI is concurrent with speech and language development, indeed with all aspects of sensory and cognitive development. In these children with CIs, central auditory patterns will develop with electric hearing, rather than in relation to templates developed during previous AH, as in postlingual adult CI users. While the mechanisms and trajectories of passive adaptation may differ between postlingual adults and prelingual children with CIs, it is likely that adaptation will be incomplete (e.g., Sagi et al. 2010). Active training may further adaptation to CIs and may be especially beneficial for challenging, unfamiliar listening conditions.

9.3 Active Training in CI Users

Passive learning will occur to some degree in all CI recipients, provided they regularly wear their device. However, passive adaptation may plateau during the first year of CI use. Passive adaptation may also be limited for difficult listening conditions or for individuals with severe tonotopic mismatch. Active training may provide further adaptation. Early studies (e.g., Busby et al. 1991; Dawson and Clark 1997) showed limited but durable improvements in CI users' speech and word recognition with active training. Since then, improvements in computer hardware and software have allowed for different training approaches.

9.3.1 Clinical CI Rehabilitation

Children with CIs are likely to receive in-person auditory rehabilitation throughout their childhood. This is important for auditory and cognitive development, as well as for academic and social skills. However, in the USA, there are few clinical auditory rehabilitation resources available to postlingual adults after CI activation. This is partly because of the initial period of passive adaptation during the first year of CI use. This is unfortunate, because clinical rehabilitation during this period may increase and/or accelerate adaptation to the CI. Clinical rehabilitation services for adult CI users are limited by resources in terms of clinician availability, insurance coverage, patient availability, etc.

Boothroyd (2007) outlined four essential components that comprise a holistic approach to auditory rehabilitation in adults: (1) Sensory management, (2) Instruction, (3) Perceptual training, and (4) Counseling. These components would generally be provided in a clinical setting requiring in-person interactions, but are generally cost prohibitive. Moberly et al. (2020) conducted a pilot study in newly activated postlingual adult CI users that compared "comprehensive auditory rehabilitation" (CAR) to standard of care (no post-implantation auditory rehabilitation) or to an active control where pre-implant counseling was provided. The CAR group received pre-implant counseling, 8 h of in-person auditory training in the clinic, and instructions to perform supplemental training (e.g., computer-based auditory training). While CI speech performance and quality of life improved over the 6-month study period, there were no significant differences among the three groups. The lack of benefit in the CAR group may have been due to strong concurrent passive adaptation during the first 6 months of CI use. Bernstein et al. (2021) evaluated short-term auditory rehabilitation in adult CI users with 3 months to 3 years of experience with their device. In the randomized controlled trial, half the participants were assigned to the training group (six weekly 90-min in-person training sessions that involved phonemic contrasts, sentences, etc., as well as counseling) and half were assigned to the control group, who completed nonauditory puzzles (e.g., Sudoku). Results showed significantly better speech performance in the training than in the control group; the advantage was retained 2 months after training had stopped.

While in-person auditory training may be beneficial, the costs and time commitment are generally high, which most likely will limit accessibility in adult CI users. Also, because CI outcomes are so variable, training needs to be tailored to individual CI users. On the one hand, this is desirable, as it will optimize the training for individual needs. On the other hand, there may be inconsistencies in training approaches across providers and clinics, even for CI patients with similar baseline capabilities. There may also be limited speech training resources; in-person training is often provided with the same voice from session to session, rather than using multiple voices. Computer-based auditory training may be a useful alternative or supplement to in-person training, as it can address many of the aforementioned limitations.

9.3.2 Psychophysical Training

Auditory training has been used to improve CI users' perception of psychophysical stimuli. For example, Goldsworthy and Shannon (2014) trained stimulation rate discrimination in CI listeners. Baseline rate discrimination was measured for reference stimulation rates ranging from 110 to 3520 pulse per second (pps). Participants received 32 h of rate discrimination training with feedback. The ratio between the reference and probe rates was held constant, and the reference rate was adapted according to the correctness of each response, thereby adjusting the level of difficulty. The authors found that the training improved rate discrimination by a factor of 2. Similar training benefits were observed in a follow-up study by Bissmeyer et al. (2020).

Fu and Galvin (2008) trained electrode discrimination in an adult prelingually deaf CI user. After measuring baseline electrode discrimination, training was provided in which relatively strong contrasts (5 h, week 1) and then relatively weak electrode contrasts (5 h, week 2) were compared. Two types of training were provided. First, stimulus preview was implemented in each training trial, in which the reference and probe stimuli were presented with visual cues, and then randomized and represented without the visual cues. Second, feedback was provided after each trial. Vowel and consonant recognition were also tested before and after the electrode discrimination training. Figure 9.3 shows heat maps for d′ values between the probe and reference electrodes at baseline, after training with the strong contrasts, after training with the weak contrasts, and at follow-up 2 weeks after training was stopped. Electrode discrimination progressively improved after training with the strong and weak contrasts, and improvements were largely retained 2 weeks after training was stopped. Improvements in vowel recognition (from 56% to 62% correct) and consonant recognition (from 38% to 51% correct) were observed after the electrode discrimination training. It is possible that some sort of procedural learning occurred for phoneme recognition that was unrelated to the improved electrode discrimination. The participant commented that the phonemes sounded different (i.e., discriminable), but that they were unable to identify the phonemes due to the limited speech perception experience given prelingual deafness. This observation led directly to Fu and colleagues developing computer-based auditory training (CBAT) for speech, first for research-based studies and eventually for home use by CI recipients.

9.3.3 Computer-Based Auditory Training (CBAT)

9.3.3.1 Speech Recognition in Quiet

CBAT has been shown to improve English vowel recognition in quiet for CI users (Fu et al. 2004; Miller et al. 2008, 2016; Stacey et al. 2010; Oba et al. 2011; Zhang et al. 2012; Ingvalson et al. 2013; Schumann et al. 2015; Barlow et al. 2016; Green et al. 2019; Reis et al. 2021; Kerneis et al. 2023; Magits et al. 2023) and for normal-hearing participants listening to CI simulations (e.g., Chavant et al. 2021; Faulkner et al. 2006, 2012; Fu et al. 2005; Fu and Galvin 2003; Nogaki et al. 2007; Rosen et al. 1999; Stacey and Summerfield 2007, 2008). Figure 9.4 shows post-training gains for vowel, consonant, and sentence recognition in quiet after CBAT for numerous studies. The mean post-train performance gains were 14 percentage points for vowels and consonants, and 19 points for sentences.

Studies conducted by Fu and colleagues (e.g., Fu et al. 2004, 2005; Kerneis et al. 2023; Zhang et al. 2012) generally employed a phonemic contrast training protocol.

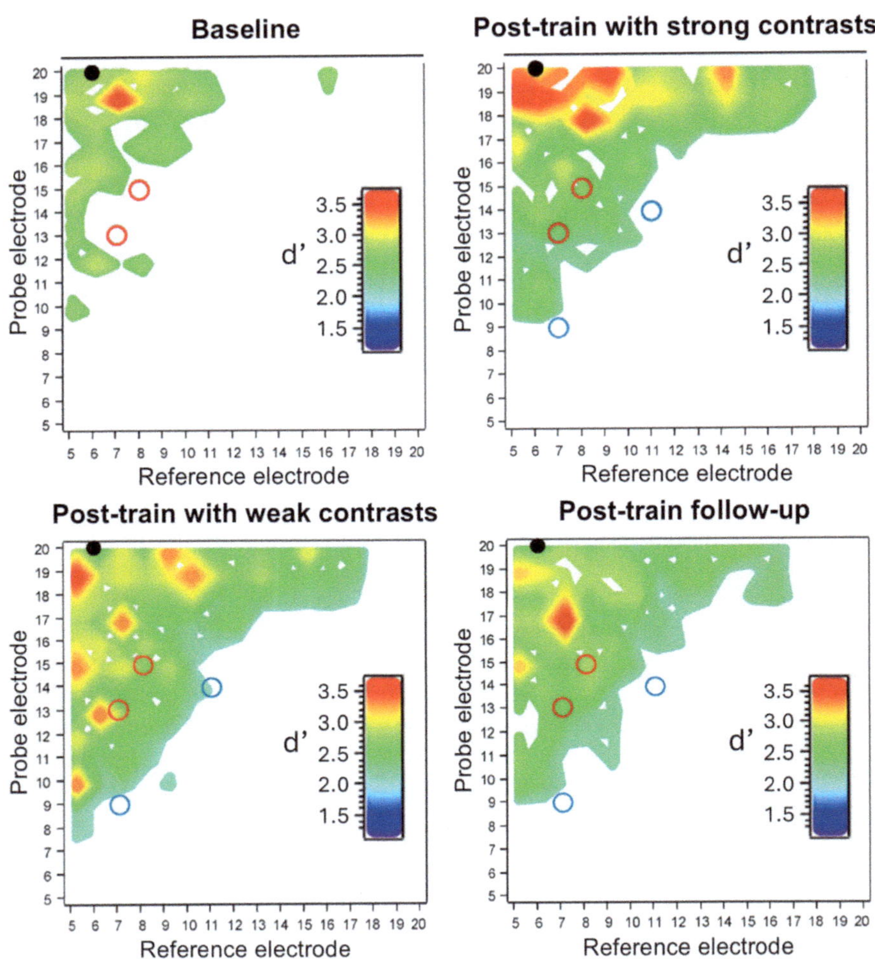

Fig. 9.3 Heat maps for electrode discrimination data from an adult CI user in terms of d´ values. Data were collected at baseline, after training with strong electrode contrasts (open red circles), after training with weak electrode contrasts (open blue circles), and at follow-up 2 weeks after training was stopped. (Figure from Fu and Galvin (2008))

In general, baseline speech recognition performance (vowel, consonant, and sentence recognition) was collected in CI users who had at least 6 months of experience with their device. Baseline performance was repeatedly measured until achieving asymptotic performance to reduce procedural learning effects. Next, participants trained at home using custom software installed on their home computers. Training was based on phonemic contrasts using a large set of multi-talker monosyllable words and v-C-v stimuli that were recorded for the training studies. The level of difficulty was automatically adjusted for each participant as training progressed by changing the number of and/or similarity among response choices (i.e., relatively

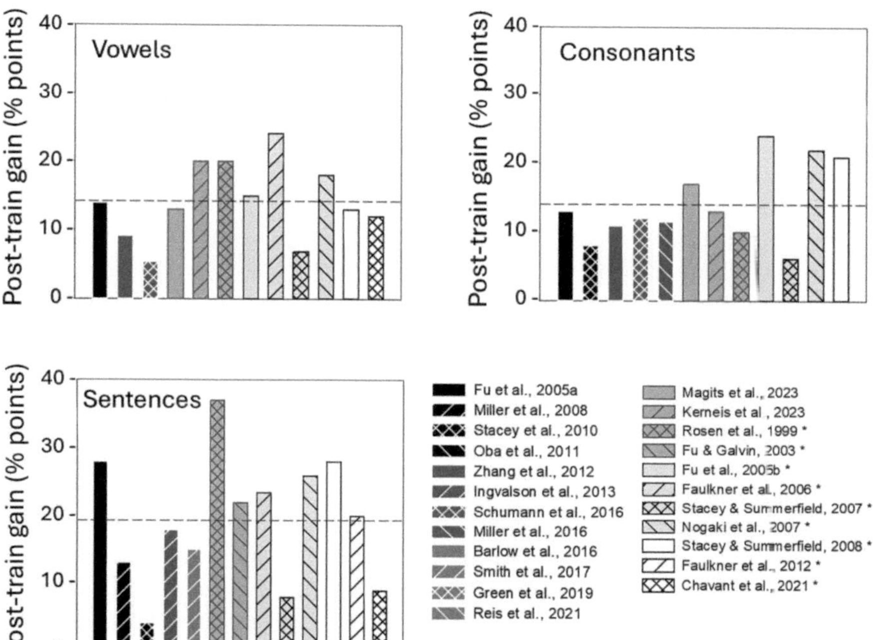

Fig. 9.4 Post-training performance gains for vowels, consonants, and sentences in quiet among different CI studies. The dashed line shows the mean performance gain across studies. The asterisks indicate CI simulation studies in normal-hearing listeners

strong or weak phonemic contrasts). Participants were given auditory and visual feedback, allowing them to compare their answer to the correct response. Participants trained for 1 h per day, 5 days per week, for 1 month or more, returning to the lab for retesting. Post-training performance consistently showed significant improvements in vowel and consonant recognition scores, measured using stimuli not used for training (h-V-d for vowels, a-C-a for consonants). Follow-up measures 1 month after training was stopped showed that post-training improvements were largely retained.

Improvements in CI speech performance after CBAT observed in English-speaking CI users have also been observed in other languages, using the phonemic contrast protocol. Wu et al. (2007) found significantly improved vowel, consonant, and lexical tone recognition in Mandarin-speaking pediatric CI users following similar training protocols as in Fu et al. (2004). Kerneis et al. (2023) found significantly improved vowel and consonant recognition in French-speaking CI users with phonemic contrast training. For each of these studies, large custom multi-talker training databases in Mandarin or French language were recorded by Fu and colleagues.

9.3.3.2 Speech Recognition in Noise

While auditory training may improve performance in ideal conditions, CI users regularly encounter challenging listening conditions in their everyday lives. Performance gains with CBAT in quiet may not necessarily generalize to speech recognition in noise. While training may not help CI users regain "lost" acoustic cues masked by interfering noise, training may help CI users to develop superior coping strategies (e.g., better utilization of context cues).

Several studies have investigated the benefits of CBAT using speech recognition in noise in CI listeners. For example, a pilot CBAT study by Fu and Galvin (2008) showed better sentence recognition in noise after 1 month of training with a keyword-in-sentence protocol than with a phonemic contrast protocol, both conducted in competing speech babble. The advantage with the keyword-in-sentence protocol may have been due to the greater context cues in the carrier sentences. Oba et al. (2011) performed digits-in-noise training in adult CI users; participants trained at home for ~10 h over a 1-month period. Mean speech recognition thresholds (SRTs) for digit-in-noise recognition, defined as the signal-to-noise ratio (SNR) that produces 50% correct performance, improved by 3 dB in competing babble (trained condition) and 4 dB in steady noise (untrained condition). Mean SRTs for sentences (untrained) improved by ~2 dB in babble and ~1 dB in noise. Barlow et al. (2016) used a short psychophysical training protocol that showed an increase in speech perception in noise and spectral discrimination performance in postlingual adult CI users. Green et al. (2019) used a training protocol in which narrative texts were divided into short phrases. Trainees listened to a phrase and were then asked to identify a target word from the phrase that was presented in competing speech which shared (in most cases) at least two phonemes with the target speech. After 12 h of self-administered, at-home training over a 4-week period, sentence recognition in noise significantly improved by an average of ~2 dB. However, there were no significant post-training improvements in phoneme identification. Figure 9.5

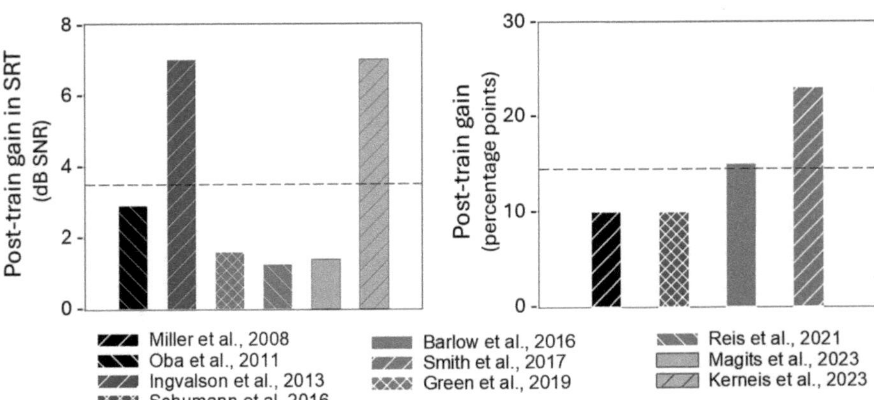

Fig. 9.5 Post-training performance gains for sentence recognition in noise for various CI studies. The dashed line shows the mean performance gain across studies

shows post-training gains for sentence recognition in noise after CBAT for numerous CI studies. The mean post-train performance gain in SRTs was 3.5 dB, and the gain at fixed SNRs was 14 percentage points.

9.3.3.3 Music Perception

Improving music perception is a high priority for many CI users (e.g., Looi and She 2010). Music perception differs greatly from speech perception in many ways. Due in part to the nature of speech communication—including redundant acoustic cues and contextual cues—central speech patterns are quite robust to spectral distortion. While trained musicians may develop central patterns related to melodies, pitch, timbre, rhythm, etc., such patterns may be less discrete in nonmusicians, and more susceptible to the spectro-temporal distortions associated with CI processing. In particular, pitch and timbre are poorly represented in CIs, which contributes to the generally poor music sound quality ratings by CI users. However, familiar music may form central patterns that can be retrieved by CI users. Gfeller et al. (2000) reported that previous familiarity with a musical piece could positively or negatively affect a CI user's appreciation of that piece (i.e., previous experience shapes current expectations and/or experience). Galvin et al. (2008, 2009) found that CI users with extensive music experience, both before and after implantation, performed better in difficult music perception tasks than musically inexperienced CI users.

As such, music training for CI users may be beneficial. Structured music training has been shown to improve participants' music timbre perception. Gfeller et al. (2002) trained musical timbre perception in CI users, finding significant improvements in timbre perception and appraisal after 12 weeks of training at home with CBAT. Galvin and colleagues have performed several CBAT studies training melodic contour identification (MCI) in CI users. In the MCI task, listeners are asked to identify 5-note melodic contours in which the direction of pitch changes and/or the semitone interval between successive notes is varied. As such, MCI requires listeners to use pitch information, rather than just discriminate musical interval or identify familiar melodies. In their training paradigm, after baseline testing of MCI, listeners were trained to identify contours for note ranges and instruments not used for testing; auditory and visual feedback were provided, allowing listeners to repeatedly compare their response to the correct response (Fig. 9.6). Galvin et al. (2007, 2009) found significant post-training improvements in MCI performance and familiar melody identification, which was not trained. Galvin et al. (2012) used CBAT to train MCI in CI users with or without a competing contour; baseline performance was much poorer with than without the competing contours. Post-training gains were significantly higher after training with than without the competing contour, suggesting that more challenging music training may be more effective. Fuller et al. (2018) compared CBAT MCI training, group-based music training, and an unrelated nonmusical training task (woodworking) in adult CI users. They found that while the group-based music training was rated as most enjoyable, only the MCI training significantly improved MCI performance.

Given the importance of pitch cues to perception of both lexical tones and music, it is not surprising that associations between Mandarin speech perception and

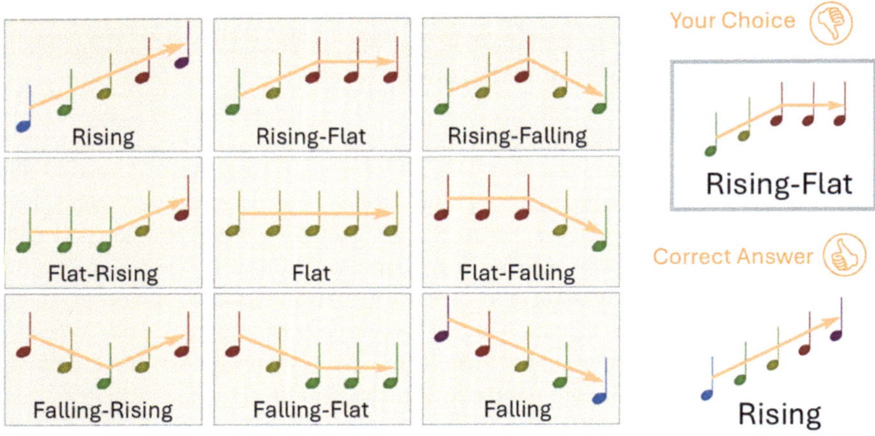

Fig. 9.6 MCI CBAT training paradigm. The left panel shows the response screen. During training, a contour stimulus is presented, and listeners select which contour best matches the stimulus. The right panel shows the audio and visual feedback provided in cases of incorrect response, in which the listener can repeatedly compare the correct answer to their (incorrect) response

musical pitch perception have been observed in Mandarin-speaking CI users (e.g., Wang et al. 2011, 2012). The MCI training paradigm has been used in several studies with Mandarin-speaking CI users. Fu et al. (2015) found that MCI CBAT training significantly improved MCI performance in pediatric Mandarin-speaking CI users. Similarly, Cheng et al. (2018) found that 8 weeks of MCI CBAT training significantly improved MCI performance in Mandarin-speaking CI users, as well as untrained lexical tone and sentence recognition.

Other approaches besides CBAT have been used to train music perception. Firestone et al. (2020) offered “attentive music listening” to adult CI users, who were instructed to listen to music directly streamed to their CI 40 min per day, 5 days per week, for a month or more. Post-training results showed improved sensitivity to changes in frequency, along with improved speech understanding in quiet and in noise; interestingly, cortical responses to changes in frequency also improved after the music training. Jiam et al. (2019) conducted a randomized clinical trial of music training in CI users. In the cross-over study, participants first trained by listening to an audiobook and then completed a series of CBAT music exercises, or vice versa. The authors found that both pitch and timbre discrimination improved with the audiobook and music training, with a small advantage for music training for timbre discrimination. Good et al. (2017) offered 6 months of piano lessons to pediatric CI users, finding significantly improved perception of melodic contours, rhythm, and perception; the music training also generalized to improved perception of vocal emotion. Yang et al. (2019) enrolled Mandarin-speaking bimodal and bilateral pediatric CI users into a choir (the “Little Dolphins”), participating in choir practices several times per week and supplementary home training for singing. The authors found that participation in the choir resulted in highly accurate pitch production;

earlier participation in the choir activities was significantly associated with better accuracy.

It should be noted that the above studies largely represent acute training benefits that are not similar to long-term music experience associated with formal music training. Such "musician advantages" have been associated with better perception of speech in noise or vocal emotion identification when speech is spectro-temporally degraded as in the CI case (e.g., Fuller et al. 2014a, b; Nie et al. 2018). While post-training gains were largely retained in many of the above studies 1 month or more after training was stopped, such training benefits may not necessarily translate into better overall music sound quality for CI users. Also, the limited CI population (and the variability in CI outcomes) and limited availability for music training results in many studies being underpowered, giving rise to some skepticism as to whether music training is worthwhile for CI users (e.g., McKay 2021). However, such skepticism is often levied against music training in general for failing to show robust benefits outside of music, contributing to the increased closures of music education in schools. It is worthwhile to remember that music is its own reward, and any improvements in CI signal processing and/or training that can improve music perception and/or enjoyment are likely to be rewarding to CI users.

9.4 Objective Measures of Adaptation and Training Benefits

There is some evidence that passive adaptation and/or active training (and their attendant performance improvements) may be reflected by objective, noninvasive measures of cortical activity and/or neural connectivity and structure. For example, cortical auditory evoked potentials (CAEPs) constitute cortical correlates of speech processing and can be measured via electroencephalogram (EEG; Hyde 1997). Such measures may reflect cortical reorganization after cochlear implantation, pre-attentive discrimination of stimulus features such as frequency, and/or post-training improvements in stimulus discrimination. These measures may help us identify the time course and trajectory of passive adaptation, perceptual limits of CIs, what auditory training approaches are the most beneficial, and what training protocols specifically benefit which aspects or modes of auditory perception.

9.4.1 Evidence of Cortical Plasticity

Reception of sensory input during developmental critical periods is crucial. During these periods, the cortex undergoes significant organizational changes as a direct result of the sensory input that it receives. Prolonged periods of auditory deprivation can result in lifelong deficits in speech understanding that are reflected in measurable cortical deficits (Kral 2013). Thus, in children with congenital deafness, it is crucial to restore bilateral sensory input before the critical period has passed (Gordon et al. 2011, 2013; Kral et al. 2001, 2006; Kral and O'Donoghue 2010). Failure to do so will interfere with normal maturation of the auditory cortex, potentially resulting in reduced cochleotopy and deficits in corticocortical and corticothalamic loops. Kral et al. (2001) found that congenitally deaf cats who received CIs in adulthood (i.e., after the sensitive period of auditory development) exhibited widespread cortical deficits, including longer CAEP latencies and reduced synaptic activity in infragranular cortical layers.

During the critical period, relatively small differences in age at implantation can affect long-term outcomes (Tyler et al. 1997a; Papsin et al. 2000; Geers et al. 2002; Hassanzadeh et al. 2002; Kirk et al. 2002). Sharma et al. (2002) showed that late-implanted congenitally deaf children (age 7 or older) exhibited abnormal CAEP latencies in response to speech (even after years of stimulation and adaptation), while early-implanted children (age 3.5 or younger) displayed normal latencies within 6 months post-implantation. Sharma et al. (2005) later demonstrated fundamentally different CAEP morphologies between early- and late-implanted children. Multiple studies in both adults and children with CIs provide evidence that with long-term device experience, changes in CAEPs were accompanied by improvements in speech perception (e.g., Jordan et al. 1997; Kaga et al. 1991; Ponton et al. 1996, 1999, 2000). Similarly, pediatric CI users displayed increased amplitude and decreased latency of evoked middle latency response in the year after cochlear implantation (Gordon et al. 2005a, b). The measurement of CAEPs in children with hearing loss may be used to monitor cortical improvements that may precede or be reflected in behavioral improvements.

Although the brain is most plastic during the critical period, some plasticity is retained through adulthood. While the specialization of high-order auditory areas is maintained, prolonged auditory deprivation in adulthood may result in metamodal reorganization (e.g., prioritization of visual tasks in shared visual-auditory cortical areas) and/or cross-modal take-over (e.g., recruitment of former auditory areas for visual processing; Lazard et al. 2014). As a result, prolonged duration of deafness may limit the time course and/or degree of adaptation to electric hearing in postlingual CI users (e.g., Sandmann et al. 2015). Cross-modal reorganization has also been observed in the auditory cortex of adults with mild-to-moderate hearing loss (Campbell and Sharma 2014), suggesting a continuous lifelong effort by the brain to reorganize according to the regularity, quality, and/or intensity of stimuli.

9.4.2 Improved Cortical Response with Passive Adaptation to the CI

Passive adaptation to electric hearing can highlight plasticity in the auditory cortex. As discussed above, cochlear implantation in congenitally deaf children can lead to normal or near-normal patterns and levels of cortical activation. Postlingual adult CI users similarly show changes in cortical organization and activation with passive adaptation that correlate with improved speech perception performance. During the first 6–12 months of implant use, speech performance continues to improve in postlingual adult CI users (Lenarz et al. 2012). During the same period, CI users show increasing N1 amplitudes and decreasing N1 latencies (Sandmann et al. 2015), decreasing P3 latencies (Jacquemin et al. 2019), increasing and progressively stimulus-specific recruitment of the visual cortex for auditory tasks (Giraud et al. 2001; Stropahl et al. 2017), and increased activation in the left inferior prefrontal and right temporal cortices (Mortensen et al. 2006). Previous experience with AH may also affect the trajectory of passive adaptation reflected by cortical measures. For example, increased activation in Broca's area was found in postlingual but not prelingual adult CI users (Petersen et al. 2013).

Much recent electrophysiological work has been conducted with single-sided deaf (SSD) CI users looking at cortical reorganization after cochlear implantation of the deaf ear. SSD CI users offer the unique opportunity to compare cortical responses and speech performance between acoustic and electric hearing. After receiving the implant, adult SSD CI users have been shown to exhibit increased P2 area, increased N1 amplitude, and bilateral activation (Legris et al. 2018, 2019; Purdy and Kelly 2016). These changes in cortical response have been associated with improved speech performance (Purdy and Kelly 2016; Legris et al. 2018). In pediatric SSD CI users, cortical responses improve in the CI ear and become less asymmetric relative to the AH ear with ongoing experience with the CI (Polonenko et al. 2017; Lee et al. 2020). Legris et al. (2020) also found increased N1 amplitude for stimuli presented in noise in adult SSD CI users that was associated with improved speech understanding in noise. However, such improvements in cortical responses and behavioral performance generally fall short of those observed in normal-hearing listeners. For example, Weglage et al. (2022) found that auditory event-related potentials (ERPs) in response to speech stimuli were significantly different between the AH and CI ears. ERP data suggested that syllable processing was attenuated and delayed in the CI ear compared to the AH ear. Van Heteren et al. (2022) compared the acoustic change complex (ACC) between the AH and CI ears in SSD listeners and in bilaterally deaf CI users. They found longer N1 latencies and smaller N1-P2 amplitudes in the CI ear than in the AH ear in SSD CI users. However, the responses in the CI ear were similar between SSD and bilaterally deaf CI users, suggesting fundamental limits with electric hearing due to the relatively poor spectro-temporal resolution. The cortical changes associated with passive adaptation suggest that active training may increase and/or accelerate adaptation to electric hearing.

9.4.3 Improved Cortical Response with Active Training in CI Users

One important question is whether auditory training improves performance on the trained task/stimuli or whether it extends to other perceptual tasks and/or domains. Cortical recordings have been used to assess task-specific or generalized learning in normal-hearing humans and/or animal models. In Tremblay et al. (1998), neurophysiological changes—in this case, in mismatch negativity (MMN)—occurred immediately following the first day of training with a stimulus discrimination task and, for many participants in the study, prior to any improvement in behavioral performance. This and a host of imaging and EEG studies of auditory training in normal-hearing humans (see Irvine 2018) provide a clear basis for the use of cortical-level measures in identifying training benefits for CI users.

Thus far, very few training studies have been conducted in adult CI users. Miller et al. (2016) studied the benefits of training with high-variability speech stimuli on categorical perception of speech contrasts (i.e., /ba/–/da/ and /wa/–/ja/). They found that the training significantly improved sensitivity to the phonetic boundary that was associated with increased MMN amplitude for the speech contrasts. Barlow et al. (2016) provided extensive psychophysical training to experienced adult CI users, including gap detection, frequency discrimination, spectral ripple perception, temporal modulation detection, etc. After training, only spectral ripple perception significantly improved, and there was no significant improvement in speech performance. However, they found that N1P2 amplitude significantly increased after training. The lack of association between improved cortical response and speech perception after training may have been partly due to the long-term CI experience and sensitivity of the speech outcome measure to the training provided. Firestone et al. (2020) assessed quality of life, frequency change detection thresholds (FCDTs), speech understanding, and cortical responses to tone stimuli in CI users after 1 month or more of guided music listening training. They found significant post-training improvements in FCDTs, speech performance, and cortical responses, suggesting that music may be a worthwhile training approach in CI users.

The dearth of auditory training/electrophysiological studies in adult CI users is unfortunate, as objective measures may provide evidence of what may be learned with auditory training (e.g., improved onset of auditory processes, improved attention, etc.). In children with CIs, it can be difficult to parse developmental changes in speech performance from the benefits of auditory training. Objective measures may help to characterize training benefits in children with CIs, provided that studies are well-controlled and sufficiently powered. Importantly, objective measures may show pre-attentive discrimination of stimuli that may be used to guide training and rehabilitation approaches.

9.5 Considerations for CI Training Studies

Conducting training studies in CI users is challenging for a variety of reasons. First, there is great variability in CI outcomes, even after long-term adaptation to the device (Finley et al. 2008; Lazard et al. 2012). It is difficult to establish control groups with sufficient numbers of CI users with which to compare training outcomes. One approach adopted by many training studies (e.g., Oba et al. 2011) is to use "within-subject" controls, where training benefits are calculated relative to baseline performance for each participant. When recording baseline measures, it is important to account for task-related learning. Figure 9.7 shows performance on successive baseline administrations of various tasks (Oba et al. 2011). It is clear that performance on all measures improved with each successive test, possibly due to procedural learning. It is important not to conflate procedural with perceptual learning. Repeated baseline testing until achieving asymptotic performance is important to evaluate training benefits. Some procedural (and perceptual learning) may be

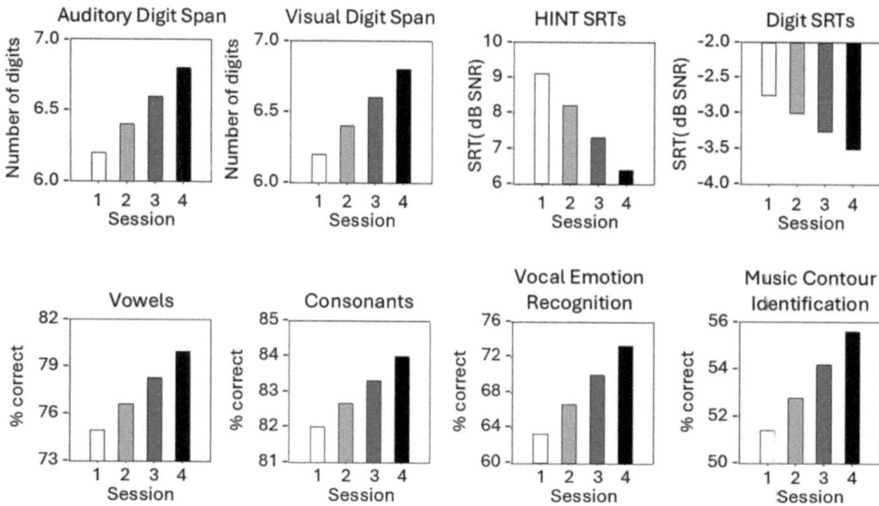

Fig. 9.7 Mean performance over four baseline (pre-training) sessions for various outcome measures from Oba et al. (2011)

due to cognitive factors. As such, cognitive measures may also help to distinguish perceptual learning. Oba et al. (2011) found that training with a nonauditory task (visual digit span) did not generalize to improved auditory performance. Such nonauditory training may be a good control condition with which to evaluate auditory perceptual learning.

Second, the time commitment is substantial. While CI users may benefit from auditory training, it can be difficult to make the time to participate in training research, where training often occurs over a 1-month period, or to perform CBAT outside of research, given lifestyle and work demands. One study examining the potential impact of training rates (e.g., 5 vs. 3 vs. 1 training sessions per week) showed that significant benefit could be reaped from even one training session per week (Nogaki et al. 2007). Mobile computing may help to make training more accessible and fit within CI users' daily lives.

Third, training approaches can differ greatly across studies, making it difficult to compare outcomes. In a systematic review of training studies in postlingual adult CI users, Cambridge et al. (2022) showed that most training studies lacked control groups, post-training follow-up measures, consistent feedback, and/or ecological validity, suggesting that there is a pressing need for high-quality experimental research regarding the effectiveness of auditory training in CI users.

It is generally difficult to conduct randomized controlled trials (RCTs) for training in CI users, due to time commitment, intersubject variability, etc. As such there have been relatively few RCT training studies in CI users. Schumann et al. (2015) randomly assigned 27 experienced CI users to a training group or control group. The training group performed computer-based phoneme discrimination training twice per week for 3 weeks. Results showed significant post-training gains in phoneme identification as well as sentence recognition in noise (untrained condition). Follow-up testing at 6 months post-training showed maintenance of performance gains. Magits et al. (2023) randomized 40 CI users into personalized ($n = 20$) and non-personalized ($n = 20$) training groups. Participants completed 16 weeks of CBAT. The personalized training group exhibited significant task-related learning (digits in noise, phoneme recognition), while both groups exhibited significant improvements for sentence recognition in noise (untrained task). Again, post-training gains were maintained at follow-up testing 8 months after training was stopped.

Given the difficulties and complexities of conducting training studies in CI users, it is important that procedural learning be controlled (i.e., repeated baseline testing), that there be some age-appropriate measure of nonauditory cognition, that there be outcome measures that are not directly related to the trained task, and that, if possible, the training be fun and engaging.

9.6 New Challenges and New Training Approaches

Auditory training may not offer many surprising results; one trains, one improves, no surprise. However, the changing demographics for CI users, especially adults, require greater consideration of the most effective and efficient training approaches. Without such considerations, one might train, but not improve, contrary to expectations.

Inclusion criteria for cochlear implantation have progressively expanded to include increasing amounts of AH in the implanted or non-implanted ears. The amount of AH in the non-implanted ear limits not only integration of acoustic and electric hearing, but also adaptation to electric hearing. In the case of SSD CI users, the AH ear will dominate bilateral perception with little contribution from the CI; in this case, there will be little adaptation to the CI. For bimodal CI users, who have limited AH in the non-implanted ear, the CI will contribute more to bilateral perception, with performance often greater in the CI than in the AH ear. In the bimodal CI case, adaptation to the CI will be greater than that in SSD CI users.

Performance asymmetry across ears has been shown to limit bimodal benefit in CI users (Yoon et al. 2015), with the greatest benefit observed when asymmetry was least. Theoretically, reducing performance asymmetry by training the poorer ear alone may help to increase bimodal benefit. For example, Kerneis et al. (2023) showed significant improvements in binaural performance after training with the CI ear alone in bimodal CI users or with the poorer ear alone in bilateral CI users. However, Chavant et al. (2021) showed that training with both acoustic and electric hearing simultaneously provided a greater binaural benefit than with training the CI ear alone. Many CI clinicians recommend watching videos while listening with the CI alone, where lipreading and/or subtitles might provide additional support. Any and all of these methods may be beneficial, with no downside to any of them. Indeed, variability in training approaches may help CI users maintain interest and learn from different stimuli and designs. While software such as AngelSound can offer a wide range of stimuli and training exercises, CBAT may offer limited emotional engagement, which may ultimately reduce patient participation. As discussed earlier, although the community music-making training in Fuller et al. (2018) did not improve MCI performance as much as with CBAT, CI patients greatly enjoyed the social aspect of the multimodal training. Approaching CBAT from a gaming perspective may also be advantageous by increasing engagement such that users do not feel as if they are actually performing training. Gaming approaches may also be more culturally relevant to young people, improving chances of engagement. Virtual reality may be another approach to auditory training with multisensory inputs. For example, a pediatric CI user could train in a simulated multisensory classroom environment, while an adult CI user could train talking on the telephone in a simulated coffee shop. Using a hear-through setting as found in many modern headphones, training protocols could be designed to take advantage of live ambient noise as a masker with CI users completing training on their phone or tablet. Especially when incorporating newer technologies into training designs, it is crucial to keep a focus on accessibility and usability. If possible, those designing new training paradigms/software/methodologies should solicit the input of individual CI users (e.g., Veltman et al. 2023).

Ideally, training should target multiple challenging listening conditions. For example, pitch cues are poorly transmitted and perceived by CI users, therefore limiting perception of music, vocal emotion, talker identity, etc. Auditory training could focus attention on pitch cues (or their covarying cues such as intensity, duration, tempo, etc.) to improve perception of pitch cues in speech and music. Noise is challenging for all CI users. Auditory training could target top-down processes to develop listening strategies to cope with noise. Spatial perception is also challenging for bilateral, bimodal, and SSD CI users. Head-related transfer functions (HRTFs) could be used to train perception of interaural loudness differences (ILDs) and interaural time differences (ITDs) in virtual acoustic space. For example, Valzolgher and colleagues recently developed virtual reality-based localization training paradigms for both individuals with normal hearing (Valzolgher et al. 2020) or with a CI (Valzolgher et al. 2023). Audiovisual virtual space could further enhance training approaches. CI patients should also have more immediate input into their auditory training. Allowing CI users to make their own adventure may further their progress on their training journey.

This chapter has focused on training for adult CI users. Children have much greater plasticity than adults do, and childhood development depends on multisensory input. For early-implanted pediatric CI users, electric hearing contributes to childhood development, not just to auditory development. There are many more rehabilitation approaches available to children than to adults. However, auditory training for CI children could also be optimized to improve or accelerate development of electric hearing. For example, there are few resources to train emotion perception in pediatric CI users. Emotion communication is crucial to childhood development. However, the CI does not transmit adequate pitch information to support good vocal emotion perception or production (Chatterjee et al. 2023 and others). Very young children tend to learn the emotional content of speech (e.g., "motherese") well before they learn to perceive or produce words. As such, it is important to train production and perception of pitch cues to emotional prosody in pediatric CI users.

Acknowledgments The authors declare that no Generative AI was used in the creation of this manuscript.

References

Barlow N, Purdy SC, Sharma M et al (2016) The effect of short-term auditory training on speech in noise perception and cortical auditory evoked potentials in adults with cochlear implants. Semin Hear 37:84–98. https://doi.org/10.1055/s-0035-1570335

Bernstein CM, Brewer DM, Bakke MH et al (2021) Maximizing cochlear implant outcomes with short-term aural rehabilitation. J Am Acad Audiol 32:144–156. https://doi.org/10.1055/s-0041-1722987

Bissmeyer SRS, Hossain S, Goldsworthy RL (2020) Perceptual learning of pitch provided by cochlear implant stimulation rate. PLoS One 15:e0242842. https://doi.org/10.1371/journal.pone.0242842

Boothroyd A (2007) Adult aural rehabilitation: what is it and does it work? Trends Amplif 11:63–71. https://doi.org/10.1177/1084713807301073

Brown AD, Jones HG, Kan A et al (2015) Evidence for a neural source of the precedence effect in sound localization. J Neurophysiol 114:2991–3001. https://doi.org/10.1152/jn.00243.2015

Busby PA, Roberts SA, Tong YC, Clark GM (1991) Results of speech perception and speech production training for three prelingually deaf patients using a multiple-electrode cochlear implant. Br J Audiol 25:291–302. https://doi.org/10.3109/03005369109076601

Buss E, Flaherty MM, Leibold LJ (2017) Development of frequency discrimination at 250 Hz is similar for tone and /ba/ stimuli. J Acoust Soc Am 142:EL150. https://doi.org/10.1121/1.4994687

Cambridge G, Taylor T, Arnott W, Wilson WJ (2022) Auditory training for adults with cochlear implants: a systematic review. Int J Audiol 61:896–904. https://doi.org/10.1080/14992027.2021.2014075

Campbell J, Sharma A (2014) Cross-modal re-organization in adults with early stage hearing loss. PLoS One 9:e90594. https://doi.org/10.1371/journal.pone.0090594

Chatterjee M, Feller A, Kulkarni AM, Galvin JJ (2023) Emotional prosody perception and production are linked in prelingually deaf children with cochlear implants. JASA Express Letters 3:127601. https://doi.org/10.1121/10.0023996

Chavant M, Hervais-Adelman A, Macherey O (2021) Perceptual learning of vocoded speech with and without contralateral hearing: implications for cochlear implant rehabilitation. J Speech Lang Hear Res 64:196–205. https://doi.org/10.1044/2020_JSLHR-20-00385

Cheng X, Liu Y, Shu Y et al (2018) Music training can improve music and speech perception in pediatric mandarin-speaking cochlear implant users. Trends Hear 22:233121651875921. https://doi.org/10.1177/2331216518759214

Dawson PW, Clark GM (1997) Changes in synthetic and natural vowel perception after specific training for congenitally deafened patients using a multichannel cochlear implant. Ear Hear 18:488–501. https://doi.org/10.1097/00003446-199712000-00007

Donaldson GS, Dawson PK, Borden LZ (2011) Within-subjects comparison of the HiRes and Fidelity120 speech processing strategies: speech perception and its relation to place-pitch sensitivity. Ear Hear 32:238–250. https://doi.org/10.1097/AUD.0b013e3181fb8390

Dorman MF, Loizou PC (1997) Changes in speech intelligibility as a function of time and signal processing strategy for an Ineraid patient fitted with continuous interleaved sampling (CIS) processors. Ear Hear 18:147–155. https://doi.org/10.1097/00003446-199704000-00007

Drennan WR, Pfingst BE (2006) Current-level discrimination in the context of interleaved, multichannel stimulation in cochlear implants: effects of number of stimulated electrodes, pulse rate, and electrode separation. J Assoc Res Otolaryngol 7:308–316. https://doi.org/10.1007/s10162-006-0045-8

Faulkner A, Rosen S, Norman C (2006) The right information may matter more than frequency-place alignment: simulations of frequency-aligned and upward shifting cochlear implant processors for a shallow electrode array insertion. Ear Hear 27:139–152. https://doi.org/10.1097/01.aud.0000202357.40662.85

Faulkner A, Rosen S, Green T (2012) Comparing live to recorded speech in training the perception of spectrally shifted noise-vocoded speech. J Acoust Soc Am 132:EL336–EL342. https://doi.org/10.1121/1.4754432

Fetterman BL, Domico EH (2002) Speech recognition in background noise of cochlear implant patients. Otolaryngol Head Neck Surg 126:257–263. https://doi.org/10.1067/mhn.2002.123044

Finley CC, Holden TA, Holden LK et al (2008) Role of electrode placement as a contributor to variability in cochlear implant outcomes. Otol Neurotol 29:920–928. https://doi.org/10.1097/MAO.0b013e318184f492

Firestone GM, McGuire K, Liang C et al (2020) A preliminary study of the effects of attentive music listening on cochlear implant users' speech perception, quality of life, and behavioral and objective measures of frequency change detection. Front Hum Neurosci 14:110. https://doi.org/10.3389/fnhum.2020.00110

Firszt JB, Holden LK, Reeder RM, Skinner MW (2009) Speech recognition in cochlear implant recipients: comparison of standard HiRes and HiRes 120 sound processing. Otol Neurotol 30:146–152. https://doi.org/10.1097/MAO.0b013e3181924ff8

Fu Q-J, Galvin JJ (2003) The effects of short-term training for spectrally mismatched noise-band speech. J Acoust Soc Am 113:1065–1072. https://doi.org/10.1121/1.1537708

Fu Q-J, Galvin JJ (2007) Perceptual learning and auditory training in cochlear implant recipients. Trends Amplif 11:193–205. https://doi.org/10.1177/1084713807301379

Fu Q-J, Galvin JJ (2008) Maximizing cochlear implant patients' performance with advanced speech training procedures. Hear Res 242:198–208. https://doi.org/10.1016/j.heares.2007.11.010

Fu Q-J, Nogaki G (2005) Noise susceptibility of cochlear implant users: the role of spectral resolution and smearing. JARO 6:19–27. https://doi.org/10.1007/s10162-004-5024-3

Fu Q-J, Shannon RV, Galvin JJ (2002) Perceptual learning following changes in the frequency-to-electrode assignment with the Nucleus-22 cochlear implant. J Acoust Soc Am 112:1664–1674. https://doi.org/10.1121/1.1502901

Fu Q-J, Galvin J, Wang X, Nogaki G (2004) Effects of auditory training on adult cochlear implant patients: a preliminary report. Cochlear Implants Int 5(Suppl 1):84–90. https://doi.org/10.1179/cim.2004.5.Supplement-1.84

Fu Q-J, Nogaki G, Galvin JJ (2005) Auditory training with spectrally shifted speech: implications for cochlear implant patient auditory rehabilitation. J Assoc Res Otolaryngol 6:180–189. https://doi.org/10.1007/s10162-005-5061-6

Fu Q-J, Galvin JJ, Wang X, Wu J-L (2015) Benefits of music training in mandarin-speaking pediatric cochlear implant users. J Speech Lang Hear Res 58:163–169. https://doi.org/10.1044/2014_JSLHR-H-14-0127

Fuller CD, Galvin JJ, Free RH, Başkent D (2014a) Musician effect in cochlear implant simulated gender categorization. J Acoust Soc Am 135:EL159–EL165. https://doi.org/10.1121/1.4865263

Fuller CD, Galvin JJ, Maat B et al (2014b) The musician effect: does it persist under degraded pitch conditions of cochlear implant simulations? Front Neurosci 8:179. https://doi.org/10.3389/fnins.2014.00179

Fuller CD, Galvin JJ, Maat B et al (2018) Comparison of two music training approaches on music and speech perception in cochlear implant users. Trends Hear 22:233121651876537. https://doi.org/10.1177/2331216518765379

Galvin JJ, Fu Q-J (2005) Effects of stimulation rate, mode and level on modulation detection by cochlear implant users. JARO 6:269–279. https://doi.org/10.1007/s10162-005-0007-6

Galvin JJ, Fu Q-J (2009) Influence of stimulation rate and loudness growth on modulation detection and intensity discrimination in cochlear implant users. Hear Res 250:46–54. https://doi.org/10.1016/j.heares.2009.01.009

Galvin JJ, Fu Q-J, Nogaki G (2007) Melodic contour identification by cochlear implant listeners. Ear Hear 28:302–319. https://doi.org/10.1097/01.aud.0000261689.35445.20

Galvin JJ, Fu Q-J, Oba S (2008) Effect of instrument timbre on melodic contour identification by cochlear implant users. J Acoust Soc Am 124:EL189–EL195. https://doi.org/10.1121/1.2961171

Galvin JJ, Fu Q-J, Shannon RV (2009) Melodic contour identification and music perception by cochlear implant users. Ann N Y Acad Sci 1169:518–533. https://doi.org/10.1111/j.1749-6632.2009.04551.x

Galvin J, Eskridge E, Oba S, Fu Q-J (2012) Melodic contour identification training in cochlear implant users with and without a competing instrument. Semin Hear 33:399–409. https://doi.org/10.1055/s-0032-1329227

Geers A, Brenner C, Nicholas J et al (2002) Rehabilitation factors contributing to implant benefit in children. Ann Otol Rhinol Laryngol Suppl 189:127–130. https://doi.org/10.1177/00034894021110s525

George CR, Cafarelli Dees D, Sheridan C, Haacke N (1995) Preliminary findings of the new Spectra 22 speech processor with first-time cochlear implant users. Ann Otol Rhinol Laryngol Suppl 166:272–275

Gfeller K, Christ A, Knutson JF et al (2000) Musical backgrounds, listening habits, and aesthetic enjoyment of adult cochlear implant recipients. J Am Acad Audiol 11:390–406

Gfeller K, Witt S, Adamek M et al (2002) Effects of training on timbre recognition and appraisal by postlingually deafened cochlear implant recipients. J Am Acad Audiol 13:132–145

Giraud AL, Price CJ, Graham JM et al (2001) Cross-modal plasticity underpins language recovery after cochlear implantation. Neuron 30:657–663. https://doi.org/10.1016/s0896-6273(01)00318-x

Goldsworthy RL, Shannon RV (2014) Training improves cochlear implant rate discrimination on a psychophysical task. J Acoust Soc Am 135:334–341. https://doi.org/10.1121/1.4835735

Good A, Gordon KA, Papsin BC et al (2017) Benefits of music training for perception of emotional speech prosody in deaf children with cochlear implants. Ear Hear 38:455–464. https://doi.org/10.1097/AUD.0000000000000402

Gordon KA, Papsin BC, Harrison RV (2005a) Effects of cochlear implant use on the electrically evoked middle latency response in children. Hear Res 204:78–89. https://doi.org/10.1016/j.heares.2005.01.003

Gordon KA, Tanaka S, Papsin BC (2005b) Atypical cortical responses underlie poor speech perception in children using cochlear implants. Neuroreport 16:2041–2045. https://doi.org/10.1097/00001756-200512190-00015

Gordon KA, Wong DDE, Valero J et al (2011) Use it or lose it? Lessons learned from the developing brains of children who are deaf and use cochlear implants to hear. Brain Topogr 24:204–219. https://doi.org/10.1007/s10548-011-0181-2

Gordon KA, Jiwani S, Papsin BC (2013) Benefits and detriments of unilateral cochlear implant use on bilateral auditory development in children who are deaf. Front Psychol 4:719. https://doi.org/10.3389/fpsyg.2013.00719

Gray RF, Wareing MJ, Court I (1995) Cochlear implant results: a comparison of live-voice and videotaped tests. Laryngoscope 105:1001–1004. https://doi.org/10.1288/00005537-199509000-00022

Green KMJ, Julyan PJ, Hastings DL, Ramsden RT (2005) Auditory cortical activation and speech perception in cochlear implant users: effects of implant experience and duration of deafness. Hear Res 205:184–192. https://doi.org/10.1016/j.heares.2005.03.016

Green T, Faulkner A, Rosen S (2019) Computer-based connected-text training of speech-in-noise perception for cochlear implant users. Trends Hear 23:2331216519843878. https://doi.org/10.1177/2331216519843878

Harnsberger JD, Svirsky MA, Kaiser AR et al (2001) Perceptual "vowel spaces" of cochlear implant users: implications for the study of auditory adaptation to spectral shift. J Acoust Soc Am 109:2135–2145. https://doi.org/10.1121/1.1350403

Hassanzadeh S, Farhadi M, Daneshi A, Emamdjomeh H (2002) The effects of age on auditory speech perception development in cochlear-implanted prelingually deaf children. Otolaryngol Head Neck Surg 126:524–527. https://doi.org/10.1067/mhn.2002.125110

Hyde M (1997) The N1 response and its applications. Audiol Neurootol 2:281–307. https://doi.org/10.1159/000259253

Ingvalson EM, Lee B, Fiebig P, Wong PCM (2013) The effects of short-term computerized speech-in-noise training on postlingually deafened adult cochlear implant recipients. J Speech Lang Hear Res 56:81–88. https://doi.org/10.1044/1092-4388(2012/11-0291)

Irvine DRF (2018) Auditory perceptual learning and changes in the conceptualization of auditory cortex. Hear Res 366:3–16. https://doi.org/10.1016/j.heares.2018.03.011

Jacquemin L, Mertens G, Schlee W et al (2019) Literature overview on P3 measurement as an objective measure of auditory performance in post-lingually deaf adults with a cochlear implant. Int J Audiol 58:816–823. https://doi.org/10.1080/14992027.2019.1654622

Jiam NT, Deroche ML, Jiradejvong P, Limb CJ (2019) A randomized controlled crossover study of the impact of online music training on pitch and timbre perception in cochlear implant users. J Assoc Res Otolaryngol 20:247–262. https://doi.org/10.1007/s10162-018-00704-0

Jordan K, Schmidt A, Plotz K et al (1997) Auditory event-related potentials in post- and prelingually deaf cochlear implant recipients. Am J Otolaryngol 18:S116–S117

Kaga K, Kodera K, Hirota E, Tsuzuku T (1991) P300 response to tones and speech sounds after cochlear implant: a case report. Laryngoscope 101:905–907. https://doi.org/10.1288/00005537-199108000-00017

Kerneis S, Galvin JJ, Borel S et al (2023) Preliminary evaluation of computer-assisted home training for French cochlear implant recipients. PLoS One 18:e0285154. https://doi.org/10.1371/journal.pone.0285154

Kirk KI, Miyamoto RT, Lento CL et al (2002) Effects of age at implantation in young children. Ann Otol Rhinol Laryngol Suppl 189:69–73. https://doi.org/10.1177/00034894021110s515

Kral A (2013) Auditory critical periods: a review from system's perspective. Neuroscience 247:117–133. https://doi.org/10.1016/j.neuroscience.2013.05.021

Kral A, O'Donoghue GM (2010) Profound deafness in childhood. N Engl J Med 363:1438–1450. https://doi.org/10.1056/NEJMra0911225

Kral A, Hartmann R, Tillein J et al (2001) Delayed maturation and sensitive periods in the auditory cortex. Audiol Neurootol 6:346–362. https://doi.org/10.1159/000046845

Kral A, Tillein J, Heid S et al (2006) Cochlear implants: cortical plasticity in congenital deprivation. Prog Brain Res 157:283–313. https://doi.org/10.1016/s0079-6123(06)57018-9

Kreft HA, Donaldson GS, Nelson DA (2004) Effects of pulse rate and electrode array design on intensity discrimination in cochlear implant users. J Acoust Soc Am 116:2258–2268. https://doi.org/10.1121/1.1786871

Lazard DS, Giraud A-L, Gnansia D et al (2012) Understanding the deafened brain: implications for cochlear implant rehabilitation. Eur Ann Otorhinolaryngol Head Neck Dis 129:98–103. https://doi.org/10.1016/j.anorl.2011.06.001

Lazard DS, Innes-Brown H, Barone P (2014) Adaptation of the communicative brain to post-lingual deafness. Evidence from functional imaging. Hear Res 307:136–143. https://doi.org/10.1016/j.heares.2013.08.006

Lee H-J, Smieja D, Polonenko MJ et al (2020) Consistent and chronic cochlear implant use partially reverses cortical effects of single sided deafness in children. Sci Rep 10:21526. https://doi.org/10.1038/s41598-020-78371-6

Legris E, Galvin J, Roux S et al (2018) Cortical reorganization after cochlear implantation for adults with single-sided deafness. PLoS One 13:e0204402. https://doi.org/10.1371/journal.pone.0204402

Legris E, Roux S, Aoustin J-M et al (2019) Cortical auditory responses according to hearing rehabilitation in unilateral hearing loss. Eur Ann Otorhinolaryngol Head Neck Dis 136:439–445. https://doi.org/10.1016/j.anorl.2019.08.001

Legris E, Galvin J, Roux S et al (2020) Development of cortical auditory responses to speech in noise in unilaterally deaf adults following cochlear implantation. PLoS One 15:e0239487. https://doi.org/10.1371/journal.pone.0239487

Lenarz M, Sönmez H, Joseph G et al (2012) Long-term performance of cochlear implants in postlingually deafened adults. Otolaryngol Head Neck Surg 147:112–118. https://doi.org/10.1177/0194599812438041

Li T, Fu Q-J (2007) Perceptual adaptation to spectrally shifted vowels: training with nonlexical labels. J Assoc Res Otolaryngol 8:32–41. https://doi.org/10.1007/s10162-006-0059-2

Lin FR, Chien WW, Li L et al (2012) Cochlear implantation in older adults. Medicine (Baltimore) 91:229–241. https://doi.org/10.1097/MD.0b013e31826b145a

Lin FR, Yaffe K, Xia J et al (2013) Hearing loss and cognitive decline in older adults. JAMA Intern Med 173:293. https://doi.org/10.1001/jamainternmed.2013.1868

Lin FR, Pike JR, Albert MS et al (2023) Hearing intervention versus health education control to reduce cognitive decline in older adults with hearing loss in the USA (ACHIEVE): a multicentre, randomised controlled trial. Lancet 402:786–797. https://doi.org/10.1016/S0140-6736(23)01406-X

Litovsky RY, Colburn HS, Yost WA, Guzman SJ (1999) The precedence effect. J Acoust Soc Am 106:1633–1654. https://doi.org/10.1121/1.427914

Loeb GE, Kessler DK (1995) Speech recognition performance over time with the Clarion cochlear prosthesis. Ann Otol Rhinol Laryngol Suppl 166:290–292

Loizou PC (1999) Introduction to cochlear implants. IEEE Eng Med Biol Mag 18:32–42. https://doi.org/10.1109/51.740962

Looi V, She J (2010) Music perception of cochlear implant users: a questionnaire, and its implications for a music training program. Int J Audiol 49:116–128. https://doi.org/10.3109/14992020903405987

Magits S, Boon E, De Meyere L et al (2023) Comparing the outcomes of a personalized versus nonpersonalized home-based auditory training program for cochlear implant users. Ear Hear 44:477–493. https://doi.org/10.1097/AUD.0000000000001295

McKay CM (2021) No evidence that music training benefits speech perception in hearing-impaired listeners: a systematic review. Trends Hear 25:2331216520985678. https://doi.org/10.1177/2331216520985678

Middlebrooks JC (2008) Cochlear-implant high pulse rate and narrow electrode configuration impair transmission of temporal information to the auditory cortex. J Neurophysiol 100:92–107. https://doi.org/10.1152/jn.01114.2007

Miller JD, Watson CS, Kistler DJ et al (2008) Preliminary evaluation of the speech perception assessment and training system (SPATS) with hearing-aid and cochlear-implant users. Proc Meet Acoust 2:1–9. https://doi.org/10.1121/1.2988004

Miller S, Zhang Y, Nelson P (2016) Neural correlates of phonetic learning in postlingually deafened cochlear implant listeners. Ear Hear 37:514–528. https://doi.org/10.1097/AUD.0000000000000287

Moberly AC, Vasil K, Baxter J et al (2020) Comprehensive auditory rehabilitation in adults receiving cochlear implants: a pilot study. Laryngoscope Investig Otolaryngol 5:911–918. https://doi.org/10.1002/lio2.442

Mortensen MV, Mirz F, Gjedde A (2006) Restored speech comprehension linked to activity in left inferior prefrontal and right temporal cortices in postlingual deafness. NeuroImage 31:842–852. https://doi.org/10.1016/j.neuroimage.2005.12.020

Nicholas JG, Geers AE (2003) Personal, social, and family adjustment in school-aged children with a cochlear implant. Ear Hear 24:69S–81S. https://doi.org/10.1097/01.AUD.0000051750.31186.7A

Nie Y, Galvin JJ, Morikawa M et al (2018) Music and speech perception in children using sung speech. Trends Hear 22:233121651876681. https://doi.org/10.1177/2331216518766810

Nogaki G, Fu Q-J, Galvin JJ (2007) Effect of training rate on recognition of spectrally shifted speech. Ear Hear 28:132–140. https://doi.org/10.1097/AUD.0b013e3180312669

Oba SI, Fu Q-J, Galvin JJ (2011) Digit training in noise can improve cochlear implant users' speech understanding in noise. Ear Hear 32:573–581. https://doi.org/10.1097/AUD.0b013e31820fc821

Papsin BC, Gysin C, Picton N et al (2000) Speech perception outcome measures in prelingually deaf children up to four years after cochlear implantation. Ann Otol Rhinol Laryngol Suppl 185: 38–42. https://doi.org/10.1177/0003489400109s1216

Pelizzone M, Cosendai G, Tinembart J (1999) Within-patient longitudinal speech reception measures with continuous interleaved sampling processors for ineraid implanted subjects. Ear Hear 20:228–237. https://doi.org/10.1097/00003446-199906000-00005

Petersen B, Gjedde A, Wallentin M, Vuust P (2013) Cortical plasticity after cochlear implantation. Neural Plast 2013:318521. https://doi.org/10.1155/2013/318521

Pfingst BE, Xu L, Thompson CS (2007) Effects of carrier pulse rate and stimulation site on modulation detection by subjects with cochlear implants. J Acoust Soc Am 121:2236–2246. https://doi.org/10.1121/1.2537501

Plant K, Holden L, Skinner M et al (2007) Clinical evaluation of higher stimulation rates in the nucleus research platform 8 system. Ear Hear 28:381–393. https://doi.org/10.1097/AUD.0b013e31804793ac

Polonenko MJ, Gordon KA, Cushing SL, Papsin BC (2017) Cortical organization restored by cochlear implantation in young children with single sided deafness. Sci Rep 7:16900. https://doi.org/10.1038/s41598-017-17129-z

Ponton CW, Don M, Eggermont JJ et al (1996) Auditory system plasticity in children after long periods of complete deafness. Neuroreport 8:61–65. https://doi.org/10.1097/00001756-199612200-00013

Ponton CW, Moore JK, Eggermont JJ (1999) Prolonged deafness limits auditory system developmental plasticity: evidence from an evoked potentials study in children with cochlear implants. Scand Audiol Suppl 51:13–22

Ponton CW, Eggermont JJ, Kwong B, Don M (2000) Maturation of human central auditory system activity: evidence from multi-channel evoked potentials. Clin Neurophysiol 111:220–236. https://doi.org/10.1016/s1388-2457(99)00236-9

Purdy SC, Kelly AS (2016) Change in speech perception and auditory evoked potentials over time after unilateral cochlear implantation in postlingually deaf adults. Semin Hear 37:62–73. https://doi.org/10.1055/s-0035-1570329

Reis M, McMahon CM, Távora-Vieira D et al (2021) Effectiveness of computer-based auditory training for adult cochlear implant users: a randomized crossover study. Trends Hear 25: 233121652110259. https://doi.org/10.1177/23312165211025938

Rosen S, Faulkner A, Wilkinson L (1999) Adaptation by normal listeners to upward spectral shifts of speech: implications for cochlear implants. J Acoust Soc Am 106:3629–3636. https://doi.org/10.1121/1.428215

Sagi E, Fu Q-J, Galvin JJ, Svirsky MA (2010) A model of incomplete adaptation to a severely shifted frequency-to-electrode mapping by cochlear implant users. J Assoc Res Otolaryngol 11: 69–78. https://doi.org/10.1007/s10162-009-0187-6

Sandmann P, Plotz K, Hauthal N et al (2015) Rapid bilateral improvement in auditory cortex activity in postlingually deafened adults following cochlear implantation. Clin Neurophysiol 126:594–607. https://doi.org/10.1016/j.clinph.2014.06.029

Santarelli R, De Filippi R, Genovese E, Arslan E (2008) Cochlear implantation outcome in prelingually deafened young adults. Audiol Neurotol 13:257–265. https://doi.org/10.1159/000115435

Sarant JZ, Harris DC, Galvin KL et al (2018) Social development in children with early cochlear implants: normative comparisons and predictive factors, including bilateral implantation. Ear Hear 39:770–782. https://doi.org/10.1097/AUD.0000000000000533

Schumann A, Serman M, Gefeller O, Hoppe U (2015) Computer-based auditory phoneme discrimination training improves speech recognition in noise in experienced adult cochlear implant listeners. Int J Audiol 54:190–198. https://doi.org/10.3109/14992027.2014.969409

Shader MJ, Gordon-Salant S, Goupell MJ (2020) Impact of aging and the electrode-to-neural interface on temporal processing ability in cochlear-implant users: amplitude-modulation detection thresholds. Trends Hear 24:2331216520936l6. https://doi.org/10.1177/2331216520936160

Sharma A, Dorman MF, Spahr AJ (2002) A sensitive period for the development of the central auditory system in children with cochlear implants: implications for age of implantation. Ear Hear 23:532–539. https://doi.org/10.1097/00003446-200212000-00004

Sharma A, Dorman MF, Kral A (2005) The influence of a sensitive period on central auditory development in children with unilateral and bilateral cochlear implants. Hear Res 203:134–143. https://doi.org/10.1016/j.heares.2004.12.010

Spahr AJ, Dorman MF, Loiselle LH (2007) Performance of patients using different cochlear implant systems: effects of input dynamic range. Ear Hear 28:260–275. https://doi.org/10.1097/AUD.0b013e3180312607

Spivak LG, Waltzman SB (1990) Performance of cochlear implant patients as a function of time. J Speech Hear Res 33:511–519. https://doi.org/10.1044/jshr.3303.511

Stacey PC, Summerfield AQ (2007) Effectiveness of computer-based auditory training in improving the perception of noise-vocoded speech. J Acoust Soc Am 121:2923–2935. https://doi.org/10.1121/1.2713668

Stacey PC, Summerfield AQ (2008) Comparison of word-, sentence-, and phoneme-based training strategies in improving the perception of spectrally distorted speech. J Speech Lang Hear Res 51:526–538. https://doi.org/10.1044/1092-4388(2008/038)

Stacey PC, Raine CH, O'Donoghue GM et al (2010) Effectiveness of computer-based auditory training for adult users of cochlear implants. Int J Audiol 49:347–356. https://doi.org/10.3109/14992020903397838

Stropahl M, Chen L-C, Debener S (2017) Cortical reorganization in postlingually deaf cochlear implant users: intra-modal and cross-modal considerations. Hear Res 343:128–137. https://doi.org/10.1016/j.heares.2016.07.005

Svirsky MA, Silveira A, Neuburger H et al (2004) Long-term auditory adaptation to a modified peripheral frequency map. Acta Otolaryngol 124:381–386

Tremblay K, Kraus N, McGee T (1998) The time course of auditory perceptual learning: neurophysiological changes during speech-sound training. Neuroreport 9:3557–3560. https://doi.org/10.1097/00001756-199811160-00003

Tyler RS, Fryauf-Bertschy H, Kelsay DM et al (1997a) Speech perception by prelingually deaf children using cochlear implants. Otolaryngol Head Neck Surg 117:180–187. https://doi.org/10.1016/s0194-5998(97)70172-4

Tyler RS, Parkinson AJ, Woodworth GG et al (1997b) Performance over time of adult patients using the Ineraid or nucleus cochlear implant. J Acoust Soc Am 102:508–522. https://doi.org/10.1121/1.419724

Uchida Y, Sugiura S, Nishita Y et al (2019) Age-related hearing loss and cognitive decline—the potential mechanisms linking the two. Auris Nasus Larynx 46:1–9. https://doi.org/10.1016/j.anl.2018.08.010

Valzolgher C, Verdelet G, Salemme R et al (2020) Reaching to sounds in virtual reality: a multisensory-motor approach to promote adaptation to altered auditory cues. Neuropsychologia 149:107665. https://doi.org/10.1016/j.neuropsychologia.2020.107665

Valzolgher C, Bouzaid S, Grenouillet S et al (2023) Training spatial hearing in unilateral cochlear implant users through reaching to sounds in virtual reality. Eur Arch Otorrinolaringol 280: 3661–3672. https://doi.org/10.1007/s00405-023-07886-1

van Heteren JAA, Vonck BMD, Stokroos RJ et al (2022) The acoustic change complex compared to hearing performance in unilaterally and bilaterally deaf cochlear implant users. Ear Hear 43: 1783–1799. https://doi.org/10.1097/AUD.0000000000001248

Van Hoesel RJM, Jones GL, Litovsky RY (2009) Interaural time-delay sensitivity in bilateral cochlear implant users: effects of pulse rate, modulation rate, and place of stimulation. JARO 10:557–567. https://doi.org/10.1007/s10162-009-0175-x

Vandali AE, Whitford LA, Plant KL, Clark GM (2000) Speech perception as a function of electrical stimulation rate: using the Nucleus 24 cochlear implant system. Ear Hear 21:608–624. https://doi.org/10.1097/00003446-200012000-00008

Veltman J, Maas MJM, Beijk C et al (2023) Development of the Musi-CI training, a musical listening training for cochlear implant users: a participatory action research approach. Trends Hear 27:23312165231198368. https://doi.org/10.1177/23312165231198368

Waltzman SB, Cohen NL, Shapiro WH (1986) Long-term effects of multichannel cochlear implant usage. Laryngoscope 96:1083–1087. https://doi.org/10.1288/00005537-198610000-00007

Wang W, Zhou N, Xu L (2011) Musical pitch and lexical tone perception with cochlear implants. Int J Audiol 50:270–278. https://doi.org/10.3109/14992027.2010.542490

Wang S, Liu B, Dong R et al (2012) Music and lexical tone perception in Chinese adult cochlear implant users. Laryngoscope 122:1353–1360. https://doi.org/10.1002/lary.23271

Weglage A, Müller V, Layer N et al (2022) Side-of-implantation effect on functional asymmetry in the auditory cortex of single-sided deaf cochlear-implant users. Brain Topogr 35:431–452. https://doi.org/10.1007/s10548-022-00902-3

Wooi Teoh S, Pisoni DB, Miyamoto RT (2004) Cochlear implantation in adults with prelingual deafness. Part I. Clinical results. Laryngoscope 114:1536–1540. https://doi.org/10.1097/00005537-200409000-00006

Wu J-L, Yang H-M, Lin Y-H, Fu Q-J (2007) Effects of computer-assisted speech training on Mandarin-speaking hearing-impaired children. Audiol Neurootol 12:307–312. https://doi.org/10.1159/000103211

Xie Z, Gaskins CR, Shader MJ et al (2019) Age-related temporal processing deficits in word segments in adult cochlear-implant users. Trends Hear 23:233121651988668. https://doi.org/10.1177/2331216519886688

Yang J, Liang Q, Chen H et al (2019) Singing proficiency of members of a choir formed by prelingually deafened children with cochlear implants. J Speech Lang Hear Res 62:1561–1573. https://doi.org/10.1044/2019_JSLHR-H-18-0385

Yeo BSY, Song HJJMD, Toh EMS et al (2023) Association of hearing aids and cochlear implants with cognitive decline and dementia: a systematic review and meta-analysis. JAMA Neurol 80:134–141. https://doi.org/10.1001/jamaneurol.2022.4427

Yoon Y-S, Shin Y-R, Gho J-S, Fu Q-J (2015) Bimodal benefit depends on the performance difference between a cochlear implant and a hearing aid. Cochlear Implants Int 16:159–167. https://doi.org/10.1179/1754762814Y.0000000101

Zhang T, Dorman MF, Fu Q-J, Spahr AJ (2012) Auditory training in patients with unilateral cochlear implant and contralateral acoustic stimulation. Ear Hear 33:e70–e79. https://doi.org/10.1097/AUD.0b013e318259e5dd

Artificial Intelligence for Hearing Outcome Prediction and Parameter Fitting in Cochlear Implants

10

Behnam Molaee-Ardekani

Abstract

This chapter explores the potential use of artificial intelligence (AI) in cochlear implant (CI) science, focusing on two distinct applications: postoperative hearing outcome prediction from preoperative data and CI parameter fitting. A comprehensive overview of the current state of the art, recent advancements, and the challenges of adopting AI in these areas is presented. Regarding outcome prediction, various preoperative data features and machine learning (ML) models (e.g., linear models, artificial neural networks, random forest models, gradient boosting models, and support vector machines) are explored in both postlingual and prelingual adults and children. Despite some improvements provided by ML-based models, they are not yet widely used in practical applications due to generalization issues and a lack of data acquisition standardization. Concerning parameter fitting, which is a crucial process in maximizing the individual performance of CI recipients, researchers are looking for AI-based fitting methods to offer faster and potentially more effective solutions than manual fittings. A few AI-based fitting methods, such as genetic algorithms, reinforcement learning, and stochastic approaches like the one used in the fitting to outcome expert (FOX) method, are detailed and linked together as different members of a broader generic family. Despite the promising progress in genetic algorithms and some interest in using FOX in certain clinics, AI has yet to find its solid footing in practical CI fittings.

B. Molaee-Ardekani (✉)
Advanced Innovation Department, Cochlear Ltd, Vallauris, France

Research and Technology Department, Oticon Medical, Vallauris, France
e-mail: bmolaee@cochlear.com

B. Molaee-Ardekani, M. Segovia-Martinez (eds.), *Advancing Cochlear Implants*,
https://doi.org/10.1007/978-3-032-15719-5_10

Keywords

Cochlear implant fitting · Hearing outcomes prediction · Artificial intelligence · Machine learning · Performance outcome · Fitting to outcome expert (FOX) · Genetic algorithm fitting · Reinforcement learning fitting

10.1 Introduction

The artificial intelligence (AI) has found many applications in cochlear implant (CI) science (Crowson et al. 2020b; Essaid et al. 2024). This chapter explores the potential use of AI in focusing on two distinct CI applications: postoperative hearing outcome prediction from preoperative data and CI parameter fitting. Each application is examined in dedicated sections, providing an overview of the current state of the art, recent advancements, and the challenges that prevent or delay the widespread adoption of AI in these areas.

Investigation into hearing outcome prediction indicates that many researchers have utilized linear models and generalized linear models to identify the most relevant preoperative features and to establish a mapping function between these features and hearing outcomes (Gantz et al. 1993; Lazard et al. 2012; Blamey et al. 2012; Plant et al. 2016; Kraaijenga et al. 2016; Favaretto et al. 2019; Goudey et al. 2021). There are some recent studies attempting to extend these models to more elaborate machine learning (ML) models (Tan et al. 2015; Kim et al. 2018; Han et al. 2019; Crowson et al. 2020a; Shafieibavani et al. 2021; Lu et al. 2022; Song et al. 2022). Section 10.2 investigates the characteristics of these models (e.g., input features, topology, and performance outcome variables) and their advantages and limitations.

Section 10.3 explores how AI may optimize the process of CI fitting (i.e., parameter tuning) to improve user satisfaction and/or auditory performance. It describes various types of fitting methods, including behavioral mapping, objective measure-based fitting using electrically evoked compound action potential (eCAP), and stapedius reflex threshold (eSRT) (de Vos et al. 2018; Gärtner et al. 2021; Molaee-Ardekani et al. 2025), fitting to outcome expert (FOX) (Govaerts et al. 2010; Vaerenberg et al. 2010, 2014; Bermejo et al. 2013; Vaerenberg 2014), evolutionary or genetic algorithm fitting (Choi 2001; Bourgeois-Republique et al. 2005; Başkent et al. 2007; Eiler et al. 2008; Lineaweaver and Wakefield 2011; Saadoun et al. 2022), and ML agent-based (e.g., reinforcement learning) fitting (Van Opstal and Noordanus 2023). These fitting methods are linked as different members of a broader and generic family.

The future prospects and challenges that AI-based methods face in predicting hearing outcomes and performing CI fitting (in the author's opinion) are discussed separately for each of these two topics.

10.2 The Application of AI in CI Performance Outcomes

While CIs have enabled many individuals to regain their hearing, there is significant variability in outcomes among patients. In addition to the quality of the surgery and related complications such as trauma, patient-specific factors contribute to this variability. Typically, the recommendation for cochlear implantation is based on the clinical team's experience regarding expected performance postimplantation. Accurately predicting postoperative performance from preoperative clinical data can assist practitioners in evaluating candidates for implantation and setting realistic performance expectations. This challenge is even greater in prelingual pediatric patients with cochlear abnormalities or nerve deficiencies compared to postlingual adults. Researchers aim to use AI-based algorithms to predict the outcomes of cochlear implantation with preoperative data. This section presents a survey of the methodologies designed for predicting postoperative speech performance using preoperative data. The models and results reported for adults (both postlingually and prelingually hearing-impaired patients) are presented first, followed by a few predicting methodologies developed for children.

10.2.1 Prediction Models

Models for predicting postoperative hearing outcomes may be categorized into regression and classification models. Regression models predict numerical values of future outcomes or their changes after CI surgery. Classification models categorize patient performance: Binary models assess if a patient will benefit from implantation, while multiclass models provide detailed categorizations, such as dividing performance into quartiles.

Linear models (LM) and its variant form, generalized linear models (GLM), are the most widely used approaches in the literature for CI performance prediction (Lazard et al. 2012; Blamey et al. 2012; Plant et al. 2016; Kraaijenga et al. 2016; Favaretto et al. 2019; Goudey et al. 2021). More recently, a broader range of machine learning models has been introduced to address this task, including artificial neural networks (Crowson et al. 2020a; Shafieibavani et al. 2021), random forests (Kim et al. 2018; Shafieibavani et al. 2021), gradient boosting (Crowson et al. 2020a; Shafieibavani et al. 2021), and support vector machine (Tan et al. 2015; Lu et al. 2022). These models can be adapted for both regression and classification tasks with minor adjustments to their prediction and objective (or cost function) equations. Below is a brief overview of some models commonly used in CI outcome prediction. For more detailed information, readers may consult comprehensive resources on machine learning, such as the textbook by Aurélien Géron (2019).

Multivariable GLMs can be expressed as $\hat{y} = f(\sum_i w_i x_i + w_0) = f(\mathbf{w} \cdot \mathbf{x} + w_0)$, where $\widehat{y}$ is the estimation of the outcome variable, $\mathbf{x} \in \mathbb{R}^m$ is the feature vector with m elements, and $\mathbf{w}$ is the coefficient vector that should be trained. A univariable model is a model where $m = 1$, and the feature vector contains only one feature. A multivariate model is a model with multiple dependent outcomes. An LM is a special form of the GLM where $f(x) = x$ is an identity function. Artificial neural network (ANN) with multilayer perceptron topology consists of L layers of interconnected neurons as if L multivariate (i.e., multiple multivariable) GLMs are sequentially connected to each other (Matloff 2017).

Decision tree (DT) is a model that splits data into subsets based on feature values, creating a tree structure. Each node in the tree represents a decision based on a feature that minimizes variance (for regression) or maximizes class separation (for classification). The final prediction is determined by the mode (for classification) or average (for regression) of the data samples in each leaf node.

Random forests (RF) and gradient boosting (GB) are both ensemble methods that improve upon the basic DT. Random forest builds multiple DTs and combines their outputs using averaging (regression) or mode (classification) methods. Gradient boosting constructs an ensemble of trees sequentially, where each new tree corrects errors made by the previous ones. XGBoost (XGB) is a specific implementation of GB with optimization and regularization enhancements to prevent overfitting. XGBs may use different types of models in their tree structure. For example, XGB-Lin uses linear functions and XGB-RF uses RF as base learners.

A support vector machine (SVM) is a model that either finds an optimal hyperplane $\mathbf{w}\cdot\mathbf{x} + w_0 = 0$ to separate data points of different classes (classification) or finds a function $\mathbf{w}\cdot\mathbf{x} + w_0$ that deviates from the target values by a value no greater than a specified margin ϵ (regression). In SVN, usually a kernel (e.g., radial basis function kernel) maps input features into a higher-dimensional space $K(\mathbf{x}_i,\mathbf{x}_j)$ where the linear hyperplane can separate the data.

10.2.2 Predictors and Outcomes

The predictors are the input features to the models. They are mostly categorized to (i) demographic data, (ii) patient clinical history, (iii) audiological metrics, (iv) etiology, and (v) anatomical and neurophysiological characteristics of the patient's hearing system. Table 10.1 lists some of the most popular predictors employed in various studies.

The performance outcomes of cochlear implantation are mostly evaluated by speech perception scores following the implantation. Trained models aim to either predict the numerical values of these speech perception metrics or categorize them into performance levels, such as low and high. Many studies use word recognition scores (WRS) (Kim et al. 2018) and other variants such as consonant-vowel-consonant (CVC) (Kraaijenga et al. 2016) or consonant-nucleus-consonant (CNC) (Roditi et al. 2009) scores. Some studies have reported sentence recognition scores in quiet and noisy environments (James et al. 2019; Crowson et al. 2020a). Performance outcomes are typically obtained unilaterally, with the other ear masked if needed. However, the advantages and complexities of bilateral measurement have also been studied (Plant et al. 2016).

Table 10.1 A list of predictor features used in various outcome prediction studies

	Blamey et al.	Lazard et al.	Tan et al.	Kraaijenga et al.[a]	Kraaijenga et al.[b]	Plant et al.	Kim et al.	Favaretto et al.	Crowson et al.	Goudey et al.
Year of study	2012	2012	2015	2016	2016	2016	2018	2019	2020	2021
Nr of participation in the study	2251	2251	23	370	58	65	120	50	1604	2735
Prediction horizon in years	<12	<12	2	1	1	1	2	1	1	1
Nr evaluated/input features	19	19	658	13/9	13/6	7/7	6	9/5	282	20/14
Nr selected features	4	9	2	9/3	3/1	2/3	3	4/1	15	NA
Demographic										
Gender		⊡		⊡	⊡			⊡	⊡	⊡
Age at implantation				⊟	⊡	⊟	⊠	⊟	⊠	⊟
Age at spHL onset	⊠	⊠								
Age at onset of deafness				⊠	⊡					
Patient clinical history										
Duration of spHL or DoD for the CI ear	⊠	⊠		⊟	⊡	⊠	⊠		⊠	⊠
spHL or DoD contralateral ear				⊡	⊡	⊟				⊟
spHL or DoD bilateral ears						⊟				
Duration of hearing loss before implantation										⊟
Duration of mHL		⊠								
Duration of using hearing aid		⊠					⊠			⊟
Duration of using CI	⊠	⊠								
Audiological metrics										
Preop monosyllabic WRS implanted ear				⊠	⊠				⊠	⊟
Preop monosyllabic WRS contralateral ear									⊠	⊡

(continued)

Table 10.1 (continued)

	Blamey et al.	Lazard et al.	Tan et al.	Kraaijenga et al.[a]	Kraaijenga et al.[b]	Plant et al.	Kim et al.	Favaretto et al.	Crowson et al.	Goudey et al.
Preop monosyllabic WRS both ears										⊟
Preop best aided sentence recognition score		⊟					⊡		⊠	
PT or PTA levels of the CI ear		⊡		⊟	⊟	⊡	⊡	⊡	⊠	⊠
PT or PTA levels of the contralateral ear				⊟	⊟	⊠	⊡	⊡	⊠	⊟
PT or PTA levels of the better or both ears		⊠		⊟	⊟					
Unaided hearing threshold at 500 Hz of the CI ear		⊡							⊠	
Unaided hearing threshold at 500 Hz of the better ear		⊟							⊠	
Ear for implantation (better or worse/left/right)		⊡		⊡	⊡				⊡	⊟
Etiology										
Congenital, labyrinth malformation, meningitis, infections, progressive, trauma, otosclerosis, sudden deafness, Meniere's disease, iatrogenic, etc.	⊠	⊠		⊠	⊡			⊠	⊡	⊠

The importance of each feature is coded with three symbols: ⊡ for unimportant features, ⊟ for features significant in one of the investigated models (e.g., a uni- or multivariable model) but not necessarily in the final or all models in the study, and ⊠ for important features. To reduce table size, some features have been grouped as a same feature even though they are not exactly identical across studies. In this table: Fields marked "v1/v2" indicate values for univariable (v1) and multivariable (v2) models

spHL, severe-to-profound hearing loss; *DoD*, duration of deafness; *mHL*, moderate hearing loss; *PT*, pure-tone audiogram (i.e., single frequency); *PTA*, pure-tone average (i.e., over multiple frequencies)

[a]Model for postlingual HI patients

[b]Model for prelingual HI patients

Other types of continuous variable measures such as clinical evaluation of language fundamentals (CELF), meaningful auditory integration score (MAIS), and the meaningful use of speech scale (MUSS) have also been used to evaluate hearing and speech in patients (Tan et al. 2015; Lu et al. 2022). Categorical variables like the categories of auditory performance (CAP), which scores between 0 and 7 indicating hearing rehabilitation, and the speech intelligibility rating (SIR), which scores between 1 and 5 indicating speech rehabilitation, have been used (Tan et al. 2015). The hearing in noise threshold (HINT) score has been used as both a continuous variable and categorical (e.g., High: 80–100, Mid: 50–79, Low: 0–49) (Crowson et al. 2020a). Another categorical example can be found in Favaretto et al. (2019), where outcomes are classified as positive or negative based on comparing the postoperative pure-tone average (PTA) to 38 dBHL or reaching a speech recognition threshold (SRT) of 50%. In some multicenter retrospective studies, each clinic may have used a different speech perception test in quiet or noise, with a different language. In these studies, the scores may need to be normalized first (Lazard et al. 2012).

10.2.3 Linear Models: Feature Selection and Outcomes

One of the early studies contributing to the development of methodologies for predicting hearing-outcome performance was conducted by Gantz et al. (1993). Since then, researchers have tried to determine which preoperative features may better predict postoperative outcome performance. Earlier studies primarily focused on LM and GLM, with significant efforts made to reduce the complexities of these models by identifying the minimum number of input features needed to achieve robust results.

Proper feature selection requires access to a substantial CI population, in terms of both recipients and clinical centers. This is evident in the dataset used by Lazard et al. (2012) and Blamey et al. (2012) in a multicenter study involving 2251 patients across 15 centers. Using a multivariable GLM model, Blamey et al. (2012) selected four factors (from 19 available factors) that had been shown to effectively describe speech outcomes in a previous study (Blamey et al. 1996). These factors were the duration of severe-to-profound hearing loss (spHL), age at spHL onset, duration of using CI, and 15 types of etiology (see Table 10.1). The total degrees of freedom of the model were 32. The first two factors negatively impacted outcome performance, indicating that patients with longer spHL and higher age are less likely to improve. The third factor had a positive impact. The etiology had smaller effect and varied based on the type of etiology. Sudden idiopathic, genetic, Ménière's disease, and otosclerosis etiologies provided better scores than average, whereas auditory neuropathy spectrum disorder, acoustic neuroma, temporal bone fracture, meningitis, and chronic otitis media had lower scores.

Lazard et al. (2012) used the same dataset and proposed a nine-factor GLM model by adding five new factors to the ones presented in Blamey et al. (2012). These new factors were selected from 15 five-factor GLMs, which included the examined factor

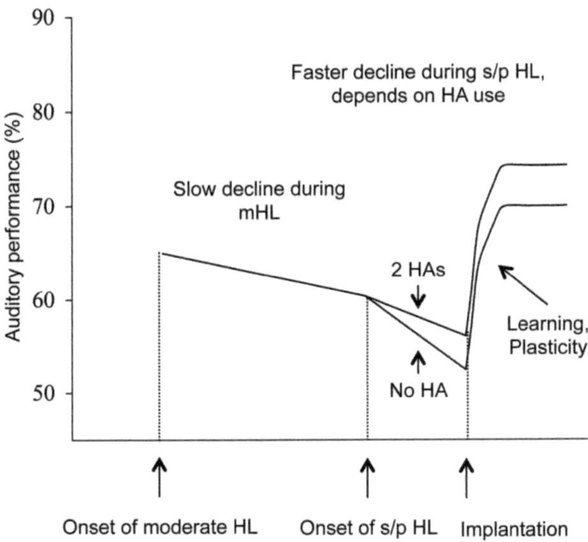

Fig. 10.1 A three-stage model illustrating the mean expected auditory performance ranking over time for a hypothetical "average CI recipient." In the GLM analysis, the durations of spHL and mHL were treated as continuous covariates to assess the rates of decline (slopes) in auditory performance over time. The duration of spHL was nested within HA use, and the results indicated that the negative impact of spHL may be influenced by HA use. [Figure reproduced from Lazard et al. (2012) under the CC BY license]

and the four aforementioned factors. The new added factors were the duration of moderate hearing loss (mHL), use of hearing aids during spHL, unaided pure-tone average (PTA) levels of the better ear, CI brand, and the percentage of active electrodes. It is worth noting that the last two factors are not strictly preoperative, but they can typically be selected or estimated prior to CI surgery. The preoperative speech score factor was eliminated from the list (although significant) as it was significantly influenced by other factors. The 500 Hz hearing aid (HA) pure-tone factor was substituted by PTA for a larger F factor. The nine-factor GLM increased the explained variance of speech perception tests from 10% [as obtained by Blamey et al. (2012)] to 22%. Figure 10.1 illustrates mean expected auditory performance ranking over time for a hypothetical average CI patient using a three-stage model. It shows how duration of mHL and spHL may drop the outcome with time, and how wearing a hearing aid (HA) may reduce the inclination.

Approximately a decade later, the next major multicenter study was conducted by Goudey et al. (2021), involving a large cohort of 2735 patients. This period between studies is particularly significant, as it saw the emergence of new insights and features identified by researchers working with small- to medium-sized CI populations. These developments laid the groundwork for Goudey et al. (2021), whose study is discussed following the review of these earlier contributions.

Kraaijenga et al. (2016) proposed two different models for postlingual ($n = 370$) and prelingual ($n = 58$) adults. Out of 13 features, the multivariable LM identified three significant features for postlingual patients: preimplant speech perception, age at onset of deafness, and etiology (with meningitis and otosclerosis having lower performance scores). For prelingual patients, this reduced to one feature: preimplant speech perception. Additionally, for postlingual patients, univariable analyses highlighted PTA, duration of spHL, and age at implantation as significant factors.

Plant et al. (2016) suggested that separate models should be developed for unilateral and bilateral speech perception scores, although using a unilateral model remains more practical in clinical settings. In recent years, there has been a growing number of CI candidates with asymmetric or single-sided hearing loss. As a result, the duration of hearing loss (HL)—a key feature in most predictive models—may be less relevant for patients requiring unilateral implantation. The authors hypothesized that hearing ability in the contralateral ear could serve as a key predictive factor. To explore this, they recruited 65 postlingually hearing-impaired (HI) adults with significant residual acoustic hearing in one or both ears. Both ipsilateral and contralateral features were analyzed using various multivariable linear models. The models identified two key predictors: the duration of spHL in the implanted ear and the unaided PTA of the contralateral ear. A shorter duration of spHL in the implanted ear was associated with better unilateral and bilateral outcomes. Interestingly, better contralateral hearing was linked to poorer unilateral word recognition scores with the implanted ear alone, but improved overall bilateral speech recognition.

Favaretto et al. (2019) studied 50 postlingually HI adults and demonstrated that input features are linked differently to various speech outcome metrics. They applied both a univariable Fisher test and a multivariable logistic model (i.e., GLM) to analyze the data. In the univariable analysis of nine features, none showed a statistically significant association with a binarized postoperative PTA at 38 dBHL (PTA > 38). When using a binarized postoperative monosyllabic WRS (SRT $> 50\%$), only the etiology of HL was significant. For more complex speech comprehension tests—such as two-syllable word recognition among four similar options SRT $> 50\%$)—the most predictive factors were etiology, CI brand, family history of HL, comorbidities, and age at implantation. In the multivariable analysis for speech comprehension, etiology was the only significant factor. The family HL factor was dropped from the multivariable analysis because it consistently showed good outcome performance. Comorbidities were excluded due to a lack of data, and the implant brand was dropped as it correlated with etiology. These findings suggest that more complex speech comprehension tests, such as two-syllable word recognition, may be more informative in clinical research.

Before turning to the large-scale study by Goudey et al. (2021), it is important to note that these smaller studies provided valuable insights into the role of specific features and modeling strategies.

Goudey et al. (2021) conducted a major three-center study involving 2735 hearing-impaired adults, both postlingual and prelingual with 71%, 18%, and 11% of participants from each center. A key finding was the substantial variability in results across centers, largely due to inconsistencies in data collection—such as differing definitions of hearing loss and spHL. Linear models across all centers consistently identified three key predictors out of 16 variables: PTA of the implanted ear, duration of spHL, and etiology. Other features were either unavailable across all centers or only significant in one, likely due to inconsistent data collection practices. The study emphasized the importance of several parameters, including preoperative performance (e.g., WRS of the implanted and both ears), audiological measures (e.g., PTA of both ears), duration of any type of hearing loss or deafness (including spHL), and hearing aid use duration.

Consistent with findings from Kraaijenga et al. (2016), the study also confirmed that prelingually hearing-impaired patients perform significantly worse than postlingual patients, underscoring the need for models to account for hearing loss onset. Despite the large sample size, the multivariable linear modeling approach explained only 12% of the variance in outcome performance across the full dataset and 21% within the dominant center.

10.2.4 Machine Learning Models: Adults Performance Outcomes

Given the limited utility of linear models, recent years have seen a few studies exploring the use of ML models (i.e., more advanced decision-making models than LM/GLM) to predict postoperative hearing outcomes from preoperative data. These studies have mostly utilized random forest (Kim et al. 2018; Shafieibavani et al. 2021), XGBoost (Crowson et al. 2020a; Shafieibavani et al. 2021), and artificial neural networks (ANN) (Crowson et al. 2020a; Shafieibavani et al. 2021).

Kim et al. (2018) trained a random forest regression (RFR) model using data from 120 postlingual HI adults with normal inner ear anatomy and fully inserted CI electrodes. The RFR model included 50 small decision trees, each grown to have at least three terminal nodes (leaves). Input features were randomly selected using a bootstrap method, ensuring that each tree received a unique subset of features.

The authors introduced six features to the models and demonstrated that the RFR outperformed the GLM. The WRS was predicted with mean absolute errors (MAE) of 15.6 and 6.1, and Pearson's correlation coefficients (i.e., between true and estimated WRS values) of 0.6 and 0.96 for the GLM and RFR, respectively. These results were cross-validated using data from two other centers with 21 and 16 patients. After applying bias correction to account for differences in WRS test types across centers, the combined dataset yielded an MAE of 9.7 and a correlation coefficient of 0.9.

Feature importance was assessed by measuring the increase in MAE when each feature was omitted from the prediction process. All features had a greater impact in the RFR model compared to the GLM, indicating that the RFR's nonlinear structure allowed for more effective utilization of the input data. The most to least impactful features on postoperative WRS were: duration of deafness (DoD), duration of hearing aid use, age at CI implantation, preoperative sentence recognition score, and pure-tone average (PTA) of the ipsilateral and contralateral ears.

Crowson et al. (2020a) compared ANN and XGB models using a single-site study involving 1604 adults. The ANN was used to predict the HINT score both as a percentage and as categorical outcomes (low, mid, high), while the XGB model was applied only to predict HINT percentage scores (i.e., regression model). Overall, both models demonstrated comparable performance in predicting HINT outcomes. However, the authors did not disclose the architecture or topology of the models, making replication of the results impossible. The study utilized a total of 282 input features: 272 numerical and 10 categorical. The numerical features included 11 audiometric measures (PTA, hearing thresholds at 500, 1000, 2000 Hz for both ears, word discrimination scores for both ears, and sentence test), three demographic features, and various questionnaire-derived features (e.g., health conditions, hearing disease factors, tinnitus situation). The categorical features included parameters such as CI brand, tinnitus severity, gender, and etiology.

The ANN was trained using two feature sets: one with only numerical features and another with all features. The model trained on only numerical features achieved a remarkably low HINT% root mean square error (RMSE) of 0.57% and a classification accuracy of 94.5%. In contrast, the model trained on all features had an RMSE of 25.0% and an accuracy of 73%. The near-perfect performance of the numerical-only model may not be reproducible with similar ANN models such as the one that Shafieibavani et al. (2021) train on multicenter data.

The XGB model achieved an RMSE of 25.3%, closely matching the ANN's performance when using all features. Feature importance analysis, conducted using the SHAP Python library (Lundberg and Lee 2017), identified key predictors such as preoperative sentence test performance, age at surgery, and patient-reported outcomes related to tinnitus and quality of life.

Shafieibavani et al. (2021) carried out a large-scale study using the same three-center dataset as Goudey et al. (2021), but excluded prelingual patients, resulting in a cohort of 2489 postlingual adults. The study compared the performance of seven models using both numerical and categorized WRS, with the latter divided into four quartiles.

The ML models proposed in the study included a shallow ANN with two hidden layers for regression and three for classification, a random forest (RF), and two XGB models—one using a linear base learner (XGB-Lin) and the other using a random forest base learner (XGB-RF). Additionally, three previously published models were re-implemented as baselines (reference) for comparison: GLMs from Lazard et al. (2012) and Blamey et al. (2012), and the RFR model from Kim et al. (2018). The reference models were retrained on the same dataset for fair comparisons.

The results for the regression models (using a 10-fold cross-validation method) showed that the best MAEs were obtained by RF and XGB-RF models (median values: 20.76 and 20.81, respectively), and other models (i.e., ANN, XGB-Lin, and baselines) performed slightly worse than these two models with about 1% increased MAE, however significantly (see Fig. 10.2).

The generalizability of the XGB-RF model to unseen data was assessed using three types of cross-validation: (i) single clinic (training and testing on the same center), (ii) multiclinic (training on all centers, testing on each center), and (iii) external (training on two centers, testing on the third). The results were consistent across single- and multiclinic validations, with only slight variations in the external

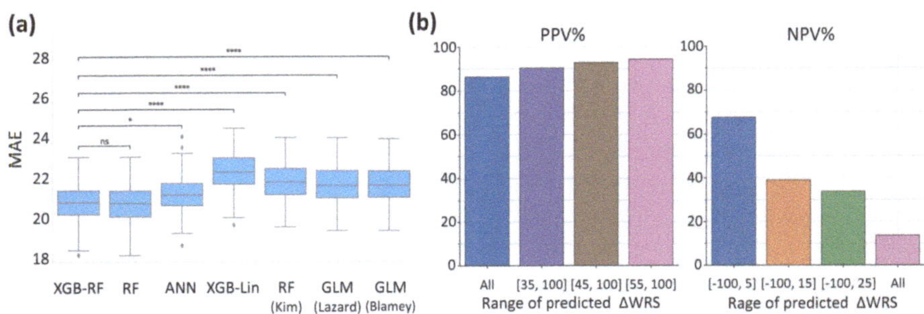

Fig. 10.2 (**a**) Comparison of the MAE in predicting WRS across seven models. Statistical significance of performance differences relative to the XGB-RF model is indicated by lines above the bars. Significance levels are denoted as follows: ns: $p > 0.05$, *: $p \leq 0.05$, **: $p \leq 0.01$, ***: $p \leq 0.001$, ****: $p \leq 0.0001$. (**b**) Positive and negative predictive values (PPV and NPV), expressed as percentages, for achieving a postimplantation ΔWRS of 10 points across different prediction-interval partitions. When no filtering is applied to select the predicted values (i.e., All), PPV and NPV are 86% and 24%, respectively. Restricting the predicted values to intervals [55, 100] and [−100, 5] (i.e., high and low values) increases these metrics to 94% and 68%, respectively. (Figure adapted from Shafieibavani et al. (2021) with permission from Trends in Hearing, Sage Publications)

validation, indicating strong generalizability. However, MAEs varied between centers, reflecting cohort differences and inconsistencies in data acquisition—an issue also highlighted by the LM models in Goudey et al. (2021).

The relatively high MAE values observed across models were hypothesized to stem from the inherent variability in WRS measurements, which can fluctuate by up to 30% between tests (Moulin et al. 2017). To address this, the study explored whether predicting categorized WRS outcomes—rather than continuous values—would improve model performance. However, classification results remained modest: Among the seven models tested, the best accuracy was approximately 31% for the ANN model, while the lowest was around 28% for the LM baselines. These findings suggest that although ML models outperformed linear models, their overall predictive accuracy remained limited on this dataset (likely with inconsistent data collection practices).

The study also investigated whether models could be trained to predict changes in WRS after cochlear implantation (ΔWRS), rather than absolute postoperative scores. This approach could help identify which patients are most or least likely to benefit from CI surgery. Shafieibavani et al. (2021) trained an XGB-RF model to predict ΔWRS and evaluated its performance using positive and negative predictive values (PPV and NPV). These metrics reflect the model's ability to correctly predict whether ΔWRS exceeds a specified positive or negative threshold. Using a threshold of 10 points, the model achieved a PPV of 86% and an NPV of 24%. This indicates that the model was more reliable in predicting substantial improvements in WRS, but less effective at identifying patients at risk of poorer outcomes. Only 24% of patients whose WRS declined by at least 10 points were correctly identified.

10.2.5 Machine Learning Models: Children Performance Outcomes

In the previous sections, the features, models, and speech outcomes discussed primarily focused on prelingually or postlingually hearing-impaired adults. However, some of these features are not informative or applicable to prelingually hearing-impaired children. For instance, presurgical speech performance is not relevant for a 12-month-old child with congenital sensorineural hearing loss (SNHL). Instead, more universally applicable features that are suitable for all children should be considered. Additionally, speech performance metrics like WRS and CVC are challenging to measure in pediatrics. Recently, a few studies have aimed to identify suitable features, models, and outcomes for children. This section explores some of these studies. In these studies, the functional connectivity of specific brain regions plays a crucial role in predicting outcomes. It has also been shown that the connectivity and interaction between some specific brain areas significantly influence reading and language development in children with cochlear implants. Additionally, the duration of deafness and the timing of early intervention are critical factors in improving outcomes and strengthening neural network connections (Koirala et al. 2025).

Tan et al. (2015) used the clinical evaluation of language fundamentals—Preschool, Second Edition (CELF-P2), suitable for children aged 3–7, to measure the primary language outcome. Features were extracted from preoperative fMRI, and an SVM classifier was employed to predict whether CELF-P2 > 300 (indicating a high-performing patient). The fMRI data were acquired under three conditions: silence, speech, and audio noise approximately 10–15 dB above the hearing threshold. Three contrast cortical activation t-maps were computed: speech vs. silence, noise vs. silence, and speech vs. noise. For each activation map, brain region clusters (a "bag of words") representing 14 levels of t-map activations were identified. Activation boundaries corresponded to seven p-values ranging from 0.05 to 0.0005 in a two-sided significance level analysis.

Results from 23 SNHL children showed that the best activation t-map was the silence versus speech map, forming 658 brain clusters (features). This large number of features did not clearly indicate which brain regions were involved in the CELF-P2 prediction process. The authors used a recursive feature elimination (RFE) approach for feature selection. The algorithm started by training an SVM model with all 658 features. Features were ranked by their absolute weights, and those with the lowest weights were discarded. This process was repeated until only two features remained, indicating these brain clusters: (1) left superior and middle temporal gyri and (2) right cerebellum (Fig. 10.3). Cross-validation analysis showed an accuracy of 68.8% and an AUC (area under receiver operating characteristic curve) of 0.71.

Song et al. (2022) assessed CI outcomes in 68 SNHL children using the categories of auditory performance (CAP) score, which ranges from no sound awareness (score: 0) to using a telephone with an unknown speaker (score: 7). The study proposed two tasks: (1) using ML to determine if children belong to the SNHL or healthy control (HC) population and (2) estimating the CAP score for SNHL-identified children (Fig. 10.4). Task 1 used three ML models and aggregated their

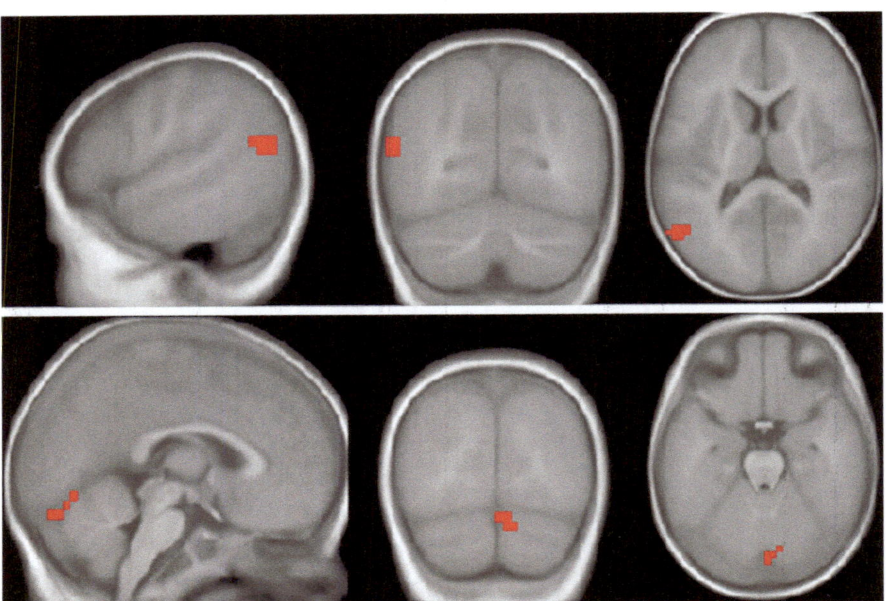

Fig. 10.3 Two discriminative brain regions identified from the contrast between speech and silence: A region located in an area corresponding to the left superior and middle temporal gyri (top); a region located in the right cerebellum (bottom). (Figure adapted from Tan et al. (2015), under the CC BY license)

results, while task 2 used multiple logistic regression (MLR). Both tasks utilized features representing functional connections in resting-state fMRI. Unlike the methodology developed by Tan et al. (2015), the brain regions in this study were fixed and predetermined. The features were based on the correlation between region activations rather than contrasted activation maps (e.g., speech vs. silence).

Features were obtained as follow: MRI/fMRI pairs were normalized to the Montreal Neurological Institute (MNI) space, and average fMRI signals were calculated for 246 brain regions defined in human Brainnetome Atlas. Pearson's correlation coefficients among these signals created a functional connectivity (FC) matrix, vectorized (flattened) into 30,135 independent values. This vectorized FC was obtained for 97 subjects (63 patients and 34 control children), and a kernel PCA reduced this matrix's complexity. New FC vectors were projected onto the most important eigenvectors of the kernel PCA.

For task 1, three classifiers (SVM, KNN, and logistic regression) were used, with hyperparameters and feature numbers optimized via a nested grid-search algorithm. The classifiers' accuracies were 0.72, 0.76, and 0.72, with AUC values of 0.78, 0.79, and 0.80, respectively. Aggregating the classifiers increased accuracy to 0.77 and AUC to 0.84.

For task 2, the MLR model used 10 features from kernel PCA, age at implantation, gender, and a binary auditory brainstem response (ABR) flag. The average

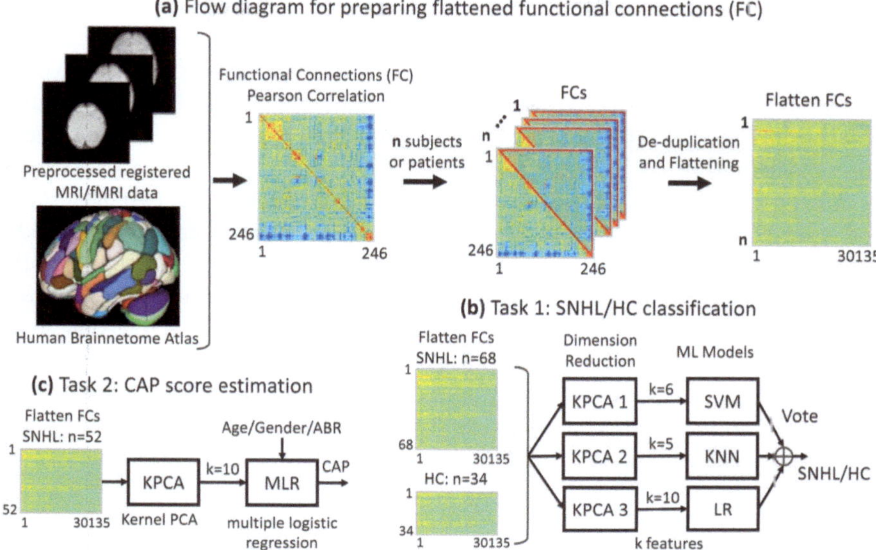

Fig. 10.4 Overall procedure for SNHL/HC classification and CAP score estimation using MRI images. (**a**) Flow diagram showing that preprocessed MRI images are used to determine 246 brain region activity levels based on the Brainnetome Atlas. Pearson correlation is applied to measure functional connectivity (FC) between these regions. The connectivity matrix is de-duplicated by taking the values in the upper triangular portion, followed by flattening these values. Repeating this process for n subjects yields a matrix of size $n \times 30{,}135$. (**b**) In Task 1, flattened FCs are fed into kernel PCA to reduce the number of features. The outputs of three different machine learning models are combined in a voting system for SNHL/HC classification. (**c**) In Task 2, a multiple logistic regression model is used to predict CAP scores for patients who underwent CI surgery, using features provided by kernel PCA along with three additional features (age, gender, and ABR). (Image inspired from Song et al. (2022), under the CC BY license)

accuracy of predicted CAP scores was 82.7%, with 100% accuracy for the four worst-performing children (CAP score: 0) and 54.5% for the 11 best-performing children (CAP score: 3). This indicated that the MLR classifier could identify children with potentially low benefit from implantation.

Children with cochlear nerve deficiency (CND) often experience variable hearing outcomes from cochlear implantation, likely due to nerve deficiencies that may not be clearly visible in fMRI. Some studies have attempted to extract anatomical features from MRI to predict postoperative auditory outcomes for CND patients. Han et al. (2019) used an LM to predict the CAP score from preoperative ABR and the area ratio of the vestibulocochlear nerve (VCN) to the facial nerve (FN). The model is mathematically expressed as $\text{CAP} = 0.7 + 1.9\ \text{ABR} + 1.2\ (\text{VCN/FN})$, where ABR is 1 or 0 depending on whether an ABR response with a distinct wave V can be recorded in response to a click sound at 90 dBHL.

Lu et al. (2022) developed multiclass SVM classifiers to predict CAP and SIR in 70 patients 2 years after CI surgery. The models incorporated 12 features, including five continuous radiological features—VCN area, internal auditory canal (IAC) diameter, bony cochlear nerve canal (BCNC) diameter, nerve bundle number, and VCN to FN area ratio; five discrete audiological features—40-Hz auditory-evoked related potential (AERP), ABR, cochlear microphonics (CM), distortion product otoacoustic emission (DPOAE), and acoustic immittance, as well as PTA and age at CI implantation.

A three-step feature reduction procedure was used to identify the two most stable and representative predictors: VCN area and nerve bundle number. The final model showed a 71% accuracy for predicting CAP and a 93% accuracy for predicting SIR, with AUC values of 0.71 and 0.74, respectively. The feature reduction steps were: (1) removing features with low variance across classes (ABR, CM, DPOAE), visual inspection removal (40-Hz AERP, acoustic immittance), and high-correlation removal (VCN to FN area ratio); (2) measuring stability scores for the remaining features and sorting them; and (3) adding features one by one to the SVM model and measuring accuracy and AUC, finding that two features yielded the best results.

10.2.6 Limitations and Opportunities in AI-Based Outcome Predictions

The reviewed studies highlight the evolution of methodologies for predicting speech performance outcomes in CI patients. Early research focused on LM and GLM, identifying essential preoperative features for robust predictions. Recent advancements have utilized sophisticated ML models like random forest, XGB, and ANN, which handle complex, nonlinear relationships and demonstrate improved predictive performance over traditional models.

When predicting hearing outcomes in adults, models should consider the inclusion of the following factors:

- Details about auditory system degradation in different phases (e.g., prelingual or postlingual, duration of HA use, deafness onset time, duration of spHL, age at implantation, etc.)
- Preoperative speech and audiometric performance (e.g., speech-in-quiet/noise tests, residual hearing evaluation, PTA, etc.)
- Etiology (e.g., sudden idiopathic, genetic, and Ménière's disease provide better scores than auditory neuropathy spectrum disorder, acoustic neuroma, temporal bone fracture, and meningitis).

These feature groups are not necessarily independent of each other. For instance, preoperative speech performance is impacted by etiology (Czurda et al. 2024). This correlation has led LM/GLM-based studies to suggest different lists of finalized features, making it challenging to determine a unique list of the best features while limiting the model complexity. Shafieibavani et al. (2021) showed that complementary information in these features suggests using all of them. Although the improvement by advanced ML models (such as ANN and XGB-RF) was modest, these models may lay the foundation for future studies utilizing subtle complementary information and nonlinearities.

Most studies are based on small populations or low-complexity linear models, limiting their practical clinical use. Few studies use large multicenter data. The database by Lazard et al. (2012) is promising but outdated, analyzed only with simple GLM models. Goudey et al. (2021) used a large dataset from three centers, revealing inconsistencies in data collection, highlighting the need for standardized data.

Different centers use different speech tests with various types and numbers of words and sentences, making it difficult for the model to provide robust predictions. Speech performance also changes over time, so retrospective studies should account for the duration of CI use. Another important factor often neglected by researchers is the need for two types of models for unilateral and bilateral performance estimation as suggested by Plant et al. (2016).

For SNHL children, fMRI data is valuable for outcome prediction. Although speech tests in children are less accurate, metrics like CELF-P2 and CAP provide necessary information. Studies with small populations (<100) show high accuracy in detecting SNHL patients and estimating CAP. For CND patients, MRI features like VCN area and nerve bundle number are crucial, with SVM models predicting CAP with ~70% accuracy.

To address limitations and improve predictive methodologies, future research may focus on:

- Standardized Data Collection: Implement standardized protocols across clinical centers to reduce variability and improve model reliability. Remove ambiguities in data collection terms (e.g., "duration of deafness") in multicenter collaborations.
- Multicenter Data Collection: Secure funding for a large international research project to collect diverse datasets globally.
- Open Access: Provide open access to datasets or developed models (or both) to enable reuse and improve performance.
- Feature Engineering: Develop advanced feature engineering to identify key features for predicting speech outcomes. For example, combine MRI and fMRI features for better ML performance in CND children.
- Longitudinal Studies: Conduct longitudinal studies to track speech performance changes over time, refining predictive models and improving accuracy.

By addressing these limitations and exploring new methodologies, researchers can continue to advance the field of predictive modeling for CI outcomes, ultimately improving the quality of care for patients with hearing impairments.

10.3 The Application of AI in CI Fitting

Parameter fitting is a crucial step in maximizing the performance of CI. This process involves fine-tuning various parameters, such as stimulation dynamic—namely, the threshold (T-level) and comfort (C-level)—and frequency range for each electrode, to improve speech perception for each individual patient. Finding the optimal parameter settings can be difficult due to numerous interacting features in CIs and the different listening preferences of patients. Manufacturers offer default parameter settings based on clinical and electro-acoustic data, with final adjustments typically made through manual ad hoc tuning methods that highly depend on the experience of the audiologist. In this traditional method, only a few features can be investigated, so it does not consistently yield the best outcomes due to the complex nature of auditory processing and the different pathologies present in the auditory system. This is where AI-based fitting methods may provide faster or even more effective solutions. AI-based methods can readily utilize and integrate various types of data ranging from low-level CI stimulation factors like electrode impedance and current distribution to mid-level neural processing such as eCAP and high-level patient intelligibility outcomes. Doing such, AI may come up with better fitting outcomes in complex situations or may even propose new settings that are not necessarily proposed by experts during traditional fitting sessions. Another advantage of AI-based methods is their flexibility in data acquisition, which can be conducted either in clinical settings or at home during daily activities. However, implementing AI for at-home fitting may involve regulatory complexities that must be carefully considered. Additionally, AI-driven fitting approaches have the potential to progressively improve speech perception and sound quality over time by leveraging prior fitting data and continuously adapting to changes in the patient's auditory responses.

Despite these advantages, AI-based fitting solutions developed by manufacturers and researchers remain limited. Moreover, acceptance within the audiology community presents a significant barrier. In practice, current efforts in computer-assisted CI fitting are largely restricted to the use of interactive evolutionary algorithms (IEA) for semiautomated fittings and a rule-based system known as fitting to outcome expert (FOX).

Recently, Van Opstal and Noordanus (2023) introduced an AI-based fitting framework aimed at advancing the development of more versatile automatic fitting methods for cochlear implants. Inspired by their work, this chapter presents both published and practically implemented fitting methods as members of a broader, more generic fitting family. The following subsections describe the overarching characteristics of this fitting family and highlight several of its key members, including interactive evolutionary algorithms (IEAs), fitting to outcome expert (FOX), and the Van Opstal and Noordanus (VON) fitting approach.

10.3.1 A Generic Fitting Family

Figure 10.5 illustrates the overall block diagram of a generic fitting method. This method incorporates two types of input: direct electrical stimulation and acoustic stimulation. The goal is to enable the auditory system to process pulsatile stimulation information in two distinct pathways. Direct electrical stimulation bypasses the sound processor—including the coding strategy algorithm and its parameters—and directly stimulates the auditory nerve. This provides valuable insights into the signal

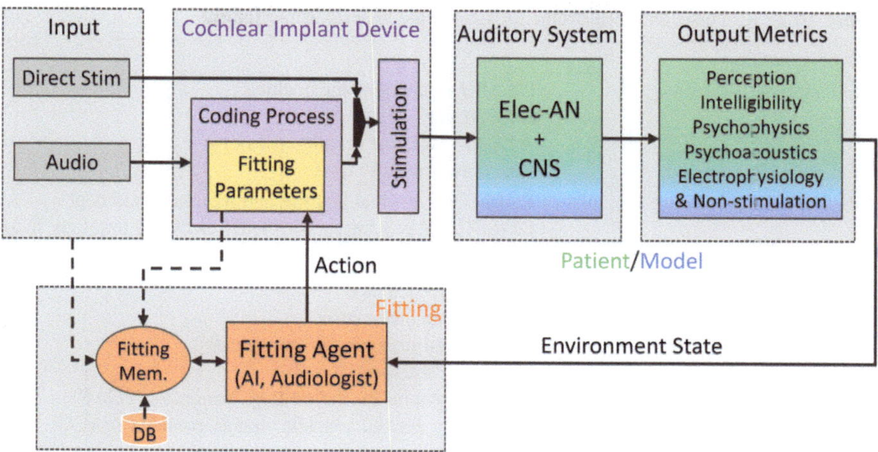

Fig. 10.5 Block diagram of a generic fitting framework. Direct electrical stimulation is used to characterize the properties of the ascending auditory pathways by analyzing various output metrics. This evaluation begins at the Elec–AN interface and extends to higher processing centers in the CNS, bypassing the influence of CI fitting parameters. In contrast, performing similar assessments using acoustic stimulation incorporates the effects of the CI's speech processor and its parameters. A fitting agent—such as an AI system or an expert audiologist—integrates multiple output metrics to adjust CI parameters for optimized outcomes. The fitting process can also be informed by historical data, either from the same patient or from a broader patient population stored in a database (DB)

transmission characteristics of the CI system to the neural level. In contrast, acoustic stimulation engages the entire CI system, including the sound processor and its parameters, allowing assessment of the system's overall effect on the patient's auditory perception.

The effects of each type of stimulation (i.e., direct electrical or acoustic) can be assessed at multiple levels of the auditory system. These range from the cochlear level, specifically at the electrode–auditory nerve (Elec–AN) interface, to higher-order cognitive functions within the central nervous system (CNS). A variety of tests and measurements are used to evaluate these effects, including assessments of perception and intelligibility, psychophysical and psychoacoustic evaluations, and electrophysiological recordings. In addition to stimulation-based assessments, other no-stimulation-based data modalities—such as preoperative audiograms and pre- and postoperative CT scans of the cochlea and electrode placement—can also be integrated.

The rationale behind this multimodal approach is that combining data from diverse testing protocols can yield unique, quantifiable insights into the different processing levels of a patient's auditory system under current CI settings. Tests that are independent of fitting parameters help reveal how CI stimulation is transmitted across various auditory layers. In contrast, tests that depend on fitting parameters provide information on how those parameters influence the effects and outcomes of

Table 10.2 Key tests and measurements used in the generic fitting family

No CI stimulation	
Preoperative	CT/MRI imaging of the cochlea Audiograms Acoustic SRT test in quiet and noise Reaction-time (RT) test
Per- and postoperative	EcochG for residual hearing and trauma assessment CT/MRI/x-Ray imaging for the CI electrode tonotopy map
Direct CI electrode stimulation	
Electro-physics/ electrophysiology	Electrode impedance Cross-impedance matrix Electrode stimulation current spread eCAP characteristics Spread of neural excitation (e.g., panoramic eCAP) Electrically evoked acoustic change complex (eACC) Electrically evoked auditory steady-state responses (eASSRs)
Psychoacoustic/psychophysics	Stimulation loudness evaluation (e.g., T- and C-levels) Masking level difference (MLD) test Interaural time (ITD) and level (ILD) difference RT to sudden electrode change RT to a sudden modulation of electrode pulse frequency
Full-system stimulation	
Psychoacoustic/psychophysics	Audiograms‡ Loudness evaluation (narrow band noise or speech) ‡ Reaction times to spectrotemporal ripples Pitch and frequency discrimination thresholds Interaural time (ITD) and level (ILD) difference Free-field sound localization Masking level difference (MLD) test
Intelligibility/perception	Phoneme recognition tests‡ Speech recognition tests (e.g., CVC-words test) ‡ Speech perception tests (e.g., SRT in quiet and noise) Sentence recognition tests (e.g., QuickSIN test; HINT)
Electrophysiology (cognitive level)	Acoustic change complex (ACC) ASSRs in the EEG with acoustic tone complexes

The tests are categorized into three main groups, with further subcategories based on the type of outcome data. The items indicated by ‡ are the outcome measures used in the FOX method (Sect. 10.3.3).

stimulation in a fully operational CI. It is also important to note that integrating these two categories of tests remains a complex and ongoing challenge.

Table 10.2 outlines some key tests and measurements to quantify the effects of stimulations on CI patients and their outcomes. The tests in the table are categorized according to their corresponding data modality. Data modalities not derived from CI stimulation are typically measured once per patient, while data derived from stimulation may be measured multiple times across various fitting sessions. The stimulating-derived data can either be from direct electrode or full-system stimulation.

Once a set of multimodal data is collected from the patient, it should be combined into a coherent decision-making model that enables the clinician to predict the perceptual outcomes of a CI patient after adjusting specific fitting parameters. However, integrating all these interdependent modalities of the acquired responses (such as anatomy, electrode impedance, stimulation dynamic range, eCAP, ECochG, phoneme recognition, speech performance, reaction time, and EEG) is quite complex. This complexity exceeds the capacity of a simple regression method or even a human to manage effectively. This is where advanced AI-based methods can help to uncover the underlying structure within these datasets. Their goal is to use this uncovered knowledge to fine-tune the parameters of the CI in a way that optimizes speech and music perception performance in both quiet and noisy environments.

The AI agent that receives the outcome multimodal data and conducts a fitting action (e.g., modifying a fitting parameter, requesting a complementary test with a given stimulation input, etc.) can be either a rule-based deterministic model, a machine learning (ML) stochastic model, or a combination of the two. Depending on the algorithm used by the AI agent, it may have either a fixed preprogrammed fitting flow or a flow that is dynamically determined by the agent. Both fixed and dynamic methods can alter the data modalities introduced to the agent at different steps of the fitting process. The fitting flow determines which tests and measurements should be performed and at what point. It may consist of several iterative loops during which new data acquisitions can be requested. In the dynamic flow, the outcome of each loop determines the execution details of the next loop. For a new action to be performed, the AI agent needs access to the history (memory) of fitting parameter values in the current and previous fitting sessions (if any), patient outcome data and corresponding inputs, normal hearing and CI-user population database, and previously performed actions by the agent.

Training the ML algorithm within the AI agent requires access to various types of data, including no CI stimulation, direct-electrode stimulation, and full-system stimulation data. The ML algorithm should learn how to capture the complex relationships between fitting parameters and perceptual outcomes for a particular input sound. Firstly, the ML requires access to patient-specific auditory system characteristics. This involves performing no and direct stimulation tests (as outlined in Table 10.2) and incorporating the results from both modalities. These tests are crucial for reflecting the ascending transfer function of the patient's auditory system. Subsequently, the ML should learn how to provide the optimal interface (mapping) between the acoustic input and identified properties of the patient's auditory system that are available by full-system stimulation tests (i.e., audio-based stimulations).

Lastly, in the generic fitting family, computational models may be used to simulate or predict patient outcomes. Although this adds another level of uncertainty to the entire system, it can be beneficial for training the AI agent when the number of patients is insufficient.

10.3.2 Traditional Fittings

Traditional fittings can be considered the simplest forms of the generic fitting family. Initially, audiologists employ single or multielectrode direct stimulations to elicit certain subjective and objective outcomes. These outcomes may vary based on factors such as patient communication abilities, etiology, and surgical complexities.

Electrode impedance is primarily used to determine which electrodes should be activated or deactivated. Electrode frequency mappings are typically configured based on manufacturer recommendations and the number of inserted electrodes.

The thresholds and dynamic ranges for electrical stimulation—such as T- and C-levels—are determined either subjectively through patient feedback or objectively using physiological measures like eCAP and eSRT. Among these, eCAP is often favored because it can be recorded directly through the implant system without the need for additional external equipment. This makes it more convenient and generally more reliable to obtain. In contrast, eSRT requires an external tympanometry device to detect the reflexive contraction of the stapedius muscle. Accurate measurement also depends on the integrity and normal function of the middle ear, including proper stiffness of the ossicular chain. These additional requirements can make eSRT more challenging to use consistently in clinical practice. To this, other complexities such as variabilities of eSRT response to electrical stimulation characteristics should be added (Martins et al. 2024).

However, objective measures are typically used as supportive guidelines rather than as the primary basis for fitting—except in cases where subjective feedback is unavailable, such as with young children. For a fast and safe fitting, even in infants and children, it is recommended to use a combination of objective and behavioral tests to avoid setting C-levels excessively high (de Andrade et al. 2014).

A meta-analysis by de Vos et al. (2018) reported inconsistent correlations between eCAP thresholds and behavioral T- and C-levels, with weak pooled correlation coefficients of $r = 0.58$ and $r = 0.61$, respectively. For eSRT, correlations with behavioral C-levels also vary widely, ranging from approximately $r = 0.6$ (Bergeron and Hotton 2015) to $r = 0.9$ (Raghunandhan et al. 2015). Additionally, there is often an offset between eCAP or eSRT thresholds and the T- and C-levels, which may vary over time, across cochlear regions, and with different stimulation characteristics (Björsne and Magnusson 2025; Molaee-Ardekani et al. 2025). This reduces the practicality of relying solely on objective measures for CI fitting.

Traditional CI fitting—whether based solely on subjective feedback or guided by objective measures—aims to provide safe and balanced loudness levels across different frequencies to the patient's auditory system. To insure this, acoustic stimulation is used following direct electrical stimulation to perform full-system testing, incorporating all fitting parameters. Some overall loudness adjustments may be necessary. This ensures that the perceived loudness remains comfortable and is not bothersome to the patient. However, in this fitting approach, there is no systematic methodology for optimizing speech performance in quiet or noisy environments. The audiologist acts as the fitting agent (see Fig. 10.5), and the success of the fitting process relies a lot on their knowledge and clinical experience. Through a process of

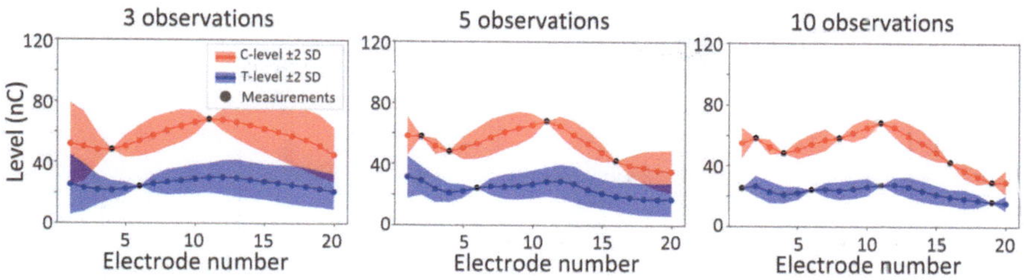

Fig. 10.6 Traditional fitting can be supported by predicting unmeasured T- and C-levels using redundancies in their patterns across electrodes and patients. Increasing the number of observations reduces prediction uncertainty, shown as shaded regions

trial and error, the audiologist adjusts the cochlear implant's fitting parameters—particularly those that are conventionally tuned—in an effort to improve the patient's hearing performance.

Additional enhancements can be introduced to traditional CI fittings to accelerate the process and reduce reliance on the audiologist's expertise for parameter adjustments. For instance, there are notable similarities between T-levels and C-levels across the cochlear implant patient population. This redundancy can be exploited using dimensionality reduction techniques such as principal component analysis (PCA) (Migliorini et al. 2024). By applying PCA, or alternatively ML models, T- and C-levels can be estimated across all electrodes using only a limited number of measurements taken from a few selected electrodes. This significantly streamlines the fitting process. This approach does not necessarily aim to improve fitting accuracy but can offer robust estimations—especially beneficial for less-experienced audiologists—by utilizing large datasets to inform predictions. Figure 10.6 illustrates this concept, showing example fittings with varying numbers of observations from the same patients. As more observations are provided, the uncertainty in the predicted values decreases.

10.3.3 Fitting to Outcome eXpert (FOX)

Fitting to outcome expert (FOX) is the name of a methodology and software for automatic fitting of CIs (Vaerenberg et al. 2010; Vaerenberg 2014). Figure 10.7 depicts the simplified block diagram of the FOX fitting method. It mostly uses the following four types of patient outcomes belonging to the full-system stimulation mode (see Table 10.2):

- Free-field audiometry (from 250 Hz to 8 kHz)
- Loudness scaling (at 250 Hz, 1 kHz, and 4 kHz)
- Phoneme discrimination (2 × 20 phonemes)
- CVC-word speech recognition test (scores at 40, 55, 70, and 85 dB SPL)

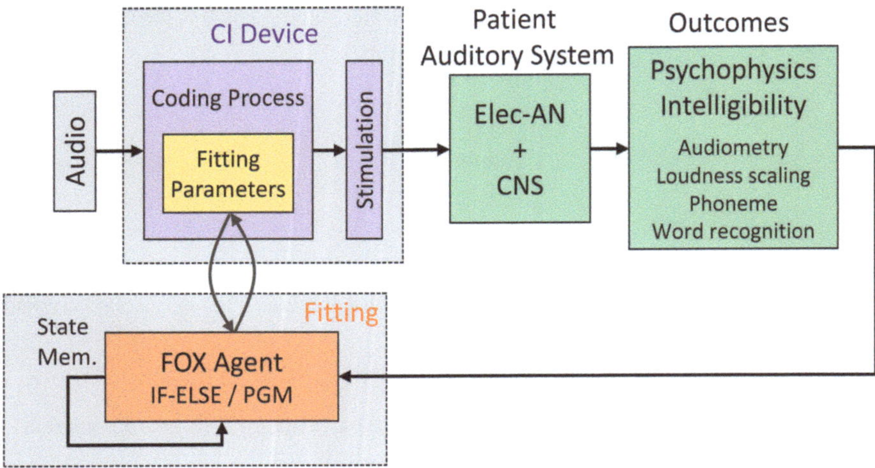

Fig. 10.7 Various patient outcome measurements, in response to electrical stimulations originating from acoustic inputs and passing through the auditory system—from the electrode-neuron interface to the CNS—are evaluated iteratively by FOX (in either deterministic or stochastic versions). This process may lead to parameter adjustments. FOX may initiate its status and decision-making based on the parameter values set during current and past fitting sessions

The software, however, potentially has the capability to use other types of data such as eCAP, eSRT, questionnaires, etc., belonging to no and direct-electrode stimulation modalities.

Unlike traditional fittings that try to provide loudness-balanced stimulations inside a determined stimulation boundary (T- and C-level) without direct guarantying the speech performance, FOX targets the speech enhancement. Depending on the ongoing fitting session number after implantation, FOX asks for some or all of the aforementioned patient outcomes. FOX has emerged with two different fitting agents: deterministic and stochastic (Vaerenberg 2014).

The deterministic version of the FOX includes hundreds of IF-ELSE deterministic rules. Audiologists may add supplementary rules and/or new data modalities to the existing features. The stochastic agent works based on a probabilistic graphical model (PGM). Topology and weights of PGM model have been determined by an expert team composed of audiologists, engineers, and mathematicians. Both versions of FOX agents have been designed based on the expertise of the design team, acquired from over 600 CI patients. This implies that no classical mathematical methods have been used to train FOX agents; instead, they have been designed based on expert knowledge.

FOX fitting strategy is based on a stepwise fitting procedure. Initially, a series of 10 automaps, obtained from good performant CI patient population, with incrementally increasing stimulation levels are created. Then the patient is instructed to start with a MAP (CI stimulation program) and to switch to an incrementally louder MAP every day as long as the sound percept remained tolerable. After about 1 week, the loudest MAP that the patient has tried successfully is used as a reference MAP. Some electrodes may consider for deactivation based on the electrode impedance analysis. From now on, the patient is instructed to find the automap MAP (with possibly deactivated electrodes) that provides the most comfortable loudness stimulations. About 1 month later, it is time to individualize the most comfortable MAP using two types of psychoacoustic outcomes: audiometry and phoneme recognition. Using these outcomes, FOX can assess if the patient has difficulty in discriminating the contrasts in different frequencies. Based on the frequency contents of the unrecognized phonemes and audiometry thresholds, FOX adjusts the gains in different frequency channels to maximize the frequency resolving power of the CI. About 2–3 months after, patient's everyday MAP is modified based on results from loudness scaling and speech audiometry. These tests are identification tasks and are subject to learning effects and change over time; therefore, some adaptation to the device is needed before performing these tests. Loudness scaling provides a complementary lower-level identification task than the speech test.

It is noteworthy that by default, FOX does not divide the outcome data modalities to direct stimulation and full-system stimulation as described in Sect. 10.3.1. As a result, it does not try to find any correspondence between direct effect of stimulation on patient's auditory system and the effect of CI parameters on the patient's auditory perception. Instead, it starts with full-system low-level outcomes (i.e., audiometry and phoneme recognition) to optimize the MAP, followed by full-system high-level outcomes (i.e., loudness scaling and speech recognition) to improve the speech understanding performance.

10.3.4 Opti-FOX: Stochastic FOX Agent

As mentioned earlier, FOX has been introduced with two different agents. The rule-based agent is constructed on parameterized deterministic rules. There are hundreds of IF-ELSE statements that receive patient's outcomes to decide how the fitting parameters should modify. The rule-based agent presents several limitations, including the difficulty to maintain the knowledge base when the number of rules increases and its inability to learn from data. Another limitation is the generalization of the rules from one CI device to another. Opti-FOX tries to mitigate these limitations by substituting the rule-based agent with a Bayesian stochastic agent (Bermejo et al. 2013; Vaerenberg 2014).

10.3.4.1 Tuning Network

The Opti-FOX stochastic agent is constructed by a special type of probabilistic graphical model (PGM) named tuning network. In this tuning network, variables/nodes have three discrete states $\{-, 0, +\}$ indicating for *decrease*, *no change*, and *increase*, respectively.

The tuning network represents how a change in parents' variables affects a child variable. The network consists of three horizontal layers where links go from one layer to the ones below it as shown in Fig. 10.8a. The top layer is the one formed by the tunable parameters of the CI. Tunable parameters are spanned into two

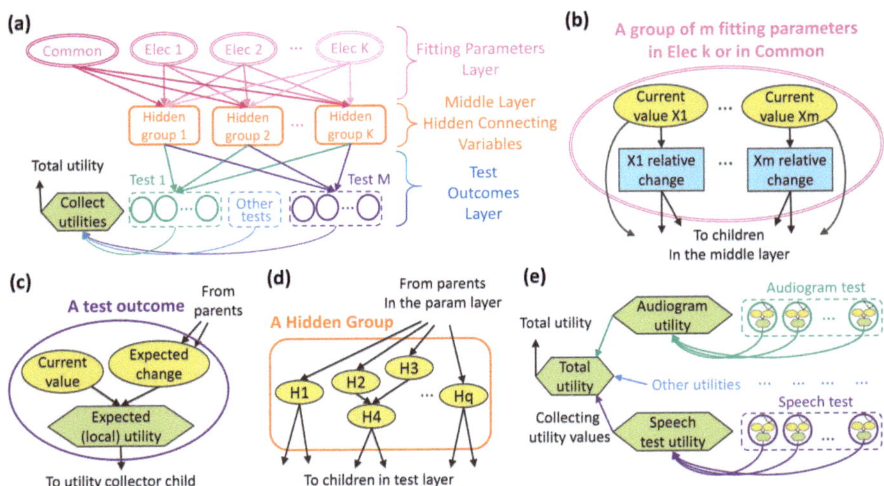

Fig. 10.8 The stochastic FOX model. (**a**) The tuning (Bayesian) model used in Opti-FOX for K electrodes and M tests includes three layers. (**b**) In the fitting parameter layer, each group of common parameters or electrode may include m parameters (e.g., T- and C-levels, etc.). Each of the m parameters is usually represented by a chance node (yellow) and a decision node (blue). (**c**) Each single measurement in each of the M tests (i.e., outcome) is usually represented by two chance nodes and a utility node (green). (**d**) The hidden layer is composed of chance nodes, connecting the parameter and test layers. (**e**) For each given test, a utility node (green) measures the utility value of the current fitting status for that test. A total utility value is calculated over all test utility values. Each node in the tuning network has three output states based on the values that it received from its parents' nodes

categories: Electrode-dependent parameters and CI parameters common to all electrodes. The middle layer consists of hidden (unobservable) nodes. For each given electrode, a group of hidden parameters is formed. The hidden layer acts as a preprocessor to reduce the amount of connectivity between the fitting parameter layer and the test layer. The test layer is the bottom layer in the tuning network. It contains the nodes representing the results (outcomes) of the tests and measurements (i.e., audiogram, phoneme recognition, loudness scaling, and speech test).

There are three types of nodes in the tuning network: decision nodes drawn as blue rectangles, chance nodes drawn as yellow ovals, and value nodes drawn as green hexagons. Decision nodes represent choices available to the decision-maker algorithm. Chance nodes represent all the statistical properties of the network, and they are not under the decision-maker's direct control. Value nodes represent the utility of the decisions (i.e., proposed changes by the decision-maker algorithm). Improved outcomes are assigned to higher utility values.

The tuning network represents each tunable fitting parameter (for example, T-level, C-level, or stimulation pulse width in i_{th}electrode) by two nodes: a chance node representing the current value of the parameter and a decision node representing the relative change proposed by the decision-maker in the network (Fig. 10.8b). Each test outcome is modeled with three different nodes: a chance node representing the current value of the test outcome, a chance node representing the expected change in the result of the outcome given a set of adjustments to the tunable fitting parameters, and a value node modeling the expected utility based on the other two (Fig. 10.8c).

The hidden layer is only composed of chance nodes. Each group of hidden variables (Fig. 10.8d) is responsible to connect the fitting layer variables to each of the test outcomes in the test layer. For example, Opti-FOX includes audiogram, loudness scaling, phoneme discrimination, and CVC speech recognition tests in its test layer. Noteworthy to mention that each given test may include several outcome variables. For instance, the phoneme recognition test includes 20 outcomes in Opti-FOX. Therefore, local utility values at the level of each outcome should be merged to form a broader utility value at the level of each test. Figure 10.8e shows how local utility values are merged toward producing a total expected utility value for all the tests in the network. It is necessary to have a utility function, for example, a weighted sum function that measures the global expected gain in performance in response to changes proposed by decision nodes.

10.3.4.2 Tuning Model

A unique feature of the tuning network is that it assumes that every variable V involved has three values: decreased (v^-), not-changed (v^0), and increased (v^+). In other words, all the variables have the same domain $\{-, 0, +\}$. This provides the possibility that the tuning network is constructed from several canonical models named as tuning models (Bermejo 2012; Bermejo et al. 2013). The term canonical has the meaning that the model forms an elementary unit that can be used in the construction of more complicated models. The tuning model is composed of a child node Y (a chance or a decision node) and its parents $\boldsymbol{X} = \{X_1, X_2, \ldots, X_n\}$ (chance or decision nodes). The set $\{Y\} \cup \boldsymbol{X}$ is called a family. When all the variables in a family are discrete (which is the case in $\{-, 0, +\}$ states), the conditional probability $P(Y | \boldsymbol{X})$ can be expressed in form of a conditional probability table (CPT), which size grows exponentially with the number of nodes (here: 3^{n+1}). In the tuning model, parents (i.e., causes) apply their effects on the child independently. This independence of causal interaction (ICI) simplifies the CPT as it splits a single n-dimensional table to n two-dimensional tables of size 3^2.

Figure 10.9 sketches how a tuning model is formed from a generic model of a family. The tuning model includes n independent small CPTs that are formed between each X_i and Z_i variables. Z_i is an auxiliary stochastic variable that represents the effect of each parent X_i on child Y as if other parents do not exist. Each X_i and Z_i variables have three states $\{x_i^-, x_i^0, x_i^+\}$ and $\{z_i^-, z_i^0, z_i^+\}$. The CPT is characterized with maximally four parameters $p_i^{--}, p_i^{-+}, p_i^{+-}, p_i^{++}$ indicating the probability for input-output state mapping. Putting extra assumptions about mapping properties between input and output states, such as monotonical increase ($p_i^{--} = p_i^{+-} = 0$), monotonical decrease ($p_i^{-+} = p_i^{++} = 0$), or symmetrical mapping ($p_i^{--} = p_i^{++}$), reduces the number of free parameters even to a single parameter p_i. At last, the tuning model combines isolated effect of each parent using the function $y = f_{tunning}(z)$.

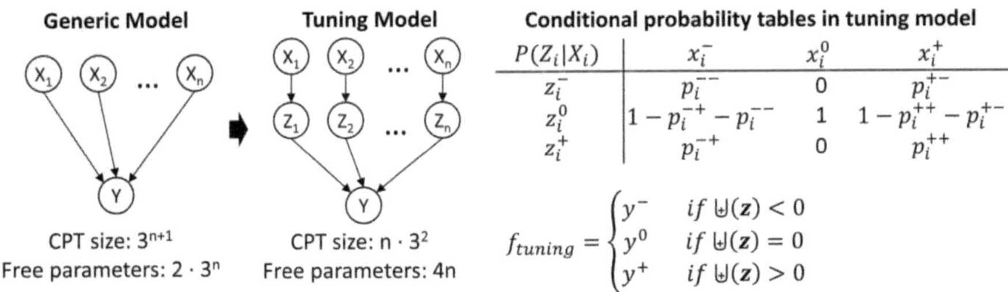

$P(Z_i \mid X_i)$	x_i^-	x_i^0	x_i^+
z_i^-	p_i^{--}	0	p_i^{+-}
z_i^0	$1 - p_i^{-+} - p_i^{--}$	1	$1 - p_i^{++} - p_i^{+-}$
z_i^+	p_i^{-+}	0	p_i^{++}

$$f_{tuning} = \begin{cases} y^- & if \uplus(\mathbf{z}) < 0 \\ y^0 & if \uplus(\mathbf{z}) = 0 \\ y^+ & if \uplus(\mathbf{z}) > 0 \end{cases}$$

Fig. 10.9 Nodes connection in a tuning model. (Left) A child node Y receives inputs from n parent nodes X. The CPT has a size of $3^{(n+1)}$, considering each parent, and the child has three states. (Middle) The size of the CPT is reduced if Y receives independent massages from the parents using auxiliary variables Z_i. (Right) A CPT indicating input/output probabilities for a given child-parent (Z_i,X_i). The child Y sums all the independent massages it receives to determine its final three-state output. Note: If the node Y is a decision node, its output determines how the fitting parameter should modify

This function applies a voting method to the states $\{-, 0, +\}$ corresponding to the auxiliary variable z. The output state value of the child (i.e., $\{y_i^-, y_i^0, y_i^+\}$) is determined by positivity of the accumulated votes $\uplus(z)$.

There are some situations where some parent nodes cannot be declared as independent causes for a child. The solution is to use a common auxiliary node for those dependent parents. For example, if X_1 and X_2 are two dependent causes, $P(Z_{1,2} | X_1, X_2)$ should be expressed by a CPT of size 3^3. Then the mutual effect of these two parameters (i.e., $z_{1,2}$ with three states) can be inserted into the f_{tunning} function.

10.3.4.3 Network Construction and Inference

Building a tuning network for CI fitting begins by deciding its morphology and determining its parameters (i.e., tuning parameter, hidden parameters, tests outcomes, and utilities). The domain of each node (i.e., decision, chance, and value nodes) corresponding to these parameters will be {−, 0, +}, which implies a discretization of a continuous variable to three levels. The lower and upper thresholds between which a continuous variable is considered as "no change" should be determined. For instance, the threshold might be ±5% of the maximum range value for a variable and ±10% for another variable. The second step is to draw causal links between the nodes (e.g., based on the three-layer scheme presented in Fig. 10.8a) and determine if a canonical model can be used for each family in the network. The next step, which is a challenging task, is to set the conditional probability values ($p_i^{--}, p_i^{-+}, p_i^{+-}, p_i^{++}$) for each family in the tuning network. This task is performed once by the designers (a team composed of mathematicians, engineers, and audiologists) of the tuning network.

Inference in a tuning network consists in looking for the optimal set of proposed changes in the tunable parameters (via the decision nodes) that maximize the expected total utility. An exhaustive search is computationally unaffordable. More practical algorithms such as "search and score" algorithms should be used. In these methods, the utility scores to single changes are evaluated iteratively. The "search" is performed using an inference algorithm. Given the high number of variables and complexities in the tuning network, exact inference algorithms are not affordable. Approximate inference algorithms such as the likelihood weighting method proposed by Shachter and Peot (1990) may be used. The computation cost in this algorithm increases linearly with the number of nodes instead of exponentially. However, the main drawback is that the accuracy decreases with extremely unlikely parameter setting configurations.

10.3.5 Evolutionary and Genetic Algorithm CI Fitting

Genetic algorithms (GAs) and their broader class, evolutionary algorithms (EAs), offer a powerful approach to CI parameter fitting. Inspired by principles of biological genetics and natural evolution, respectively, GAs and EAs optimize a vast number of parameter combinations, fine-tuning them to achieve desirable solutions quickly and efficiently.

The EA uses a mathematical search routine, inspired by biological concepts like natural selection, mutation, and crossover, to find an acceptable (or even optimal) solution. For example, tuning C-levels, T-levels, and frequency allocation table in a 20-channel CI device with just eight possible setting levels per parameter results in $8^{(3 \times 20)} = 1.5 \times 10^{54}$ possible settings, making a complete search impractical. Even if the setting is put at a more global parameters interactions level, there will be $8 \times 3 \times 20 = 480$ possible settings, still too many to evaluate exhaustively.

The EA fitting is a subcategory of the generic fitting family, where only audio signals are used for input stimulation, and the AI agent is the EA. The EA directly receives feedback from the CI patient and employs an interactive method to modify the fitting parameters with minimal evaluations.

The EA can be applied in both research and routine CI fittings (Eiler et al. 2008). Engineers and audiologists use EA to optimize new CI coding strategies and fitting algorithms. Audiologists can also fine-tune CI parameters in the field using EA, aided by a remote programming device.

Numerous EAs with diverse optimization techniques have been created for CI fitting (Bourgeois-Republique et al. 2005; Saadoun et al. 2022). Typically, in these methods, the fitting parameters (the so-called "variables" in EA methods) put together to form a sequence named "individual." The CI patient evaluates a "population" of these "individuals" based on their sound quality or speech intelligibility. The highest-rated individuals advance to the next generation, where they undergo mutation and/or crossover for further enhancement. In GAs, the equivalents of "variables," "individuals," and "population" are "genes," "chromosomes," and "population," respectively.

Bourgeois-République et al. (2004) developed software for EA-based CI fitting, tested on a patient with a 15-electrode MXM Digisonic implant (Bourgeois-Republique et al. 2005). The genome, limited to T- and C-values per electrode, improved the speech understanding score from 48.5/100 to 91.5/100. The study showed EA could suggest effective fitting parameters that may be overlooked by audiologists. For this patient, the EA recommended significantly reducing the dynamic range of about six electrodes to avoid large drops in speech recognition due to stimulation interactions between electrodes, which can be contra intuitive of what an audiologist may recommend. The software evaluation on two additional patients confirmed that EA fitting yields comparable or slightly improved results compared to manual fittings. Both patients expressed a preference for EA fitting (Legrand et al. 2007). Başkent et al. (2007) used GA in a study on normal-hearing subjects. Their genome included the number of vocoder channels and the frequency shift and compression-expansion difference between input and synthesis frequency ranges. Using sentence recognition scores, they showed that GA could produce results comparable to the best configuration of the vocoder where the vocoder runs at maximum channels with no frequency shift or compression-expansion difference.

In a more recent study, Saadoun et al. (2022) used EA to modify electrode frequency allocation in bimodal CI patients. Improvements in global sound quality and music perception were evidenced by higher ratings on questionnaires and sound quality scales. In this study, the genome of each "individual" included electrode frequency bandwidth. The exploration domain for upper limit of ith frequency band was set to the interval $[f_i^L, 1.2f_i^H]$ (see Fig. 10.10a for an example with five

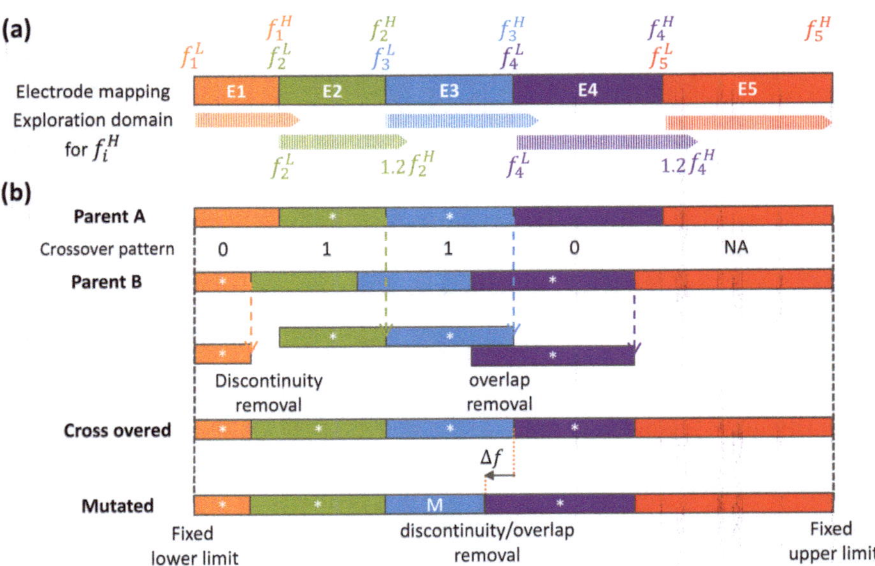

Fig. 10.10 Genome (i.e., "individual") representation of electrode bandwidths. (**a**) A genome represents the bandwidths of five electrodes. Each gene has a limited exploration domain, indicated by the hashed arrows below each electrode. (**b**) Two parent genomes can undergo crossover and mutation to produce a new child. The crossover pattern is determined by a random binary number (0/1: Take the segment from parent B/A). The crossover process requires discontinuity and overlap removal to achieve a continuous electrode map. Mutation involves randomly modifying a bandwidth by Δf followed by discontinuity removal

electrodes). A new child (generation) was generated from its parents A and B by performing crossover and mutation operations on the "variables" presented in "individuals." As shown in Fig. 10.10b, a crossover binary pattern can determine which variable should be taken from which parent. The selected variables are concatenated to make a new "individual." Discontinuity and overlaps between frequency bands should be removed from the individual by setting the f_i^L of each frequency band equal to the f_{i-1}^H of the previous band. During a mutation, the f_i^H is modified with a given probability value (e.g., 0.2) in a stochastic manner (e.g., Gaussian distribution) limited to the exploration domain $\left[f_i^L, 1.2f_i^H\right]$.

Figure 10.11 shows the EA procedure that Saadoun et al. (2022) designed for creating 13 individuals over three generations for frequency fitting. The individual with the highest word recognition score (WRS) is chosen as the final winner. In case of a tie, the patient's preferred individual is selected. Initially, four parents are randomly generated in the exploration domain from a manual fitting with default factory frequency setting. These parents produce three children in the first generation through crossover and mutation. In the second generation, the four individuals with the best WRS are selected from the seven available, and three new children are created similarly to the first generation. The third generation repeats this process, selecting the four best individuals from the 10 available before generating new children.

The EA algorithm suggested a different MAP than the default setting. The EA yielded an enlargement of the bands in the low frequencies and shifted their center frequencies toward higher values. In patients with a higher number of electrode

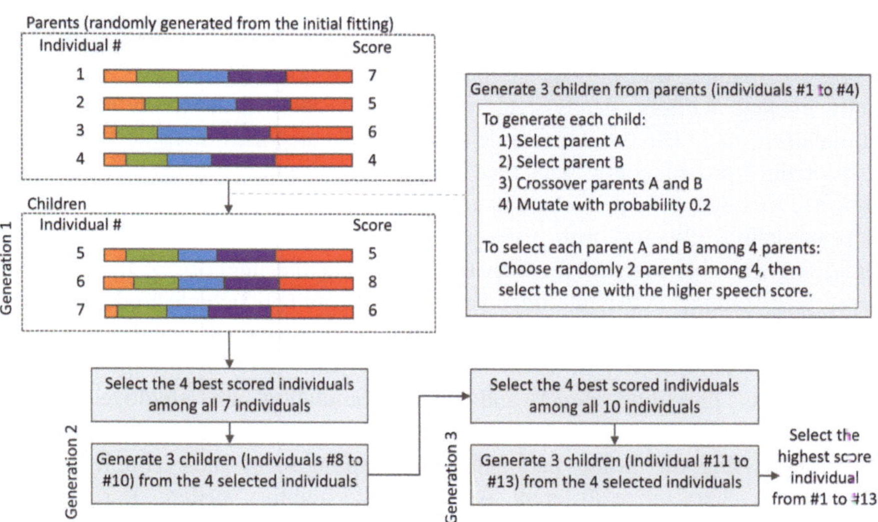

Fig. 10.11 Fitting steps using the EA for electrode frequency allocation. In total, 13 individuals (i.e., "genomes") are formed across three generations. The individual that yields the best speech test score will be selected. The process indicated in Fig. 10.10 is used to form each new child from two parents

contacts (e.g., Cochlear Ltd. or Advanced Bionics LLC), several frequency bands were dramatically narrowed, while the bands allocated to the neighboring electrodes were significantly widened, suggesting the detection and the resolution of channel interferences (similar to Bourgeois-Republique et al. (2005)). At the final evaluation of 27 CI patients, WRS in noise was significantly improved with EA (6.46 ± 1.63) compared to default MAP (4.17 ± 0.97). WRS in quiet and free-field audiometry did not show significant differences.

10.3.6 Van Opstal and Noordanus (VON) Fitting Method

Van Opstal and Noordanus (2023) presented the concept of a fitting method that can be considered the broadest member in the generic fitting family. This fitting method to which we refer as VON differs from FOX, OptiFOX, and EA methods in two main aspects: Firstly, it uses both direct and audio stimulations, connecting them via their corresponding test outcomes; secondly, the agent is automatically trained on a population of CI patients, with the possibility of using simulated data as well.

The fitting policy of the VON involves using a variety of experimental tests that provide essential information about the characteristics of different processing stages in the ascending auditory pathways. It aims to optimize the interface between the acoustic input and the patient-specific auditory system. The experimental tests should specifically assess the quality of the electrode-to-auditory nerve (EL-AN) interface (i.e., signal transfer properties) and the integrity of the ascending auditory pathways.

The EL-AN quality (e.g., neural health) assessment can be performed using various relevant measurements and tests such as analysis of cochlea anatomy using CT imaging, transimpedance matrix, spread of excitation, eCAP, and electrically evoked auditory steady-state response (eASSR), as listed in the "no CI stimulation" and "Direct CI electrode stimulation" categories in Table 10.2. These tests mainly provide static data that are not significantly affected by the fitting process, so they can be performed once for the EL-AN quality assessment. Various data modalities obtained from these tests can be used in a rule-based approach, providing valuable insights for decisions regarding the integrity of the electrode array. For example, eCAP and impedance can determine which electrodes to use or inactivate, while eASSR can estimate near-optimal values for C-levels.

To assess the integrity of ascending auditory pathways, two groups of experimental tests providing direct and audio stimulation are employed. Direct stimulations are used to characterize the properties of ascending auditory pathways while bypassing the fitting parameters. The stimulations have simple patterns designed to characterize different aspects of the auditory system. For example, some temporal and spectral aspects of the auditory system can be investigated by shifting the stimulation rate of an electrode by Δf and changing the stimulating electrode from one to another, respectively. The effect of these simple stimulation patterns can be quantified using various psychophysics or electrophysiology tests such as reaction time (RT), eASSR, and electrically evoked auditory change complex (eACC) in the EEG (see Table 10.2). Performing similar tests using audio stimulations adds the impacts of the CI's speech processor (which includes the fitting adjustments) on the same data modalities. There are two approaches to optimize this impact: (1) By the analysis of audio-related test outcomes, (2) by the comparison between relevant data modalities acquired with direct electrode and audio stimuli. In either case, a reward or cost value should be calculated from the outcomes. Any improvement in the integrity of the ascending auditory pathways, measured for example as, more similarity between direct electrode and audio-related data modalities, shorter reaction times, and shorter N1 latency of the ACC (that corresponds to better SRT) should increase (decrease) the reward (cost) value. To maximize the reward, one needs to find the functional relationship between the fitting parameters and different test outcomes (e.g., RT when the stimulating electrode changes, or RT when the rate increases by Δf). However, finding such a functional relationship is difficult as it includes many complex, highly interdependent, and nonlinear subsystems from the EL-AN interface to high-level auditory processing levels. Advanced ML methods may be of great help to figure out how the fitting parameters are mapped to various tests outcomes.

Van Opstal and Noordanus (2023) suggested a network architecture and a few data modalities for monaural and binaural fittings. These data modalities are mostly acquired by various RT and ASSR/ACC tests. Below it is explained as an example, how RT tests can determine a reward value for the fitting procedure. The ML network may use a combination of rewards coming from different experimental tests, integrated into a total reward value.

As mentioned earlier, RT data modalities can be measured with direct electrode and audio stimuli. For the direct stimulation, one can measure RT, for example, to sudden electrode change (e.g., from contacts 1&2 to contacts 3&2, where electrode 2 acts as a background noise electrode) or rate change (e.g., from 180 to 220 Hz). It has been shown by Noorani and Carpenter (2016) that the distribution of RT responses can be modeled by a reciprocal Gaussian function (i.e., 1/RT has a Gaussian distribution). If expressed in a reciprobit plot, a linear function (i.e., straight line) is formed between RT values and their cumulative probability values (Fig. 10.12). In a reciprobit plot, the probit value of cumulative probability of RT is plotted versus RT values in a reciprocal scale (which yields a Gaussian distribution). This straight reciprobit line shifts to left (i.e., lower reaction time) if the central nervous system (CNS), including auditory pathways, has more sensory evidence to decide at a faster rate. Repeating the direct-electrode RT test for different electrode contacts generates a set of reciprobit lines represented by a gray zone in the figure. On the other hand, for the audio stimulation, one may use different audio stimuli having different spectrotemporal patterns (ST ripples) that are found in different parts of speech signals. For a given ST ripple, a greater reward should be generated if its reciprobit line is closer to those of direct-electrode stimuli (i.e., the gray zone). The fitting parameters should be tuned so that they increase this reward. The RT response of each ST ripple may be different from the others based on the amount of sensory evidence that is provided to the CNS for that ST ripple. A good fitting is the

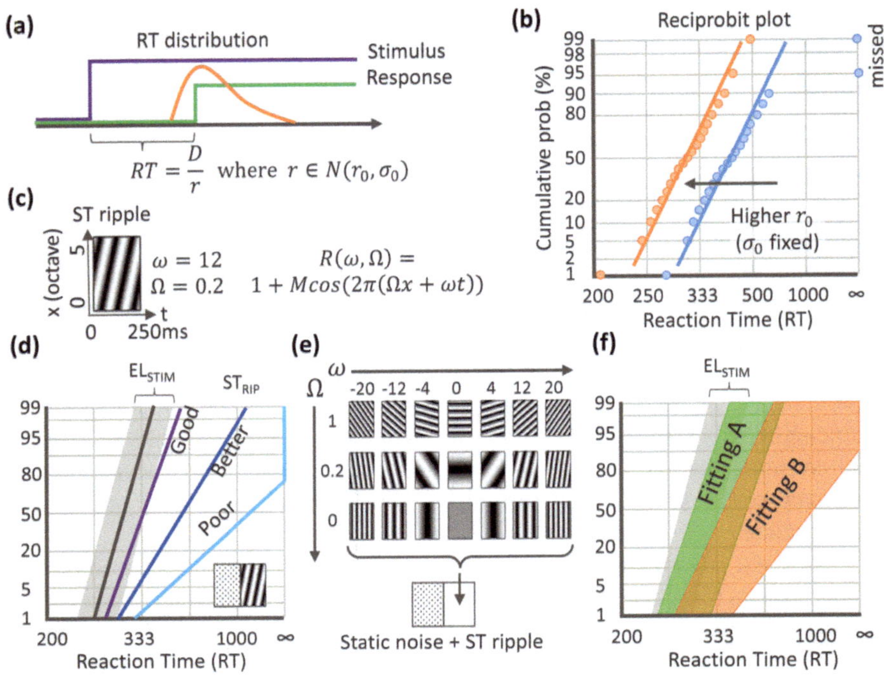

Fig. 10.12 Use of RT convergence in response to direct and acoustic stimulation for CI fitting. (**a**) RT can be modeled by a reciprocal Gaussian function. (**b**) The reciprobit plot of RT has a linear form. (**c**) Various spatiotemporal modulations in acoustic stimulations can be used to stimulate different parts of the auditory system. (**d**) Reciprobit plots of RT can be compared between direct electrical stimulation and acoustic stimulation. The closer the distance between the RT responses, the better the auditory system is at analyzing the spatiotemporal content of the auditory stimulation. (**e**) Various spatiotemporal stimulations used in a fitting. (**f**) Fitting A is better than fitting B as it produces reciprobit plots closer to those of electrical stimulations

one that generates reciprobit lines for all ST ripples as close as possible to those of direct-electrode stimulations (Fig. 10.12).

Several possible ML architectures can be considered for the fitting agent. In the training phase, the agent should find the mapping between fitting parameters and the outcomes. In the testing phase (i.e., fitting mode), it should result a list of fitting parameters that maximize the reward (e.g., obtained from different RT, ASSR, ACC outcomes). One of the promising networks for this kind of problems is reinforcement networks. They can continuously learn new decision policies when more data or learning time is provided to them. Figure 10.13 illustrates a fitting agent based on an actor-critic reinforcement method. In this method two ML networks are used as the actor and the critic networks. The actor is responsible for selecting actions (i.e., proposing fitting modifications) based on a current policy coded in its network structure. The critic evaluates the actions by estimating cumulative reward values in future time (the so-called the value function V) and provides feedback to the actor

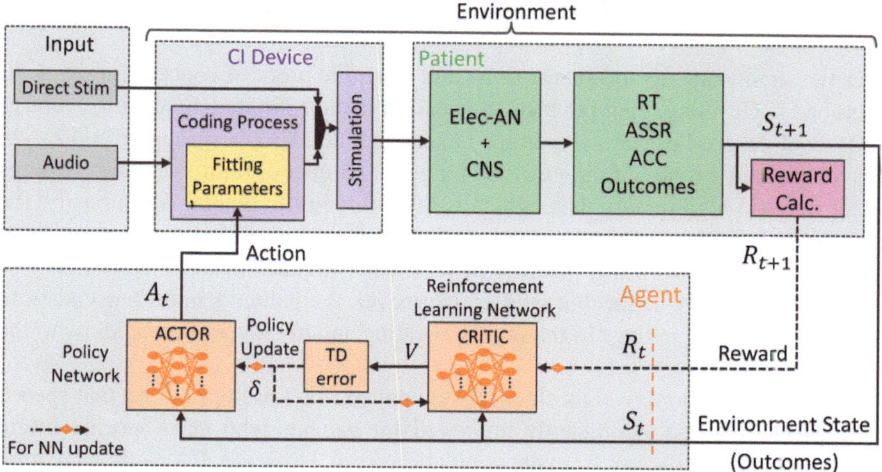

Fig. 10.13 Suggested Van Opstal and Noordanus (VON) fitting method which incorporates both direct and audio stimulations, and a reinforcement ML agent that optimizes the fitting parameters using a reward mechanism calculated on various outcomes including RT, ASSR, and ACC responses obtained from the patient in response to stimulations

to update its policy. The feedback is a time difference (TD) error representing the difference between estimated value function and obtained value after the last action. Below is an event summary that happens in an actor-critic reinforcement model:

1. Both the actor and critic networks receive the current state S_t as input. The state is a union of experimental test outcomes.
2. The actor network outputs the probabilities of each action, and an action A_t is sampled from this distribution.
3. The chosen action A_t is executed in the environment (i.e., CI device, EL-AN interface, CNS, and experimental tests) resulting in a new state S_{t+1} and a reward R_{t+1}.
4. The critic network evaluates the current state S_t and the next state S_{t+1} to estimate value functions $V(S_t)$ and $V(S_{t+1})$, where $V(S_t) = \mathrm{E}(R_{t+1} + \gamma^1 R_{t+2} + \gamma^2 R_{t+3} + \ldots)$ is $\mathrm{E}(R_{t+1} + \gamma^1 R_{t+2} + \gamma^2 R_{t+3} + \ldots)$ is the expected value of future estimated rewards with discount factor $\gamma \in [0, 1]$ starting from state S_t.
5. The TD error δ is computed using the reward R_{t+1} and the value estimates as $\delta = R_{t+1} + \gamma V(S_{t+1}) - V(S_t)$.
6. Network updates in the context of deep learning:
   - The actor network finds the best action policy using the policy gradient (backpropagation) method with the TD error as a feedback signal.
   - The critic network is updated to reduce the TD error by minimizing the difference between the estimated and actual value functions.

10.3.7 Perspectives of AI-Based Methods in CI Fitting

Despite significant advancements in AI and scientific proof of concept for using AI methods in CI fitting, their practical application remains limited. Scalability, robustness, and ease of use are important factors that need to be properly addressed. Developing an AI-based fitting method for a vast number of fitting variables requires extensive and high-quality data, which is often difficult to obtain. Additionally, the variability in patient responses, the complexity of individual auditory systems, and the changing list of fitting parameters with each implant make it challenging to develop universally applicable models. Moreover, the patient's brain must adapt to the new parameter settings in the implant, adding another layer of complexity to the fitting process.

The few evolutionary algorithms (EAs) applied in CI fitting indicate that speech performance can be significantly improved for patients who need specific fitting parameter values different from the population norm. However, EA methods require active participation from the patient to guide the formation of genomes and evaluate their performance over time. This necessitates voluntary participation from CI manufacturers to develop sophisticated hardware and software to modify CI parameters and perform the necessary tests during the patients' daily activities. Standards and safety aspects must be addressed to facilitate this promising path. As initial steps, hybrid methods combining traditional and EA fitting methods can be considered, allowing the EA to fine-tune some of the more important fitting parameters over time in a chronic manner.

The FOX method relies on the expertise of a few specialists who incorporate their knowledge into a deterministic or stochastic model for a given implant. Consequently, this model faces challenges with automatic scalability and can only achieve the performance level of expert audiologists. FOX is the most widely used AI-based fitting method within the CI community, demonstrating effectiveness comparable to human performance in several clinical studies (Vaerenberg et al. 2014; Battmer et al. 2015; Buechner et al. 2015; Meeuws et al. 2017; Wathour et al. 2023). The studies show while FOX can improve outcomes for some individuals, it may not significantly enhance speech perception scores for all patients due to varying individual responses. Additionally, FOX requires some patient-specific outcome data to optimize the fitting process, which can be difficult to obtain quickly in clinical settings. Although the initial fitting time may be reduced, over longer periods, it may not differ significantly from conventional methods. Moreover, FOX is not an open-access project, limiting the CI community's ability to contribute to its improvement or develop it into a fully computerized, trainable model.

The VON fitting framework can inspire the CI community to collaborate on multicenter projects to achieve a universal fitting method. This framework emphasizes objective measures such as eCAP and eASSR, which modern implants can capture, and easy-to-capture subjective responses like reaction time to direct and acoustic stimulations. These measurements aim to assess various aspects of the auditory processing chain, from the electrode-neuron interface to the auditory cortex. However, realizing such an ambitious framework requires close collaboration among CI manufacturers and research centers to share knowledge, data, and infrastructure.

Developing the agent (e.g., actor-critic reinforcement) is challenging and shares issues with EAs such as needing large, high-quality datasets. Additionally, for the actor-critic method to function correctly, the patient's brain must adapt to the new implant settings, making the training process of the agent lengthy. Researchers may need to expedite this process by introducing hybrid methods, incorporating prior information into the agent, using computational models to simulate patient outcomes, or developing smaller agents specialized in subsets of fitting parameters.

The AI fitting models discussed in this chapter are designed to handle complex, multimodal outcomes—potentially more efficiently than audiologists—leading to improved fitting solutions. Alternatively, these models can serve as tools to help audiologists reach fitting outcomes more quickly, as briefly noted in Sect. 10.3.2. However, this supportive role was not the primary focus of this chapter.

Acknowledgments The author acknowledges the use of Microsoft Copilot for proofreading and enhancing the readability of the manuscript. All suggested changes were carefully reviewed and edited. The author assumes full responsibility for the content of the final publication.

References

Başkent D, Eiler CL, Edwards B (2007) Using genetic algorithms with subjective input from human subjects: implications for fitting hearing aids and cochlear implants. Ear Hear 28:370–380. https://doi.org/10.1097/AUD.0b013e318047935e

Battmer RD, Borel S, Brendel M et al (2015) Assessment of 'fitting to outcomes expert' FOXTM with new cochlear implant users in a multi-centre study. Cochlear Implants Int 16:100–109. https://doi.org/10.1179/1754762814Y.0000000093

Bergeron F, Hotton M (2015) Comparison of eSRTs and comfort levels in users of digisonic SP cochlear implants. Cochlear Implants Int 16:110–114. https://doi.org/10.1179/1754762814Y.0000000092

Bermejo Delgado I (2012) Probabilistic graphical models for the tuning of systems. Master thesis, Dept. Artificial Intelligence, UNED, Madrid, Spain.

Bermejo I, Díez FJ, Govaerts P, Vaerenberg B (2013) A probabilistic graphical model for tuning cochlear implants, Lecture notes in computer science (including subseries lecture notes in artificial intelligence and lecture notes in bioinformatics) 7885 LNAI, pp 150–155. https://doi.org/10.1007/978-3-642-38326-7_23

Björsne A, Magnusson L (2025) The relationship between AutoNRT thresholds and subjective programming levels revisited. Cochlear Implants Int. https://doi.org/10.1080/14670100.2025.2455891

Blamey P, Arndt P, Brimacombe J et al (1996) Factors affecting auditory performance of postlinguistically deaf adults using cochlear implants. Audiol Neurootol 1. https://doi.org/10.1159/000259212

Blamey P, Artieres F, Başkent D et al (2012) Factors affecting auditory performance of postlinguistically deaf adults using cochlear implants: an update with 2251 patients. Audiol Neurotol 18:36–47. https://doi.org/10.1159/000343189

Bourgeois-République C, Chabrier JJ, Collet P (2004) Automatic fitting of cochlear implants with evolutionary algorithms. Proc ACM Symp Appl Comput 1:296–300. https://doi.org/10.1145/967900.967963

Bourgeois-Republique C, Frachet B, Collet P (2005) Using an interactive evolutionary algorithm to help fitting a cochlear implant. In: GECCO 2005—genetic and evolutionary computation conference 2005, pp 133–139. https://doi.org/10.1145/1102256.1102287

Buechner A, Vaerenberg B, Gazibegovic D et al (2015) Evaluation of the 'fitting to outcomes eXpert' (FOX®) with established cochlear implant users. Cochlear Implants Int 16:39–46. https://doi.org/10.1179/1754762814Y.0000000085

Choi CTM (2001) Shape optimization of cochlear implant electrode array using genetic algorithms. In: Annual international conference of the IEEE engineering in medicine and biology—proceedings, vol 2, pp 1445–1448. https://doi.org/10.1109/iembs.2001.1020475

Crowson MG, Dixon P, Mahmood R et al (2020a) Predicting postoperative cochlear implant performance using supervised machine learning. Otol Neurotol 41:E1013–E1023. https://doi.org/10.1097/MAO.0000000000002710

Crowson MG, Lin V, Chen JM, Chan TCY (2020b) Machine learning and cochlear implantation—a structured review of opportunities and challenges. Otol Neurotol 41:E36–E45. https://doi.org/10.1097/MAO.0000000000002440

Czurda R, Wesarg T, Aschendorff A et al (2024) Investigation of maximum monosyllabic word recognition as a predictor of speech understanding with cochlear implant. J Clin Med 13:646. https://doi.org/10.3390/jcm13030646

de Andrade KCL, Leal M de C, Muniz LF et al (2014) A the importance of electrically evoked stapedial reflex in cochlear implant. Braz J Otorhinolaryngol 80:68–77. https://doi.org/10.5935/1808-8694.20140014

de Vos JJ, Biesheuvel JD, Briaire JJ et al (2018) Use of electrically evoked compound action potentials for cochlear implant fitting: a systematic review. Ear Hear 39(3):401–411. https://doi.org/10.1097/AUD.0000000000000495

Eiler C, Başkent D, Recker K, Edwards B (2008) Genetic algorithms: are they the future of hearing aid fittings? Hear J 61:16–21. https://doi.org/10.1097/01.HJ.0000343021.84357.78

Essaid B, Kheddar H, Batel N et al (2024) Advanced artificial intelligence algorithms in cochlear implants: review of healthcare strategies, challenges, and perspectives. IEEE, 12:119015–119038.https://doi.org/10.1109/ACCESS.2024.3429524

Favaretto N, Marioni G, Brotto D et al (2019) Cochlear implant outcomes in the elderly: a uni- and multivariate analyses of prognostic factors. Eur Arch Otorrinolaringol 276:3089–3094. https://doi.org/10.1007/s00405-019-05613-3

Gantz BJ, Woodworth GG, Knutson JF et al (1993) Multivariate predictors of audiological success with multichannel cochlear implants. Ann Otol Rhinol Laryngol 102:909–916. https://doi.org/10.1177/000348949310201201

Gärtner L, Spitzer P, Lauss K et al (2021) Optimized SNR-based ECAP threshold determination is comparable to the judgement of human evaluators. PLoS One 16:1–21. https://doi.org/10.1371/journal.pone.0259347

Géron A (2019) Hands-on machine learning with Scikit-Learn, Keras and TensorFlow: concepts, tools, and techniques to build intelligent systems, 2nd edn. O'Reilly Media Inc., p 856. ISBN: 9781492032649

Goudey B, Plant K, Kiral I et al (2021) A multicenter analysis of factors associated with hearing outcome for 2,735 adults with cochlear implants. Trends Hear 25:23312165211037524. https://doi.org/10.1177/23312165211037525

Govaerts PJ, Vaerenberg B, De Ceulaer G et al (2010) Development of a software tool using deterministic logic for the optimization of cochlear implant processor programming. Otol Neurotol 31:908–918. https://doi.org/10.1097/MAO.0b013e3181dd160b

Han JJ, Suh MW, Park MK et al (2019) A predictive model for cochlear implant outcome in children with cochlear nerve deficiency. Sci Rep 9:1154. https://doi.org/10.1038/s41598-018-37014-7

James CJ, Karoui C, Laborde M-LL et al (2019) Early sentence recognition in adult cochlear implant users. Ear Hear 40:905–917. https://doi.org/10.1097/AUD.0000000000000670

Kim H, Kang WS, Park HJ et al (2018) Cochlear implantation in postlingually deaf adults is time-sensitive towards positive outcome: prediction using advanced machine learning techniques. Sci Rep 8:18004. https://doi.org/10.1038/s41598-018-36404-1

Koirala N, Manning J, Neumann S et al (2025) The neural characteristics influencing literacy outcome in children with cochlear implants. Brain Commun 7. https://doi.org/10.1093/braincomms/fcaf086

Kraaijenga VJCC, Smit AL, Stegeman I et al (2016) Factors that influence outcomes in cochlear implantation in adults, based on patient-related characteristics—a retrospective study. Clin Otolaryngol 41:585–592. https://doi.org/10.1111/coa.12571

Lazard DS, Vincent C, Venail F et al (2012) Pre-, per- and Postoperative factors affecting performance of postlinguistically deaf adults using cochlear implants: a new conceptual model over time. PLoS One 7:e48739. https://doi.org/10.1371/journal.pone.0048739

Legrand P, Bourgeois-Republique C, Péan V et al (2007) Interactive evolution for cochlear implants fitting. Genet Program Evolvable Mach 8:319–354. https://doi.org/10.1007/s10710-007-9048-4

Lineaweaver SKR, Wakefield GH (2011) Psychometric augmentation of an interactive genetic algorithm for optimizing cochlear implant programs. In: Genetic and evolutionary computation conference, GECCO'11, pp 1755–1762. https://doi.org/10.1145/2001576.2001812

Lu S, Xie J, Wei X et al (2022) Machine learning-based prediction of the outcomes of cochlear implantation in patients with cochlear nerve deficiency and normal cochlea: a 2-year follow-up of 70 children. Front Neurosci 16:1–13. https://doi.org/10.3389/fnins.2022.895560

Lundberg SM, Lee SI (2017) A unified approach to interpreting model predictions. In: Advances in neural information processing systems. NIPS-2017, pp 4765–4774. https://doi.org/10.48550/arXiv.1705.07874

Martins PG, Luís L, Segovia-Martinez M, Molaee-Ardekani B (2024) Assessment of electrically evoked stapedius reflex threshold (eSRT) for auditory mapping in pulse width modulation cochlear implants: the effect of multielectrode stimulation. Otol Neurotol 45(10):1128–1134. https://doi.org/10.1097/MAO.0000000000004354

Matloff N (2017) Statistical regression and classification: from linear models to machine learning. CRC Press, Taylor & Francis Group, New York. ISBN 9781138066465

Meeuws M, Pascoal D, Bermejo I et al (2017) Computer-assisted CI fitting: is the learning capacity of the intelligent agent FOX beneficial for speech understanding? Cochlear Implants Int 18:198–206. https://doi.org/10.1080/14670100.2017.1325093

Migliorini E, van Dijk B, Philips B et al (2024) The relation between cochlear implant programming levels and speech perception performance in post-lingually deafened adults: a data-driven approach. Eur Arch Otorrinolaringol 281:1163–1173. https://doi.org/10.1007/s00405-023-08195-3

Molaee-Ardekani B, Goiana-Martins P, Segovia-Martinez M, Luís L (2025) The effect of current amplitude and multi-electrode stimulation on eSRT for auditory fitting in cochlear implants with pulse-width loudness coding. Cochlear Implants Int. https://doi.org/10.1080/14670100.2025.2457200

Moulin A, Bernard A, Tordella L et al (2017) Variability of word discrimination scores in clinical practice and consequences on their sensitivity to hearing loss. Eur Arch Otorrinolaringol 274:2117–2124. https://doi.org/10.1007/s00405

Noorani I, Carpenter RHS (2016) Neuroscience and biobehavioral reviews the LATER model of reaction time and decision. Neurosci Biobehav Rev 64:229–251. https://doi.org/10.1016/j.neubiorev.2016.02.018

Plant K, McDermott H, van Hoesel R et al (2016) Factors predicting postoperative unilateral and bilateral speech recognition in adult cochlear implant recipients with acoustic hearing. Ear Hear 37(2):153–163. https://doi.org/10.1097/AUD.0000000000000233

Raghunandhan S, Ravikumar A, Kameswaran M et al (2015) Electrophysiological correlates of behavioral comfort levels in cochlear implantees: a prospective study. Indian J Otolaryngol Head Neck Surg 67:210–222. https://doi.org/10.1007/s12070-013-0679-x

Roditi RE, Poissant SF, Bero EM, Lee DJ (2009) A predictive model of cochlear implant performance in postlingually deafened adults. Otol Neurotol 30(4):449–454. https://doi.org/10.1097/MAO.0b013e31819d3480

Saadoun A, Schein A, Péan V et al (2022) Frequency fitting optimization using evolutionary algorithm in cochlear implant users with bimodal binaural hearing. Brain Sci 12:253. https://doi.org/10.3390/brainsci12020253

Shachter RD, Peot MA (1990) Simulation approaches to general probabilistic inference on belief networks. In: Henrion M, Shachter RD, Kanal LN, Lemmer JF (eds) Machine intelligence and pattern recognition. North-Holland, New York, pp 221–231

Shafieibavani E, Goudey B, Kiral I et al (2021) Predictive models for cochlear implant outcomes: performance, generalizability, and the impact of cohort size. Trends Hear 25:1–14. https://doi.org/10.1177/23312165211066174

Song Q, Qi S, Jin C et al (2022) Functional brain connections identify sensorineural hearing loss and predict the outcome of cochlear implantation. Front Comput Neurosci 16:825160. https://doi.org/10.3389/fncom.2022.825160

Tan L, Holland SK, Deshpande AK et al (2015) A semi-supervised support vector machine model for predicting the language outcomes following cochlear implantation based on pre-implant brain fMRI imaging. Brain Behav 5:1–25. https://doi.org/10.1002/brb3.391

Vaerenberg B (2014) Programming cochlear implants for auditory performance (FOX). PhD thesis, University of Antwerpen

Vaerenberg B, Govaerts PJ, De Ceulaer G et al (2010) Experiences of the use of FOX, an intelligent agent, for programming cochlear implant sound processors in new users. Int J Audiol 50:50–58. https://doi.org/10.3109/14992027.2010.531294

Vaerenberg B, De Ceulaer G, Szlávik Z et al (2014) Setting and reaching targets with computer-assisted cochlear implant fitting. Sci World J 2014:646590. https://doi.org/10.1155/2014/646590

Van Opstal AJ, Noordanus E (2023) Towards personalized and optimized fitting of cochlear implants. Front Neurosci 17:1–20. https://doi.org/10.3389/fnins.2023.1183126

Wathour J, Govaerts PJ, Derue L et al (2023) Prospective comparison between manual and computer-assisted (FOX) cochlear implant fitting in newly implanted patients. Ear Hear 44: 494–505. https://doi.org/10.1097/AUD.0000000000001314

Part III

Emerging Technologies and Future Directions

Computational Modeling in Cochlear Implants: Physical Modeling

11

M. A. Callejón-Leblic and Tania Hanekom

Abstract

Computational modeling provides a methodical approach to deconstructing the complexity of biological systems into realizable, definable, interconnected subsystems. In this chapter, modeling of the electrical behavior of the peripheral auditory system under electrical stimulation with a cochlear implant (CI) is explored and unpacked. An overview is provided of the different components required to create a three-dimensional volume electric model of the electrically stimulated cochlea, including extracochlear components such as the head, which affects current pathways between the stimulating and the return electrodes, and the facial nerve and vestibular systems that may also be affected by stimulation. A generalized workflow for developing volume electrical models of the cochlea is presented. It includes image acquisition, three-dimensional reconstruction techniques, meshing, applying the appropriate physics and boundary conditions, using numerical methods to solve the system of equations, obtaining meaningful predictions, and validating and interpreting predictions. The chapter concludes

M. A. Callejón-Leblic and Tania Hanekom have contributed equally to this chapter.

M. A. Callejón-Leblic (✉)
Biomedical Engineering Group, University of Seville, Seville, Spain

Otorhinolaryngology Unit, Virgen Macarena University Hospital, Seville, Seville, Spain

Oticon Medical, Alcobendas, Madrid, Spain
e-mail: mcallejon@us.es

T. Hanekom
Bioengineering, Department of Electrical, Electronic and Computer Engineering, University of Pretoria, Pretoria, South Africa
e-mail: tania.hanekom@up.ac.za

B. Molaee-Ardekani, M. Segovia-Martinez (eds.), *Advancing Cochlear Implants*,
https://doi.org/10.1007/978-3-032-15719-5_11

with an overview of existing volume electrical models, their contributions to our understanding of CI functioning, challenges and limitations of a computational modeling approach, and their current and future applications.

Keywords

Volume electrical model · Cochlear implant · Finite element method · Head model · Tissue electrical properties

11.1 Introduction

Computational modeling provides a comprehensive and methodical approach to deconstructing the complexity of biological systems into realizable, definable, interconnected subsystems. The scope of each subsystem is determined by the function that needs to be emulated and the level at which the specific biological processes involved may be disentangled. The model of each subsystem comprises a mathematical description of the structure and function of the anatomy and physiology.

The mechanism of hearing may be divided into peripheral and central processes. The first involves the transduction of sound pressure waves to a neural code in the cochlea (e.g., Goldsworthy 2022), and the second involves the central nervous system's processing of the neural code to invoke a perception (e.g., Gervain and Geffen 2019; Drakopoulos et al. 2025). For this reason, computational models of hearing are also typically categorized as peripheral, i.e., describing the sound wave-to-neural code transduction mechanism, or central, i.e., describing the processing pathways in the brain to make sense of the peripheral neural code. Both types of models may use a physiologically based approach, a phenomenological approach, or a combination of the two. Because of the complexity of the complete auditory system, a large number of models have been created to describe distinct characteristics of this system. Often, several different models, each describing a specific component or aspect, are combined to capture the characteristics of the auditory system more completely.

At the periphery, different physical principles are required to describe normal and prosthetic cochlear function. A model of normal cochlear function requires a description of the mechano-electro transduction process that captures the transformation of mechanical sound waves into neural responses that can be interpreted by the brain, thus enabling the perception of sound. A vast number of cochlear mechanics models have been developed to describe the mechanisms, whereby the basilar membrane responds to sound waves, subsequently deflects hair cell stereocilia, and finally causes ion channel activation in the auditory nerve (e.g., Reichenbach and Hudspeth 2014). Physiologically based models that aim to capture the *physical nature* of the hearing system (e.g., Ni et al. 2014) require an accurate representation of the structure of the cochlea, the mechanical properties of these structures, and the fluid dynamics that govern the system's response. Conversely, a model of *prosthetic cochlear function*, specifically electrical stimulation of the cochlea (e.g., Goldwyn et al. 2010), must capture the mechanisms that allow the auditory nerve to be artificially excited by passing an electric current through it. This requires a volumetric description of the cochlear morphology that allows electrical characteristics to be assigned to the various tissues and implant components. The model can then be used to predict the effect of electrical excitation of the system on peripheral neural activation patterns. Such models may also find application in diagnostic tool development, drug delivery research, cochlear implant (CI) device optimization, electrophysiological studies of the mechanism of hearing, investigation of alternative stimulation techniques such as optical stimulation, and educational and training applications.

As this book focuses on CI devices, we explore physiologically based models of electrical stimulation in CIs. Accurately modeling the auditory periphery's response to electrical stimulation requires at least two types of models: a volume conduction (VC) or volume electric field (VEF) model—collectively referred to here as volume electric (VE) models—and a neural model that describes the auditory nerve's response to the injected current. These two model types form the core topics of this chapter and the following one (Chap. 12).

VE models attempt to capture a detailed three-dimensional (3D) electrical description of the cochlea, auditory nerve, and, in many cases, the surrounding tissues and structures (Callejón-Leblic 2024; Frijns et al. 2000; Rattay et al. 2001a; Malherbe et al. 2015). They are used to simulate the distribution of electric potential, field, or current through the geometrically complex tissue volume. The parameters that describe the anatomy and geometry of a particular CI user's cochlear structures are extracted from medical imaging data, generally computed tomography (CT) and/or magnetic resonance imaging (MRI) scans. Different tissues in the model are assigned different electrical properties, for which the values are derived from experimental data. The finite element method (FEM) is a commonly employed numerical method to solve the partial differential equations governing the underlying physics. The magnitude and distribution of the electrical quantities provide insight into the operation of the biophysical interface as affected by, among others, electrode design and placement. It also provides an indication of which structures, including the auditory neurons, may be affected by the stimulation because they are in the current paths between the stimulation and return electrodes.

Predictions of neural excitation behavior require integration of the output of the VE model with a neural model. Neural models may account for both spatial and temporal excitation dynamics. A generalized overview of the typical model creation workflow is illustrated in Fig. 11.1 and will be detailed in the different subsections of this chapter. Note that the workflow components outlined in this chapter may present different peculiarities or vary across research groups and modeling software packages depending on the CI application.

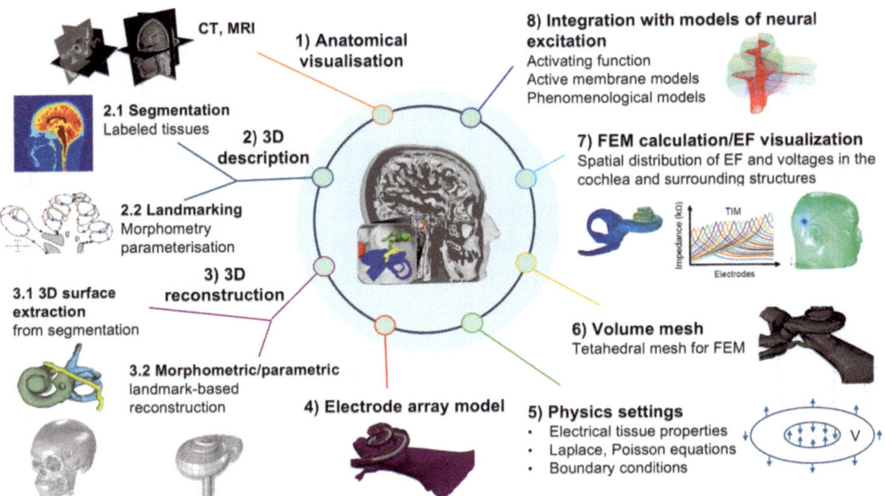

Fig. 11.1 Illustrative overview of the computational workflow for the generation of volume conduction or electric field models of the cochlea and surrounding head tissues. (1) The workflow usually starts with the visualization of the anatomy of the CI recipient through CT and/or MRI scans. (2) Two modeling paths can be used to provide a 3D description of the model: (i) segmentation of the image into several tissue types (cochlea, auditory nerve, and other surrounding tissues such as scalp, skull, and brain tissues) and (ii) description of the cochlear anatomy through landmarking and morphological parameterization. (3) 3D surface reconstruction of these tissues is performed based on either segmented tissue masks or parametric landmarks. (4) The CI electrode array is coupled into the model. (5) Physics equations, electrical properties of domain tissues, and boundary conditions are then set, including the current at each electrode depending on the CI stimulation mode used, and (6) a final volume mesh is generated. (7) The equations are solved using a numerical technique, such as the FEM. The solution, i.e., the voltage and EF distribution through the VE model, is displayed and analyzed. (8) Finally, the voltage values modeled can be integrated into a model of the auditory periphery to predict neural excitation

11.2 Anatomical Modeling of the Cochlea, Electrode Array, and Surrounding Structures

When the behavior of the auditory system must be mathematically described, it is important to determine which tissues and structures need to be included to obtain an acceptable representation of the physical environment. There is generally a trade-off between the accuracy of the anatomical representation, i.e., the amount of anatomical detail that may be included, and the solvability of the model, which relates to the time required for the numerical solver to converge to a solution (Thiselton and Hanekom 2024). The primary challenge when constructing models of the electrically stimulated cochlea is that there are three to four orders of magnitude size differences among the various structures that need to be modeled. For example, Reissner's membrane, located inside the cochlea, is approximately 10 μm thick, compared to the cochlea, with an average length of approximately

10 mm, compared to the human head, if included in the model, with a length of around 24 cm. Due to the geometric scale variations and complexity of the anatomy, the description of the peripheral auditory system's electrical behavior is limited to tissue volumes that are practical to model, given computing hardware and software limitations.

11.2.1 Cochlear Anatomy from an Electrical Perspective

The cochlea is a very complex organ consisting of a plethora of cells and tissue types ranging from micrometer to millimeter dimensions. To simulate its response to electrical stimulation, it is necessary to include anatomical details that may affect the distribution of the electrical potential, current, and field throughout the volume, particularly at the location of the target neural elements, i.e., the auditory nerve fibers (ANFs). A description of the various tissues as volumes with specific impedances is at the core of VE models of the cochlea. Figure 11.2 shows a schematic representation of a mid-modiolar section through a human cochlea, outlining the macromorphology of the cochlear turns. The micromorphology, such as the detailed

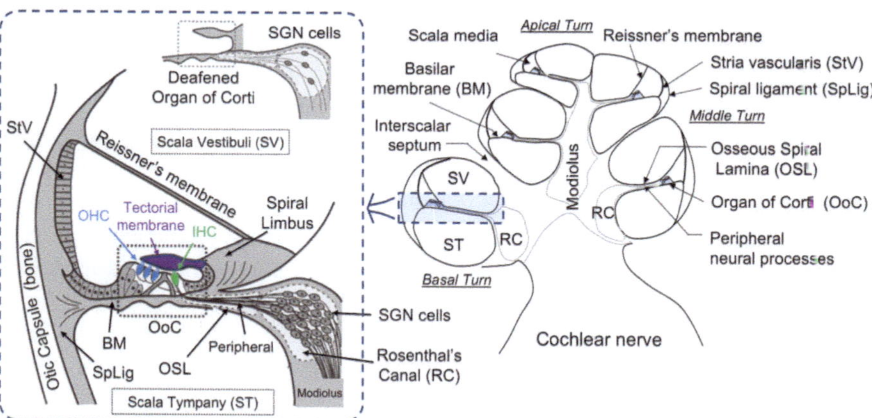

Fig. 11.2 Diagrammatic representation of a mid-modiolar section through a human cochlea (right). The cochlear canal spirals from the basal to the middle, and apical turns. The three scalae are indicated on the lower basal turn, while Reissner's membrane and the basilar membrane (BM) that divides the bony labyrinth into the three scalae are labeled on the lower middle turn. An enlargement of the area on the diagram, covering the scala media, is shown on the left. The peripheral processes of the spiral ganglion (SGN) cells in Rosenthal's canal traverse the osseous spiral lamina (OSL) to connect to the inner hair cells (IHCs) and three rows of outer hair cells (OHCs) in the Organ of Corti (OoC). The OoC is located on the BM, while the cilia of the hair cells connect to the tectorial membrane. The spiral limbus supports the OoC and provides an anchor site for the tectorial membrane. The lateral wall of the scala media is lined with the stria vascularis (StV), while the spiral ligament (SpLig) anchors the BM and membranous labyrinth (scala media) to the otic capsule. On the superior side, the scala media is separated from the scala vestibuli by Reissner's membrane that extends from the SpLig laterally to the spiral limbus medially. A deafened OoC is depicted at the top of the left figure, showing the absence of the OoC structures and a decrease in the density of the SGN cells. (The left part of the figure was adopted, with modification, from Hurley et al. (2007) with permission from John Wiley and Sons)

structure of the organ of Corti, is typically too small to include realistically in a VE model of the cochlea (Thiselton and Hanekom 2024) and is therefore omitted.

The cochlea consists of an outer bony labyrinth and an inner membranous labyrinth. The bony labyrinth forms the boundary of the cochlear space in the temporal bone, while the membranous labyrinth separates the bony labyrinth into three canals: the scala media, enclosed within the membranous labyrinth, and the scala vestibuli and scala tympani, superior and inferior from the scala media, respectively. The scala media is isolated and insulated from the scala tympani by the basilar membrane and from the scala vestibuli by Reissner's membrane. It contains endolymph, which has a different electrochemical composition from the perilymph in the bony labyrinth. The scala media extends to the apex of the cochlea but is not attached at the apical end so that the perilymph in the scala vestibuli and scala tympani is electrically connected via an opening called the helicotrema.

The basilar and Reissner's membranes provide high-impedance barriers to stimulation currents that may be injected (Johnstone et al. 1966) along the cochlear canal. The electrical impedance of the scala media may also be affected by the location of the membranes relative to one another, since it will determine the size and shape of the duct (Black et al. 1981; Thiselton and Hanekom 2024) (see Fig. 11.2). The spiral lamina, a perforated bony shelf extending from the modiolus, is an important impedance boundary between the scala tympani and the scala vestibuli. Its location determines the relative sizes of these two canals with respect to each other. Furthermore, the peripheral processes of the ANFs protrude from the perforations (habenula perforata) into the scala media, making this structure important for predicting neural excitation behavior. While neural degeneration in CI users often causes the peripheral processes to die, leaving only the spiral ganglion in the modiolus as excitable tissue, an increasing number of individuals with partial hearing loss have been implanted recently (Matin et al. 2021). These individuals have functional numbers of intact ANFs in some sections of the cochlea, which may be included in models that include the spiral lamina (Kalkman et al. 2015; Malherbe et al. 2016).

The modiolus is a conical-shaped structure around which the cochlear canal spirals and forms the inner boundary of the bony labyrinth. It consists of porous bone and contains the spiral ganglion, i.e., the cell bodies of the peripheral ANFs, which are often the main excitable structure targeted by electrical stimulation once the peripheral processes have degenerated. The modiolar wall of the scala tympani and scala vestibuli is important since it outlines the region where spiral ganglia exist. The location of this boundary, specifically in relation to the electrode array, has an important effect on neural excitation and should thus be included in VE models of the cochlea.

The spiral ligament, with the stria vascularis on its inner boundary lining the scala media, is also important as it is known to act as a current leakage pathway out of the scala tympani (Briaire and Frijns 2000).

Up to this point, the important anatomical structures that describe normal cochlear morphology have been considered. In this case, hearing impairment is mostly due to extensive damage to the hair cells and limited damage to the auditory neurons—neither of these components has an effect on the general construction of a

VE model of the cochlea. However, cochlear implantation is also considered for a variety of etiologies that do not present with normal cochlear morphology. These include ossification of the cochlea, which usually presents as a result of meningitis or otosclerosis, as well as cochlear malformations. Modeling of these cases may necessitate a deviation in the general constitution of a cochlear model. In the case of ossification of the cochlea, additional detail in the form of a bone volume in the basal section of the scala tympani may have to be added (Frijns et al. 2009; Badenhorst et al. 2021; Van Der Westhuizen et al. 2022). If a malformation is present, the dimensions and structural details of the cochlea may have to be adjusted, e.g., in the case of a common cavity malformation, no inner ear structures are present.

11.2.2 Anatomy of Surrounding Structures

Because of the conductive nature of the internal volume of the human body, a VE model of the cochlea needs to be enveloped by a conductive medium that simulates the extracochlear environment. Models that simplify the geometry and tissue constitution of the extracochlear environment limit realistic simulation of the return current path in monopolar (MP) stimulation mode. In MP mode, the current flows between the cochlea and one or more external return electrodes in the lateral part of the head. These electrodes are typically placed subcutaneously between the scalp and the skull layer in the mastoid area behind the ear. Several other structures, such as the facial nerve (FN) and vestibular system (VS), may also be affected by or may affect the extracochlear return current.

11.2.2.1 Head

Despite their greater complexity and computational cost, full-head models better emulate the return current path and improve numerical predictions of extracochlear currents (Malherbe et al. 2015; Tran et al. 2015; Callejón-Leblic et al. 2024). The challenge with constructing a VE head model is the amount and relevance of detail that should be included. The range of detail of full-head models varies in the literature from simple homogeneous volumes to basic three-compartment volumes that include the scalp, skull, and brain (Malherbe et al. 2015; Mangado et al. 2018a) to detailed models that include the scalp, skull, brain, CSF, ventricles, eyeballs, and sinus cavities (Tran et al. 2015; Callejón-Leblic et al. 2024).

Structural details about the temporal bone, such as the boundaries of the high-density otic capsule, highly individual pneumatization region, and foramina that may exist between the stimulating electrodes in the cochlea and the external return electrode located in the mastoid, are arguably the most important to include as these will affect the predicted current pathways.

Figure 11.3 shows different components of a detailed 3D head model. The surrounding head tissues were segmented based on MRI, while the cochleo-vestibular system, the FN, the ossicles, the carotid, and the sigmoid sinus were segmented from a cone-beam CT (CBCT) scan of the temporal bone.

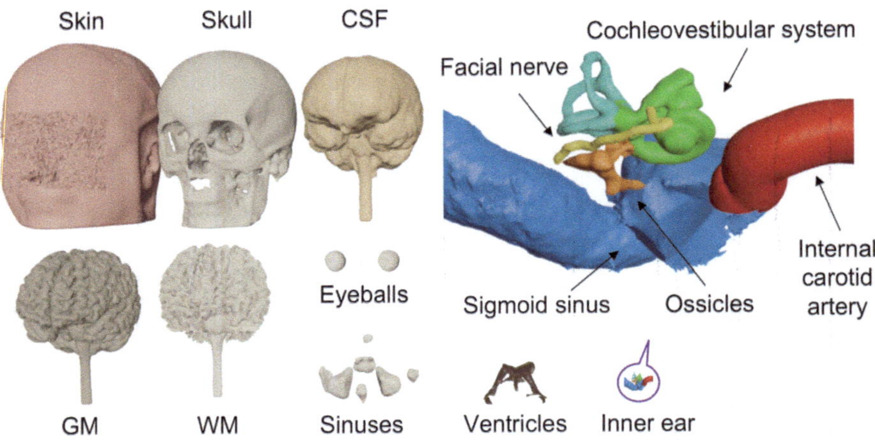

Fig. 11.3 Full-head model containing different tissues: skin, skull (bone), cerebrospinal fluid (CSF), sinuses, eyeballs, gray matter (GM), white matter (WM), and ventricles. The inner ear model contains the cochleo-vestibular system, the FN, the ossicles, the internal carotid artery, and the sigmoid sinus

11.2.2.2 Facial Nerve

The FN, of which the labyrinthine section runs near the cochlea, is a mixed nerve with motor fibers (which control the muscles of facial expression) and parasympathetic fibers (which regulate autonomous processes such as salivation and taste). It is an important structure in CI models as it may constitute a leakage path for the electric current (Vanpoucke et al. 2004), leading to undesired side effects such as facial nerve stimulation (FNS).

FNS is often solved by reducing or canceling the electric charge supplied to specific electrodes or using alternative pulse waveforms such as triphasic pulses. In rare cases, refractory FNS may require CI removal. Although a relatively low overall FNS rate of about 5% is reported, some subjects are at higher risk depending on individual cochlear pathology and anatomy (Van Horn et al. 2020). Individual bone characteristics such as otosclerosis, malformations, fractures, and dehiscence have been identified as risk factors (Bunne et al. 2024). Cochlear dehiscences are bony defects caused by the loss or absence of the otic capsule area and may be present at different sites along the cochlea, for instance, through the carotid or through the facial nerve. In CI recipients, individual bone properties and state may lead to spurious current leakage and undesired FNS. Figure 11.4 shows an anatomical view of the FN and its different segments.

11.2.2.3 Vestibular System

The vestibular system detects the location and movement of the head in space to facilitate proprioception and balance. It comprises the semicircular canals and the otolith organs (utricle and saccule). The semicircular canals are sensitive to rotational movements, while the otolith organs detect linear movement, acceleration, and head position relative to gravity. The vestibular nerve joins the cochlear nerve to

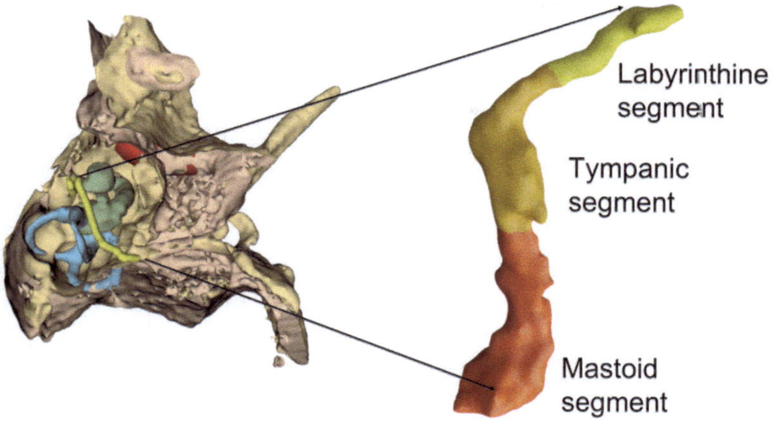

Fig. 11.4 Segmentation-based model of the temporal bone with cochleo-vestibular and FN structures. The three main portions of the FN are described as the labyrinthine segment, the tympanic segment, and the mastoid segment

form the vestibulocochlear nerve (cranial nerve VIII). The structure of the vestibular system is similar to that of the cochlea, i.e., it consists of a bony labyrinth filled with perilymph and, within it, a membranous labyrinth filled with endolymph. The membranous labyrinth contains the sensory organs. The saccule is connected to the cochlear duct via the ductus reuniens, which means that the cochlear and vestibular systems are connected allowing for movement and exchange of endolymphatic fluid between the two structures.

Since the vestibular system and the cochlea are interconnected anatomically and functionally, sharing common fluids and neural pathways, cochlear implants can sometimes elicit vestibular responses (Orlando et al. 2024).

The complexity of models of the vestibular system depends on the extent to which vestibular function needs to be predicted. Models that primarily focus on intracochlear stimulation and the effect that external structures have on current paths typically consider a simplified perilymph-filled structure. However, if the effect of electrical stimulation on vestibular function is investigated, a more complex model consisting of endolymphatic and perilymphatic spaces and the various nerve tissue components (posterior nerve, anterior nerve, lateral nerve, utricular nerve, and saccular nerve) is required. The otolith organs are the closest to the cochlea. Consequently, the nerve fibers in these organs are more likely to be excited during intracochlear stimulation than nerve fibers in the ampullae of the semicircular canals.

11.2.3 Imaging Techniques for Quantifying Anatomical Parameters

Several imaging techniques are available to quantify the geometry of the cochlea and surrounding structures for modeling purposes. Models can incorporate different levels of individualization based on anatomical variance among CI recipients.

Generalized models aimed at answering particular research questions can use high-resolution imaging techniques such as micro-CT (μCT) images, histology, dissection, and photography or advanced techniques like scanning thin-sheet laser imaging microscopy. These imaging modalities allow high-fidelity morphometry that supports the development of highly detailed models, though they may only be used on cadavers due to high radiation output or the requirement for dissection. Lower resolution clinical imaging, e.g., CT-based modalities and MRI, is required when developing models for healthcare settings.

Histology provides image resolutions of 20–200 nm (Sahl et al. 2017) that capture the detailed internal structures of the cochlea and other relevant structures. It has been used to visualize the electrode in situ, to assess electrode location (Rau et al. 2018), insertion trauma (Geerardyn et al. 2024), and tissue reactions (Ishai et al. 2017), to parameterize the internal structures of the cochlea for modeling purposes (Thiselton and Hanekom 2024), and to derive parameters for modeling the fibrous tissue layer around intracochlear electrodes (de Rijk et al. 2023). The method also allows for preparing high-resolution 3D image stacks, e.g., through cryosectioning. The disadvantage is that histology is destructive and may thus only be applied to cadavers.

Planar X-ray film-screen radiography has a typical resolution of 100–200 μm with lower resolutions for digital image acquisition (Huda and Brad Abrahams 2015). The visibility of the cochlea and neighboring structures in the dense temporal bone is limited because of the superimposition of structures in the 2D projections. The primary advantages of planar X-ray imaging are that live CI recipients can be imaged, and the electrode array may be clearly visualized without beam-hardening artifacts typical of CT reconstructions. It is routinely used to assess the positioning of the electrode array within the cochlea after surgery. X-ray images may thus be used to inform modeling of electrode array insertion depth and misplacement of an array, such as tip-fold over or buckling. In the absence of postoperative 3D imaging, it may be possible to reconstruct the radial trajectory of the outer wall of the cochlea from the trajectory of a lateral wall electrode array, allowing some person-specificity to be added to an existing generalized cochlear model.

Computed tomography methods, e.g., μCT, conventional, helical, and multidetector CT, as well as cone-beam CT (CBCT), form the backbone of most CI-related VE computational modeling. The dimensions of the bony labyrinth can be quantified in 3D from CT image stacks, thereby providing a framework for constructing the model's geometry. The resolution of these methods depends on the specific technology. Micro-CT imaging has a typical in-plane resolution of 50–100 μm, although some techniques may render a voxel size down to 200 nm (Hipsley et al. 2020). Conventional and helical CT images have a typical in-plane resolution of between 500 and 1000 μm, while that of multidetector CT images varies from 300 to 600 μm. The in-plane resolution of CBCT is between 50 and 600 μm (Kaasalainen et al. 2021). Due to the high radiation required by μCT, this modality is not suitable for use in live individuals. Because of its high resolution, the intracochlear membranes are sometimes visible (though soft tissue visibility is limited without a contrast agent), while the spiral lamina and the interscalar septa are usually visible. For this reason, μCT images are often used to create highly detailed individualized models (e.g., Potrusil et al. 2020) or to derive analytical (Pietsch et al. 2017) or statistical shape models of cochlear structures (Noble et al. 2012). For clinical CT modalities suitable for imaging live individuals, only the outline of the cochlear canals is visible, while none of the internal structures required to construct a detailed electrical impedance model of the cochlea are discernable. Internal structure data typically have to be augmented by other means, such as prediction through analytical models (Thiselton and Hanekom 2024) or template morphing (Malherbe et al. 2016), to render a detailed model. CT-based imaging is subject to metal beam-hardening artifacts, which may obscure some anatomical structures in proximity to the electrode array. Notwithstanding, it can be used to quantify the intracochlear trajectory of the electrode array, which is a crucial parameter in VE models of the electrically stimulated cochlea.

Magnetic resonance imaging provides a detailed view of soft tissues and fluid cavities, which is useful for describing the macro electrical impedance environment of the structures around the inner ear. MRI imaging is considered safe and is routinely used during the preoperative workup of CI candidates to verify, among others, the patency of the cochlear canal, the presence, and caliber of the cochlear nerve and to assess possible malformations that may affect CI candidacy or the surgical approach. It is possible to discern the spiral lamina and basilar membrane complex, the interscalar septa, and the modiolus on MRI images obtained using scanning parameters allowing an in-plane resolution of 50 μm (Benson et al. 2020). MRI imaging of the cochlea is not possible postoperatively due to metal artifacts caused by the electrode array and implant, disqualifying the method as the primary source for deriving geometric parameters for 3D modeling. Even so, preoperative MRI data could be used to augment models derived from postoperative CT imaging through co-registration (Athanasopoulos et al. 2024). MRI images of non-implanted individuals have been used to create highly detailed 3D models of the extracochlear electrical environment (Mangado et al. 2018a; Callejón-Leblic et al. 2024) through segmentation. Such models may then be used as template models to which cochlear models constructed from other datasets may be registered.

Several other imaging techniques may be used to quantify cochlear morphology, e.g., optical coherence tomography (OCT) and synchrotron radiation-based imaging. However, these technologies have not yet found broad clinical or research application due to limited accessibility but may, in future, provide highly detailed morphometric data for modeling purposes.

11.2.4 Imaging Postprocessing and 3D Reconstruction

The methods that are typically employed to reconstruct the 3D anatomy of the cochlea and implanted electrode array from medical images are explored below. The same methods may be extended to the reconstruction of the extracochlear environment, e.g., the FN, vestibular system, and head structures.

11.2.4.1 Reconstruction of Anatomy

Two techniques—landmarking and segmentation—form the basis for reconstructing the cochlea from medical images. Landmarking involves measuring distinct and identifiable points on the cochlea within a 3D coordinate system (Wärmländer et al. 2019; Human-Baron and Hanekom 2024). The landmarks are typically selected on the boundaries of important structures so that a lattice of lines connecting the landmarks demarcates distinct tissue areas. Segmentation, on the other hand, involves discretizing and filtering the color gradient on either 2D or 3D images and assigning each to specific tissue volumes (Fredieu et al. 2015).

A landmark-based approach to describing the morphology is common in cochlear modeling (e.g., Malherbe et al. 2015; Badenhorst et al. 2021; Lei et al. 2021; Ramos-De-Miguel et al. 2022). The method lends itself to observing correspondence among individuals (Wärmländer et al. 2019), allowing human observers to approximate unclear landmarks (Malherbe et al. 2016). This makes it more robust against blurred tissue boundaries on low-quality images than segmentation methods. Model generation from landmarks is computationally efficient since it involves only a limited number of points to define the geometry, thereby supporting relatively simple mathematical processing and manipulation of the extracted geometry (Schurzig et al. 2023). The mathematical elegance of landmark-based modeling is also evident in the development of parametric models that require measurement of only a few landmarks to arrive at a relatively accurate estimation of a cochlea's person-specific 3D spiraling trajectory (Pietsch et al. 2017). However, landmarking requires significant manual input, making it time-consuming, although progress has been made toward automating this process (Wang et al. 2023). It may also fail to capture fine anatomical details on low-resolution images, which is why studies have been aimed at predicting the location of such landmarks (Thiselton and Hanekom 2024).

Measuring landmarks involves preprocessing the image stack using software such as Fiji (Schindelin et al. 2012), by orienting the cochlea according to the consensus cochlear coordinate system (Verbist et al. 2009). This is followed by radial reslicing of the image stack through the modiolar axis so that landmarks can be measured on mid-modiolar slices through the cochlea (Thiselton and Hanekom 2024) as shown in Fig. 11.5a. Three-dimensional reconstruction is performed by connecting in-plane landmarks to outline the different cochlear tissues, after which volumes may be constructed by directly extruding the 2D surfaces or by extruding the line lattice into surfaces enclosing the various tissue volumes (Fig. 11.5b). Additional processing may involve filling in inner-structure details through prediction or morphing and smoothing to remove measurement noise. Figure 11.5c shows a detailed cochlear model created through this process.

Segmentation-based methods are well suited to model irregular and complex shapes that may be difficult to describe with a limited number of landmarks. Much work has recently been done in applying machine learning to develop automated segmentation algorithms to combat the labor-intensive process of manually outlining tissue areas (e.g., Siebrecht et al. 2024). The method is computationally costly (due to pixel- and voxel-level processing) and mathematical complex when the detailed geometry needs to be manipulated.

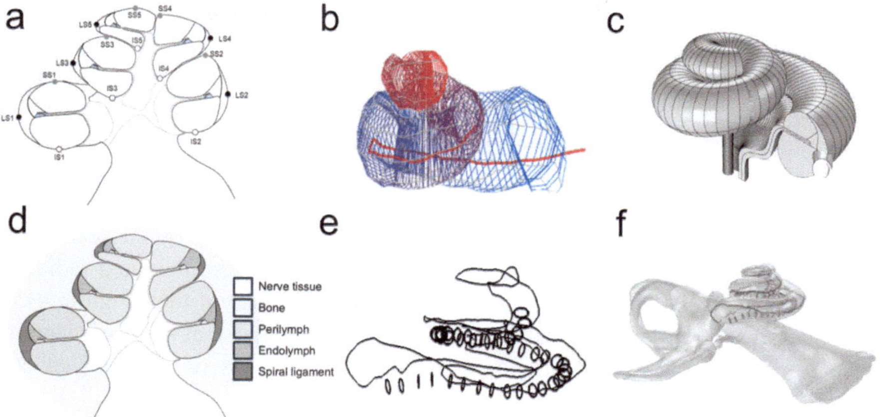

Fig. 11.5 Landmarking involves determining the coordinates of specific points on the anatomy. Landmarks that describe the lateral-most (LSn), inferior-most (ISn), and superior-most (SSn) points on a mid-modiolar section through the cochlear canal are indicated in (**a**). These landmarks measured on consecutive radial slices through the modiolus form the basis for reconstructing the cochlear spirals. A wireframe reconstruction of a moving average smoothed version of a cochlea measured from a CT scan using eight landmarks is shown in (**b**). The thick red line shows the electrode trajectory, which enters the cochlea from a cochleostomy. A reconstruction of the cochlea after more processing and smoothing of the measured landmarks, which includes an estimation of the inner structures, is shown in (**c**). Segmentation is illustrated in (**d**) where areas (or volumes) with similar gray scale values are grouped and assigned specific electrical properties. The gray scale in the image is for illustrative purposes and does not reflect gray scale values in, for example, a CT image (refer to Fig. 11.2 for a typical gray scale reference). Segmentation of the electrode array is illustrated in (**e**) with a full cochlear and electrode reconstruction from segmentation shown in (**f**)

Segmentation-based modeling also requires preprocessing of the image stack to provide an optimal view of these structures. The segmentation process may be manual, semi-automatic where an observer supervises automatic segmentation processes (Sakellarios et al. 2017), or fully automatic where machine learning algorithms are employed (Vaidyanathan et al. 2021). Fig. 11.5d illustrates segmentation of the cochlea. Reconstruction is performed by stacking the 2D surfaces that delineate each tissue type in 3D. A voxel-based reconstruction can then be done by redefining each pixel as a voxel, or a surface extraction method may be employed to approximate the boundaries among tissue types (Sakellarios et al. 2017). Figure 11.5e shows the results of a segmentation of an electrode array, while Fig. 11.5f shows a model of the inner ear, internal auditory canal, and electrode that was created using a segmentation approach. Software tools such as 3D Slicer,[1] ITK-SNAP,[2] or MATLAB[3] are available for image processing applications. For

[1] https://www.slicer.org/

[2] https://www.itksnap.org/pmwiki/pmwiki.php

[3] https://es.mathworks.com/products/matlab.html?s_tid=hp_products_matlab

segmentation of head and brain tissues, dedicated software such as Freesurfer,[4] FSL,[5] and CAT-12[6] exist.

The two methods have somewhat different use cases in CI modeling. Landmarking, because of its computational and data efficiency, is effective for creating person-specific VE models for clinical application within a limited timeframe. It also facilitates simple morphometric analysis to quantify the effect of interpersonal variance in anatomy on various CI performance measures. Its robustness allows the use of retrospective clinical imaging data where quality cannot be controlled—an important factor for clinical applications where cost and radiation exposure need to be considered. On the other hand, because of their high level of detail, segmentation-based models are preferred for the purpose of surgical planning and "once-off" construction of highly detailed VE models of the cochlea and extracochlear structures.

11.2.4.2 Reconstruction of the Electrode Array

Neural excitation is sensitive to the electrode neural separation distance (Malherbe et al. 2016; Zhou et al. 2023), and hence, an accurate estimation of the intracochlear electrode array location relative to the surviving nerve fibers is crucial. In cases where destructive imaging methods such as sectioning are undesirable, the 3D trajectory of the electrode array needs to be approximated from CT images and, in particular, from low-resolution clinical CT images in the case of live individuals. The challenge in estimating the 3D electrode trajectory and exact electrode contact locations is rooted in the beam-hardening artifacts caused by the electrode contacts, and partial volume effects due to the small size of the electrode contacts relative to the resolution of the scans (Zhao et al. 2019). For person-specific modeling, several methods have been used to estimate the intracochlear trajectory of the electrode array such as calculating the centroid of the pixels with the highest absorption rate on successive mid-modiolar slices through the cochlea (e.g., Malherbe et al. 2015, 2016), expert manual localization (Schurzig et al. 2023), and automated methods (Zhao et al. 2019).

Preknowledge of the electrode design and dimensions is essential for model construction irrespective of the modeling approach, firstly because the design reflects a specific neural targeting strategy that must be captured in a model, and secondly because it is difficult to pinpoint the exact size and location of electrode contacts from CT images. Table 11.1 provides a summary of relevant modeling parameters for the electrode arrays offered by MED-EL, Advanced Bionics (AB), Cochlear Ltd., and Oticon Medical at the time of writing this chapter.

[4] https://surfer.nmr.mgh.harvard.edu/

[5] https://fsl.fmrib.ox.ac.uk/fsl/docs/#/install/index

[6] https://neuro-jena.github.io/cat12-help/

Table 11.1 Electrode array modeling parameters as provided by manufacturers or derived from technical documentation (MED-EL, Oticon Medical, AB, Cochlear) and the literature (Dhanasingh and Jolly 2017; Ertas et al. 2022)

Electrode array name (type)	Active length (mm)	*N*° of contacts	Type of contact	Contact shape	Contact length (mm), dimensions ($w \times h$) (mm), or area* (mm^2)	Interelectrode distance (mm)	Carrier cross-sectional ($w \times h$) dimensions or diameter (mm)
Cochlear: Slim Straight CI622 and Slim 20 (straight)	25 (19.10)	22	Single	Half-band	0.25 (A: 12 contacts) 0.3 (B: 10 contacts)	0.85	0.35 × 0.25 (A) 0.60 × 0.50 (B)
Cochlear: Contour Advance Perimodiolar CI612 (precurved)	19.00 (14.25)	22	Single	Half-band	0.25 (A: 12 contacts) 0.3 (B: 10 contacts)	0.6	0.5 (A) 0.8 (B)
MED-EL: Standard (straight)	31.50 (26.40)	24	Double	Rounded rectangular	0.14*	2.40	0.3
MED-EL: Flex 28 (straight)	28 (23.1)	19	7 double (B) 5 single (A)	Rounded rectangular	0.13* (A) 0.14* (B)	2.10	0.35 × 0.45 (A) 0.3 (B)
AB: HiFocus Slim J (straight)	23 (20)	16	Single	Planar	0.14*	1.30	0.55 × 0.26 (A) 0.76 × 0.56 (B)
AB: HiFocus MidScala (precurved)	18.5 (15)	16	Single	Planar	0.14*	0.975	0.50 × 0.50 (A) 0.70 × 0.70 (B)
Oticon Medical: EVO (straight)	25.0 (24.0)	20	Single	281° partial band	0.46* (A: 10 contacts) 0.60* (B: 10 contacts)	1.20	0.40 (A: 10 contacts) 0.50 (B: 10 contacts)

Apical and basal dimensions are indicated by (A) and (B), respectively

11.3 Mathematical and Physical Foundations of Volume Conduction Models

11.3.1 Volume Electric Governing Equations

The physics that underlies CI stimulation can be described as that of a conducting volume containing an array of electrodes delivering electrical stimuli to it. It is commonly accepted to assume a static or a quasi-static assumption to formulate the problem (Gaugain et al. 2023), which can be solved through Poisson or Laplace equations. These equations are fundamental and appear in a wide variety of applications, such as electrostatics, magnetostatics, and current conduction. The formulations below are suitable for describing electric potentials in conductive biological tissues driven by electrical currents and represent the bioelectric forms of the Poisson and Laplace equations.

$$\text{Poisson}: \nabla\cdot((\sigma + \varepsilon\partial_t)\nabla\phi) = -I_s \overset{\text{EQS}}{\rightarrow} \nabla\cdot(\sigma\nabla\phi) = -I_s \tag{11.1}$$

$$\text{Laplace}: \nabla\cdot((\sigma + \varepsilon\partial_t)\nabla\phi) = 0 \overset{\text{EQS}}{\rightarrow} \nabla\cdot(\sigma\nabla\phi) = 0 \overset{\text{Homogeneous}}{\rightarrow} \nabla^2\phi = 0 \tag{11.2}$$

where I_s is the impressed (volume) current source density in (A/m^3), ϕ is the electric potential in (V), σ is the electrical conductivity in (S/m), and $\varepsilon = \varepsilon_0\varepsilon_r$ is the permittivity in (F/m): ε_r the relative permittivity and ε_0 the permittivity of free space. The electric currents injected through the intracochlear electrodes cause a net flow of ions through the cochlear and surrounding head tissues, which can be described in terms of an electric field, $\mathbf{E} = -\nabla\phi$ (V/m), or by Ohm's law, through a current density, $\mathbf{J} = \sigma\mathbf{E}$ (A/m^2). Notice that Poisson's equation is a generalized version of Laplace's equation with a nonzero term $-I_s$. The Laplace formulation represents sources through boundary conditions, whereas the Poisson formulation represents distributed volume sources.

In a homogeneous medium, the values of ε and σ are spatially constant and can be taken outside the divergence operation ($\nabla\cdot$). The electro-quasi-static (EQS) approximation assumes that displacement currents are negligible compared to conduction currents (i.e., $\varepsilon\partial t \ll \sigma$). In the frequency domain, this assumption holds when $\varepsilon\omega \ll \sigma$, with $\omega = 2\pi f$ denoting the angular frequency of the applied stimulus. For human living tissues, it is generally accepted that the EQS approximation is valid in the frequency range between 100 Hz and 10 kHz (Gaugain et al. 2023). For cochlear implants, the EQS assumption has been justified based on early experimental data by Spelman et al. (1982), who reported resistive impedances inside and outside of the scala tympani between 8 Hz and 12.5 kHz. Under this assumption, the solution of the model is time-independent, and the input time-varying current source can be substituted by independent DC sources at each given time.

Finally, Poisson and Laplace equations are closed with the appropriate boundary conditions for the specific problem under study. Additionally, solving the physics in a 3D geometry requires a numerical solver and a mesh that allows operating with small surface and/or volume elements, addressed in the next sections.

11.3.2 Source Modeling and Boundary Conditions

Source and boundary conditions complete the definition of the physics in VE models. For Laplace (or Poisson) equations, the continuity of the electric potential ($V_1 = V_2$) and the continuity of the normal component of the current density ($(\mathbf{J}_1 - \mathbf{J}_2) \cdot \mathbf{n} = 0$) across the internal boundaries are applied (Makarov et al. 2016).

At the external surfaces, electric isolation ($\mathbf{J} \cdot \mathbf{n} = 0$) is usually applied, depending on the specific implementation of the surrounding cochlear model. When models are restricted to the cochlea, grounded ($V = 0$) spherical, cubic, or ellipsoid bone domains are commonly used to describe the extracochlear environment (see Fig. 11.6a). Previous studies have investigated the impact of different geometries (Malherbe et al. 2015) and boundary conditions (Wong et al. 2016) on current spread and potential distribution. Current flows were noticeably different for different grounding schemes beyond the scala tympani. To overcome such limitations, full-head models have also been proposed in the literature (Malherbe et al. 2015; Tran et al. 2015; Mangado et al. 2018a; Bai et al. 2019; Badenhorst et al. 2021; Callejón-Leblic et al. 2024).

Boundary conditions at the location of active and ground electrodes allow the definition of voltage values and source and sink currents at these locations. Normally, CI devices apply a constant current to the active contacts ($+I_0$). Mathematically, this can be defined as the integral of the normal electric current density through the electrode surface being equal to the current I_0, i.e., $\int_{\partial\Omega} \mathbf{n}.\ \mathbf{J}\ dS = I_0$. At the reference(s) sink electrode(s), the same current with opposite sign $-I_0$ is applied, depending on the CI stimulation mode. Different stimulation modes are characterized by the way that the electric current returns to one or several electrodes, either located inside the cochlea or in the temporal bone. In monopolar (MP) mode, the return electrode is usually an extracochlear plate that forms part of the implant casing or a rod-shaped electrode placed lateral to the cochlea in the head (Fig. 11.6f). Current focusing strategies and multipoles, such as bipolar (BP), tripolar (TP), and partial tripolar (pTP) modes, aim to reduce current spread by using intracochlear returning electrodes in the same array (see Chaps. 1 and 2).

11.3.3 Electrical Properties of Tissues

The electrical properties of tissues relevant to VE models of the cochlea and surrounding structures are a collection of values obtained from various species and measuring techniques over an extended period. It is difficult to measure the values of

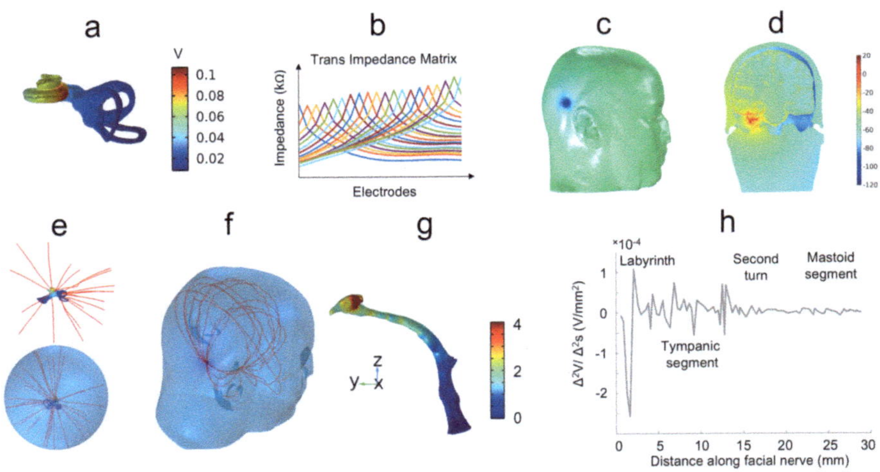

Fig. 11.6 Examples of different types of visualization and interpretation of the simulated results for a medial electrode, stimulated with a current of 100 μA. The predicted voltage distribution (V) through the cochleo-vestibular region is shown in (**a**). Simulated transimpedances along the electrode array are shown in (**b**) in units of V per unit current (Ω). Current density J (Am^{-2}) at the scalp is shown in (**c**) and at a coronal section of the head near the cochlea. The results are represented on a dB scale (0 dB indicates 1 Am^{-2}) (**d**). Current streamlines from an intracochlear active electrode to a grounded surrounding sphere are shown in (**e**) and to a return electrode placed in the head in (**f**). EF distribution over the facial nerve in units of V/m is shown in (**g**). In (**h**), the activating function calculated through the outer surface of a modeled FN fiber is shown

cochlear tissues because of the small size of the inner ear and its complex structure. The electrical properties that have been used in VE models of the cochlea and external model domain are given in Table 11.2 as compiled from the literature (Misrahy et al. 1958; Strelioff 1973; Frijns et al. 2000, 2001, 2009; Hanekom 2001, 2005; Rattay et al. 2001b; Saba 2012; Kalkman et al. 2014; Malherbe et al. 2015; Tran et al. 2015; Wong et al. 2016; Nogueira et al. 2016, 2017; Dang 2017; Mangado et al. 2018a, b; Bai et al. 2019; Potrusil et al. 2020; Badenhorst et al. 2021; Lei et al. 2021; Ramos-De-Miguel et al. 2022; Cakmak et al. 2023; Callejón-Leblic et al. 2024). The temporal bone and otic capsule are two key structures that have been studied extensively (e.g., Malherbe et al. 2015). The otic capsule is a dense and hard bone that provides a high-impedance barrier to current flow from the cochlear canal. Table 11.2 bulks the resistivity of the otic capsule with that of the temporal bone in studies where the otic capsule was not specifically delineated.

Membranes can also be modeled as contact impedances (Thiselton and Hanekom 2024) instead of volumes. These are specified through the values reported in Table 11.2 in combination with the thicknesses of the membranes: Basilar membrane is 0.004 mm, Reissner's membrane is 0.0001 mm, and stria vascularis is 0.04 mm. Thin bone structures, such as the spiral lamina, may also be modeled in this way.

Table 11.2 Tissue electrical properties used in purely resistive VE modeling studies

Material or tissue	Common (S/m)	Other values (S/m)
Perilymph	1.43	1.42, 1.79
Endolymph	1.67	
Ossified cochlea	0.154	
Fibrous/encapsulation tissue	0.159	
Organ of Corti	0.012	0.33
Spiral ligament	1.67	
Basilar membrane	0.0125	0.0267, 0.0625[a]
Stria vascularis	0.0053	
Reissner's membrane	9.8×10^{-5}	0.01
Round window	0.001	
Neural tissue (spiral ganglion; auditory nerve)	0.33	0.2, 0.3, 0.35
Anisotropic: axial	0.33	
Anisotropic: transversal	0.067	
Modiolus	0.2	0.01, 0.0334
Spiral lamina/modiolar bone		0.2, 0.0334, 0.085
Temporal bone (external medium)	0.016	0.00118, 0.01, 0.013, 0.014, 0.0156, 0.2, 0.23, 0.3, 0.0728
Otic capsule (normal)	0.156	0.008, 0.013, 0.014, 0.016
Otic capsule (demineralized; otosclerotic)		0.0192, 1.6
Facial nerve	0.33	0.3
Ventricles (CSF)	1.79	
Eyeballs		0.5, 1.5
Brain: gray matter	0.33	
Brain: white matter	0.15	
Brain: gray matter and white matter	0.2	
Cerebellum	0.16	
Brainstem/spinal cord	0.16	
CSF	1.79	
Skull/bone	0.013	0.02
Scalp/skin	0.33	0.42
Sinuses	6.25×10^{-5}	
Blood		0.63, 0.66
Electrode carrier	1.00	0, 0.1, 0.99, 10^{-5}, 10^{-7}, 10^{-15}
Electrode contact	1000	100, 10^{6}, 9.43×10^{6}

The most commonly used (or only) value, as well as other values (or different values if there is no commonly used value), is indicated

[a]The value of 0.0625 commonly used for the basilar membrane is a fivefold scaled value originating from studies where the thickness of the basilar membrane was scaled up for meshing purposes

While most studies use a quasi-static approximation ($\varepsilon = 1$), a few investigated capacitive effects as a result of the dielectric properties of tissues on the predicted electric fields. Inguva et al. (2015) approached frequency-domain simulations by using a biphasic square pulse input current, decomposed it into its main spectral components through Fourier transform, to solve for each harmonic component. The reconstructed output was the sum of the recombined solutions for each frequency. The authors found significant differences between the predictions based on a quasi-static assumption and those obtained through frequency-domain simulations, mainly because of the capacitive effects in neural tissue. Other works have also attempted to model the time-dependent properties of the electrode–electrolyte interface through equivalent basic RC circuits (Choi and Wang 2014) and Cole models (Dang 2017). Molaee-Ardekani and Donahue (2023) have suggested that the frequency characteristics of this interface are important to accurately model different transient charge mechanisms arising at it.

11.3.4 Meshing and Discretization

Solving the physics equations in complex 3D anatomical models requires the discretization of the geometry of the model into a mesh. The first step usually involves the creation of a surface mesh that discretizes the surface boundaries of the anatomical model into a set of small elements (e.g., triangles). The second step is the discretization of the internal volume domains with volume elements (e.g., tetrahedra). The nodes of this set of triangles and tetrahedra constitute an array of points where the electric field (EF) is calculated through numerical techniques such as the FEM. Specialized simulation software programs usually provide in-built meshing tools to generate and manipulate meshes based on model geometry and specific physics.

Image segmentation software programs usually provide the output of the segmentation as a triangle surface mesh (e.g., through a stereolithography STL file), which contains the arrays of triangles, edges, and nodes that form the surface mesh. This surface mesh should not contain intersecting, inverted, or missing triangles, and these should be small enough to accurately describe the variation of physical variables (such as the EF). Additionally, the mesh should be smooth enough and not contain irregularities that may produce spurious EF values (Makarov et al. 2016).

Complex geometries containing anatomically adjacent or contacting structures, such as the scala tympani and vestibuli and the membranes, as well as the inserted electrode array that may touch the cochlear walls or even penetrate some of the structures, can cause meshing errors (Wong et al. 2016; Mangado et al. 2018a). A manifold boundary mesh, with no intersecting edges or faces, is often a prerequisite for FE analysis. For this reason, the geometry of the anatomical structures often has to be manipulated to allow correct meshing. However, such corrections may constrain the modeling of realistic anatomical intersections between tissues creating a trade-off between anatomical accuracy and meshability. Two different approaches coexist to handle tissue intersections: (i) a surface-based approach, in which a non-manifold mesh suitable for FE analysis needs to be constructed; these contain "cleaned" intersected faces that are shared by two intersecting domains simultaneously (Callejón-Leblic et al. 2024), and (ii) a volume-based approach, in which the creation of a surface mesh is omitted and the volume mesh is directly built from the segmented image using tetrahedra, cubes, or hexahedra (Pietroni et al. 2022). Although the latter approach also presents the advantage that tissue properties can be more easily assigned to separated domain tissues, the lack of smooth boundaries between tissues may produce higher EF discontinuities (Miranda et al. 2018).

When the cochlea is embedded within a full-head model or surrounding domain, fine mesh elements are needed to adequately capture the geometry of small tissue details, while coarse elements are adequate for larger head structures. Adaptive mesh refinement preserves numerical accuracy while minimizing computational cost (Tran et al. 2015). For the sake of validation, metrics of mesh quality provide information about the size, resolution, and uniformity of the mesh elements in the model (Pepper and Heinrich 2017).

11.3.5 Numerical Methods and Simulation

As explained in the previous section, the mesh divides the modeling domain into elements representing small sections of the domain over which the variables are assumed to vary in a computationally simple way, e.g., linearly. Numerical methods, such as the finite element method (FEM) and boundary element method (BEM), are required to solve the partial differential equations discussed in Sects. 11.3.1 and 11.3.2. The FEM is a domain-based numerical method, i.e., it operates over the entire model domain using, for example, triangular elements in 2D or tetrahedra in 3D, while the BEM is a boundary-based numerical method that operates on edges in 2D and surfaces in 3D.

The FEM is well suited for solving VE problems with complex geometries involving irregular shapes because it discretizes the entire domain. It can also accommodate heterogeneous material properties. Its main disadvantage is that the method can be computationally expensive when a large number of elements are required to discretize the entire domain. On the other hand, the integral formulation of the BEM becomes challenging to implement for complex geometries and material inhomogeneities, making it more suitable for problems with simple boundaries and homogeneous materials. Because only the boundaries are discretized, the problem dimensionality and, therefore, the system of equations that need to be solved (for problems with uncomplicated geometries and material properties) are reduced relative to those of the FEM, making it more computationally efficient.

Although computationally typically more expensive than the BEM, the majority of VE cochlear models use the FEM. The primary reason for this is the advantage that the FEM offers in solving the partial differential equations (PDEs) over a relatively accurate representation of the irregular anatomy of the cochlea. Furthermore, both the surface-based and volume-based discretization approaches are suitable for FE analysis in contrast to the BEM, which can only be used with a surface-based approach.

Implementing numerical solvers for these models typically involves specialized software tools. Commonly used FEM software programs are Comsol Multiphysics and Ansys, which can handle the complex geometries and boundary conditions of the cochlear implant and surrounding tissues. Matlab is frequently used to script and process simulation data or manage FEM workflows with interactive links that may be set up with packages such as Comsol[7] and Ansys.[8] BEM is often custom-written in software programs such as Matlab; however other packages such as Bempp (Betcke and Scroggs 2021) are also available.

After geometry definition, discretization of the domain, and assignment of material properties and boundary conditions, the type of study must be selected, typically a volume conduction analysis (Malherbe et al. 2015) or an electrostatic analysis (Callejón-Leblic et al. 2024). These studies require setting up appropriate sources and boundary conditions, as explained in Sect. 11.3.2.

11.3.6 Visualizing and Interpreting the Results

11.3.6.1 Volume Electric Model Outputs

Predictions from VE models may be visualized based on direct outputs of the models, e.g., electric potential distributions, EF and current density patterns, or by further modeling to represent neural responsiveness, e.g., activating functions and neural excitation profiles.

Electric potential (voltage) distributions show the spatial distribution of electrical potentials within the cochlea resulting from CI stimulation (Fig. 11.6a). They illustrate how voltage levels vary across different cochlear regions as a function of the electrical parameters (geometric, material, and stimulation properties) of the model (Hanekom 2005; Nogueira et al. 2016). An alternative way of visualizing the voltage spread is through transimpedances or voltage matrices that correspond to clinical measurements provided by CI software. When an electric current is injected through an active intracochlear electrode, it spreads throughout the cochlear fluids and can be measured on other nearby electrodes (Fig. 11.6b). Purely resistive models have been shown to reasonably predict voltages at the off-stimuli contacts, while the reactive components of the electrode–electrolyte interface need to be considered at the stimulating electrode (Molaee-Ardekani and Donahue 2023).

EF distributions may be derived from the electric potential distributions by applying the gradient operator (∇) to represent the spatial rate of change of the potential field, while current density is related to the EF through the medium's resistivity. These vector fields provide insight into the flow and hence the spread patterns of electrical potential and current as a function of the properties and composition of the cochlear and surrounding tissues (e.g., Tran et al. 2015; Kalkman

[7] https://www.comsol.com/livelink-for-matlab

[8] https://www.mathworks.com/matlabcentral/fileexchange/53032-interactive-link-between-matlab-and-ansys

et al. 2016; Badenhorst et al. 2017). Callejón-Leblic et al. (2024) provided current density plots through the complete head (Fig. 11.6c, d) and integrated the current density over different tissues on a coronal plane through the head between the cochlea and MP return electrode to predict the proportion of current that is shunted through each of the tissues given a specific electrode configuration and stimulation protocol. Current streamlines show the main current pathways through cochlear and head tissues from the active to the MP return electrodes. Figure 11.6e, f, illustrates the differences in current streamlines for a model consisting of a simplified grounded surrounding spherical domain versus a realistic full-head model with a return electrode placed in the lateral head. Figure 11.6g shows the EF distribution over the facial nerve upon stimulation with a medial electrode in MP mode.

11.3.6.2 Peripheral Neural Excitation

If a neural model is not integrated with the VE model, the activating function (Rattay 1986) provides a means of estimating the likelihood that neurons would be excited at specific spatial locations. The activating function quantifies the influence of an external electric field on a neuron by calculating the second spatial derivative of the extracellular potential along the length of a neuron, indicating regions where the neuron is likely to be excited. Figure 11.6h shows the activating function calculated through a modeled nerve fiber in the outer surface of the facial nerve for MP stimulation on a medial electrode contact (Callejón-Leblic et al. 2024).

VE models of the cochlea are often combined with ANF models to convert the output of the VE model to predicted neural excitation patterns (Frijns et al. 1995; Hanekom 2001; Rattay et al. 2001b). Voltage, field, and current distributions in VE models lend themselves to integration with biologically based ANFs that are able to take a distributed input. Typical models that have been used in this context are the Hodgkin–Huxley-based compartmental model from Rattay et al. (2001a), the generalized Swartz–Eikhof–Frijns model of Frijns et al. (1994), and variations on these, e.g., Smit et al. (2009) and Badenhorst et al. (2017). However, biophysical models formulated as leaky-integrator and ion channel gating models are computationally demanding, which limits the number of neurons that may be realistically implemented, as well as the duration over which neural responses may be calculated. This, in turn, restricts the predictions that can be made with these models to single-fiber responses, which can be combined into simple population response models such as neural excitation profiles (Badenhorst et al. 2017) and electrically evoked compound action potentials (ECAPs) (Briaire and Frijns 2005). To predict more complex neural responses, such as psychoacoustic behavior or the time-varying excitation behavior for a segment of speech, based on the 3D VE model outputs, the computational load of the neural models must be drastically reduced. De Nobel et al. (2024) developed a biophysics-inspired phenomenological ANF model that may be combined with 3D VE models and is significantly faster than the traditional models. Ramos-De-Miguel et al. (2022) also investigated the use of a phenomenological computational model to predict the ECAP from the output of a 3D VE model. This model was not aimed at improving the speed of the simulations but rather at providing an accurate, user-specific prediction of the ECAP as a function of time.

The modeling of the neural response to electric stimulation of the cochlea is overviewed in the next chapter.

11.3.7 Validation and Calibration

Volume electric models may be validated directly or indirect against different clinical and experimental measures. Physical models that aim to represent the intracochlear environment have been used to investigate the characteristics of the electric field around stimulation electrodes. Black et al. (1981) used a special measurement electrode to measure ground currents returned from an active electrode during CI surgery in humans, while Ifukube and White (1987) performed direct measurements of electric fields around CI electrodes in saline-filled test tubes using a probe electrode. Kral et al. (1998) used a similar approach to compare electric field measurements along the length of a Nucleus-22 electrode array in a saline bath, feline cadaveric cochleae, and live cats. They found that the measurements made on the inactive electrodes of the electrode array matched the measurements made with a separate probe. Studies like these formed the basis for present-day clinical measures that determine intracochlear potential spread by exciting one active electrode and measuring the electric potential at the inactive electrodes (Vanpoucke et al. 2004; Stahl et al. 2022). It is easy to simulate this setup in VE models, allowing validation of the model through a comparison with measurements made in human CI users.

Impedance measurements, where the voltage measured at an inactive electrode contact is divided by the stimulus current, are also available in clinical software and provide a direct method of validation. Purely resistive models have been shown to reasonably predict impedances at off-stimuli electrode positions, in the range between 0.2 and 4 kΩ (Rattay et al. 2001a; Malherbe et al. 2015; Tran et al. 2015; Mangado et al. 2018a; Bai et al. 2019; Lei et al. 2021; Jiang et al. 2021; de Rijk et al. 2022; Kopsch et al. 2022; Cakmak et al. 2023). However, at the active electrodes, the electrode contact impedance due to reactive components needs to be modeled (Molaee-Ardekani and Donahue 2023) to fit higher impedances reported in the range of 4–10 kΩ (Cheng et al. 2022; Hrncirik et al. 2023). Measurements of the impact of fibrosis on transimpedance matrices (Leblans et al. 2022) may also be useful as a validation measure. Malherbe et al. (2015) and Kalkman et al. (2014) used clinical data to adjust the bone conductivity that should be used in VE models of cochlear implants, thereby adding to the validation of the models.

Another measure, the attenuation constant of potential spread, may be calculated by fitting the simulated potential profiles to the exponential function $V(x) = Ae^{-x/\lambda} + C$ (Malherbe et al. 2015; Lei et al. 2021; Jiang et al. 2021), where A is the amplitude, x is the distance from the active electrode, λ is the attenuation constant in millimeter, and C is the baseline offset of the voltage profile at the stimulating electrode. Other measures, such as the half-bandwidth of simulated transimpedances (defined as the width where the profile is at 50% of its peak amplitude), may also be used as validation measures against which to compare experimental data. In the literature, models predict asymmetric profiles with broader voltage decays toward the base than toward the apex (Bai et al. 2019; Callejón-Leblic et al. 2024), possibly explained due to anatomical differences in the tapering of the cochlea, with reduced volume and thus less longitudinal current spread around apical electrodes.

More recently, full-head model approaches have provided the possibility to validate extracochlear current flow, for instance, through simulated scalp and facial nerve potential and EF against surface measurements such as EEG data (Tran et al. 2021; Callejón-Leblic et al. 2024) and the stimulus artifact observed in EMG data measured from the facial muscles (Van Der Westhuizen et al. 2022).

Measurements of neural excitation behavior can also be used as validation measures. These measures are indirect as they require a neural model to be added to the VE model, as described in the previous section. Inaccuracies in the neural model may thus obscure the true fidelity of the VE model and vice versa. Spatial tuning curve measurements on single fibers in the inferior colliculus of a guinea pig (Rebscher et al. 2007) have been used to validate the place and magnitude of stimulation thresholds in a VE-neural compound model (Malherbe et al. 2013), while single-fiber data measured in cat (Javel and Shepherd 2000) have been used to validate the neural response predictions of a VE-neural model (Badenhorst et al. 2017). Briaire and Frijns (2005) validated their model against ECAP data measured in CI users. Malherbe et al. (2016) and Snel-Bongers et al. (2013) directly compared predicted neural excitation thresholds with perceptual thresholds, although care should be taken in interpreting peripheral predictions against perceptual data. Facial nerve stimulation thresholds relative to electrical stimulation thresholds for individual CI users also provide a valuable means of model validation (Badenhorst et al. 2021; Van Der Westhuizen et al. 2022).

However, despite the availability of these measures, validation remains scarce and challenging in the CI literature. The main reasons for this are that direct measurements of voltage, EF, and current distributions in live CI users are not possible, and errors may accumulate if additional models, such as a neural model, need to be added to a VE model to simulate a measurable response, e.g., ECAP or perceptual thresholds. This also adds an additional layer of complexity and computational load to the models.

11.4 Models in the Literature, Applications, and Future Challenges

11.4.1 Models in the Literature, Findings, and Applications to Date

This section summarizes relevant models published in the literature, highlighting their main findings to date. It is important to note that, despite the coexistence of heterogeneous modeling approaches in the literature—characterized by variations in imaging techniques, levels of anatomical detail, boundary conditions, numerical implementations, and electrical conductivity values—generalized findings have consistently emerged across studies regarding the influence of individual cochlear morphology, the exact position and configuration of electrodes, the effect of different CI stimulation modes, and tissue resistivity properties. Models have also provided relevant insights about extracochlear current pathways and side effects such as facial nerve stimulation. These findings support the capacity of VE models to reasonably predict CI electrical stimulation behavior and their potential applicability as clinical tools. The main insights gained on these topics are summarized below.

11.4.1.1 Influence of Individual Cochlear Morphology

The degree of anatomical detail and personalization has been shown to significantly influence modeled outcomes and their capability to fit clinical data (Malherbe et al. 2016; Nogueira et al. 2016). For example, models have demonstrated the impact of spiral cochlear anatomy on the asymmetric distribution of the electric field along the cochlea, with higher intracochlear peak voltages observed in the apex compared to the base during monopolar (MP) stimulation (Bai et al. 2019). Additionally, faster voltage decay toward the basal turn, potentially due to its larger cross-sectional area, has been reported (Kalkman et al. 2014; Lei et al. 2021). This asymmetric voltage pattern has been found to be dependent on individual parameters such as cochlear canal shape (Kalkman et al. 2014; Malherbe et al. 2015; Mangado et al. 2018a), lumen diameter, and taper ratio (Lei et al. 2021). Thus, the relationship between individual cochlear morphology and spatial electric field (EF) selectivity—which seems to be crucial for effective speech perception—could be further understood and optimized by employing subject-specific VE models.

11.4.1.2 Influence of Electrode Position and Configuration

VE models have also shown the dependence of EF spread on the exact position and geometrical configuration of the electrode inside the scala tympani and its distance to the modiolus (Briaire and Frijns 2000; Hanekom 2001; Kalkman et al. 2014, 2015). The effect of electrode encapsulation and fibrous scar tissue on excitation patterns has also been evidenced for perimodiolar and lateral electrodes (Hanekom 2001, 2005). Generally, these models support the observation that thresholds decrease when the electrode is closer to the modiolus (Briaire and Frijns 2006) and that current can be focused and steered using multipolar electrode configurations (Bonham and Litvak 2008). Thus, models can help guide the design of new optimized electrode designs that provide more focused EF patterns over targeted ANFs.

11.4.1.3 Influence of Different CI Stimulation Modes

Models prove to be useful *in silico* tools to analyze and test new stimulation modes (Dang 2017; Nogueira et al. 2017; Cakmak et al. 2023). Through a better characterization of the electrical properties and behavior of the auditory periphery, computational models have allowed us to gain further knowledge of the EF profiles generated by different CI stimulation modes (Briaire and Frijns 2000; Hanekom 2001; Rattay et al. 2001b; Kalkman et al. 2014; Callejón-Leblic et al. 2024). Simulations show a trade-off between EF localization and stimulation current required, with more localized EF profiles being achieved by current focusing modes such as BP and TP, however, at a cost of lower intracochlear voltages (Hanekom 2001; Rattay et al. 2001b; Saba 2012; Kalkman et al. 2014; Callejón-Leblic et al. 2024). The models therefore support the generalized observation that higher current thresholds are required in BP and TP than in MP to reach the same loudness sensation in CI patients (Landsberger et al. 2012; Vellinga et al. 2017).

11.4.1.4 Influence of Tissue Resistivities

Models predict a large influence of tissue conductivity properties, especially the bone (otic capsule), on EF magnitude and distribution through the cochlea and other nearby structures. Malherbe et al. (2015, 2016), Mangado et al. (2018a), Badenhorst et al. (2021), and Lei et al. (2021) showed that bone resistivity has the highest influence on the prediction of CI outcome measures such as frequency selectivity, cross-turn stimulation, and auditory threshold levels relative to other tissue resistivities. Simulations have also served to provide recommendations on the use of bone conductivity values for modeling purposes: a mean in vivo bone tissue resistivity value of 6.6 kΩ cm was calculated to better fit transimpedance clinical measurements in Lei et al. (2021), in agreement with the 6.5 kΩ cm previously reported in Malherbe et al. (2015). Therefore, models can be used to conduct sensitivity analysis to facilitate parameterization and calibration of model predictions against experimental data.

11.4.1.5 Extracochlear Current Pathways

VE models have helped characterize extracochlear current pathways through realistic full-head models. The first full-head model published in the literature (Tran et al. 2015) allowed the prediction of the main paths for intra- and extracochlear currents in MP mode. They quantified the percentage of current leaving the cochlea toward the modiolus (14%), the basal end of the cochlea (22%), and through the cochlear walls (64%). After leaving the cochlea, the returning current was broadly classified as flowing through the scalp or the cranial cavity. Callejón-Leblic et al. (2024) computed the percentage of extracochlear current passing through a transversal plane in the head midway between an active intracochlear electrode and an extracochlear return electrode located in the lateral part of the head accounting for a classical MP configuration mode. In addition to the skin (37%), other relevant current pathways through the CSF (27%), GM (23%), and WM (10%) were revealed. The prevalence of the different current pathways depended on skin and bone conductivity. The prediction of the true extracochlear current pathways is relevant for optimizing the focal pattern and power efficiency of the induced EF over targeted neurons while minimizing non-desired current leakage to other nontargeted areas.

11.4.1.6 Facial Nerve Stimulation

General and user-specific CI models have demonstrated their clinical suitability in assessing and managing the progress of CI-derived side effects and complications such as FNS. Modeling studies have found higher FNS thresholds for more conductive bone, such as osteosclerotic (Frijns et al. 2009) and ossified post-meningitic cochleae (Badenhorst et al. 2021). These models explained the mechanism of FNS as being driven by different slopes in FN and cochlear threshold changes as a function of bone conductivity. As bone conductivity increases, both ANF and FN excitation thresholds increase, but the latter do so less than ANF thresholds, which leads to a higher risk of FNS. Frijns et al. (2009) also reported a dependence on electrode site and type, with medial and lateral wall electrodes exhibiting an increased probability for FNS, possibly due to their proximity to the FN. Badenhorst et al. (2021) reported that ossified cochlear canals worsen such an increase in thresholds due to the decreased resistivity of the otic capsule. Callejón-Leblic et al. (2024) predicted maximum EFs and excitation functions at the labyrinthine segment of the FN for a medial stimulating electrode upon MP stimulation. They also observed that bone conductivity significantly impacted both the magnitude and distribution of EFs over the facial nerve, with higher and more widely distributed EFs through both the tympanic and mastoid FN segments as bone conductivity decreased. Analyses and insights from modeling studies may thus guide the design of interventions to decrease the probability of side effects such as spurious FNS.

11.4.1.7 Diagnostics and Care

Because of the diverse factors that may be manipulated, investigated, and tested *in silico* by computational models of the auditory periphery, they form an important component of CI modeling pipelines that allow complex analysis, simulation, and prediction of clinically relevant cochlear implant phenomena. Mangado et al. (2018a) created a model framework intended for predicting the prognosis of cochlear implantation using a detailed model of the cochlea. Leclère et al. (2023), on the other hand, presented a framework for assessing information transmission in cochlear implants but used a simplified model of the electrode–neuron interface. Hernández-Gil et al. (2025) used a patient-specific model of the cochlea and CI electrode to predict a multipolar stimulation strategy that would produce the most localized current density profile possible at target neurons, thereby seeking to predict optimal device programming parameters. Hanekom et al. (2023) explained the role of computational modeling in person-centered care of CI users suffering from nonauditory stimulation (NAS) based on detailed models of the cochlea and FN. The approach includes model-based diagnostics and *in silico* investigation of interventions. The principles could also be extended to include other forms of NAS such as vestibular co-stimulation. The models are therefore ideally suited to support clinical decision-making, device optimization, diagnostics, therapeutic interventions, and person-centered care.

11.4.2 Challenges, Limitations, and Future Work

Computational models have consistently evolved for two decades, increasing their complexity to more accurately reproduce the behavior of the electrically stimulated cochlea. However, outstanding challenges and limitations still exist, which have yet to be solved. For instance, the complex and time-consuming task of creating accurate anatomical models of the cochlea and inner ear structures is dependent on available image acquisition parameters and/or resolution. The fact that different modeling approaches coexist in the literature, which involve different levels of anatomical detail, and the lack of software harmonization make comparisons among modeling studies difficult. While studies on modeling of specific human cochleae derived from high-resolution imaging of cadavers are fairly common (Hanekom and Hanekom 2016), person-specific models of live subjects are less prevalent. CI manufacturers have developed software tools that facilitate presurgical analysis and selection of optimal electrodes upon conventional medical images of patient-specific cochlear anatomy, e.g., OTOPLAN (MED-EL) (Paouris et al. 2023), Nautilus (Oticon Medical and Cochlear Ltd) (Margeta et al. 2022), and CImago (Advanced Bionics) (Geiger et al. 2023). Based on both threshold segmentation and parametric statistical models, these are useful clinical tools to predict the position and length of the electrode array inside the estimated tympanic and vestibular scala and the probability of trauma and electrode translocation. However, these tools do not incorporate VE models of cochlear EF spread. The unavailability of clinically relevant person-specific models for live implant users is primarily because of the low resolution of such medical images; most modeling techniques rely on high-resolution imaging to extract the volumes or landmarks that define the anatomical structure. The lack of person-specific models of CI patients prevents the validation of simulated results, which remains very scarce in the literature. Lack of validation remains one of the most relevant challenges the CI research community faces.

While there is consensus on the reliability of purely resistive models in predicting common electric responses to CI electrical stimulation, it is also agreed that they still fail to accurately reproduce *in vivo* and clinical data (Nogueira et al. 2016). This failure is probably in part due to uncertainty at different levels, e.g., electrical tissue conductivities, boundary conditions, fibrosis encapsulation phenomena, position of the electrode contacts in the scala tympani, the healthy or degenerated state of the ANFs, as well as uncertainty on the exact excitation mechanisms of spiral ganglion neurons, among others.

Future work may be directed toward the development of new techniques and algorithms for better characterization of individual electrical conductivities, as well as a more comprehensive characterization of time-dependent properties of tissues, primarily those of the electrode–electrolyte interface, which may play a pivotal role when predicting clinical data using CI user-specific models (Molaee-Ardekani and Donahue 2023). For developing personalized anatomical models, the creation of harmonized semi-automated modeling pipelines that can be systematically used in clinical practice is crucial. Currently, open repositories with images from CI patients and open VE models are scarce (Gerber et al. 2017; Sieber et al. 2019; Callejón-Leblic et al. 2024). The creation of open repositories with images from CI patients would increase the number of models and measures, thus expanding the number of novel metrics for accuracy and validation.

More recently, some authors have proposed phenomenological and artificial intelligence (AI)-driven approaches to match simulated results to in vivo data (Ramos-De-Miguel et al. 2022; de Nobel et al. 2023, 2024). While these approaches often lack interpretability, making identifying the specific physical mechanisms responsible for the observed predictions challenging, they may open the way toward a cost-saving computational paradigm which accurately fits clinical data.

In conclusion, advances in computational modeling of the peripheral auditory system have greatly improved our understanding of the functioning of CIs. These models capture the characteristics of the cochlea, extracochlear environment, and device to describe the biophysical interface through which electric hearing is realized. They provide valuable information about how stimulation currents and fields distribute throughout the tissue volumes, eventually giving rise to specific spatiotemporal excitation patterns of the ANFs. Based on these insights, the models provide a means to optimize CI electrode design, which may lead to the customization of an implant and sound coding strategies for specific individuals.

With the rapid development of technology and strong transdisciplinary teams, the future of modeling holds excellent promise. Integrating these models with advanced imaging techniques and machine learning algorithms is envisioned to support innovative personalized hearing rehabilitation. With the continuous evolution of computational models, we will progressively unravel the mysteries of the hearing system, ultimately reaching a level of understanding that will allow us to provide natural-sounding hearing for all.

Acknowledgements Research was supported in part by the *Programa Estatal para la Investigación y el Desarrollo Experimental*, under the *Plan Estatal de Investigación Científica, Técnica y de Innovación* 2024–2027; project funded by PID2024-157499OA-C32 and MICIU/AEI/10.13039/501100011033/FEDER, Spain.

The authors acknowledge the use of Microsoft Copilot for proofreading and enhancing the readability of the manuscript. All suggested changes were carefully reviewed and edited. The authors assume full responsibility for the content of the final publication.

References

Advanced Bionics HiRes ™ Ultra 3D. https://www.advancedbionics.com/content/dam/advancedbionics/Documents/Regional/en-gb/Professional/Ultra-3D-Tech-Specs-Brochure.pdf. Accessed 3 May 2025

Athanasopoulos M, Samara P, Athanasopoulos I (2024) Advances in 3D inner ear reconstruction software for cochlear implants: a comprehensive review. Methods Protoc 7:1–13. https://doi.org/10.3390/mps7030046

Badenhorst W, Hanekom T, Hanekom JJ (2017) Analysis of a purely conductance-based stochastic nerve fibre model as applied to compound models of populations of human auditory nerve fibres used in cochlear implant simulations. Biol Cybern 111:439–458. https://doi.org/10.1007/s00422-017-0736-8

Badenhorst W, Hanekom T, Gross L, Hanekom JJ (2021) Facial nerve stimulation in a post-meningitic cochlear implant user: using computational modelling as a tool to probe mechanisms and progression of complications on a case-by-case basis. Cochlear Implants Int 22:68–79. https://doi.org/10.1080/14670100.2020.1824431

Bai S, Encke J, Obando-Leitón M et al (2019) Electrical stimulation in the human cochlea: a computational study based on high-resolution micro-CT scans. Front Neurosci 13. https://doi.org/10.3389/fnins.2019.01312

Benson JC, Carlson ML, Lane JI (2020) MRI of the internal auditory canal, labyrinth, and middle ear: how we do it. Radiology 297:252–265. https://doi.org/10.1148/RADIOL.2020201767

Betcke T, Scroggs MW (2021) Bempp-cl: a fast Python based just-in-time compiling boundary element library. J Open Source Softw 6:2879. https://doi.org/10.21105/JOSS.02879

Black RC, Clark GM, Patrick JF (1981) Current distribution measurements within the human cochlea. IEEE Trans Biomed Eng BME 28. https://doi.org/10.1109/TBME.1981.324668

Bonham BH, Litvak LM (2008) Current focusing and steering: modeling, physiology, and psychophysics. Hear Res 242. https://doi.org/10.1016/j.heares.2008.03.006

Briaire JJ, Frijns JHM (2000) Field patterns in a 3D tapered spiral model of the electrically stimulated cochlea. Hear Res 148. https://doi.org/10.1016/S0378-5955(00)00104-0

Briaire JJ, Frijns JHM (2005) Unraveling the electrically evoked compound action potential. Hear Res 205:143–156. https://doi.org/10.1016/J.HEARES.2005.03.020

Briaire JJ, Frijns JHM (2006) The consequences of neural degeneration regarding optimal cochlear implant position in scala tympani: a model approach. Hear Res 214. https://doi.org/10.1016/j.heares.2006.01.015

Bunne M, Myhrum M, Iftikhar M et al (2024) Cochlear-facial dehiscence—the most common cause of facial nerve stimulation from a cochlear implant? A case–control study. Cochlear Implants Int 25:154–164. https://doi.org/10.1080/14670100.2024.2316452

Cakmak O, Pal S, Sahin M (2023) Improving the stimulation selectivity in the human cochlea by strategic selection of the current return electrode. Front Audiol Otol 1. https://doi.org/10.3389/fauot.2023.1259852

Callejón-Leblic MA (2024) A full-head model to investigate intra and extracochlear electric fields in cochlear implant stimulation. Phys Med Biol 69(15). https://doi.org/10.5281/ZENODO.8191036

Callejón-Leblic MA, Lazo-Maestre M, Fratter A et al (2024) A full-head model to investigate intra and extracochlear electric fields in cochlear implant stimulation. Phys Med Biol 69. https://doi.org/10.1088/1361-6560/AD5C38

Cheng Q, Yu H, Liu J et al (2022) Design and optimization of auditory prostheses using the finite element method: a narrative review. Ann Transl Med 10:715–715. https://doi.org/10.21037/ATM-22-2792

Choi CTM, Wang SP (2014) Modeling ECAP in cochlear implants using the FEM and equivalent circuits. IEEE Trans Magn 50:49–52. https://doi.org/10.1109/TMAG.2013.2282640

Cochlear Ltd (2022) Cochlear implant electrode comparison sheet. https://assets.cochlear.com/api/public/content/FUN1142-Electrode-Comparison-Sheet.pdf. Accessed 3 May 2025

Dang K (2017) Electrical conduction models for cochlear implant stimulation. HAL Open Science

de Nobel J, Kononova AV, Briaire JJ et al (2023) Optimizing stimulus energy for cochlear implants with a machine learning model of the auditory nerve. Hear Res 432. https://doi.org/10.1016/j.heares.2023.108741

de Nobel J, Martens SSM, Briaire JJ et al (2024) Biophysics-inspired spike rate adaptation for computationally efficient phenomenological nerve modeling. Hear Res 447:109011. https://doi.org/10.1016/j.heares.2024.109011

de Rijk SR, Hammond-Kenny A, Tam YC et al (2022) Detection of extracochlear electrodes using stimulation-current-induced non-stimulating electrode voltage recordings with different electrode designs. Otol Neurotol 43:E548–E557. https://doi.org/10.1097/MAO.0000000000003512

de Rijk SR, Boys AJ, Roberts IV et al (2023) Tissue-engineered cochlear fibrosis model links complex impedance to fibrosis formation for cochlear implant patients. Adv Healthc Mater 12:1–15. https://doi.org/10.1002/adhm.202300732

Dhanasingh A, Jolly C (2017) An overview of cochlear implant electrode array designs. Hear Res 356:93–103. https://doi.org/10.1016/J.HEARES.2017.10.005

Drakopoulos F, Pellatt L, Sabesan S et al (2025) Modelling neural coding in the auditory midbrain with high resolution and accuracy. Nat Mach Intell 7:1478–1493. https://doi.org/10.1038/s42256-025-01104-9

Ertas YN, Ozpolat D, Karasu SN, Ashammakhi N (2022) Recent advances in cochlear implant electrode array design parameters. Micromachines (Basel) 13:1081. https://doi.org/10.3390/mi13071081

Fredieu JR, Kerbo J, Herron M et al (2015) Anatomical models: a digital revolution. Med Sci Educ 25:183–194. https://doi.org/10.1007/s40670-015-0115-9

Frijns JHM, Mooij J, ten Kate JH (1994) A quantitative approach to modeling mammalian myelinated nerve fibers for electrical prosthesis design. IEEE Trans Biomed Eng 41:556–566. https://doi.org/10.1109/10.293243

Frijns JHM, de Snoo SL, Schoonhoven R (1995) Potential distributions and neural excitation patterns in a rotationally symmetric model of the electrically stimulated cochlea. Hear Res 87: 170–186. https://doi.org/10.1016/0378-5955(95)00090-Q

Frijns JHM, Briaire JJ, Schoonhoven R (2000) Integrated use of volume conduction and neural models to simulate the response to cochlear implants. Simul Model Pract Theory 8:75–97. https://doi.org/10.1016/S0928-4869(00)00008-2

Frijns JHM, Briaire JJ, Grote JJ (2001) The importance of human cochlear anatomy for the results of modiolus-hugging multichannel cochlear implants. Otol Neurotol 22(3):340–349

Frijns JHM, Kalkman RK, Briaire JJ (2009) Stimulation of the facial nerve by intracochlear electrodes in otosclerosis: a computer modeling study. Otol Neurotol 30:1168–1174. https://doi.org/10.1097/MAO.0B013E3181B12115

Gaugain G, Quéguiner L, Bikson M et al (2023) Quasi-static approximation error of electric field analysis for transcranial current stimulation. J Neural Eng 20:016027. https://doi.org/10.1088/1741-2552/ACB14D

Geerardyn A, Zhu MY, Klabbers T et al (2024) Human histology after structure preservation cochlear implantation via round window insertion. Laryngoscope 134:945–953. https://doi.org/10.1002/LARY.30900

Geiger S, Iso-Mustajärvi M, Nauwelaers T et al (2023) Automatic electrode scalar location assessment after cochlear implantation using a novel imaging software. Sci Rep 13:1–9. https://doi.org/10.1038/S41598-023-39275-3

Gerber N, Reyes M, Barazzetti L et al (2017) Data descriptor: a multiscale imaging and modelling dataset of the human inner ear. Sci Data 4:1–12. https://doi.org/10.1038/SDATA.2017.132

Gervain J, Geffen MN (2019) Efficient neural coding in auditory and speech perception. Trends Neurosci 42(1):56–65

Goldsworthy RL (2022) Computational modeling of synchrony in the auditory nerve in response to acoustic and electric stimulation. Front Comput Neurosci 16. https://doi.org/10.3389/fncom.2022.889992

Goldwyn JH, Bierer SM, Bierer JA (2010) Modeling the electrode-neuron interface of cochlear implants: effects of neural survival, electrode placement, and the partial tripolar configuration. Hear Res 268. https://doi.org/10.1016/j.heares.2010.05.005

Hanekom T (2001) Three-dimensional spiraling finite element model of the electrically stimulated cochlea. Ear Hear 22:300–315. https://doi.org/10.1097/00003446-200108000-00005

Hanekom T (2005) Modelling encapsulation tissue around cochlear implant electrodes. Med Biol Eng Comput 43:47–55. https://doi.org/10.1007/BF02345122

Hanekom T, Hanekom JJ (2016) Three-dimensional models of cochlear implants: a review of their development and how they could support management and maintenance of cochlear implant performance. Netw Comput Neural Syst 27:67–106. https://doi.org/10.3109/0954898X.2016.1171411

Hanekom T, Hanekom J, Badenhorst W et al (2023) A tailored and transdisciplinary approach to cochlear implants. Updates on hearing loss and its rehabilitation. https://doi.org/10.5772/INTECHOPEN.1002920

Hernández-Gil M, Ramos-de-Miguel Á, Greiner D et al (2025) A computational model for multiobjective optimization of multipolar stimulation in cochlear implants: an enhanced focusing approach. Expert Syst Appl 280:127472. https://doi.org/10.1016/J.ESWA.2025.127472

Hipsley CA, Aguilar R, Black JR, Hocknull SA (2020) High-throughput microCT scanning of small specimens: preparation, packing, parameters and post-processing. Sci Rep 10:1–10. https://doi.org/10.1038/s41598-020-70970-7

Hrncirik F, Roberts I, Sevgili I et al (2023) Models of cochlea used in cochlear implant research: a review. Ann Biomed Eng 51:1390–1407. https://doi.org/10.1007/s10439-023-03192-3

Huda W, Brad Abrahams R (2015) X-ray-based medical imaging and resolution. Am J Roentgenol 204:W393–W397. https://doi.org/10.2214/AJR.14.13126

Human-Baron R, Hanekom T (2024) Unpacking the terminology used in human cochlear dimension methodologies. Transl Res Anat 35:100290. https://doi.org/10.1016/J.TRIA.2024.100290

Hurley PA, Crook JM, Shepherd RK (2007) Schwann cells revert to non-myelinating phenotypes in the deafened rat cochlea. Eur J Neurosci 26. https://doi.org/10.1111/j.1460-9568.2007.05811.x

Ifukube T, White RL (1987) Current distributions produced inside and outside the cochlea from a Scala Tympani electrode array. IEEE Trans Biomed Eng BME 34:883–890. https://doi.org/10.1109/TBME.1987.326009

Inguva C, Wong P, Sue A et al (2015) Frequency-dependent simulation of volume conduction in a linear model of the implanted cochlea. In: International IEEE/EMBS conference on neural engineering, NER 2015-July, pp 426–429. https://doi.org/10.1109/NER.2015.7146650

Ishai R, Herrmann BS, Nadol JB, Quesnel AM (2017) The pattern and degree of capsular fibrous sheaths surrounding cochlear electrode arrays. Hear Res 348:44–53. https://doi.org/10.1016/J.HEARES.2017.02.012

Javel E, Shepherd RK (2000) Electrical stimulation of the auditory nerve: III. Response initiation sites and temporal fine structure. Hear Res 140:45–76. https://doi.org/10.1016/S0378-5955(99)00186-0

Jiang C, Singhal S, Landry T et al (2021) An instrumented cochlea model for the evaluation of cochlear implant electrical stimulus spread. IEEE Trans Biomed Eng 68:2281–2288. https://doi.org/10.1109/TBME.2021.3059302

Johnstone BM, Johnstone JR, Pugsley ID (1966) Membrane resistance in endolymphatic walls of the first turn of the guinea-pig cochlea. J Acoust Soc Am 40. https://doi.org/10.1121/1.1910239

Kaasalainen T, Ekholm M, Siiskonen T, Kortesniemi M (2021) Dental cone beam CT: an updated review. Phys Med 88:193–217. https://doi.org/10.1016/j.ejmp.2021.07.007

Kalkman RK, Briaire JJ, Dekker DMT, Frijns JHM (2014) Place pitch versus electrode location in a realistic computational model of the implanted human cochlea. Hear Res 315:10–24. https://doi.org/10.1016/j.heares.2014.06.003

Kalkman RK, Briaire JJ, Frijns JHM (2015) Current focussing in cochlear implants: an analysis of neural recruitment in a computational model. Hear Res 322:89–98. https://doi.org/10.1016/j.heares.2014.12.004

Kalkman RK, Briaire JJ, Frijns JHM (2016) Stimulation strategies and electrode design in computational models of the electrically stimulated cochlea: an overview of existing literature. Netw Comput Neural Syst 27:107–134. https://doi.org/10.3109/0954898X.2016.1171412

Kopsch AC, Rahne T, Plontke SK, Wagner L (2022) Influence of the spread of electric field on neural excitation in cochlear implant users: transimpedance and spread of excitation measurements. Hear Res 424:108591. https://doi.org/10.1016/J.HEARES.2022.108591

Kral A, Hartmann R, Mortazavi D, Klinke R (1998) Spatial resolution of cochlear implants: the electrical field and excitation of auditory afferents. Hear Res 121:11–28. https://doi.org/10.1016/S0378-5955(98)00061-6

Landsberger DM, Padilla M, Srinivasan AG (2012) Reducing current spread using current focusing in cochlear implant users. Hear Res 284:16–24. https://doi.org/10.1016/j.heares.2011.12.009

Leblans M, Sismono F, Vanpoucke F et al (2022) Novel impedance measures as biomarker for intracochlear fibrosis. Hear Res 426:108563. https://doi.org/10.1016/J.HEARES.2022.108563

Leclère T, Johannesen PT, Wijetilake A et al (2023) A computational modelling framework for assessing information transmission with cochlear implants. Hear Res 432:108744. https://doi.org/10.1016/J.HEARES.2023.108744

Lei IM, Jiang C, Lei CL et al (2021) 3D printed biomimetic cochleae and machine learning co-modelling provides clinical informatics for cochlear implant patients. Nat Commun 12. https://doi.org/10.1038/s41467-021-26491-6

Makarov SN, Noetscher GM, Ara N (2016) Low-frequency electromagnetic modeling for electrical and biological systems using MATLAB. Wiley

Malherbe TK, Hanekom T, Hanekom JJ (2013) Can subject-specific single-fibre electrically evoked auditory brainstem response data be predicted from a model? Med Eng Phys 35:926–936. https://doi.org/10.1016/J.MEDENGPHY.2012.09.001

Malherbe TK, Hanekom T, Hanekom JJ (2015) The effect of the resistive properties of bone on neural excitation and electric fields in cochlear implant models. Hear Res 327:126–135. https://doi.org/10.1016/j.heares.2015.06.003

Malherbe TK, Hanekom T, Hanekom JJ (2016) Constructing a three-dimensional electrical model of a living cochlear implant user's cochlea. Int J Numer Method Biomed Eng 32:e02751. https://doi.org/10.1002/CNM.2751

Mangado N, Ceresa M, Benav H et al (2018a) Towards a complete in silico assessment of the outcome of cochlear implantation surgery. Mol Neurobiol 55:173–186. https://doi.org/10.1007/s12035-017-0731-z

Mangado N, Pons-Prats J, Coma M et al (2018b) Computational evaluation of cochlear implant surgery outcomes accounting for uncertainty and parameter variability. Front Physiol 9. https://doi.org/10.3389/fphys.2018.00498

Margeta J, Hussain R, López Diez P et al (2022) A web-based automated image processing research platform for cochlear implantation-related studies. J Clin Med 11:6640. https://doi.org/10.3390/JCM11226640/S1

Matin F, Artukarslan EN, Illg A et al (2021) Cochlear implantation in elderly patients with residual hearing. J Clin Med 10. https://doi.org/10.3390/jcm10194305

MED-EL For Professionals FLEX Electrode Arrays Right Fit for Each Cochlea. https://d1c96hlcey6qkb.cloudfront.net/96f9920d-91e7-4197-b06d-839d293f4862/4a3f43edf2c24e38ba572b2e5f18abfd?response-content-disposition=inline%3Bfilename%2A%3DUTF-8%27%2721617CE_r11_0_FLEX-Electrode-Arrays_FS-WEB.pdf&response-content-type=application%2F. Accessed 3 May 2025

Miranda PC, Callejón-Leblic MA, Salvador R (2018) Realistic modeling of transcranial current stimulation: the electric field in the brain. Curr Opin Biomed Eng 8:20–27. https://doi.org/10.1016/J.COBME.2018.09.002

Misrahy GA, Hildreth KM, Shinabarger EW et al (1958) Electrical properties of wall of endolymphatic space of the cochlea (guinea pig). https://doi.org/10.1152/AJPLEGACY.1958.194.2.396

Molaee-Ardekani B, Donahue MJ (2023) Investigating the electrode-electrolyte interface modelling in cochlear implants. Biomed Phys Eng Express 9. https://doi.org/10.1088/2057-1976/aceafb

Ni G, Elliott SJ, Ayat M, Teal PD (2014) Modelling cochlear mechanics. Biomed Res Int 2014. https://doi.org/10.1155/2014/150637

Noble JH, Gifford RH, Labadie RF, Dawant BM (2012) Statistical shape model segmentation and frequency mapping of cochlear implant stimulation targets in CT, Lecture notes in computer science (including subseries lecture notes in artificial intelligence and lecture notes in bioinformatics) 7511 LNCS, pp 421–428. https://doi.org/10.1007/978-3-642-33418-4_52

Nogueira W, Schurzig D, Büchner A et al (2016) Validation of a cochlear implant patient-specific model of the voltage distribution in a clinical setting. Front Bioeng Biotechnol 4:84. https://doi.org/10.3389/fbioe.2016.00084

Nogueira W, Litvak LM, Landsberger DM, Büchner A (2017) Loudness and pitch perception using dynamically compensated virtual channels. Hear Res 344:223–234. https://doi.org/10.1016/j.heares.2016.11.017

Orlando VR, Laércio O, Cruz M et al (2024) Postoperative complications in cochlear implant surgery and their possible risk factors. Braz J Otorhinolaryngol 90:101428. https://doi.org/10.1016/J.BJORL.2024.101428

Oticon Medical Neuro Zti Product Information. https://www.oticonmedical.com/-/media/medical/main/files/ci/products/neuro-zti/pi/en/neuro-zti-product-information%2D%2D-english%2D%2D-m80689.pdf?la=en. Accessed 3 May 2025

Paouris D, Kunzo S, Goljerová I (2023) Validation of automatic cochlear measurements using OTOPLAN® software. J Pers Med 13:805. https://doi.org/10.3390/JPM13050805

Pepper DW, Heinrich JC (2017) The finite element method: basic concepts and applications with MATLAB, MAPLE, and COMSOL. CRC Press, Taylor & Francis Group, CRC Press is an Imprint of the Taylor & Francis Group, an Informa Business

Pietroni N, Campen M, Sheffer A et al (2022) Hex-mesh generation and processing: a survey. ACM Trans Graph 42. https://doi.org/10.1145/3554920/ASSET/CD56B43A-3AFF-46BB-8938-3EC1B05CF2F2/ASSETS/GRAPHIC/TOG-21-0047-F35.JPG

Pietsch M, Aguirre Dávila L, Erfurt P et al (2017) Spiral Form of the human cochlea results from spatial constraints. Sci Rep 7:1–11. https://doi.org/10.1038/s41598-017-07795-4

Potrusil T, Heshmat A, Sajedi S et al (2020) Finite element analysis and three-dimensional reconstruction of tonotopically aligned human auditory fiber pathways: a computational environment for modeling electrical stimulation by a cochlear implant based on micro-CT. Hear Res 393. https://doi.org/10.1016/j.heares.2020.108001

Ramos-De-Miguel Á, Escobar JM, Greiner D et al (2022) A phenomenological computational model of the evoked action potential fitted to human cochlear implant responses. PLoS Comput Biol 18:1–27. https://doi.org/10.1371/journal.pcbi.1010134

Rattay F (1986) Analysis of models for external stimulation of axons. IEEE Trans Biomed Eng BME 33:974–977. https://doi.org/10.1109/TBME.1986.325670

Rattay F, Leao RN, Felix H (2001a) A model of the electrically excited human cochlear neuron. II. Influence of the three-dimensional cochlear structure on neural excitability. Hear Res 153: 64–79. https://doi.org/10.1016/S0378-5955(00)00257-4

Rattay F, Lutter P, Felix H (2001b) A model of the electrically excited human cochlear neuronI. Contribution of neural substructures to the generation and propagation of spikes. Hear Res 153: 43–63. https://doi.org/10.1016/S0378-5955(00)00256-2

Rau TS, Suzaly N, Pawsey N et al (2018) Histological evaluation of a cochlear implant electrode array with electrically activated shape change for perimodiolar positioning. Curr Dir Biomed Eng 4:145–148. https://doi.org/10.1515/CDBME-2018-0036/MACHINEREADABLECITATION/RIS

Rebscher SJ, Hetherington AM, Snyder RL et al (2007) Design and fabrication of multichannel cochlear implants for animal research. J Neurosci Methods 166:1–12. https://doi.org/10.1016/J.JNEUMETH.2007.05.013

Reichenbach T, Hudspeth AJ (2014) The physics of hearing: fluid mechanics and the active process of the inner ear. Rep Prog Phys 77. https://doi.org/10.1088/0034-4885/77/7/076601

Saba R (2012) Cochlear implant modelling: stimulation and power consumption. PhD thesis, University of Southampton

Sahl SJ, Hell SW, Jakobs S (2017) Fluorescence nanoscopy in cell biology. Nat Rev Mol Cell Biol 18:685–701. https://doi.org/10.1038/nrm.2017.71

Sakellarios AI, Tachos NS, Rigas G et al (2017) A validated methodology for the 3D reconstruction of cochlea geometries using human microCT images. Meas Sci Technol 28:054001. https://doi.org/10.1088/1361-6501/AA5F18

Schindelin J, Arganda-Carreras I, Frise E et al (2012) Fiji: an open-source platform for biological-image analysis. Nat Methods 9:676–682. https://doi.org/10.1038/nmeth.2019

Schurzig D, Repp F, Timm ME et al (2023) Virtual cochlear implantation for personalized rehabilitation of profound hearing loss. Hear Res 429:108687. https://doi.org/10.1016/J.HEARES.2022.108687

Sieber D, Erfurt P, John S et al (2019) Data descriptor: the openEar library of 3D models of the human temporal bone based on computed tomography and micro-slicing. Sci Data 6:1–9. https://doi.org/10.1038/sdata.2018.297

Siebrecht M, Briaire JJ, Verbist BM et al (2024) Automated segmentation of clinical CT scans of the cochlea and analysis of the cochlea's vertical profile. Heliyon 10. https://doi.org/10.1016/j.heliyon.2024.e35737

Smit JE, Hanekom T, Hanekom JJ (2009) Estimation of stimulus attenuation in cochlear implants. J Neurosci Methods 180:363–373. https://doi.org/10.1016/J.JNEUMETH.2009.03.024

Snel-Bongers J, Briaire JJ, Van Der Veen EH et al (2013) Threshold levels of dual electrode stimulation in cochlear implants. J Assoc Res Otolaryngol 14:781–790. https://doi.org/10.1007/S10162-013-0395-Y/FIGURES/5

Spelman FA, Clopton BM, Pfingst BE (1982) Tissue impedance and current flow in the implanted ear. Implications for the cochlear prosthesis. Ann Otol Rhinol Laryngol Suppl 98:3–8

Stahl P, Dang K, Vandersteen C et al (2022) Current distribution of distributed all-polar cochlear implant stimulation mode measured in-situ. PLoS One 17:1–16. https://doi.org/10.1371/journal.pone.0275961

Strelioff D (1973) A computer simulation of the generation and distribution of cochlear potentials. J Acoust Soc Am 54:620–629. https://doi.org/10.1121/1.1913642

Thiselton J, Hanekom T (2024) Parameterisation and prediction of intra-canal cochlear structures. Ann Biomed Eng 52:695–706. https://doi.org/10.1007/s10439-023-03417-5

Tran P, Sue A, Wong P et al (2015) Development of HEATHER for cochlear implant stimulation using a new modeling workflow. IEEE Trans Biomed Eng 62:728–735. https://doi.org/10.1109/TBME.2014.2364297

Tran P, Richardson ML, Zeng FG (2021) Input–output functions in human heads obtained with cochlear implant and transcranial electric stimulation. Neuromodulation 24:1402–1411. https://doi.org/10.1111/NER.13065/ASSET/87C40DF2-DF22-4017-B5DE-512DA915EBDA/MAIN.ASSETS/GR7.SML

Vaidyanathan A, van der Lubbe MFJA, Leijenaar RTH et al (2021) Deep learning for the fully automated segmentation of the inner ear on MRI. Sci Rep 11:1–14. https://doi.org/10.1038/s41598-021-82289-y

Van Der Westhuizen J, Hanekom T, Hanekom JJ (2022) Apical reference stimulation: a possible solution to facial nerve stimulation. Ear Hear 43:1189–1197. https://doi.org/10.1097/AUD.0000000000001170

Van Horn A, Hayden C, Mahairas AD et al (2020) Factors influencing aberrant facial nerve stimulation following cochlear implantation: a systematic review and meta-analysis. Otol Neurotol 41:1050–1059. https://doi.org/10.1097/MAO.0000000000002693

Vanpoucke FJ, Zarowski AJ, Peeters SA (2004) Identification of the impedance model of an implanted cochlear prosthesis from intracochlear potential measurements. IEEE Trans Biomed Eng 51:2174–2183. https://doi.org/10.1109/TBME.2004.836518

Vellinga D, Briaire JJ, Van Meenen DMP, Frijns JHM (2017) Comparison of multipole stimulus configurations with respect to loudness and spread of excitation. Ear Hear 38:487–496. https://doi.org/10.1097/AUD.0000000000000416

Verbist BM, Ferrarini L, Briaire JJ et al (2009) Anatomic considerations of cochlear morphology and its implications for insertion trauma in cochlear implant surgery. Otol Neurotol 30:471–477. https://doi.org/10.1097/MAO.0B013E3181A32C0D

Wang Y, Lenarz T, Kral A, John S (2023) Automatic landmark localization in 3D medical CT images: few-shot learning through optimized data pre-processing and network design. In: ACM international conference proceeding series, pp 1–7. https://doi.org/10.1145/3632047.3632048

Wärmländer SKTS, Garvin H, Guyomarc'h P et al (2019) Landmark typology in applied morphometrics studies: what's the point? Anat Rec 302:1144–1153. https://doi.org/10.1002/ar.24005

Wong P, George S, Tran P et al (2016) Development and validation of a high-fidelity finite-element model of monopolar stimulation in the implanted Guinea pig cochlea. IEEE Trans Biomed Eng 63:188–198. https://doi.org/10.1109/TBME.2015.2480601

Zhao T, Hoffman J, McNitt-Gray M, Ruan D (2019) Ultra-low-dose CT image denoising using modified BM3D scheme tailored to data statistics. Med Phys 46:190–198. https://doi.org/10.1002/MP.13252

Zhou N, Shi X, Dixit O et al (2023) Relationship between electrode position and temporal modulation sensitivity in cochlear implant users: are close electrodes always better? Heliyon 9:e12467. https://doi.org/10.1016/j.heliyon.2022.e12467

Computational Modeling of the Auditory System: Biophysical Modeling

12

Tania Hanekom and Johan J. Hanekom

Abstract

Acoustic information in the auditory system is captured in space-time patterns of neural spiking. Ideally, cochlear implants should convey information contained in sound to the central auditory nervous system through electrically elicited neural activation patterns that would result in a sound sensation indistinguishable from that provided through acoustic auditory stimulation. Computational models of nerve fiber firings elicited by acoustic or electrical stimulation of the cochlea may be used to investigate the characteristics of electrically stimulated neurons, the resulting neural spike train patterns, and how these encode acoustic information. This chapter commences by considering the response of single nerve fibers captured by models based on or related to the classical Hodgkin-Huxley nerve fiber model and proceeds to model specializations specific to electrical stimulation of nerve fibers. Next, end-to-end models, which capture processing from the acoustic input to the neural activity output of the cochlea, are discussed. The latter is in the form of a neurogram that reflects neural activity across the entire cochlea. How this population response varies for different peripheral input is considered here for normal hearing, a commercial speech processing algorithm, and a custom speech processing algorithm. This demonstrates how neurograms may be used to investigate differences between speech processors. Proceeding from this, performance measures derived from neurograms are discussed. The neural activity output from the cochlea is processed through several stages before eliciting a perception. There is no unified approach to modeling this bridge between stimulus and perception. A cochlear nucleus model to predict pitch perception is presented and modeling of processing beyond the cochlear nucleus is

T. Hanekom (✉) · J. J. Hanekom
Bioengineering, Department of Electrical, Electronic and Computer Engineering, University of Pretoria, Pretoria, South Africa
e-mail: tania.hanekom@up.ac.za; johan.hanekom@up.ac.za

B. Molaee-Ardekani, M. Segovia-Martinez (eds.), *Advancing Cochlear Implants*,
https://doi.org/10.1007/978-3-032-15719-5_12

discussed. Examples of AI approaches to modeling of psychoacoustic measures and speech intelligibility are discussed.

Keywords

Nerve fiber model · Population response · Neurograms · End-to-end auditory model · Performance measures · Models of speech perception

12.1 Introduction

Computational models of the cochlea are valuable for investigating the neural excitation behavior of both acoustically and electrically stimulated cochleae. They also provide a means to explore signal processing mechanisms within the cochlea and throughout the deeper structures of the auditory pathway. These models have been instrumental in auditory neuroscience as they allow researchers to simulate and analyze complex auditory processes without the need for invasive measurements and experimental methods.

This chapter continues the thread of computational modeling of the auditory system from Chap. 11, which presents an overview of how the response of the physical environment to electric stimulation may be investigated using volume electric models. The present chapter extends the modeling domain to the auditory nervous system to allow a holistic approach to exploring and analyzing the effects of stimulating the cochlea (acoustically or electrically) on the processing and perception of sound.

Information in the acoustic (natural) or electric (prosthetic) input to the auditory system is encoded in the temporal and spatial characteristics of nerve fiber firing or spiking patterns. Understanding these spatiotemporal spiking patterns is necessary for understanding how sound is represented and processed in the auditory nervous system (Yu and Goodrich 2014) and, hence, for developing technology to replace dysfunctional components of the natural hearing system.

The literature covers a wide variety of computational neural models, ranging from relatively simple descriptions of individual peripheral neurons innervating the Organ of Corti in the cochlea, e.g., Rattay et al. (2001) and Briaire and Frijns (2005), to more comprehensive models that predict the behavior of complex responses, such as the auditory brainstem response (Verhulst et al. 2018) and aspects of perception (Brochier et al. 2021).

Single nerve fiber models attempt to capture the response to a peripheral stimulus or synaptic input from a previous neural stage. For example, the compartmental auditory nerve fiber (ANF) model by Rattay et al. (2001) provides insight into the factors that affect neural activation patterns elicited by electrical stimulation, while the model by Spencer et al. (2012) simulates the time-sensitive response of octopus cells in the cochlear nucleus (CN) of the auditory brainstem to synaptic input from ANFs. Single nerve fiber models are often used as preprocessors for population responses as shown by, e.g., Kipping et al. (2024) who simulated the electrically evoked compound action potential (eCAP) responses in cochlear implant (CI) users with residual hearing and Verhulst et al. (2018) who explored the relationship of single neuron responses in the auditory nerve, CN and inferior colliculus (IC) to aggregate population responses such as the auditory brainstem response (ABR).

On the other end of the spectrum, end-to-end models attempt to integrate models of the constituting components of the entire auditory pathway to predict complex electrophysiological responses or perception from input stimuli (Brochier et al. 2022; Hu et al. 2023). Unlike models that isolate specific parts of the auditory system, end-to-end models represent a complete processing path from the raw sensory input (natural or prosthetic) to a final electrophysiological or perceptual output (Hu et al. 2023). This approach may capture complex interactions among the various auditory processing stages that cannot be represented by single-stage models. However, constructing an end-to-end model by combining single-stage models may not yield a comprehensive and accurate simulation of perception (Hu et al. 2023). This could, in part, explain why some recent end-to-end models replace traditional modeling of signal processing stages that represent the way in which the auditory system interprets inputs with machine learning (ML) approaches. While this phenomenological approach to emulating the behavior of the auditory system may not be biologically interpretable, it has been applied successfully in CI optimization (Gajecki and Nogueira 2024; Sinha and Azadpour 2024) and has been proposed for substituting human listeners in time-consuming experiments (Brochier et al. 2022). The use of ML is explored further in Sect. 12.6.

This chapter will explore computational auditory models, starting with a brief functional overview of the auditory nervous system. Single fiber models will subsequently be covered, followed by considering end-to-end and central auditory system models and their application.

12.2 Anatomical and Physiological Overview of the Auditory Nervous System

From a modeling perspective, the auditory nervous system is a complex system comprising several processing stages. These can be broadly classified as peripheral or central. At the periphery, the spiral ganglion neurons (SGNs), also known as the auditory nerve fibers, innervate the mechanoelectrical transducing hair cells that form part of the Organ of Corti in the cochlear duct. Their cell bodies are located in the modiolus of the cochlea in the bony Rosenthal's canal. SGNs can be divided into two main types (Zhang and Coate 2017). Type I neurons form the majority of the nerve fibers. They are large, bipolar, afferent neurons (Zhang and Coate 2017) carrying acoustic information to the brainstem. Each type I neuron's peripheral process connects to a single inner hair cell in the Organ of Corti. The soma and pre- and post-somatic regions of these neurons are unmyelinated in humans, causing significant differences in temporal behavior between human fibers and the myelinated fibers of other mammalian species, such as cat. This is important to note, as neural models are often derived from animal experiments that may not entirely mimic the human case. Type II neurons constitute around 5–10% of the ANF population. These are small and unmyelinated, and their peripheral processes connect to multiple outer hair cells in the Organ of Corti (Yu and Goodrich 2014; Zhang and Coate 2017). They are believed to affect cochlear amplification and auditory nociception (Zhang and Coate 2017). The central processes (axons) of the SGNs bundle together in the modiolus to exit the cochlea through the internal auditory canal. It joins the vestibular nerve to form the vestibulocochlear nerve (cranial nerve VIII) that enters the brainstem at the pontomedullary junction (Yu and Goodrich 2014).

Centrally, the ANFs synapse at the CN. Projections from the CN reach the superior olivary complex (SOC), considered the first location where binaural processing takes place and therefore is important for sound localization. Fibers then travel through the brainstem via the lateral lemniscus to the IC, which is involved in integrating auditory information. From here, fibers project to the primary auditory cortex (A1) through the medial geniculate nucleus of the thalamus.

Components of the central auditory system (CAS) have been modeled to explain a specific functional aspect of auditory processing, such as temporal processing in the CN, and the neurophonic response in the SOC.

12.3 Peripheral Neural Models

The primary goal of ANF models is to predict the behavior of the auditory system. ANF models are broadly categorized into two types. Phenomenological or black-box models focus on describing observable behaviors in CI users without regard for the underlying physiological mechanisms. These models are often used for practical applications where it is more important to relate the outputs to the system's inputs than to understand the mechanisms that provide these links. On the other hand, biophysical models aim to capture the nerve fibers' physical properties. The foundational Hodgkin-Huxley (HH) model (Hodgkin and Huxley 1952) provides a mathematical framework to describe how neurons generate and propagate action potentials. Biophysical models allow manipulation of physiological parameters, such as ion-channel densities, membrane potentials, and synaptic conductances, to aid understanding of the factors that affect neural responses to electrical stimulation.

12.3.1 Classical Neural Models

The classic neural membrane model of Hodgkin and Huxley (1952), based on the membrane kinetics of a giant squid axon, laid the foundation for modern neuroscience by accurately predicting membrane behavior under various stimulus conditions. Although the model was derived from a nonmammalian axon, it provided fundamental insights into the basic functioning of neurons.

Several limitations should be considered when nonmammalian models are used to describe mammalian nerve fibers, such as the operating temperature, differences

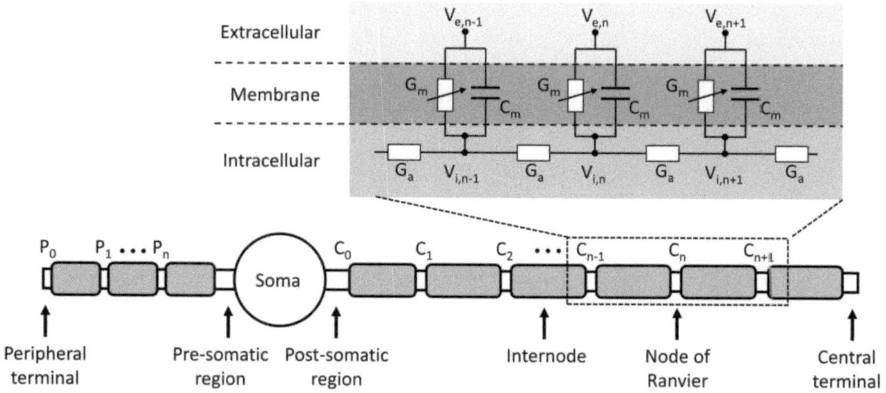

Fig. 12.1 Basic structure of biophysical models of the ANF that predict propagation of APs. The lower part of the figure illustrates the general morphology of these models, which comprises nodes of Ranvier, internodes, pre- and post-somatic regions, and the soma. The upper part of the figure shows a typical equivalent electrical circuit that represents active membrane components such as Ranvier nodes, as a parallel capacitance and voltage-controlled conductance (G_m and C_m), connected with neighboring nodes through axoplasmic conductances (G_a). The extracellular electric potentials $V_{e,n}$ are applied at each of the active membrane components, while the internal potential at each node, $V_{i,n}$, is determined by the current crossing the membrane as well as the internal resting potential of the cell

in the ion-channel composition, myelination, and the morphology of the neurons. The Hodgkin-Huxley (HH) model was therefore adapted by several researchers to improve the predictions of mammalian nerve fiber characteristics. For example, the Schwarz-Eikhof (SE) (1987) model incorporated data from mammalian nerve fibers to more closely predict mammalian excitation characteristics.

To model propagating action potentials, the membrane models have been included in cable models, initially by injecting a current into the intracellular space. Rattay (1986) performed seminal work extending these models to predict action potential generation and propagation by applying extracellular electric fields, e.g., fields produced by electrical stimulation through extracellular electrodes. The basic structure of cable models as implemented by, e.g., Cakmak et al. (2023), is shown in Fig. 12.1.

12.3.2 Human Auditory Nerve Fiber Models

The primary differences between animal and human ANFs are the lack of myelination of the spiral ganglion cells (Nadol 1988) and different ion-channel densities. When adapting animal models to predict human single fiber responses, the morphology has to be adjusted to represent human nerve fiber characteristics. Also, if nonmammalian models, such as the HH model, form the basis for the adaptation, ion-channel gating has to be accelerated by changing the temperature parameters in the equations (Macherey et al. 2007).

The HH-based model of Rattay et al. (2001) implemented these differences by simulating the neuronal substructures in separate compartments. These include unmyelinated dendritic nodes, myelinated internodes, pre- and post-somatic compartments, the soma, and central axon segments (see Fig. 12.1). These models were validated against experimental data and can estimate spike generation points, propagation delays, electrode effects, and the impact of dendrite loss (Hanekom 2001; Rattay et al. 2001).

The humanized generalized Schwarz-Eikhof-Frijns (GSEF) model (Briaire and Frijns 2005) further refines this approach by including a thin unmyelinated pre-somatic compartment modeled after Rattay et al. (2001) and scaling internodal distances (Kalkman et al. 2014). To account for nonmonotonic human ANF behavior, modifications have been proposed to incorporate additional ion-channel dynamics, either through a hyperpolarization-activated current (Mo and Davis 1997) or via a combination of slow potassium and persistent sodium channels (McIntyre et al. 2002).

Further refinement by Smit et al. (2010) introduced adjustments to potassium current kinetics and added a persistent sodium current, extending the model to a generalized human peripheral sensory nerve fiber framework. While these deterministic models predict a binary threshold for action potential generation, experimental findings indicate a stochastic relationship between stimulus intensity and discharge probability caused by stochastic fluctuations in sodium channels (Rubinstein 1995). Subsequent stochastic and hybrid models (e.g., Bruce et al. 1999a, b; Xu and Collins 2004; Macherey et al. 2007; van Gendt et al. 2016, 2017) capture this variability.

Notably, Rattay's compartmental model (Rattay et al. 2001) and the model of Badenhorst et al. (2017) remain the only purely conductance-based stochastic ANF models implemented in compound models.

12.3.3 Integration with Volume Electric Models

Volume electric (VE) models calculate the electric field or potential distribution within the cochlea due to electrical stimulation, taking into account the complex geometry of the cochlea, the electrical properties of various cochlear tissues, and the spatial arrangement of CI electrodes. Compound models integrate VE models with neural models. The primary requirement of integration between VE and neural models is that the latter must be able to accept the output of the VE model, i.e., a distributed electric potential, as input.

Biophysical models such as the GSEF model of Frijns et al. (1994) and HH-based compartmental models such as those by Rattay et al. (2001) and Smit et al. (2010) are ideal in this application as the membrane kinetics of the active nodes of Ranvier in these models are driven by the potential gradient that may be obtained from the VE models. The most general approach to integrating these neural models with VE models is to define a curved neural plane within the neural tissue volume in the VE

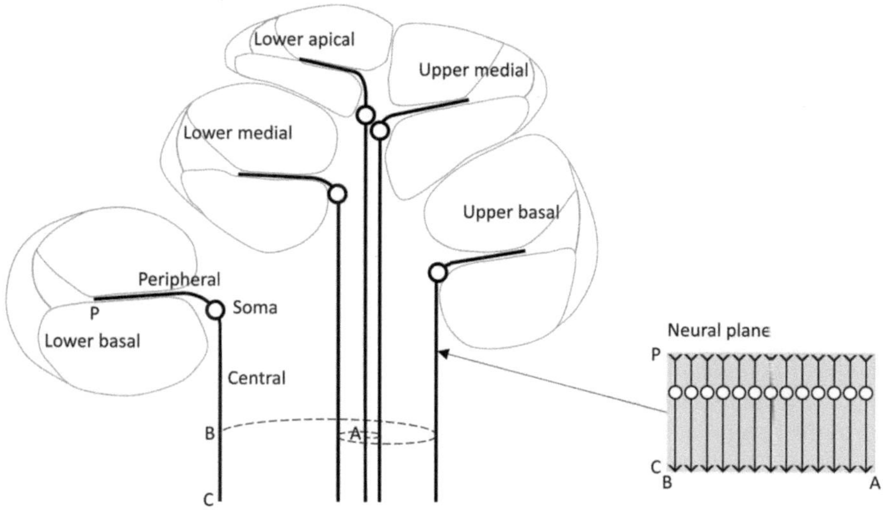

Fig. 12.2 Illustration of the interface between VE and ANF models. The figure on the left shows a mid-modiolar plane through a cochlear model indicating typical placement of ANFs. The dashed spiral shows the trajectory of the neural plane along the tonotopic axis of the cochlea (i.e., along the length of the basilar membrane) from (B)ase to (A)pex, while the heavy lines show the trajectory of the ANFs from the (P)eripheral end where they connect to the hair cells in the Organ of Corti, to the (C)entral termination point of the modeled ANFs. The open circles indicate typical placement of the modeled somas. The figure on the right illustrates the interpretation of this plane in two dimensions

model. A two-dimensional abstraction on this plane is made from the three-dimensional potential distribution predicted by the VE model (Frijns et al. 1995; Hanekom 2001; Malherbe et al. 2016; Cakmak et al. 2023) (Fig. 12.2). The ordinate runs along the length of the nerve fibers starting at the Organ of Corti at the periphery (P) to a boundary at the end of the nerve volume in the VE model and cutting through the central axons of the ANFs (C). This dimension is curved along the modeled trajectory of the nerve fibers. The abscissa runs along the cochleotopic axis along the length of the basilar membrane (BM) starting at the level of the round window at the base (B) and terminating at the apex (A). An array of adjoining independent single nerve fiber models is modeled from the basal to the apical extent of the plane. Electric potentials are extracted from the VE model output as a matrix on the neural plane, either specifically at the locations of the nodes of Ranvier in the ANF models, or on a grid that allows interpolation of the potentials at the location of the nodes of Ranvier in each of the ANF models. The inset to the right in Fig. 12.2 shows the unrolled, flattened plane with the orientation of the ANFs indicated.

The neural plane has also been extended into a volume where the spiral ganglia are distributed in a 3D representation of Rosenthal's canal. The main objective was to model neural recruitment behavior as a result of the spatial distribution of the spiral ganglia (Badenhorst et al. 2017) so that phenomena such as the eCAP and penetration of the stimulus into the spiral ganglia could be investigated.

12.4 Nerve Fiber Population Response Models

This section moves from single nerve fiber models to population response models. The neural population response (i.e., nerve fiber action potentials in response to acoustic or electrical stimulation) is a space-time representation of nerve firings. This representation is often in the form of a raster plot, also referred to as a neurogram, that shows each action potential as a dot on space-time axes. The space dimension is typically the distance along the BM. Figure 12.5a–d gives examples of neurograms that show modeled nerve firings on the auditory nerve (AN) in response to a pure tone. The particular model is described later in the corresponding subsection.

Neurograms are typically the output of an end-to-end cochlear model. End-to-end models (more details on existing models appear below) usually have an acoustic input (e.g., pure tone, speech or music) and model the different stages of processing up to the generation of action potentials on the AN fiber population. An end-to-end model may also reach deeper into the auditory processing pathway to, for example, model processing in the cochlear nucleus (CN), or may bundle CAS processing stages.

The first processing stages either model the normal hearing (NH) auditory periphery or the electrically stimulated auditory periphery. A model of NH may include an outer ear transformation and will typically include some or all of these components: (i) a middle ear (ME) model; (ii) BM filtering (either by modeling this through a filter bank or by explicitly modeling the cochlear travelling wave); (iii) an inner hair cell (IHC) model; (iv) an outer hair cell (OHC) model; (v) an inner hair cell auditory nerve (IHC-AN) synapse model; and (vi) a model of the generation of neural spike trains. Nonlinearities associated with the various processing stages are included in more recent models (Zhang et al. 2001; Zilany et al. 2009, 2014). Efferent feedback to OHCs and the modulatory effect of this feedback is an important focus in recent models, e.g., Farhadi et al. (2023).

If the electrically stimulated auditory system is modeled, the first model stages are replaced by an implementation of the CI speech processor, a model of the electrode-nerve interface which models characteristics of the biophysical interface (e.g., spread of electrical current), and an appropriate model of electrically evoked nerve fiber excitation, e.g., Bruce et al. (1999a). Models of the subsequent auditory processing stages will, however, be the same for an acoustically or electrically stimulated periphery. While the AN spike train patterns are markedly different for acoustic and electrical stimuli, the processing in the CAS will remain the same.

The AN is estimated to contain around 35,000 nerve fibers at a young age, with this number declining with age. Auditory models typically consider a subset of the total number of neurons. Using biophysical neural models (e.g., the HH model) is computationally expensive, and many researchers have opted for simpler representations of neural excitation through phenomenological models, specifically leaky integrate and fire (LIF) models.

LIF models are also employed to model the electrically stimulated cochlea. The model of Bruce et al. (1999a) is often used in this application. This single ANF model is based on cat AN data and models a stochastic threshold that determines when action potentials will be generated. Goldwyn et al. (2012) generalized this approach by creating a point process model for the electrically stimulated ANF that extends the model of Bruce et al. (1999a) by incorporating more realistic neural spiking properties, including realistic spike jitter and temporal integration.

As LIF nerve fiber models have lower computational demands, their use allows the inclusion of reasonably large numbers of neurons in a model, so that published modeling work using phenomenological neural models often includes up to 10,000 neurons (Fredelake and Hohmann 2012; Tabibi et al. 2021). While biophysical models are less typical in end-to-end models because of their computational load, simulation platforms like *Neuron* allow simulations with large numbers of neurons (Manis and Campagnola 2018).

Modeling of CAS processing may continue at a level of detail that considers the processing in each stage of the auditory pathway, but often, end-to-end models lump all of this processing together. The final model output is, generally, some predictor of perception. Predictions of perceptual outcomes are often compared for different model parameter choices. For example, model predictions of the intelligibility of vowels may be compared between NH listeners and CI users.

Saremi et al. (2016) compared seven cochlear filtering models, i.e., these are models that consider cochlear filtering only, but do not include modeling of the hair cells, the IHC-AN interface, or neural activation. The objective was to compare these models on six important characteristics of human cochlear excitation: excitation patterns; nonlinear response growth; tuning; group delays; signal in noise processing; and representation of amplitude modulation. The set of models included mostly phenomenological models, including filter bank models (parallel or cascaded) and one lumped element biophysical model (the latter is summarized in Saremi and Stenfelt 2022). The standout model regarding accuracy in prediction of data was the CARFAC model of Lyon (2011), a cascade phenomenological model. As indicated by the authors, an important criterion is that models are preferred that have fewer free parameters while having the ability to predict a variety of core observations relating to cochlear function. From the seven models, the biophysical model of Saremi and Stenfelt required the smallest number of fitted parameters, while the remaining parameters are physical parameters obtained from measurements. This demonstrates an important advantage of biophysical models: There is a direct relationship between model parameters and the underlying physical structures.

What follows is not a comprehensive review of end-to-end models, but provides an entry point into existing methods. The sections that follow consider two representative approaches used in the development of end-to-end models of AN response. The models considered model either NH or listening with a CI.

12.4.1 End-to-End Models of Normal Hearing

12.4.1.1 Phenomenological Models

A typical approach in cochlear modeling is to use filter banks to describe the mechanical filtering of the basilar membrane (BM) resulting from input sound (e.g., Carney 1993; Zhang et al. 2001; Zilany and Bruce 2006; Zilany et al. 2009,

2014). These are phenomenological models that describe input-output relationships, often without specific reference to the underlying physical parameters. A smaller number of biophysical models (models that are intended to be true to the underlying biology) are available that describe cochlear mechanics by modeling the cochlear travelling wave (Shamma et al. 1986; Eguia et al. 2010).

The modeling approach of Zhang et al. (2001) is shown in Fig. 12.3. This is a filter bank model that contains a signal path and a control path. An input stimulus is filtered into its frequency components by a gammatone filter bank in the signal path, of which the properties are controlled by a control path. Later versions (commencing with Bruce et al. 2003) of the model identified the control path specifically as the OHCs, while the IHCs are within the signal path. The gammatone filter bank contains narrowband filters that correspond to the mechanical filtering function of the BM, but that does not contain a biophysical representation of conversion from BM response to AN response. Each signal path filter in the filter bank corresponds to the tuning characteristics of a specific place on the BM.

The AN response depends on the frequency components of the stimulus, its intensity, and the characteristic frequency (CF) of each neuron. The gain and bandwidth of the filters are controlled by the control signal in the control path. These are continuously updated, so that the signal path filters capture nonlinear characteristics of cochlear filtering. Specifically, level-dependent phase characteristics and two-tone suppression are controlled in this way. The gammatone filters contain a time delay parameter, dependent on CF, that is used to capture travelling wave delay. In addition, the AN neural response that is compressive at high sound levels is captured by saturating nonlinear functions in the control path.

Outputs of the filters are inputs to the IHC model, which transduces BM mechanical responses to AN action potentials. The IHC-AN synapse model controls spontaneous rate, neural adaptation, and the neural rate-level function. Phase-locking behavior of nerve fibers in the low-frequency region of the cochlea is captured by the ac-dc ratio of the IHC response. Finally, neural spike trains are produced by a refractory time modified Poisson process from Carney (1993). The model output is an AN neural population response resulting from an input acoustic signal. The latter can be pure tones or complex sounds.

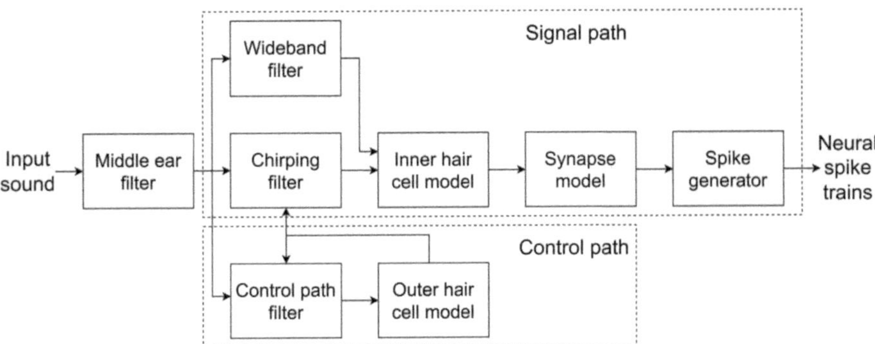

Fig. 12.3 The basic structure of the series of phenomenological models (Zhang et al. 2001; Zilany and Bruce 2006; Zilany et al. 2009, 2014)

This phenomenological model in its various iterations (Bruce et al. 2003; Zilany and Bruce 2006; Zilany et al. 2009, 2014) is a popular choice in auditory modeling work. The model can predict a wide variety of observed AN responses across a wide range of stimulus levels, including specifically nonlinear characteristics of AN responses. Modeled nonlinear response included (i) two-tone suppression and asymmetric growth of two-tone suppression for tones below and above CF, (ii) compressive gain at increasing stimulus intensity, and (iii) nonlinear temporal response properties that include discharge statistics and peak splitting in the period histogram. Nonlinear response properties may be an important aspect of encoding (Carney 2024) of speech signals in the neural response.

Filter bank models can capture the linear characteristics of the cochlear travelling wave, but modeling of nonlinear characteristics in phenomenological models requires the postulation and inclusion of further mechanisms that often do not correspond with specific biophysical characteristics. As a result, these models can be very good at summarizing observations in literature relating to linear and nonlinear response properties of the auditory system, but have limited ability to explain the origin of these observed responses and have limited predictive value. Nonetheless, some phenomenological models accurately model a large set of observations, while having the major advantage of low computational cost.

12.4.1.2 Biophysical Models

Biophysical models are intended to describe a biological system with mathematics that reflects the underlying physical properties.

This section presents an end-to-end biophysical model of normal hearing (Schmidt 2016). While a filter bank can accurately model some characteristics of a linear travelling wave, not all cochlear signal processing can be modeled by filters (Duifhuis 2012). A hydrodynamic travelling wave model can provide a more realistic representation of the signal processing in NH than filter banks as it captures the complex fluid mechanics of the cochlea. For example, the fluid movement in the cochlea includes both a long wavelength component (that corresponds to the travelling wave and that is a result of the interaction between the cochlear partition and the endolymph) and a short wavelength component (a compression pressure mode that is in phase with the stapes motion).

As will be shown, neural tuning does not have to be modeled through an explicit filter bank implementation. A hydrodynamic model of the cochlear travelling wave predicts cochlear filtering and may be preferred in auditory models that are intended to be biologically realistic, albeit at higher computational cost.

The end-to-end auditory model presented here (Fig. 12.4) (Schmidt 2016) consists of a signal conversion stage that relates the sound pressure level (in dB SPL) input to sound pressure (in Pascal units) at the tympanic membrane, a middle ear model that matches air impedance with the impedance of the fluid in the cochlea (O'Connor and Puria 2008; Duifhuis 2012), a travelling wave model that predicts the BM displacement (Duifhuis et al. 1986; Duifhuis 2012), an IHC model that relates BM displacement to an IHC membrane potential (Mountain and Hubbard 1996), a synapse model that provides neural adaptation characteristics to the auditory model

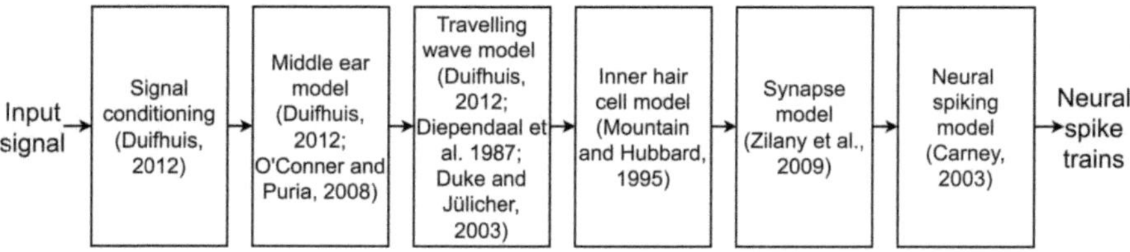

Fig. 12.4 End-to-end auditory model with an indication of the origin of each of the stages of the model (Schmidt 2016)

(the IHC-AN synapse model of Zilany et al. 2009), and a neural spiking model that generates nerve fiber spike train patterns (Carney 1993). Thus, similar to the comparable model of Eguia et al. (2010), the model is biophysical in spirit, and parameters have biological meaning, but some model sections are simplified relative to biological reality. These model sections are explained in more detail below. Model parameters are the same as used in the cited publications.

The middle ear model was that of O'Connor and Puria (2008), modified to fit human data by Duifhuis (2012), and containing as parameters middle ear stiffness, mass and air impedance. The output was the pressure at the stapes, and this was used as input to the travelling wave model by Duifhuis et al. (1986), which was modified to be critically damped (Duke and Jülicher 2003; Duifhuis 2012).

The travelling wave model captures the natural BM response in a one-dimensional, nonlinear partial differential equation. The model describes both long and short wavelengths of the cochlear fluid motion to determine BM displacement and includes the effects of the active and nonlinear parameters of cochlear mechanics. This includes active gain of the BM with low-intensity acoustic stimulation and broadening of the BM displacement area with increases in stimulation intensity (Rhode 1971).

The differential equation is solved numerically by discretization into compartments along the length of the BM. Each compartment has as parameters the BM mass, damping and stiffness along with fluid density and cochlear duct dimensions. The damping factor, described by a Van der Pol oscillator, is a function of velocity of the particular section of the BM and controls the gain observed in measurements for different intensities of stimulation.

BM displacement was nonlinearly related to the IHC voltage (Mountain and Hubbard 1996) using 800 IHCs that were equally spaced along the length of the modeled BM (35 mm). The IHC-AN synapse model of Zilany et al. (2009) was modified to accept IHC voltage as input. The AN model of Carney (1993) was used as output ANF output stage. This is not a biophysical model, but it models observed characteristics of neural discharges. It included refractory effects (absolute refractory period of 0.75 ms, and discharge history of 40 ms, which took in account the earlier response of the AN) and high, medium, and low spontaneous rate nerve fibers.

While pure tones were used as input to obtain tuning characteristics, the model is valid for complex signals (e.g., speech) as well. The model output was a neurogram

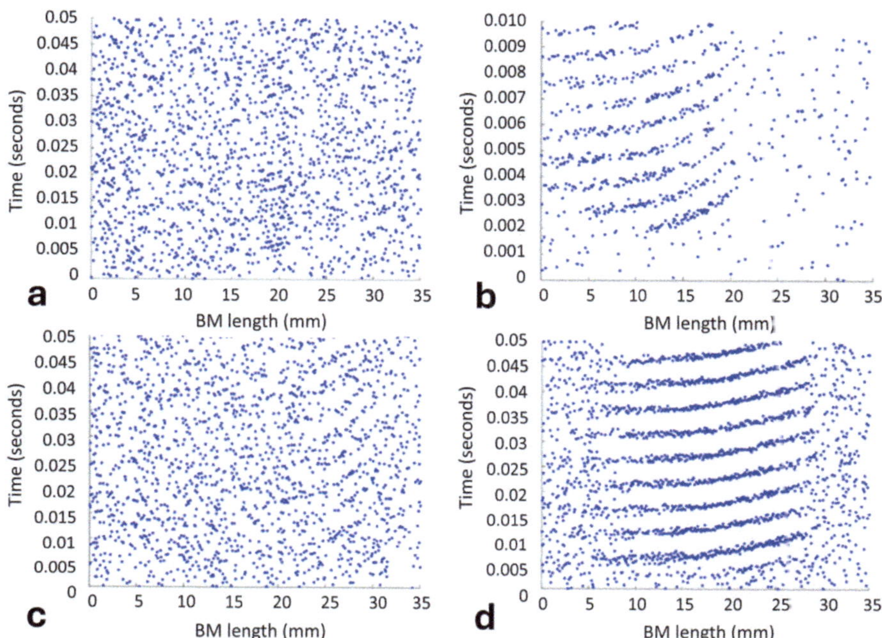

Fig. 12.5 Neurograms for pure tone stimuli for a travelling wave model of NH (Schmidt 2016). (**a**) 30 dB SPL, 1000 Hz. (**b**) 60 dB SPL, 1000 Hz. (**c**) 30 dB SPL, 200 Hz. (**d**) 60 dB SPL, 200 Hz

predicted for a given input. Example neurograms at the output of this model are shown in Fig. 12.5a–d. The neurograms reflect the cochlear travelling wave at the higher stimulation level (60 dB SPL). The travelling wave extends up to the cochlear place corresponding to the pure tone frequency according to Greenwood's equation (Greenwood 1990) (travelling deeper into the cochlea for lower frequencies) and dissipates quickly beyond this place.

Tuning curves were predicted for modeled auditory nerve fibers at different positions along the length of the cochlea as a basic validation test. Simulations with pure tone stimulation at a number of different frequencies were carried out, in which these pure tones were encoded in the modeled travelling wave, and the firing rate pattern of a selected nerve fiber was predicted. This was repeated with 2 dB intensity step size from 0 to 80 dB SPL. Pure tone frequency was varied in 100 steps per decade from 0 to 10,000 Hz, although in Fig. 12.6a, b only frequencies up to 7000 Hz are shown.

Figure 12.6a, b shows the outcome as a heat map, with maximum firing rates represented in different colors. The typical shape of a tuning curve is clearly visible in the two panels of the figure, which were derived for two high spontaneous rate nerve fibers. Superimposed on the figures are chinchilla data from Ruggero et al. (1997) and Temchin et al. (2008). The fibers in Fig. 12.6a, b were expected to have

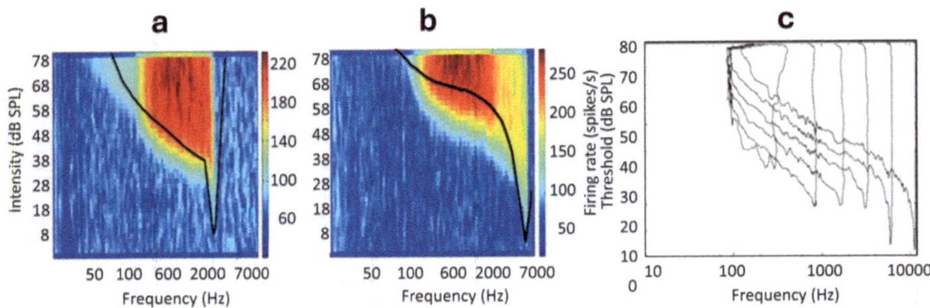

Fig. 12.6 Heat map representation of neural tuning curves (Schmidt 2016) with expected CFs of (**a**) 1715 Hz and (**b**) 6091 Hz, with frequency of pure tone stimulation on the abscissa, and intensity of pure tone stimulation on the ordinate. The superimposed solid lines represent redrawn auditory nerve fiber tuning curve data at the same CF (Ruggero et al. 1997; Temchin et al. 2008). (**c**) Model-predicted neural tuning curves for a number of cochlear positions, which may be compared with results from Temchin et al. (2008)

CFs of 1715 Hz and 6091 Hz based on their respective positions in the cochlea and the corresponding CFs predicted from Greenwood's equation (Greenwood 1990).

For a threshold of 100 spikes/second, the predicted tuning curve (Fig. 12.6) is seen to have a tip with minimum threshold of around 15 dB SPL. Both data and predictions demonstrate two stages of roll-off in the tuning curve threshold below CF: a steep increase in threshold as stimulus frequency decreases followed by a more gradual increase in threshold beyond that.

Figure 12.6c shows predicted nerve fiber tuning curves across multiple characteristic frequencies using 100 spikes/second as threshold. These are similar to chinchilla data shown in Temchin et al. (2008). While the order of magnitude of threshold drop-off was the same for model predictions and data, model-predicted threshold drop-off was shallower below CF and sharper above CF than in chinchilla data.

These model outputs show that nerve fiber tuning characteristics can be predicted with a travelling wave auditory model that does not incorporate any explicit filtering, and with a generic nerve fiber model that has no frequency selectivity. As the neural tuning curves reflect BM tuning, the specific travelling wave model used is key when modeling neural tuning curve characteristics to closely mimic data. While the one-dimensional travelling wave model used here (Duifhuis 2012) has been shown to perform well in the reproduction of global travelling wave characteristics, one-dimensional travelling wave models do not capture BM displacement with adequate accuracy, and higher dimensional models may be required (Viergever and Diependaal 1983). Nevertheless, using a biophysical traveling wave model for auditory filtering is a valid alternative to filter bank models.

In summary, this section discussed two different categories of end-to-end models. Both phenomenological models and biophysical models can predict neural response patterns. While the latter is more computationally expensive, it also captures the underlying biophysical system more accurately. The next section will consider neurograms elicited by CI electrical stimulation.

12.4.2 End-to-End Models of CI Hearing

Electrical stimulation patterns produced by CI electrodes are limited in terms of spectral resolution (CIs have a small number of electrodes, the number of active electrodes is usually smaller than the number of available electrodes, and the actual number of neural information channels may be even smaller because of current spread) and in terms of temporal resolution (the electrical stimulation rate is limited by the cochlear implant hardware). A trade-off exists between these two: Higher stimulation rates per electrode will necessarily mean that fewer electrodes can be stimulated. Because of this trade-off and the differences in focus on presentation of selected speech cues between different speech processors, there are marked differences between CI stimulation strategies across CI brands (see Chap. 2).

In computer models, an existing speech processing strategy is applied in the first processing stage, or the modeler may implement a proposed new strategy. A model of the biophysical interface provides the link between the produced electrical stimulation patterns and a model of electrically elicited neural excitation. The model output, in the form of a neurogram, is then typically used to evaluate aspects of the speech processing strategy. For example, if the speech processing strategy targeted the preservation of fine temporal structure, the modeler will check whether this has been preserved in the neural spike train patterns predicted by the model. If a new speech processing strategy is developed, the modeled neural output may be used to predict the potential benefit of the strategy, e.g., preservation of formants.

12.4.2.1 Phenomenological Models

In their cascade model for cochlear implant hearing (Fig. 12.7), Fredelake and Hohmann (2012) bundled CAS processing into what is referred to as an "internal representation." The model was then used to predict the speech reception threshold (SRT) and the slope of the speech reception function (speech intelligibility as a function of SNR). The first model stage was an implementation of a CI speech processor. The output of this modeling stage was electrical current pulses on each of the modeled electrodes. Realistic threshold (T-level) and maximum comfort level (MCL or C-level) values were used for each electrode, and a typical CI loudness

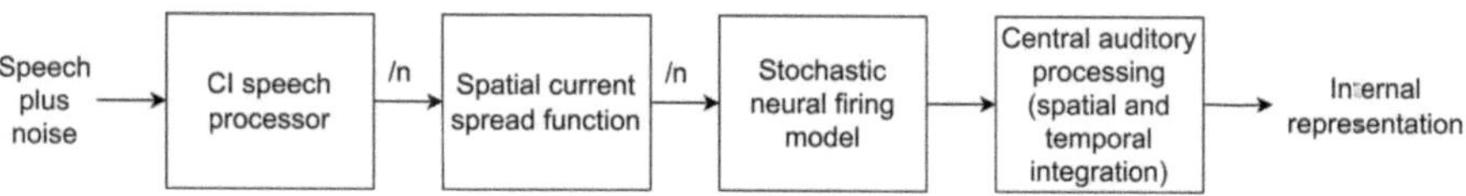

Fig. 12.7 Block diagram of the model of Fredelake and Hohmann (2012). The symbol /n means that there are n channels, where n corresponds to the number of filter channels and the number of electrodes

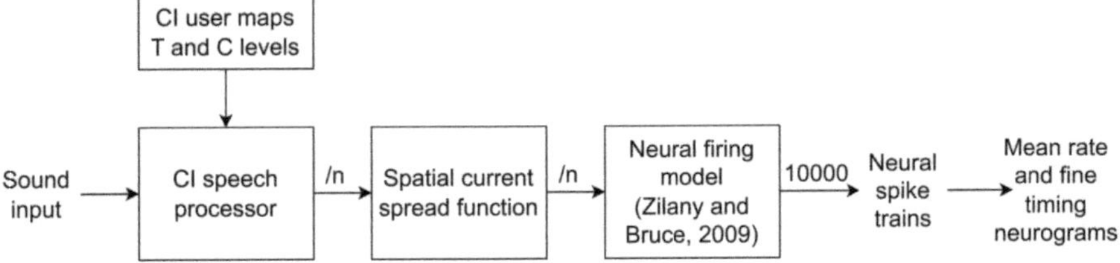

Fig. 12.8 Block diagram of the model of Tabibi et al. (2021)

growth function (LGF) was used to map acoustic amplitude to electrical pulses. In a next processing stage, spatial spread of current was modeled. A simplified one-dimensional (i.e., unrolled) cochlear model was used for this, but this modeling stage may also be far more detailed. For example, current spread was predicted by a 3D finite element (FEM) model of the implanted cochlea in Malherbe et al. (2016).

Following on this stage, Fredelake and Hohmann used a phenomenological model of neural firing (an LIF model) which determined whether a spike would be generated in response to electrical stimulation. Spiking was based on a stochastic membrane threshold by including membrane noise, similar to the neural model of Bruce et al. (1999a). The neural refractory period was included via a threshold increase after neural spiking. An "internal representation" was then determined from this neural output. This internal presentation was an "activity map" of averaged spike rate activity after spatial and temporal integration, from which speech intelligibility in noise was predicted.

Tabibi et al. (2021) developed an end-to-end phenomenological model (Fig. 12.8) to predict the performance of CI users on melodic contour identification. The authors reasoned that a model of lower complexity than a biophysical model would be acceptable to evaluate CI coding strategies. The model input was sound, which was then processed through a CI processing strategy to obtain the electrical stimulation pattern across the simulated electrode array. The model was intended to predict melodic contour identification for individual CI users, so that the speech processor implementation used T- and C-levels from individual CI users' maps.

Current spread was included in the model so that the currents from each electrode were interleaved at the simulated neural plane. Only the cathodic phase of the electrical stimulus was modeled.

An LIF nerve fiber model, an adaptation of the Zilany et al. (2009) model, was used. The modeled ANF characteristics included refractoriness, facilitation, accommodation, and spike rate adaptation. Neural thresholds were scaled according to the CI user's map in this model that simulated 10,000 nerve fibers. The neural output was then used to predict performance on melodic contour identification using "mean-rate" and "fine-timing" neurograms, and comparing these to the neurograms produced by the NH model of Zilany et al. (2009, 2014).

Leclère et al. (2023) used an end-to-end model and simulated experiments in modulation detection. For a chosen input signal, they created neural spike train patterns using the Goldwyn et al. (2012) model. This phenomenological model has stochastic spike train output. From the spike train space-time pattern, they reconstructed an estimate of the original input signal (Rieke et al. 1997). The root mean square error (RMSE) between the original (the input signal) and the reconstructed signal is calculated. Using signal detection theory, detection thresholds are predicted for different amounts of modulation.

12.4.2.2 Biophysical Models

The paragraphs that follow describe a comparison between modeled neurograms of NH, a commercial CI speech processing strategy, and a proposed travelling wave (TW) CI speech processor (Schmidt 2016). The neurograms for NH were produced with the model in Sect. 12.4.1.2 above.

CI speech processors are channel vocoders, and neurograms for NH and virtually any CI channel vocoder could have been compared. The selection here is representative of channel vocoding CI strategies, the advanced combination encoder (ACE) speech processing strategy used in Nucleus-24 CIs. For this speech processor, the per-electrode stimulation rate varies between 250 and 2400 pps, but the rate per electrode is limited by a combined maximum rate across the set of active electrodes of 14,400 pps. With a pure tone input, electrical stimuli were produced by the model, and this was followed by a current spread model (Strydom and Hanekom 2011) that determined which nerve fibers would be stimulated. Electrical current spreads away from each electrode, so that interaction occurs on the nerve fiber plane (i.e., current pulses from multiple electrodes may interact at a specific place in the neural population to give rise to an effective stimulus at this point). Finally, the stochastic model of Bruce et al. (1999b) that related electrical stimulation to generation of action potentials was modified to use currents rather than voltages as input.

The TW speech processor (Schmidt 2016) uses the high-resolution NH travelling wave model of Fig. 12.4, but as limited resolution is available in both space (limited by the number of electrodes) and time (limited by the maximum available stimulation rate of the speech processor), the output of the NH travelling wave model needs to be decimated in space and time to match the available resolution of a CI speech processor. The spatial sampling of the high-resolution NH travelling wave model was down-sampled to 22 equally spaced positions, reflecting electrode contacts that are on average 0.75 mm apart (simulating a Nucleus CI24 cochlear implant). At each of the 22 positions, temporal resolution was down-sampled to 14,400 Hz corresponding to the maximum total stimulation rate of 14,400 pps for Nucleus CI24 devices. As the maximum stimulation rate on each electrode depends on the number of electrodes used, this sampling rate was further down-sampled according to the number of active electrodes (N) to obtain a sampling rate of 14,400/N Hz on each active electrode. Electrical pulse trains were then simulated on each modeled active electrode with a stimulation pulse rate of 14,400/N pps on each. The BM displacement, as predicted at these discrete positions in the high-resolution TW model, was mapped to electrical stimulus amplitudes that are applied on the corresponding electrodes. T- and C-levels are generic, but are typical of CI users, while a CI speech processor LGF was applied to map between BM displacement

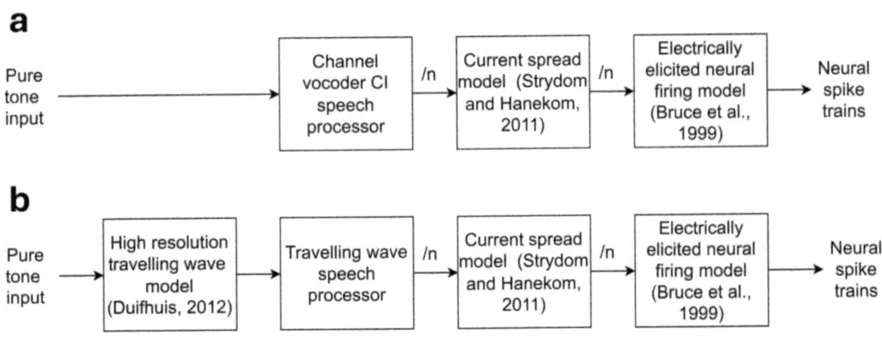

Fig. 12.9 Models of CI vocoder speech processing and a proposed travelling wave speech processor (Schmidt 2016)

amplitudes and electrical current. A trade-off between spatial and temporal resolution exists. If the spatial resolution is decreased (fewer electrodes are stimulated), the temporal resolution increases. An incorrect choice of spatial and time sampling will distort the travelling wave representation in the neural spike train patterns. After electrical stimuli have been determined, the same current spread and neural spiking model is used as for the vocoder CI speech processor model.

Figure 12.9 summarizes, in block diagram format, the two models of which neurograms were compared.

Figure 12.10 shows modeled neurograms for an NH travelling wave model, the ACE CI speech processor (representative of a typical "n-of-m" vocoder strategy in a commercial speech processor), and a proposed TW speech processor, for pure tone input of 200 Hz, 500 Hz, and 1000 Hz (Schmidt 2016). The x-axis shows the BM distance measured from the base of the cochlea. The ACE speech processor stimulates electrodes at a place corresponding to the pure tone frequency. As frequency increases, the area of neural activation shifts basally. The shift in place of stimulation is relatively small. Electrode 22 (i.e., the most apical electrode) is at 25 mm in this model. The electrode closest to the 200 Hz cochlear position is electrode 19, while the electrode closest to the 500 Hz position is electrode 16. The neural spike rate for the ACE processor reflects the electrical stimulus rate, which was 1200 pps. This is clearer from Fig. 12.11, which shows in the inter-spike interval (ISI) histograms that spikes cluster around the electrical stimulus period (0.83 ms) and multiples thereof. This is irrespective of the input pure tone frequency. For NH, the ISI histogram shows that spikes cluster around the input pure tone period (respectively, 5 ms, 2 ms, and 1 ms for the three input frequencies) and multiples thereof.

Similarly, for the TW speech processor, spikes tend to cluster around multiples of the period of the input pure tone. The neurograms in Fig. 12.10 show spike train patterns for the TW speech processor that resemble those of normal hearing, although the range of stimulation along the basilar membrane is restricted by the electrode length and the insertion depth.

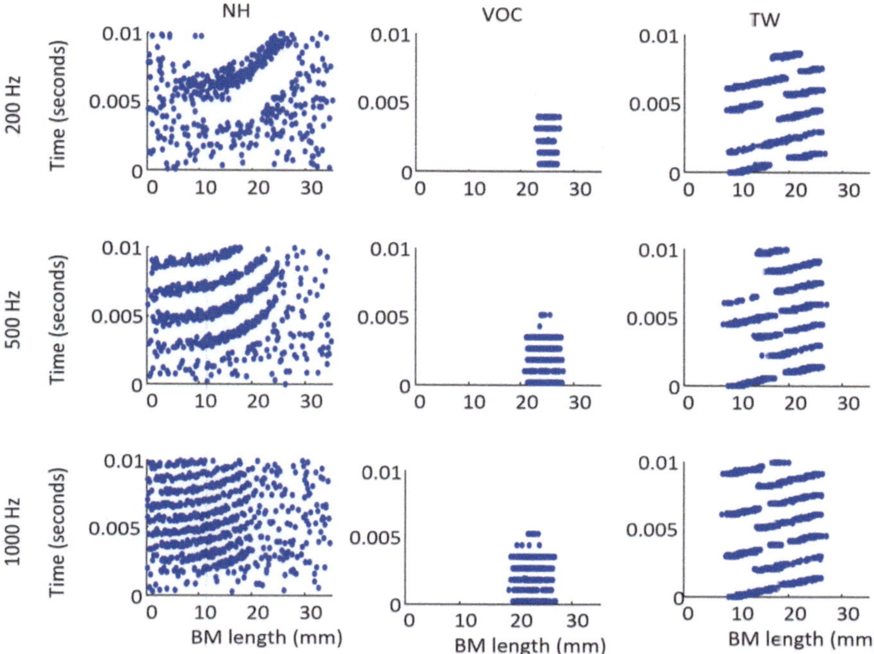

Fig. 12.10 Neurograms of normal hearing (NH), CI vocoder speech processing (VOC), and travelling wave speech processor (TW) (Schmidt 2016). Pure tone inputs were 200 Hz, 500 Hz, and 1000 Hz, respectively, and the input stimulus level was 60 dB SPL

These model predictions suggest that while the pure tone pitch is not encoded in the rate of the electrically elicited neural spike train for a typical vocoder speech processor, but only in place of stimulation, a travelling wave-based speech processor may contain spike rate information corresponding to the input pure tone's period which more closely reflects normal hearing. Of course, the trade-off between electrical pulse rate and spatial resolution will influence to what extent the neurograms of the TW processor reflect those of NH.

Objective methods are needed to characterize performance, which will be discussed in the next section.

12.4.3 Performance Measures Based on Neurograms

A number of different performance measures have been used to characterize performance based on neurograms. These are specific to the objectives of the particular study. The examples below do not intend to provide an exhaustive overview, but show the performance measures used in representative studies.

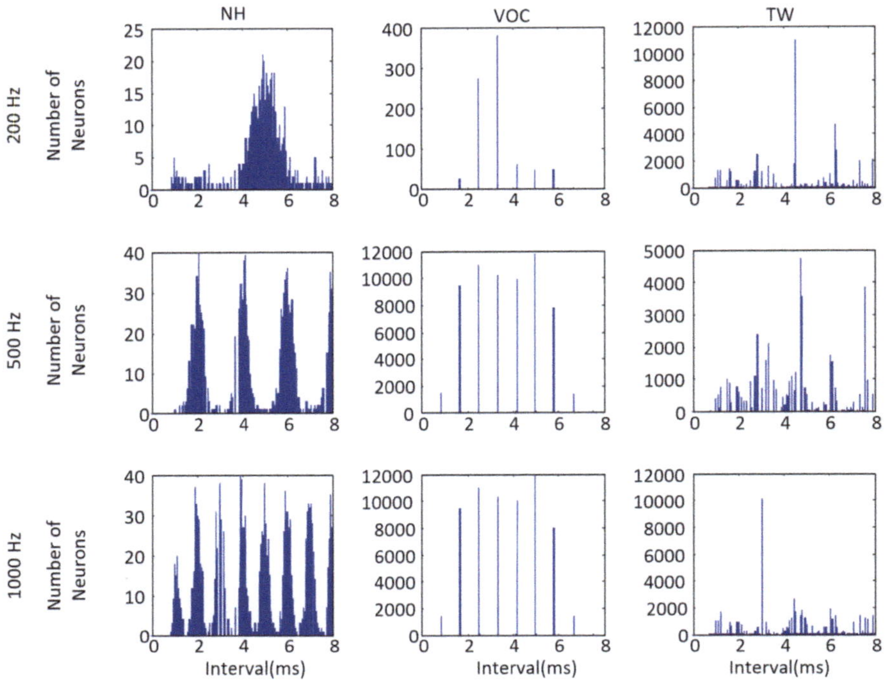

Fig. 12.11 Modeled inter-spike interval histograms from the NH model, CI vocoder speech processing (VOC) and travelling wave speech processor (TW) for the same stimuli of the previous figure (Schmidt 2016)

As a first example, the proposed TW speech processor discussed in the previous subsection had as objective improved encoding of voice pitch information. To evaluate this proposed speech processor from neurogram model output, pure tones may be used as model input to predict the pure tone pitch from the neurogram output. One approach is to use template matching. This approach extracts phase locking of nerve fibers to estimate the period of the stimulus signal (Cedolin and Delgutte 2005). An all-order ISI histogram is constructed (rather than a first-order ISI histogram) and weighted with an exponential weighting function (Cedolin and Delgutte 2005), which is then matched to templates of a specific pitch. A simple harmonic template for a specific fundamental frequency under consideration, F0, is correlated with the weighted all-order ISI histogram. Estimated pitch is extracted as the F0 for which the maximum contrast ratio (stimulus F0/predicted F0) is found in template matching. Pitch estimates based on the neurograms predicted by the NH model are accurate up to 1300 Hz (Schmidt 2016). For the vocoder CI speech processor (ACE strategy), pitch of the pure tone input is not encoded in the neural excitation pattern and is predicted from the neurograms to have a rate pitch that is equal to the electrical stimulus pulse rate irrespective of the input pure tone frequency. This holds up to 300 pps, but for higher pulse rates the predicted pitch saturates, corresponding to the observed limit in rate pitch perception in CI users

(Zeng 2002). For the TW speech processor, pitch estimates from the neurograms are accurate up to 1060 Hz and depend on the specific choices of the number of electrodes and pulse rate.

A second family of performance measures focus on speech intelligibility as predicted from neurograms. Dau et al. (1996) developed a biophysical model of normal hearing that predicted speech intelligibility and compared predictions to the speech intelligibility index (SII).

Jürgens and Brand (2009) predicted speech intelligibility of NH listeners in noise. Stadler and Leijon (2009) created a CI user-specific filter bank input stage that contained a specific CI user's map. Approximate electrode positions were used to predict neural excitation patterns with a 3D FEM model. Neural excitation was simplified to activation patterns that reflected the amount of neural excitation in a set of neural population segments. Speech plus noise was input, and statistical methods were used to predict the SRT (i.e., the SNR at which speech recognition is 50%) of CI listeners.

Fredelake and Hohmann (2012) used the output of their model, the "internal representation," to predict speech intelligibility in CI listeners through a prediction of the SRT. The SNR was varied so that the neural activity map contained noise. For a given input signal, this was then compared to several stored versions using dynamic time warping (DTW), and the most likely stored version, given the input signal plus noise, was the detected one. This process was repeated several times to obtain a predicted SRT.

In a third approach to measure performance, Tabibi et al. (2021) created "mean-rate" and "fine-timing" neurograms and applied the neurogram similarity index measure (NSIM) to compare their model outputs to the NH model of Zilany et al. (2009, 2014). NSIM (Hines and Harte 2012) is a structural similarity method based on image quality assessment.

A different approach was followed by Leclère et al. (2023), who used the neurogram spike trains to reconstruct the input signal. The RMSE between the input signal and the reconstruction is used to determine distributions from which detection thresholds were predicted. These model-predicted detection thresholds were then compared to psychoacoustic data obtained from listeners.

12.5 Models of Central Auditory System Processing

While the measures discussed in the previous section were used to predict perceptual outcomes, it is important to note that predicted neural outputs at the AN level were compared directly to perceptual outcomes, essentially ignoring further processing that takes place in the CAS, or bundling together processing stages between the AN and the primary auditory cortex (A1).

The first retro-cochlear processing stage is the cochlear nucleus (CN). The CN has been the topic of many studies, including modeling studies. In the latter, the CN has been included in models as the final processing stage of an end-to-end model (see references below). In other words, neural excitation patterns at the output of the CN were used to interpret specific observations from perceptual data. Models that consider processing in further stages of the CAS appear in publications, but these are generally not end-to-end models that consider neurograms.

12.5.1 Models of the Cochlear Nucleus

Complex quasiperiodic stimuli (like speech) evoke a pitch that is related to the fundamental frequency (F0) of the stimulus. This periodicity pitch is encoded in inter-spike intervals of auditory nerve fibers (ANFs) (Cedolin and Delgutte 2005). Voice pitch perception by CI users is notoriously poor (Gaudrain and Baskent 2018), so that CI listeners have difficulty in distinguishing between competing speakers, between questions and statements and between male and female voices.

The CN is central in the perception of voice pitch. Octopus cells (OCs) in the CN act as coincidence detectors and are thought to provide a robust estimate of pitch to higher levels of auditory processing in their inter-spike intervals (Golding et al. 1995). They need multiple coincident inputs to depolarize (Spencer et al. 2012), so that the convergence of many ANFs onto each OC provides a structural basis for the decoding of pitch information through the coincident arrival of synaptic inputs from ANFs during a specific time window (Golding et al. 1995; Ferragamo and Oertel 2002). Given the difficulty that CI users have to extract pitch from electrically elicited neural spike train patterns, it is important to consider CI activation of OCs and their role in the extraction of pitch.

Octopus cells have been modeled with compartmental Hodgkin-Huxley-type models (Levy and Kipke 1997; Cai et al. 2001; Spencer et al. 2012) as well as LIF neural model (Kalluri and Delgutte 2003). The input to the model of Spencer et al. was a modeled spike train pattern from the Zilany et al. (2009) model of the auditory periphery, while Kalluri and Delgutte's model used neural discharge patterns predicted by Carney's model (1993).

The model presented here (Blignaut 2017) models the extraction of pure tone pitch by OCs for inputs from NH and from a CI. A compartmental Hodgkin-Huxley model based on the model in Spencer et al. (2012) was implemented. The model geometry is shown in Fig. 12.12a. The model structure corresponds to anatomical measurements and previous OC models (Golding et al. 1995; Levy and Kipke 1997; Cai et al. 2001; Ferragamo and Oertel 2002; Bal and Baydas 2009). Sixty or more ANFs terminate on each OC in mice (Golding et al. 1995; Oertel et al. 2000) and cats (Bal and Baydas 2009). Assuming the same to be true for humans, synaptic inputs from ANFs were applied to all 80 dendrite compartments of each OC in the model. Basal neurons terminated distally on the dendrites, and apical neurons terminated proximally on the dendrites (Fig. 12.12a). This retained the tonotopic arrangement of the ANF termination on OCs (Oertel et al. 2000; McGinley et al. 2012).

Inputs to the OC model were derived from the AN models described earlier in Sect. 12.4.2.2. The model was tested for NH (Fig. 12.12b) with pure tones and electrical stimulation after processing a pure tone input through a CI speech

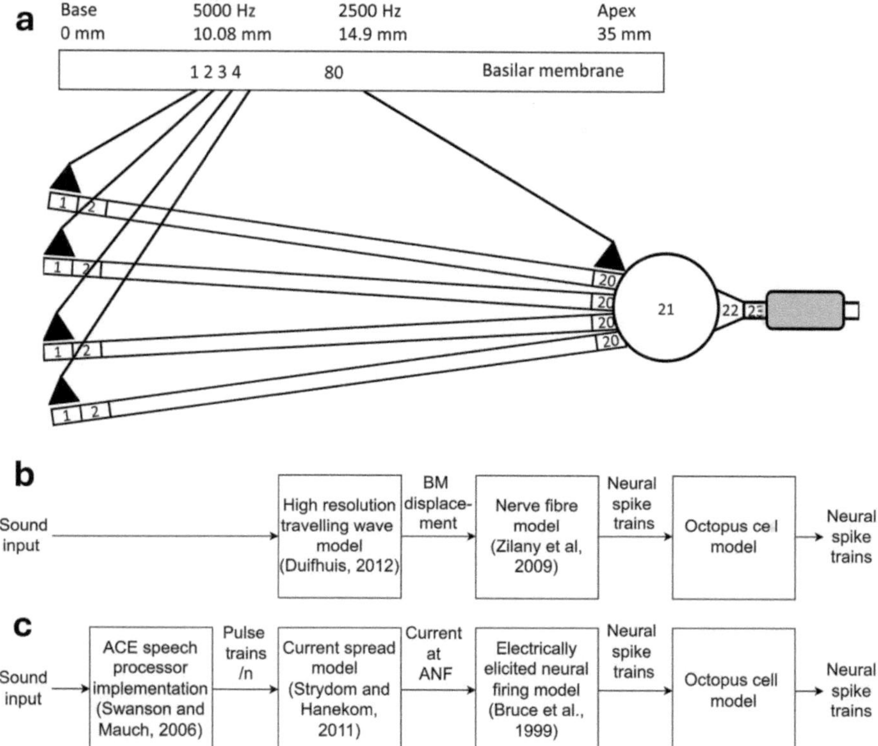

Fig. 12.12 Model used to consider NH and CI activation of octopus cells (Blignaut 2017). (**a**) Model of a single OC, containing four dendrites with 20 compartments each, a soma (compartment 21), axon hillock (compartment 22), and axon initial segment (compartment 23). Termination of ANFs on the OC is shown for ANFs between 2500 and 5000 Hz. (**b**, **c**) The constituent models that were used in the models for NH activation and CI activation of OCs, respectively

processor (Fig. 12.12c). The outputs were the discharge times of a population of ANFs spread linearly in position along the cochlea (Fig. 12.13).

Figure 12.13a shows the membrane voltage of each OC for one simulation run for stimulation of the AN with a 400 Hz pure tone acoustic signal. Individual action potentials (APs) are shown. OCs entrained to the pure tone acoustic frequency in this simulation, and the first mode of the ISI histogram (Fig. 12.13c) corresponds to the period of the pure tone (2.5 ms). The OC population response for a 200 Hz pure tone input using a CI ACE strategy is shown in Fig. 12.13d–f for three pulse rates. The range of OC activity is restricted because excitation is localized, determined by electrode position. Apical electrodes were activated for an input of 200 Hz, and only OCs receiving input from apical ANFs were activated by the CI stimulation. OCs lock onto the rate of lower pulse rates (Fig. 12.13d), and higher pulse rates result in an onset response (Fig. 12.13f). Thus, for low frequencies, electrical pulse rate,

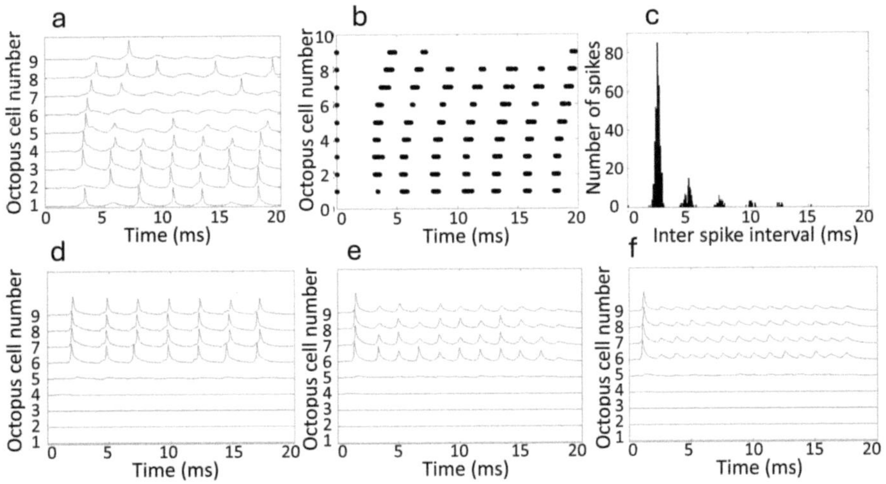

Fig. 12.13 Panels **a**, **b**, and **c**: response of the population of OCs of Fig. 12.12c to pure tone acoustic stimulation with a 400 Hz pure tone (Blignaut 2017). (**a**) Membrane voltage on each OC in the population for one simulation run. (**b**) OC firing times for 20 simulation runs of 20 ms each. (**c**) ISI histogram. Panels **d**, **e**, and **f**: OC population response to a 200 Hz pure tone at (**d**) 400 pps, (**e**) 600 pps, and (**f**) 800 pps using the ACE strategy

rather than the pure tone input frequency, is encoded in the neural spike train pattern at the output of OCs.

In summary, this OC model shows that OCs can extract pitch from acoustic pure tone inputs in the voice pitch range. However, for electrical stimulation, the OCs lock onto the stimulation pulse rate. These results explain why CI users have difficulty with tasks that require the extraction of F0 by listeners, and provide a physiological basis for the 300 Hz rate pitch discrimination limit (Zeng 2002) observed in CI users.

This analysis shows the application of neurograms at the CN level to investigate under which conditions the pitch of pure tones or voice pitch will be available to CI listeners.

12.5.2 Central Auditory System Models Beyond the Cochlear Nucleus

While neural projections in the auditory pathway are ipsilateral up to the CN, there are both ipsilateral and contralateral pathways toward higher auditory processing centers. Most fibers from the dorsal cochlear nucleus (DCN) cross the midline and ascend to the contralateral superior olivary complex and lateral lemniscus. While the CN has been the subject of a number of modeling studies, these deeper processing centers in the CAS have not been the subject of as many modeling studies. Specifically in the context of CIs, modeling of the activation of contralateral pathways by bilateral CIs is an unexplored topic. Processing stages beyond the CN that involve inputs from both ears are involved in forming auditory images by localization of sound in 3D space. For example, the medial superior olive (MSO) and lateral superior olive (LSO) are involved in sound localization. The role of the MSO has been explored in a model (Brughera et al. 2021), and the authors showed its role in localization of a sound source in a reverberant environment.

Often, CAS processing models involve bundling several auditory nuclei together in models that consider the assumed signal processing of these. As an example, Wang and Shamma (1995) separated peripheral auditory processing (the cochlea and the CN) and central auditory processing [MSO and LSO, inferior colliculus (IC), and medial geniculate body (MGB)]. In this model, the result of peripheral auditory processing is assumed to be an auditory spectrum profile which is the input to A1. In this particular model, A1 is assumed to be responsible for spectral shape analysis. This class of models, i.e., signal processing models of audition, does not consider any detail of neural spike train patterns and is not intended to be biophysical models.

ML methods provide a different class of models that also bundle together CAS processing stages. They can provide accuracy in predicting human performance on listening tasks that were previously unattainable. Kell et al. (2018) departed from a biophysical modeling approach and developed a model of CAS processing with a peripheral auditory system-inspired filter bank stage followed by a 12-layer deep neural network. The neural network was optimized to perform two real-world listening tasks, word recognition, and music genre identification, but was not trained to match human listener performance. Notwithstanding, confusion matrices predicted by the ML model for stimuli presented in noise were significantly correlated with those of listeners. The model could approximately replicate listener performance in the specific tasks for which it was optimized.

Furthermore, while this type of model is not a biophysical model and internal details of the neural network cannot be related directly to specifics of CAS neural circuitry, Kell et al. (2018) were able to show similarities between the task-optimized deep neural network structure and hierarchical organization of the auditory cortex.

12.6 The Future of Neural Models

CI models have informed our understanding of CI-elicited neural excitation patterns and how these influence listening through a CI. For example, CI speech intelligibility has been predicted from model neurograms in several modeling efforts. In future work, CI models will continue to be developed and to be used to predict outcomes for proposed speech processors. As models become more sophisticated, they will become more accurate in predicting observations across a wide variety of conditions.

One aspect of model sophistication that is increasingly seen is modeling of top-down processes. Top-down processes have largely been ignored in earlier models, but recent models have included auditory efferent feedback, and these models will continue growing in sophistication.

Investigating speech processing in noise, Yasin et al. (2020) included a filter bank cochlear model (the DRNL filter bank model, see Saremi et al. 2016), an IHC model, and an AN model, with medial olivocochlear (MOC) feedback based on the AN firing rate to control the amount of attenuation in the nonlinear path of the DRNL filter bank model. The model stages, including MOC feedback, were phenomenological. An automatic speech recognition system was used to evaluate speech recognition in noise resulting from efferent feedback.

At a deeper level of sophistication, Smalt et al. (2014) used the AN model of Zilany and Bruce (2006; Zilany et al. 2009) and included CF-dependent, time-dependent, and level-dependent MOC feedback (both ipsilateral and contralateral) to adjust OHC gain. The model demonstrated improved tone detection in background noise, similar to observed AN responses. Farhadi et al. (2023) included detail of IC feedback in the auditory efferent pathway. This model was based on the model of Zilany et al. (2014) with the addition of two feedback signals to the MOC from the IC and wide dynamic range neurons in the CN. As in other models, MOC efferent feedback was used to adjust OHC gain. This model contributed to an ongoing effort to understand the role of auditory efferent feedback from higher auditory centers. Understanding the role of auditory efferent feedback is key to understanding the signal processing in the NH auditory system.

Increased model sophistication also includes a drive toward more realistic models of auditory function, as is exemplified in the model of Verhulst et al. (2018). They focused on two core aspects in model development: (i) to constrain the different modeling stages to be modeled in terms of physical quantities where possible, and thus simultaneously to reduce the number of parameters that would need to be fitted to observations; (ii) modeling of population responses that can be related to noninvasively recorded human physiological responses. Specifically, this model can predict population responses at the AN level, the CN level, and the IC level. These can be related to human auditory brainstem response (ABR) and envelope-following response (EFR) measurements, as neuronal activity at these different levels of the auditory pathway corresponds to ABR waves I, III, and V. Both of these aspects (sophistication in biophysical models and the ability to include human physiological measurements) are important in the ongoing efforts to develop of a comprehensive model of the human auditory system.

A different type of sophistication comes from the rapidly increasing use of artificial intelligence (AI) methods. Machine learning (ML) methods, as an application of AI, have been applied to various problems within the domain of CIs. ML methods essentially learn from data to fit a model to input-output relationships, which inherently limits their application in increasing the level of detail and biological realism in auditory models.

At a high level of abstraction in the use of AI, ML has been used to model CI user outcomes from imaging data, audiological measures, patient-reported measures, and various other functional descriptors (Mo et al. 2025). This allows general predictions of whether a specific CI user is likely to have good speech recognition performance. This type of ML model is, however, not a computational model of auditory function, but rather models input-output relationships without reference to biophysical parameters. Mo et al. (2025) concluded that many ML models may be overfitting data and that they often do not generalize well. For example, they observed poor generalization of an ML model trained on a particular CI user population to populations at other institutions. Nevertheless, this phenomenological approach may be useful in grouping CI users likely to have good performance, and those likely to be at risk of poor outcomes.

At a deeper level of model sophistication, ML provides powerful tools to solve specific problems in modeling. First, increased computational cost is an issue with increased model sophistication, but computational bottlenecks may benefit from ML methods. For example, de Nobel et al. (2023) replaced the computationally intensive solution of the GSEF model, a set of differential equations that need to be solved at every model iteration and for numerous modeled nerve fibers in a biophysical AN model, by a surrogate model. The surrogate model provides a significant reduction in computational time relative to biophysical models and allows evaluation of neural responses over longer periods of time than what is computationally realistic with biophysical models, paving the way for simulation of complex phenomena such as psychophysical experiments.

Second, in some circumstances, it is of value to have a phenomenological model of the reaction of human listeners in a listening experiment. Acoustic models of CIs have long been used to model the effect of specific parameters on speech intelligibility (Svirsky et al. 2021). In this line of research, NH listeners listen to modeled CI speech processors in which specific parameters are varied (e.g., dynamic range) to model their effect on speech intelligibility. However, the response of a human listener could be predicted algorithmically. In this context, Sinha and Azadpour (2024) used deep learning to predict the outcomes when human listeners listen to acoustic models of CIs.

At a level of detail closer to biological realism, Brochier et al. (2022) modeled CI listeners' perception of vowels and consonants by training an ML model, an automatic speech recognition neural network, on neural spike trains. They used an end-to-end model to predict excitation profiles (these were obtained from an FEM model) as initial estimate of neural activation thresholds, and then adjusted neural thresholds to reflect refractoriness, adaptation, and temporal integration. Predicted neural spike trains were used as input to the ML model. The ML model in essence bundles CAS processing and models an input-output relationship between neural spike trains at the AN level and perception. The ML model predicted consonant and vowel confusions, and using these, an information transmission analysis determined the extent to which the modeled CI user received cues that underlie vowel and consonant perception. The model predicted consonant confusions that had similar confusion patterns as those of CI listeners, but was less successful in predicting confusion patterns for vowels. Similarly, the ML model of Kell et al. (2018) discussed in Sect. 12.5 bundled together CAS processing, with their focus being the careful development of a listening task-optimized neural network. This approach supports the drive toward more realistic models by approximating real-world task performance. Real-world performance increases model plausibility (Kell et al.

2018), underlining the potential value of including ML modeling stages in an end-to-end model.

Models will continue to develop in detail and biological realism, and ML methods will be used increasingly to assist in reducing computational cost and to model complex relationships that cannot yet be modeled with biophysical models (e.g., the link between biophysical models of the electrically stimulated auditory periphery and perception with a CI). More sophisticated models will increase understanding of the signal processing at the various processing stages in the auditory system. In the context of cochlear implants, from one perspective, the task is to create electrically elicited spike train patterns in the AN that can be correctly decoded by the CAS. The better our understanding of the signal processing in the CAS, the more successful will our efforts to achieve this be.

Acknowledgments The authors declare that no Generative AI was used in the creation of this manuscript.

References

Badenhorst W, Hanekom T, Hanekom JJ (2017) Analysis of a purely conductance-based stochastic nerve fibre model as applied to compound models of populations of human auditory nerve fibres used in cochlear implant simulations. Biol Cybern 111:439–458. https://doi.org/10.1007/s00422-017-0736-8

Bal R, Baydas G (2009) Electrophysiological properties of octopus neurons of the cat cochlear nucleus: an in vitro study. J Assoc Res Otolaryngol 10. https://doi.org/10.1007/s10162-009-0159-x

Blignaut GE (2017) Signal processing by octopus cells for acoustic and electric hearing: a modelling study. Masters dissertation, University of Pretoria

Briaire JJ, Frijns JHM (2005) Unraveling the electrically evoked compound action potential. Hear Res 205:143–156. https://doi.org/10.1016/j.heares.2005.03.020

Brochier T, Guérit F, Deeks JM et al (2021) Evaluating and comparing behavioural and electrophysiological estimates of neural health in cochlear implant users. J Assoc Res Otolaryngol 22: 67–80. https://doi.org/10.1007/s10162-020-00773-0

Brochier T, Schlittenlacher J, Roberts I et al (2022) From microphone to phoneme: an end-to-end computational neural model for predicting speech perception with cochlear implants. IEEE Trans Biomed Eng 69:3300–3312. https://doi.org/10.1109/TBME.2022.3167113

Bruce LC, Irlicht LS, White MW et al (1999a) A stochastic model of the electrically stimulated auditory nerve: pulse-train response. IEEE Trans Biomed Eng 46:630–637. https://doi.org/10.1109/10.764939

Bruce LC, White MW, Irlicht LS et al (1999b) A stochastic model of the electrically stimulated auditory nerve: single-pulse response. IEEE Trans Biomed Eng 46:617–629. https://doi.org/10.1109/10.764938

Bruce IC, Sachs MB, Young ED (2003) An auditory-periphery model of the effects of acoustic trauma on auditory nerve responses. J Acoust Soc Am 113. https://doi.org/10.1121/1.1519544

Brughera A, Mikiel-Hunter J, Dietz M, McAlpine D (2021) Auditory brainstem models: adapting cochlear nuclei improve spatial encoding by the medial superior olive in reverberation. J Assoc Res Otolaryngol 22:289–318. https://doi.org/10.1007/s10162-021-00797-0

Cai Y, McGee J, Walsh E (2001) Processing of pitch information in complex stimuli by a model of octopus cells in the cochlear nucleus. In: Greenberg S, Slaney M (eds) Computational models of auditory function. IOS Press, Amsterdam

Cakmak O, Pal S, Sahin M (2023) Improving the stimulation selectivity in the human cochlea by strategic selection of the current return electrode. Front Audiol Otol 1. https://doi.org/10.3389/fauot.2023.1259852

Carney LH (1993) A model for the responses of low-frequency auditory-nerve fibers in cat. J Acoust Soc Am 93. https://doi.org/10.1121/1.405620

Carney LH (2024) Neural fluctuation contrast as a code for complex sounds: the role and control of peripheral nonlinearities. Hear Res 443:108966

Cedolin L, Delgutte B (2005) Pitch of complex tones: rate-place and interspike interval representations in the auditory nerve. J Neurophysiol 94:347–362. https://doi.org/10.1152/jn.01114.2004

Dau T, Püschel D, Kohlrausch A (1996) A quantitative model of the "effective" signal processing in the auditory system. I. Model structure. J Acoust Soc Am 99:3615–3622. https://doi.org/10.1121/1.414959

de Nobel J, Kononova AV, Briaire JJ et al (2023) Optimizing stimulus energy for cochlear implants with a machine learning model of the auditory nerve. Hear Res 432. https://doi.org/10.1016/j.heares.2023.108741

Duifhuis H (2012) Cochlear mechanics: introduction to a time domain analysis of the nonlinear cochlea. https://doi.org/10.1007/978-1-4419-6117-4

Duifhuis H, Hoogstraten HW, van Netten SM et al (1986) Modelling the cochlear partition with coupled van der Pol oscillators. In: Alien JB, Hall JL, Hubbard AE, Neely ST, Tubis A (eds) Peripheral auditor mechanisms. Springer, New York, pp 290–297

Duke T, Jülicher F (2003) Active traveling wave in the cochlea. Phys Rev Lett 90. https://doi.org/10.1103/PhysRevLett.90.158101

Eguia MC, Garcia GC, Romano SA (2010) A biophysical model for modulation frequency encoding in the cochlear nucleus. J Physiol Paris 104. https://doi.org/10.1016/j.jphysparis.2009.11.014

Farhadi A, Jennings SG, Strickland EA, Carney LH (2023) Subcortical auditory model including efferent dynamic gain control with inputs from cochlear nucleus and inferior colliculus. J Acoust Soc Am 154. https://doi.org/10.1121/10.0022578

Ferragamo MJ, Oertel D (2002) Octopus cells of the mammalian ventral cochlear nucleus sense the rate of depolarization. J Neurophysiol 87:2262–2270. https://doi.org/10.1152/jn.00587.2001

Fredelake S, Hohmann V (2012) Factors affecting predicted speech intelligibility with cochlear implants in an auditory model for electrical stimulation. Hear Res 287:76–90. https://doi.org/10.1016/j.heares.2012.03.005

Frijns JHM, Mooij J, ten Kate JH (1994) A quantitative approach to modeling mammalian myelinated nerve fibers for electrical prosthesis design. IEEE Trans Biomed Eng 41:556–566. https://doi.org/10.1109/10.293243

Frijns JHM, de Snoo SL, Schoonhoven R (1995) Potential distributions and neural excitation patterns in a rotationally symmetric model of the electrically stimulated cochlea. Hear Res 87:170–186. https://doi.org/10.1016/0378-5955(95)00090-Q

Gajecki T, Nogueira W (2024) A fused deep denoising sound coding strategy for bilateral cochlear implants. IEEE Trans Biomed Eng 71:2232–2242. https://doi.org/10.1109/TBME.2024.3367530

Gaudrain E, Baskent D (2018) Discrimination of voice pitch and vocal-tract length in cochlear implant users. Ear Hear 39:226–237. https://doi.org/10.1097/AUD.0000000000000480

Golding NL, Robertson D, Oertel D (1995) Recordings from slices indicate that octopus cells of the cochlear nucleus detect coincident firing of auditory nerve fibers with temporal precision. J Neurosci 15:3138–3153. https://doi.org/10.1523/jneurosci.15-04-03138.1995

Goldwyn JH, Rubinstein JT, Shea-Brown E (2012) A point process framework for modeling electrical stimulation of the auditory nerve. J Neurophysiol 108:1430–1452. https://doi.org/10.1152/jn.00095.2012

Greenwood DD (1990) A cochlear frequency-position function for several species—29 years later. J Acoust Soc Am 87. https://doi.org/10.1121/1.399052

Hanekom T (2001) Three-dimensional spiraling finite element model of the electrically stimulated cochlea. Ear Hear 22:300–315. https://doi.org/10.1097/00003446-200108000-00005

Hines A, Harte N (2012) Speech intelligibility prediction using a neurogram similarity index measure. Speech Comm 54:306–320. https://doi.org/10.1016/j.specom.2011.09.004

Hodgkin AL, Huxley AF (1952) Propagation of electrical signals along giant nerve fibers. Proc R Soc Lond B Biol Sci 140:177–183

Hu H, Ausili SA, Williges B et al (2023) A model framework for simulating spatial hearing of bilateral cochlear implant users. Acta Acust 7. https://doi.org/10.1051/aacus/2023036

Jürgens T, Brand T (2009) Microscopic prediction of speech recognition for listeners with normal hearing in noise using an auditory model. J Acoust Soc Am 126:2635–2648. https://doi.org/10.1121/1.3224721

Kalkman RK, Briaire JJ, Dekker DMT, Frijns JHM (2014) Place pitch versus electrode location in a realistic computational model of the implanted human cochlea. Hear Res 315:10–24. https://doi.org/10.1016/j.heares.2014.06.003

Kalluri S, Delgutte B (2003) Mathematical models of cochlear nucleus onset neurons: I. Point neuron with many weak synaptic inputs. J Comput Neurosci 14. https://doi.org/10.1023/A:1021128418615

Kell AJE, Yamins DLK, Shook EN et al (2018) A task-optimized neural network replicates human auditory behavior, predicts brain responses, and reveals a cortical processing hierarchy. Neuron 98. https://doi.org/10.1016/j.neuron.2018.03.044

Kipping D, Zhang Y, Nogueira W (2024) A computational model of the electrically or acoustically evoked compound action potential in cochlear implant users with residual hearing. IEEE Trans Biomed Eng 71:3192–3203. https://doi.org/10.1109/TBME.2024.3410686

Leclère T, Johannesen PT, Wijetilake A et al (2023) A computational modelling framework for assessing information transmission with cochlear implants. Hear Res 432. https://doi.org/10.1016/j.heares.2023.108744

Levy KL, Kipke DR (1997) A computational model of the cochlear nucleus octopus cell. J Acoust Soc Am 102. https://doi.org/10.1121/1.419761

Lyon RF (2011) Cascades of two-pole–two-zero asymmetric resonators are good models of peripheral auditory function. J Acoust Soc Am 130. https://doi.org/10.1121/1.3658470

Macherey O, Carlyon RP, Van Wieringen A, Wouters J (2007) A dual-process integrator-resonator model of the electrically stimulated human auditory nerve. J Assoc Res Otolaryngol 8:84–104. https://doi.org/10.1007/s10162-006-0066-3

Malherbe TK, Hanekom T, Hanekom JJ (2016) Constructing a three-dimensional electrical model of a living cochlear implant user's cochlea. Int J Numer Method Biomed Eng 32. https://doi.org/10.1002/cnm.2751

Manis PB, Campagnola L (2018) A biophysical modelling platform of the cochlear nucleus and other auditory circuits: from channels to networks. Hear Res 360:76–91. https://doi.org/10.1016/j.heares.2017.12.017

McGinley MJ, Charles Liberman M, Bal RM, Oertel D (2012) Generating synchrony from the asynchronous: compensation for cochlear traveling wave delays by the dendrites of individual brainstem neurons. J Neurosci 32. https://doi.org/10.1523/JNEUROSCI.0272-12.2012

McIntyre CC, Richardson AG, Grill WM (2002) Modeling the excitability of mammalian nerve fibers: influence of afterpotentials on the recovery cycle. J Neurophysiol 87:995–1006. https://doi.org/10.1152/jn.00353.2001

Mo Z-L, Davis RL (1997) Heterogeneous voltage dependence of inward rectifier currents in spiral ganglion neurons. J Neurophysiol 78:3019–3027. https://doi.org/10.1152/jn.1997.78.6.3019

Mo JT, Chong DS, Sun C et al (2025) Machine-learning predictions of cochlear implant functional outcomes: a systematic review. Ear Hear. https://doi.org/10.1097/AUD.0000000000001638

Mountain DC, Hubbard AE (1996) Computational analysis of hair cell and auditory nerve processes. In: Hawkins HL, McMullen TA, Popper AN, Fay RR (eds) Auditory computation. Springer handbook of auditory research, vol 6. Springer, New York

Nadol JB (1988) Quantification of human spiral ganglion cells by serial section reconstruction and segmental density estimates. Am J Otolaryngol Head Neck Med Surg 9:47–51. https://doi.org/10.1016/S0196-0709(88)80007-3

O'Connor KN, Puria S (2008) Middle-ear circuit model parameters based on a population of human ears. J Acoust Soc Am 123. https://doi.org/10.1121/1.2817358

Oertel D, Bal R, Gardner SM et al (2000) Detection of synchrony in the activity of auditory nerve fibers by octopus cells of the mammalian cochlear nucleus. Proc Natl Acad Sci USA 97:11773–11779. https://doi.org/10.1073/pnas.97.22.11773

Rattay F (1986) Analysis of models for external stimulation of axons. IEEE Trans Biomed Eng BME 33:974–977. https://doi.org/10.1109/TBME.1986.325670

Rattay F, Lutter P, Felix H (2001) A model of the electrically excited human cochlear neuron. I. Contribution of neural substructures to the generation and propagation of spikes. Hear Res 153:43–63. https://doi.org/10.1016/S0378-5955(00)00256-2

Rhode WS (1971) Observations of the vibration of the basilar membrane in squirrel monkeys using the Mössbauer technique. J Acoust Soc Am 49. https://doi.org/10.1121/1.1912485

Rieke F, Warland D, De Ruyter Van Steveninck R, Bialek W (1997) Spikes: exploring the neural code. Comput Neurosci 20:311–315

Rubinstein JT (1995) Threshold fluctuations in an N sodium channel model of the node of Ranvier. Biophys J 68:779–785. https://doi.org/10.1016/S0006-3495(95)80252-3

Ruggero MA, Rich NC, Recio A et al (1997) Basilar-membrane responses to tones at the base of the chinchilla cochlea. J Acoust Soc Am 101. https://doi.org/10.1121/1.418265

Saremi A, Stenfelt S (2022) The effects of noise-induced hair cell lesions on cochlear electromechanical responses: a computational approach using a biophysical model. Int J Numer Method Biomed Eng 38. https://doi.org/10.1002/cnm.3582

Saremi A, Beutelmann R, Dietz M et al (2016) A comparative study of seven human cochlear filter models. J Acoust Soc Am 140. https://doi.org/10.1121/1.4960486

Schmidt L (2016) Viability of cochlear travelling wave signal processing for cochlear implants. Masters dissertation, University of Pretoria

Schwarz JR, Eikhof G (1987) Na currents and action potentials in rat myelinated nerve fibres at 20 and 37°C. Pflugers Arch 409:569–577. https://doi.org/10.1007/BF00584655

Shamma SA, Chadwick RS, Wilbur WJ et al (1986) A biophysical model of cochlear processing: intensity dependence of pure tone responses. J Acoust Soc Am 80. https://doi.org/10.1121/1.394173

Sinha R, Azadpour M (2024) Employing deep learning model to evaluate speech information in acoustic simulations of cochlear implants. Sci Rep 14. https://doi.org/10.1038/s41598-024-73173-6

Smalt CJ, Heinz MG, Strickland EA (2014) Modeling the time-varying and level-dependent effects of the medial olivocochlear reflex in auditory nerve responses. J Assoc Res Otolaryngol 15. https://doi.org/10.1007/s10162-013-0430-z

Smit JE, Hanekom T, van Wieringen A et al (2010) Threshold predictions of different pulse shapes using a human auditory nerve fibre model containing persistent sodium and slow potassium currents. Hear Res 269:12–22. https://doi.org/10.1016/j.heares.2010.08.004

Spencer MJ, Grayden DB, Bruce IC et al (2012) An investigation of dendritic delay in octopus cells of the mammalian cochlear nucleus. Front Comput Neurosci. https://doi.org/10.3389/fncom.2012.00083

Stadler S, Leijon A (2009) Prediction of speech recognition in cochlear implant users by adapting auditory models to psychophysical data. EURASIP J Adv Signal Process 2009. https://doi.org/10.1155/2009/175243

Strydom T, Hanekom JJ (2011) An analysis of the effects of electrical field interaction with an acoustic model of cochlear implants. J Acoust Soc Am 129. https://doi.org/10.1121/1.3518761

Svirsky MA, Capach NH, Neukam JD et al (2021) Valid acoustic models of cochlear implants: one size does not fit all. Otol Neurotol 42. https://doi.org/10.1097/MAO.0000000000003373

Tabibi S, Boulet J, Dillier N, Bruce IC (2021) Phenomenological model of auditory nerve population responses to cochlear implant stimulation. J Neurosci Methods 358. https://doi.org/10.1016/j.jneumeth.2021.109212

Temchin AN, Rich NC, Ruggero MA (2008) Threshold tuning curves of chinchilla auditory-nerve fibers. I. Dependence on characteristic frequency and relation to the magnitudes of cochlear vibrations. J Neurophysiol 100. https://doi.org/10.1152/jn.90637.2008

van Gendt MJ, Briaire JJ, Kalkman RK, Frijns JHM (2016) A fast, stochastic, and adaptive model of auditory nerve responses to cochlear implant stimulation. Hear Res 341:130–143. https://doi.org/10.1016/j.heares.2016.08.011

van Gendt MJ, Briaire JJ, Kalkman RK, Frijns JHM (2017) Modeled auditory nerve responses to amplitude modulated cochlear implant stimulation. Hear Res 351:19–33. https://doi.org/10.1016/j.heares.2017.05.007

Verhulst S, Altoè A, Vasilkov V (2018) Computational modeling of the human auditory periphery: auditory-nerve responses, evoked potentials and hearing loss. Hear Res 360:55–75. https://doi.org/10.1016/j.heares.2017.12.018

Viergever MA, Diependaal RJ (1983) Simultaneous amplitude and phase match of cochlear model calculations and basilar membrane vibration data. In: de Boer E, Viergever MA (eds) Mechanics of hearing. Springer, Dordrecht

Wang K, Shamma SA (1995) Spectral shape analysis in the central auditory system. IEEE Trans Speech Audio Process 3:382–395. https://doi.org/10.1109/89.466657

Xu W, Collins LM (2004) Predicting the threshold of pulse-train electrical stimuli using a stochastic auditory nerve model: the effects of stimulus noise. IEEE Trans Biomed Eng 51:590–603. https://doi.org/10.1109/TBME.2004.824143

Yasin I, Drga V, Liu F et al (2020) Optimizing speech recognition using a computational model of human hearing: effect of noise type and efferent time constants. IEEE Access 8. https://doi.org/10.1109/ACCESS.2020.2981885

Yu W-M, Goodrich LV (2014) Morphological and physiological development of auditory synapses. Hear Res 311:3–16. https://doi.org/10.1016/j.heares.2014.01.007

Zeng FG (2002) Temporal pitch in electric hearing. Hear Res 174. https://doi.org/10.1016/S0378-5955(02)00644-5

Zhang KD, Coate TM (2017) Recent advances in the development and function of type II spiral ganglion neurons in the mammalian inner ear. Semin Cell Dev Biol 65:80–87. https://doi.org/10.1016/J.SEMCDB.2016.09.017

Zhang X, Heinz MG, Bruce IC, Carney LH (2001) A phenomenological model for the responses of auditory-nerve fibers: I. Nonlinear tuning with compression and suppression. J Acoust Soc Am 109:648–670. https://doi.org/10.1121/1.1336503

Zilany MSA, Bruce IC (2006) Modeling auditory-nerve responses for high sound pressure levels in the normal and impaired auditory periphery. J Acoust Soc Am 120:1446–1466. https://doi.org/10.1121/1.2225512

Zilany MSA, Bruce IC, Nelson PC, Carney LH (2009) A phenomenological model of the synapse between the inner hair cell and auditory nerve: long-term adaptation with power-law dynamics. J Acoust Soc Am 126:2390–2412. https://doi.org/10.1121/1.3238250

Zilany MSA, Bruce IC, Carney LH (2014) Updated parameters and expanded simulation options for a model of the auditory periphery. J Acoust Soc Am 135:283–286. https://doi.org/10.1121/1.4837815

Medical Imaging and Artificial Intelligence in Cochlear Implantation

13

Jan Margeta, Paula López Diez, Behnam Molaee-Ardekani, and Thomas Demarcy

Abstract

This chapter explores the critical role of cochlear imaging as applied to the field of cochlear implants (CI), focusing on the distinct phases of the implantation process: preoperative, intraoperative, and postoperative imaging. We explore various applications of imaging in otology, detailing its use in anatomical measurements for surgical diagnosis, surgical planning, and post-implantation CI fitting. Key imaging modalities such as X-ray and fluoroscopy, computed tomography (CT), cone-beam CT (CBCT), photon counting CT (PCCT), magnetic resonance imaging (MRI), and micro-CT are discussed, along with potential future advancements in imaging technology with synchrotron imaging. The integration of artificial intelligence (AI) in cochlear imaging is examined, highlighting how AI algorithms enhance imaging processes through segmentation to obtain accurate three-dimensional (3D) models of the patient's anatomy, the fusion of information across modalities through image registration, landmark

J. Margeta (✉)
Research & Development, KardioMe, Nová Dubnica, Slovakia
e-mail: jan@kardio.me

P. López Diez
Eriksholm Research Center, Oticon A/S, Snekkersten, Denmark

Department for Applied Mathematics and Computer Science, Technical University of Denmark, Kongens Lyngby, Denmark
e-mail: ppapd@eriksholm.com

B. Molaee-Ardekani
Advanced Innovation Department, Cochlear Ltd, Vallauris, France
e-mail: bmolaee@cochlear.com

T. Demarcy
Research & Development, Volta Medical, Marseille, France
e-mail: thomas@demar.cy

B. Molaee-Ardekani, M. Segovia-Martinez (eds.), *Advancing Cochlear Implants*,
https://doi.org/10.1007/978-3-032-15719-5_13

detection, and electrode array detection for estimating the placement of the electrode and patient-specific tonotopic mapping. This chapter also addresses the detection of congenital malformations and provides insights into the types of algorithms employed, their advantages, limitations, and principles of operation. Available datasets are presented, and the evaluation of these techniques is discussed. Finally, platforms integrating these techniques, which promise the transformative impact of AI on cochlear imaging, are showcased. In conclusion, the chapter offers final thoughts emphasizing the ongoing advancements in AI and imaging resolution that promise to further revolutionize cochlear imaging technology.

Keywords

Artificial intelligence · Cochlear implants · Computer vision

13.1 Introduction

Medical imaging is pivotal throughout the entire cochlear implantation journey, providing detailed insights into ear anatomy and serving as an objective measure of electrode placement. These detailed views enable precise diagnosis and characterization of each patient's unique anatomy, promising less traumatic surgeries, faster recovery, and improved hearing outcomes. However, visualizing the cochlea and its associated structures is challenging due to its small size (7–10 mm), complex topology, and deep embedding within the dense bone of the skull. The electrode array (EA) of the cochlear implant (CI) itself is even smaller.

Once the images are acquired, experienced radiologists must perform several evaluation steps. These include the identification of critical landmarks, anatomical structures, or implanted electrodes, measuring distances and characterizing shapes, aligning pre- and postoperative images, and estimating patient tonotopy post-surgery. Given the small size of the inner ear structures and spatial resolution limitations of imaging techniques, these steps are often challenging and time-consuming. However, the latest advances in artificial intelligence (AI) techniques and overall greater access to large datasets and computational resources have enabled the training of powerful models as a result. These models can effectively handle the real-world variability between patients and significantly simplify or even automate many image analysis tasks.

In this chapter, we begin by exploring the pre-, intra-, and postoperative stages of the cochlear implantation, highlighting the distinctions among various imaging modalities and presenting the measurements and assessments expected from CI image processing (Sect. 13.2). Subsequently, we discuss the typical image processing pipeline and examine how artificial intelligence (AI) techniques contribute to automating its steps and enabling the analysis of large-scale cochlear implantation imaging datasets (Sect. 13.3). Lastly, AI and emerging technologies in CI are briefly discussed (Sect. 13.4).

13.2 Patient Operation Journey and Imaging

There are three main stages that CI recipients typically undergo, each benefiting from medical image acquisition: pre-, intra-, and postoperative operative imaging.

The preoperative imaging stage is essential for diagnostics and planning. It helps determine candidacy for cochlear implantation and supports the development of an appropriate therapy plan. This stage involves identifying key anatomical features, assessing the presence and structure of relevant middle and inner ear components, characterizing their shape and dimensions, comparing them with population statistics, and evaluating possible approaches to access the cochlea. Individualized 3D anatomical models extracted from imaging data assist in planning safe surgical approaches—especially in cases where a patient's unique anatomical geometry may present challenges in accessing the cochlea. Identifying fibrosis, ossification, or other abnormalities—such as inner ear malformations—is essential for ruling out known contraindications to CI therapy and for anticipating potential complications.

The intraoperative imaging stage supports minimally invasive approaches by enhancing the surgeon's spatial understanding of the anatomy, providing real-time feedback, and thereby reducing surgical risks and improving outcomes. The primary goals of this feedback include tracking and visualization of the EA during surgery to ensure accurate placement and prevent complications such as significant bending, tip fold-overs (TFO), kinking (Ishiyama et al. 2020), extracochlear placement, or postsurgical trauma. Intraoperative imaging techniques must also adhere to the constraints of the operating room and integrate smoothly into the surgical workflow.

The postoperative imaging stage confirms the final placement of the EA and identifies any complications resulting from surgery such as scalar translocation, trauma to intra-cochlear structures, electrode folding, or electrode misplacement. The final placement also serves as the foundation for personalized, anatomy-based tonotopic mapping. Follow-up postoperative imaging is used to detect EA migration or fibrous tissue growth.

For each of the aforementioned imaging stages, one or more specific imaging modalities may be required for cochlear implantation. The following subsections introduce these imaging modalities, followed by the types of CI-related measurements and assessments conducted at various stages of the CI patient journey.

13.2.1 Cochlear Implant Imaging Modalities

The choice of imaging modality depends on the clinical questions relevant to the patient's current phase of care, individual characteristics, tolerance to ionizing radiation, and available resources (Aschendorff 2011; Vogl et al. 2015; Widmann et al. 2020). For instance, high-resolution imaging is crucial when evaluating fine anatomical structures or constructing detailed models of the inner ear. Radiation-free imaging techniques are preferred for pediatric patients, whereas ionizing radiation is not a concern in cadaveric temporal bone studies. Some modalities may produce artifacts in the presence of an implant. Imaging technologies are continuously

Table 13.1 Overview of the most used CI imaging modalities used in clinical practice and in cadaveric research studies (indicated by ‡)

Imaging	Dim.	Visualized structures	Ionizing radiation	Suitability	Resolution
X-ray	2D	Electrode array	Low	Intra, post	Moderate
Fluoroscopy	2D +Time	Electrode array	Moderate	Intra, post	Moderate
CT	3D	Bony cavities, electrode array	High	Pre, post	Good
CBCT	3D	Bony cavities, electrode array	Medium-high	Pre, intra, post	High
PCCT	3D	Bony cavities, electrode array	Medium	Pre, post	High
MRI	3D	Soft tissues (neural structures, tumors)	None	Pre	Low
Micro-CT ‡	3D	All	Very high	Pre, post, cadaveric studies	Very high
Synchrotron ‡	3D	All	Extremely high	Pre, post, cadaveric studies	Ultra-high

evolving, with new techniques and applications being developed. In the following subsections, we present the most commonly used imaging modalities (see Table 13.1).

X-Ray Radiography

X-ray radiography is a relatively simple and inexpensive two-dimensional (2D) imaging modality based on ionizing radiation. Electromagnetic waves from an X-ray source penetrate the imaged tissue and reach the detector, forming a 2D projection image (see Fig. 13.1). Each element in the 2D image represents the amount of attenuation a ray accumulates along its path from the source to the detector. Different tissues absorb X-rays differently: Softer tissues absorb less than bones, and metals absorb the most.

This 2D projection provides a high-contrast visualization of the implanted elements, making it particularly valuable for postoperative assessment of EA placement. Achieving optimal visualization of the EA depends on properly setting the orientation of the X-ray with respect to the patient's anatomy, with the modified Stenver's view being the standard approach (McClenaghan and Nash 2020). However, such evaluations are inherently limited to two dimensions, unless multiple views are acquired and mentally reconstructed to approximate a three-dimensional (3D) representation of the EA. Additionally, the utility of X-ray imaging for preoperative planning is limited, as projections through overlying soft anatomical structures produce low contrast and do not necessarily represent the size of the tissues in the projected images accurately. Consequently, it is not possible to obtain millimetric measurements, making the delineation and measurement of critical anatomical structures rather challenging.

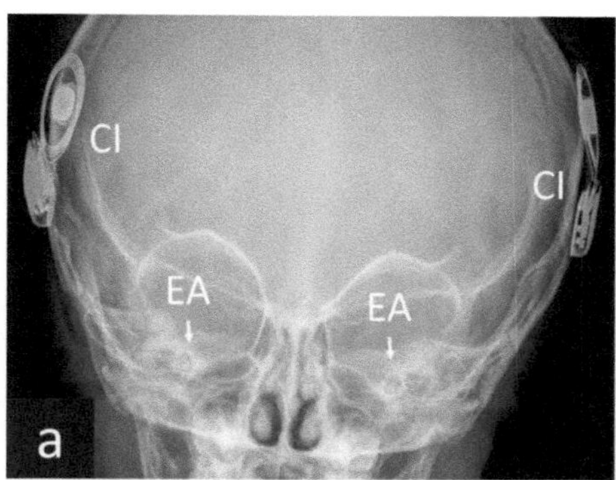

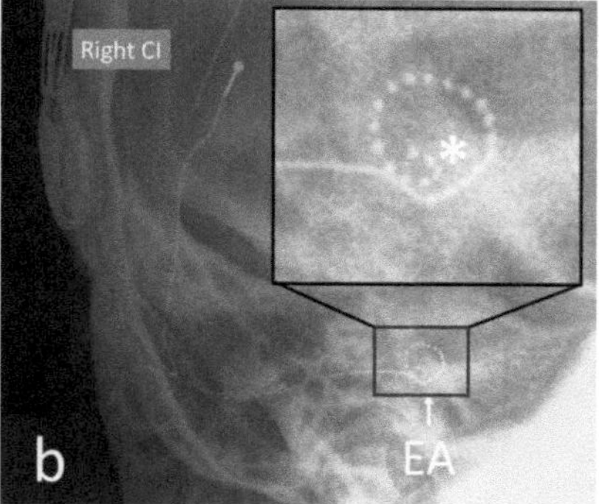

Fig. 13.1 Postoperative X-ray images: (**a**) Bilateral cochlear implants (CI) with two electrode arrays (EA) deployed. (**b**) Modified Stenver's view is suitable for visualization of the insertion complications such as tip fold-over (asterisk in the zoomed area). (Adapted with modification from McClenaghan and Nash (2020) under the CC BY license)

Fluoroscopy

Fluoroscopy is a type of real-time, low-dose X-ray that results in live video feed for immediate intraoperative feedback to the surgeon. Compared to static X-ray, this dynamic nature of fluoroscopy adds the benefit of immediate localization of the EA for intraoperative guidance (Perazzini et al. 2021), rather than for postoperative evaluation only. The downside is that fluoroscopy exposes the patient and surgical team to greater cumulative ionizing radiation. Robotic surgeries (Mom et al. 2022) can greatly reduce such risks in the future.

Computed Tomography (CT)

CT plays a crucial role in both pre- and postoperative evaluation of CI patients by providing detailed three-dimensional (3D) imaging. It reconstructs a detailed 3D view of the region by combining multiple X-ray images taken from various angles, enabling clear visualization of the anatomical structures relevant to CI therapy. Preoperative CT scans are particularly effective for imaging the bony structures of the inner ear. The resolution of these scans is defined by the size of the imaging element (voxel), which typically ranges from 0.2 to 0.5 mm. However, in some cases, the resolution may differ across dimensions, resulting in non-cubic voxels.

CT imaging of the temporal bone region is well-suited for visualizing anatomical structures (see Fig. 13.2a, b) and for generating 3D models of key components in the middle and inner ear, such as the bony labyrinth, round window (Benson et al. 2019), internal auditory canal, and facial nerve canals (Ho et al. 2015). CT scans are also used to identify potential surgical complications—such as ossification, as illustrated in Fig. 13.2c—and congenital malformations (Sennaroğlu and Demir Bajin 2017). Additionally, they are used to evaluate middle ear pathologies and determine a patient's suitability for cochlear implant therapy. The anatomical models derived from these images support surgical planning and help assess the complexity of the procedure (Jerosha et al. 2024; Vaid et al. 2015).

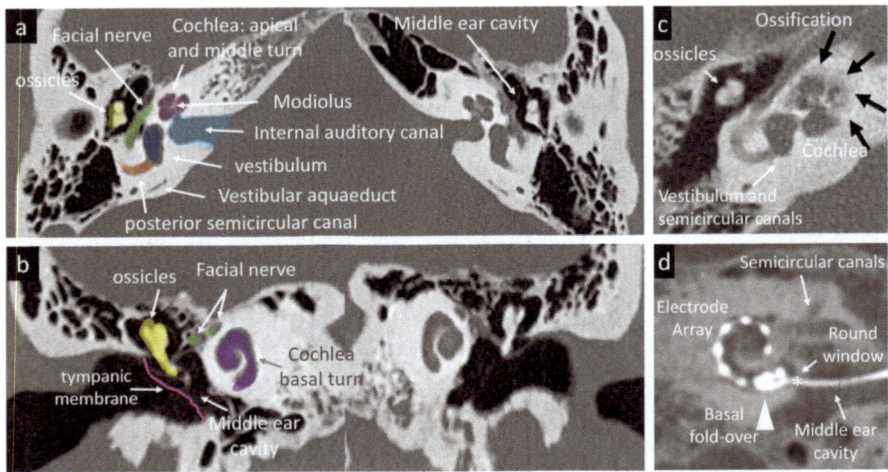

Fig. 13.2 CT imaging of the cochlea: Axial (**a**) and coronal (**b**) views demonstrating preoperative normal temporal bone structures, color-coded on the left side (patient's right side) for reference to the opposite side. (**c**) Preoperative CT image revealing an ossified cochlea. (**d**) Postoperative image showing the electrode array and a basal fold-over complication following a round window insertion technique. (Figure adapted with modification from Batts et al. (2025) (panels **a** and **b**) and Widmann et al. (2020) (panels **c** and **d**), under the CC BY license)

Postoperative CT scans (see Fig. 13.2d) are used to objectively measure the placement of the electrodes in 3D and evaluate the success of the procedure. However, these postoperative images often have metallic artifacts from the cochlear implants which obfuscate the exact boundaries of the underlying anatomical features, making them unsuitable for precise anatomical measurements.

The main limitation of CT imaging is its reduced ability to visualize soft tissues such as nerves and ligaments—typically, only their bony cavities are visible—as well as fine intra-cochlear structures like the basilar membrane. Additionally, CT involves a significantly higher dose of potentially harmful ionizing radiation compared to a single X-ray image, which limits its use in pediatric patients. To obtain the best clinical images at the lowest effective dose, it has become essential to optimize imaging parameters such as tube voltage, exposure time, reconstruction algorithms, and detector design.

Cone-Beam Computed Tomography (CBCT)

CBCT offers several advantages over conventional CT for imaging the temporal bone, particularly in intra- and postoperative settings. CBCT provides higher-resolution images while delivering significantly lower radiation doses compared to conventional CT. With a typical resolution of 0.1–0.3 mm, it enables more detailed visualization of delicate structures such as the ossicular chain, bony labyrinth, internal cochlear anatomy, and facial nerve canal. Additionally, CBCT generates fewer metal artifacts, facilitating more accurate determination of EA positioning.

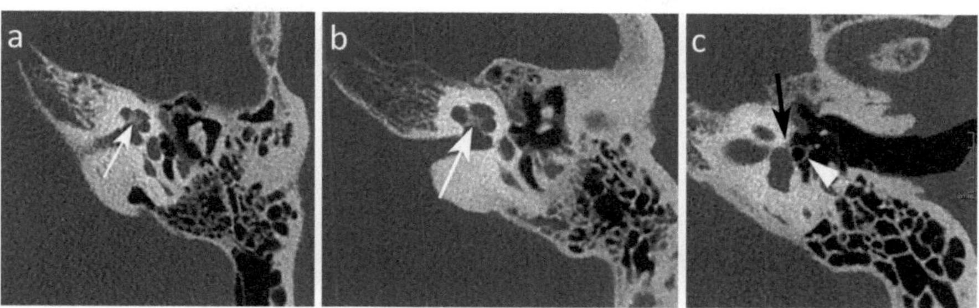

Fig. 13.3 Comparison of axially reformatted CT (**a**) and PCCT (**b**) images of the cochlear anatomy. The cochlear modiolus (white arrow) is shown slightly more clearly by PCCT despite the reduced radiation dose. (**c**) PCCT showing the fine stapedial superstructure (arrowheads). (Adapted from Hermans et al. (2023) under the CC BY license)

However, due to its cone-shaped beam geometry, CBCT has a smaller field of view and produces images that are less isotropic than those from conventional CT, which can impact the accuracy of length measurements.

Photon Counting Computed Tomography (PCCT)
PCCT is an emerging type of CT imaging combining a detector capable of counting individual photons with advanced reconstruction algorithms to significantly reduce radiation dose from imaging (see Fig. 13.3). Compared to conventional CT, PCCT offers higher spatial resolution and contrast, while minimizing noise and metal artifacts (Hermans et al. 2023; Hsieh et al. 2021). It can also be used for ultra-low-dose protocols, making it particularly suitable for patients who need frequent imaging for long-term monitoring, as well as for intraoperative imaging (Rak et al. 2024). Compared to conventional CT, PCCT has demonstrated a radiation dose reduction of 87.3% (Spahn et al. 2025).

Micro-CT (μCT)
μCT offers even higher resolution than other CT-based techniques, with voxel sizes as small as 5.9 μm (Braun et al. 2012). However, this comes at the cost of significantly increased radiation exposure, longer acquisition times, higher costs, and a much smaller field of view. These limitations make μCT unsuitable for imaging the live human ear; however, it is an excellent modality for studying cadaveric specimens. Various preparation techniques—such as freezing, fluid removal, wax infiltration, and the use of contrast agents—can further enhance image contrast or preserve the sample. μCT enables detailed visualization of extremely fine structures, such as the basilar membrane (see Fig. 13.4) and allows for clear differentiation of the cochlear scalae. Measurements obtained from μCT imaging are often considered the gold standard (Gee et al. 2021; Iyaniwura et al. 2017). As we will see later in Sect. 13.3.2, detailed models derived from μCT can be used to improve image processing methods applicable to clinical CT images.

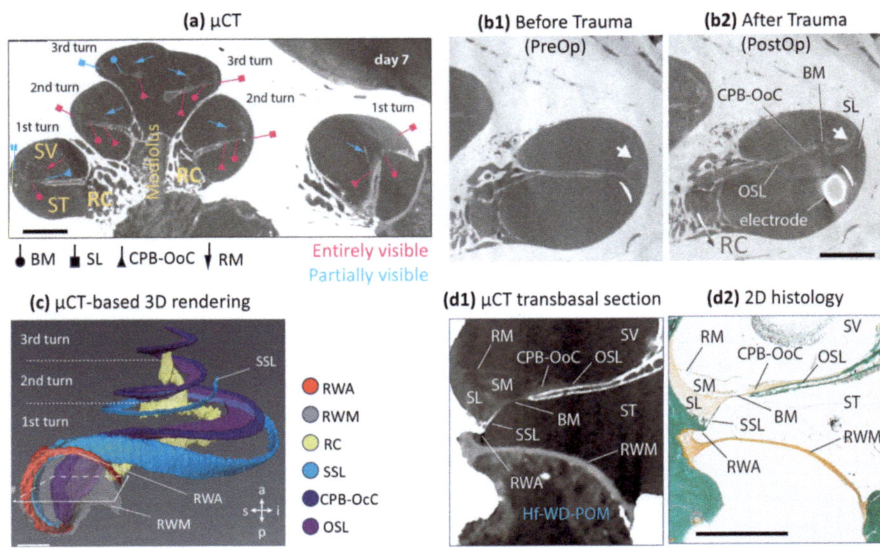

Fig. 13.4 µCT of the cochlea with contrast-enhancing staining agent. (**a**) Transmodiolar cross-section shows bony and soft intra-cochlear structures across the turns on the seventh day post-staining. Compared to CT imaging, BM, SL, CPB-OoC, or RM are observed entirely or partially. (**b**) Cross-section detail of the undisturbed cochlea preoperatively (**b1**) and postoperatively (**b2**), highlighting fractured OSL and deformed BM, CPB, and SL in the ST as a result of a traumatic electrode insertion. (**c**) 3D reconstruction of the cochlear microstructures' geometry helps to design electrode array insertion trajectories. (**d**) Detailed microstructures shown in µCT (**d1**) and 2D histology imaging at the transbasal section indicated in (**c**). (Figure adapted with modification from Starovoyt et al. (2023) under the CC BY license). *BM*, basilar membrane; *SL*, spiral ligament; *CPB-OoC*, cochlear partition bridge—organ of Corti complex; *RM*, Reissner's membrane; *OSL*, osseous spiral lamina; *ST*, scala tympani; *SSL*, secondary spiral lamina; *RWA*, round window arch; *RWM*, round window membrane

Synchrotron Radiation µCT (SRµCT)

SRµCT uses synchrotron radiation to display the finest neural structures, such as the spiral ligament, Rosenthal's canal, Reissner's membrane (Sismono et al. 2021), or dendrites of the spiral ganglion (Li et al. 2021). It also allows for segmenting the basilar membrane into corresponding frequency maps (see Fig. 13.5). Similarly to X-ray-based µCT, this technique is not suitable for in vivo imaging or routine clinical use, but it is valuable for generating detailed anatomical models.

Magnetic Resonance Imaging (MRI)

MRI is a three-dimensional, radiation-free imaging technique, making it especially suitable for pediatric patients who are more sensitive to ionizing radiation. MRI operates by using a strong magnetic field to align the spins of hydrogen protons in tissue water molecules. When radiofrequency pulses disturb this alignment, the emitted signals are captured and computationally reconstructed into detailed 3D images.

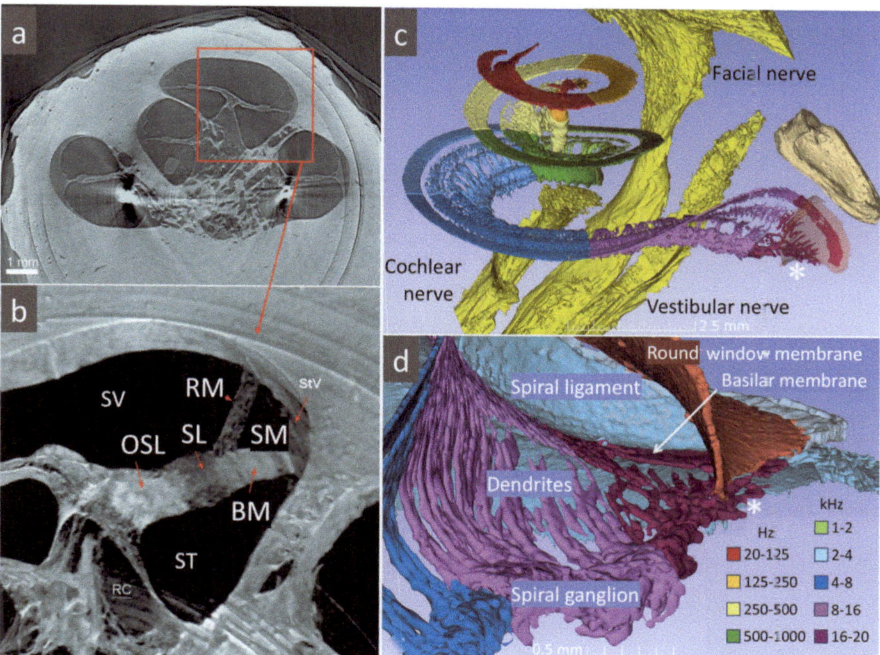

Fig. 13.5 Synchrotron imaging allows the visualization of fine structures. (**a**) Transmodiolar cross-section of the cochlea. (**b**) The microstructures can be seen with more detail than provided by X-ray-based μCT. (**c**) Detailed 3D models of cochlear structures can be segmented, and the corresponding frequency map can be made. (**d**) Individual dendrites of the spiral ganglion can be traced from the basilar membrane. (Figure panels (**a**) and (**b**) reproduced from Sismono et al. (2021), with permission from IUCr Journals; panels (**c**) and (**d**) reproduced from Li et al. (2021) under the CC BY license). *ST*, scala tympani; *SV*, scala vestibuli; *SM*, scala media; *BM*, basilar membrane; *SL*, spiral limbus; *OSL*, osseous spiral lamina; *RC*, Rosenthal's canal; *RM*, Reissner's membrane; *StV*, stria vascularis

MRI excels at visualizing soft tissues and serves as an excellent complement to CT in preoperative assessments. It is particularly effective for identifying cochlear malformations, evaluating the presence and dimensions of the cochlear nerve (Swarup et al. 2023), and detecting fibrosis, malignant tumors (Shapey et al. 2021), or other potential causes of sensorineural hearing loss. Various MRI sequences highlight different tissue properties—for example, T1-weighted MRI provides important information about tumors, and T2-weighted MRI provides clear visualization of the cochlea's general shape, as well as the vestibular and facial nerves (see Fig. 13.6) (Ho et al. 2015; Risbud et al. 2021). Diffusion-weighted imaging (DWI) can be used to evaluate tractography, enabling assessment of the central auditory pathways up to the auditory cortices (Quatre et al. 2024).

MRI has some limitations. It is relatively expensive and typically provides lower resolution (typically greater than 0.3 mm) with less detailed images of bony structures compared to CT despite adapted sequences (Puel et al. 2025). Longer

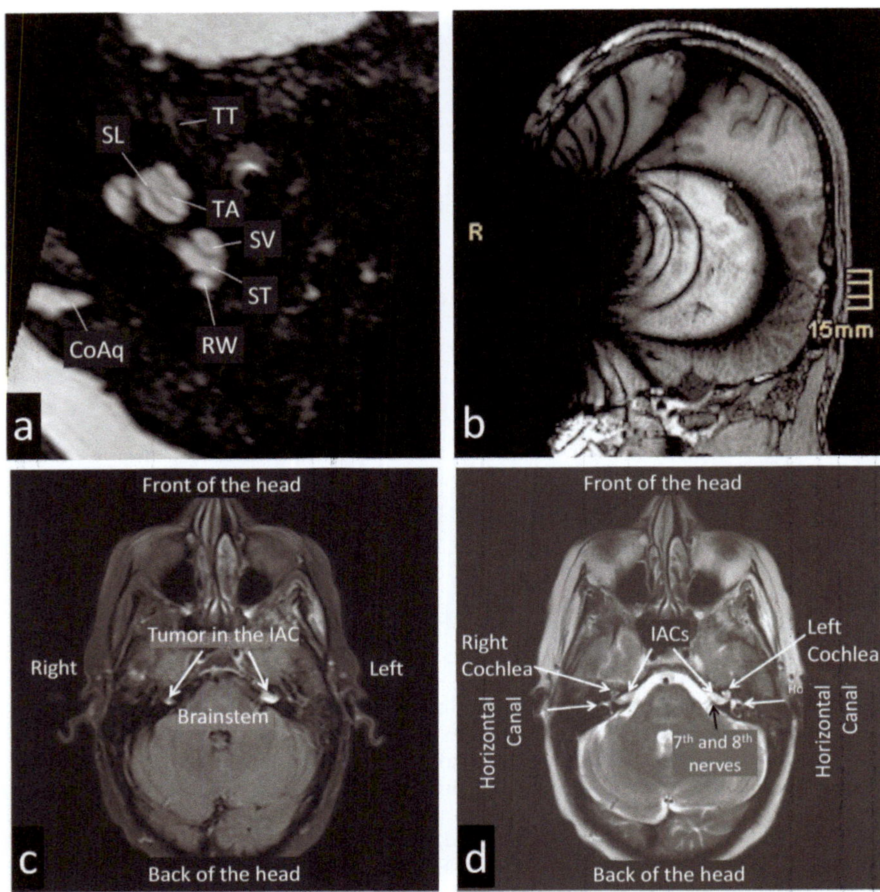

Fig. 13.6 MRI imaging of the cochlea is particularly suited for soft tissue or pediatric patients. (**a**) T2-weighted image shows the fluid-filled compartments as bright signal. (**b**) MRI is currently unsuitable for postoperative assessment due to strong metallic artifacts. (**c**) T1 post-gadolinium MRI demonstrating bilateral enhancement (white) in the internal auditory canals (IACs), indicating the presence of tumors. (**d**) Corresponding T2-weighted MRI showing no distinct tumor in the IACs, which appear fluid-filled (white). The cochlea and vestibular canals are better visualized in T2-weighted MRI due to their fluid content. (Panels (**a**) and (**b**) adapted from Santek et al. (2024) and Gottfried et al. (2022), respectively, under the CC BY license. Panels (**c**) and (**d**) adapted from Risbud et al. (2021) with permission from Wolters Kluwer Health, Inc.)

scan times often require sedation or anesthesia of pediatric patients. While recent cochlear implants are compatible with magnetic fields up to 3T (Risi et al. 2004) and advances in reconstruction algorithms exist (Wagner et al. 2015), MRI is not suitable for postoperative assessment of electrode insertion due to magnetic field distortion from metal implants (Fig. 13.6b). Preoperative MRIs are usually taken as baselines together with other preoperative imaging modalities.

13.2.2 Imaging-Based Measurements and Assessments

13.2.2.1 Preoperative Imaging

As previously discussed, one of the primary roles of image analysis is in preoperative planning, particularly through the creation of patient-specific anatomical models. The unique morphology of each human cochlea has long been studied to identify key features that could help tailor treatment for optimal outcomes. A typical cochlea has a distinct 3D spiral shape, and various approaches have been employed to characterize it.

Traditionally, surgeons and radiologists carefully analyze preoperative cochlear images to identify key anatomical structures and critical landmarks within the cochlea. A smaller-than-average cochlea may signal an abnormality or present surgical challenges. The length and diameter of the cochlear duct can also inform the selection of the electrode array—guiding decisions about its length, stiffness, and type, such as choosing between a pre-curved peri-modiolar EA and a stiffer straight EA. Consequently, measurements of cochlear geometry are routinely obtained prior to surgery. These typically include (see Fig. 13.7a):

- *Basal turn diameter, known as the A value:* The distance between round window and the opposite point on the lateral wall (LW) when passing through the modiolar axis. This typically ranges from 8 to 10 mm.
- *Width of the cochlea, known as the B value:* Measured perpendicularly to the A-value axis. This value is about 7 mm.
- *Cochlear Duct Length (CDL):* The length from the round window to the helicotrema, measured along the LW of the scala tympani. This value is around 39 mm, ranged from 32 to 45 (Alexiades et al. 2015). This measurement is

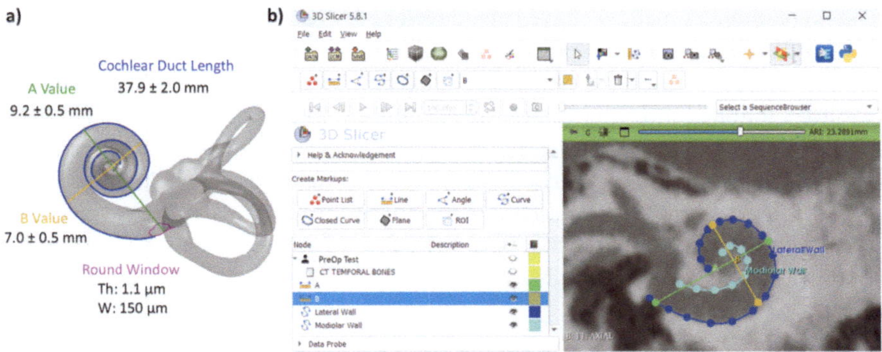

Fig. 13.7 (**a**) Illustration of the cochlear shape annotated with the main cochlear dimensions and a common measuring approach to obtain them in a medical image processing software. (Figure reproduced from Hrncirik et al. (2023) under the CC BY license). (**b**) Screenshot of 3D Slicer, a medical image processing software. The image is first reformatted to the cochlear view, and dimensions such as the A and B values can then be measured by locating two points in the 3D image. The cochlear duct length is typically obtained by tracing points along the lateral wall (blue curve). The modiolar wall is indicated in cyan

sometimes used to determine the required electrode array length and plan appropriate depth of insertion to achieve full cochlear coverage.

- *Cochlear Height (CH):* The vertical distance from the basal turn to the apex of the cochlea, contributing to the overall assessment of cochlear size. This value is about 4 mm.
- *Volume:* Determining the cochlear volume involves annotating all voxels corresponding to the cochlea while excluding non-cochlear structures. This value is mostly found between 50 and 80 mm^3, with an average around 65 mm^3.
- *Scala Tympani Cross-Sectional Measurements:* These include the diameter and cross-sectional area of the scala tympani—where the electrode is typically placed—along the cochlear duct. These measurements vary along the cochlear spiral and influence electrode selection and fit.
- *Rollercoaster:* This refers to the measurement of the largest dip in the height of the basal turn of the cochlear duct, which can indicate potential surgical electrode insertion challenges. This value is mostly found between 0.1 and 0.8 mm, with an average around 0.3 mm.

These measurements are typically performed using radiological software such as 3D Slicer[1] (Fedorov et al. 2012), which is shown in Fig. 13.7b, as well as ITK-SNAP[2] (Yushkevich et al. 2006), RadiAnt,[3] and other similar platforms. The process involves scrolling through image slices to identify anatomical landmarks, followed by placing dimension lines to obtain precise measurements. For cochlear duct length estimation, the lateral wall is usually traced from the round window to the apex. For volume and cross-sectional area measurements, relevant voxels must be annotated—a process known as image segmentation. This manual workflow requires substantial expertise, is time-consuming, and is prone to human error, making consistent and reproducible results challenging. In particular, tracing the duct and annotating voxels are among the most labor-intensive tasks.

Several indirect methods have been developed to approximate the cochlear duct length (CDL), such as the approach by Koch et al. (2017), which estimates CDL by combining simpler measurements like the A and B values. For additional methods that use simplified measurements to approximate CDL, see (Alexiades et al. 2015; Escudé et al. 2006; Schurzig et al. 2018). While these approaches offer practical advantages, more precise 3D anatomical measurements generally yield higher accuracy, albeit at the cost of increased processing time.

It is also important to note that most MRI-derived measurements differ from CT (Eser et al. 2021; Swarup et al. 2023), so extra care is needed before comparing them directly.

[1] https://www.slicer.org

[2] https://www.itksnap.org

[3] https://www.radiantviewer.com

13.2.2.2 Postoperative Imaging

In postoperative imaging, the positions of the EA contacts can be determined by placing a set of landmarks—one for each contact. These are most commonly derived from various imaging modalities. The measurements typically include:

- *Placement Confirmation:* Assessment of the electrode array's position within the cochlea, including measurements of its proximity to neural structures or cochlear walls.
- *Angular Depth of Insertion and Tonotopy:* The angular measurement indicating how far the electrode has been inserted into the cochlear spiral. A 360° insertion corresponds to one full turn. Deeper insertions cover broader frequency regions (as illustrated in Fig. 13.5c, d), helping to minimize frequency-to-place mismatch between the electrode array's tonotopic organization and that of the cochlea, thereby improving auditory outcomes.
- *Insertion Complications:* Identification of potential complications such as electrode fold-over, buckling, or extrusion, as well as estimation of surgical trauma.

Most measurements of the EA position are computed relative to the cochlear anatomy. However, metallic artifacts from the EA in standard CT images often obscure anatomical boundaries. To overcome this, postoperative images are frequently fused with segmented anatomical data derived from corresponding preoperative cochlear scans. Since pre- and postoperative images are rarely acquired in the same frame of reference, image registration is required prior to fusion. This process is typically performed using an automated pipeline that integrates classical image processing methods with AI techniques, which are described in detail in the next section.

13.3 Automating CI Image Analysis with Artificial Intelligence

Manual execution of CI image processing steps is often time-consuming and highly subjective. Artificial intelligence (AI) techniques are increasingly revolutionizing image analysis across the entire field of CI imaging (Lesica et al. 2021; Essaid et al. 2024). In this section, we introduce the key components of a generic AI-based CI image processing pipeline, highlighting the most relevant techniques and algorithms used in inner ear imaging for CI therapy—as illustrated in Fig. 13.8.

13.3.1 Anatomical Landmark Detection

Landmark detection involves identifying specific points within an image that correspond to key anatomical structures. Several of the primary measurements described previously in Sect. 13.2.2 can be derived by calculating the distance between two such landmarks.

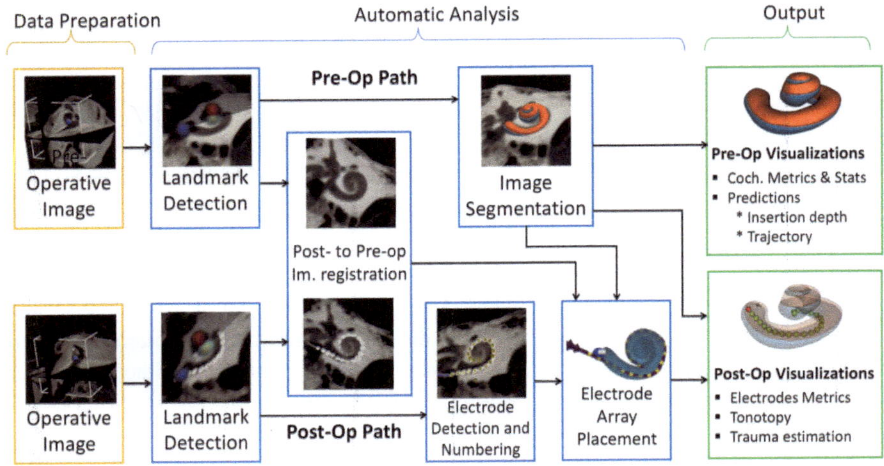

Fig. 13.8 Example of a typical cochlear image analysis steps from preoperative (top row) and postoperative (bottom row) CT imaging. Image registration aligns the two input images and allows to evaluate placement of the detected electrode array

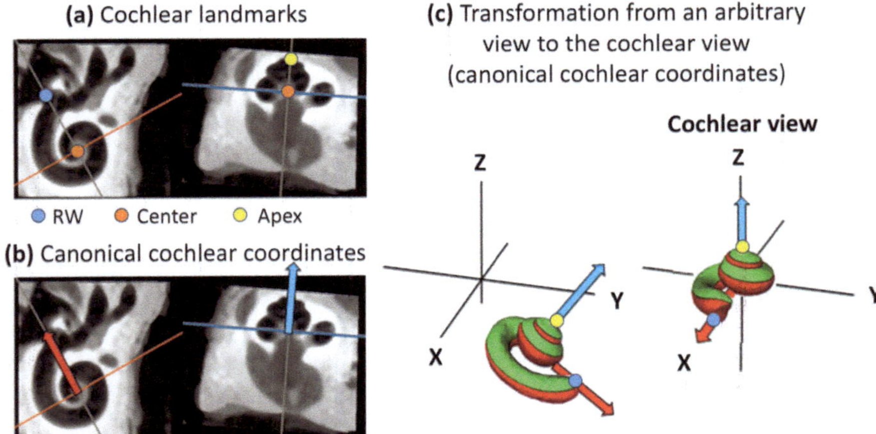

Fig. 13.9 (**a**) Cochlear landmarks: round window (RW), center, and apex as visualized in CT images. (**b**) These three landmarks are sufficient to uniquely define a coordinate frame aligned with the cochlear anatomy. (**c**) Transformation into the cochlear view enables consistent alignment of multiple cochleae for fusion or comparative analysis

In addition to measurement, landmark detection plays a crucial role in image registration—aligning multiple images to a common reference frame—provided the landmarks can be consistently identified across different scans. Ideal landmarks are anatomically unique and clearly distinguishable. Common examples include the round window (RW), the center of the basal turn of the cochlea, and apex (see Fig. 13.9). These three landmarks are sufficient to define canonical cochlear

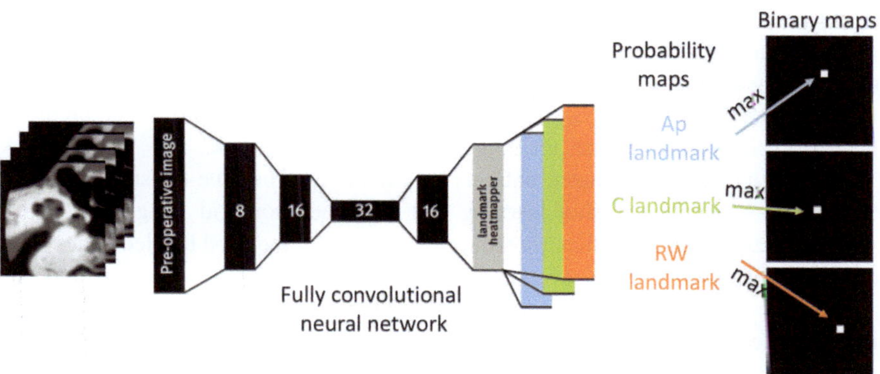

Fig. 13.10 Landmark estimation with fully convolutional neural network. The input images are passed through a set of trainable filters to produce heatmaps of the target landmarks for Apex (Ap), round window (RW) and Center (C) are estimated. In the heatmaps, maxima correspond to the landmark positions

coordinate frames and enable the alignment of multiple cochleae in the standardized cochlear view for fusion or comparative analysis.

Landmark detection is an important initial processing step in cochlear imaging. Using automated techniques to locate various anatomical landmarks helps reduce human error and improves the consistency of extracted measurements. A variety of machine learning and AI approaches have been developed for automated landmark detection in medical imaging. One of the leading methods is heatmap prediction using fully convolutional networks (Margeta et al. 2022; Stebani et al. 2023). These networks can be viewed as tools that can be trained to translate one image into another image of the same dimensions. In the landmark detection task illustrated in Fig. 13.10, the network learns to convert a series of input images (e.g., a CT or MRI scan) into three corresponding output image series with identical dimensions. Each output series represents the probability for one of the specified landmarks (i.e., apex, center, and round window). The values in these output images range from 0 to 1, where a value of 1 indicates that a voxel is at the center of a landmark. These values gradually decrease with distance from the landmark, approaching 0 in regions considered background. By identifying the voxel with the maximum probability (i.e., converting the probability maps into binary images), the position of each landmark can be determined.

In the training process, pairs of input images and corresponding landmark heatmaps (i.e., true probability maps or binary images indicating landmark locations) are used to optimize the neural network's parameters. The input image is passed through the network, and the predicted output is compared to the ground-truth heatmap. The difference between the predicted and actual outputs—known as the loss—is calculated and used to adjust the network's parameters. This adjustment process, called gradient descent, iteratively updates the weights to minimize the loss, gradually improving the network's predictions.

As the network is exposed to a diverse set of training examples, its parameters converge, and it learns to recognize the typical appearance of anatomical features surrounding the landmarks. Once trained, the model can generate accurate heatmaps for previously unseen images. The location of the maximum value in the predicted heatmap is interpreted as the position of the detected landmark.

Although creating annotated datasets for training can be time-consuming, recent techniques—such as one-shot learning that combines nonrigid image registration with prediction—can significantly reduce the amount of required labeled data (Wang et al. 2021b).

Another widely used approach for landmark detection in medical imaging involves deep reinforcement learning techniques (Alansary et al. 2019). These methods train agents to navigate through an image in search of specific landmarks. The agents observe a local image patch, which becomes smaller as they approach the target landmark. At each step, the agent selects an action—such as moving in a particular direction—and receives a reward based on its proximity to the target. Over time, the agent learns an optimal navigation strategy for locating the landmark.

This approach has been successfully applied to detect and characterize neural structures, including cochlear nerves (López Diez et al. 2021), and has demonstrated strong generalization capabilities, even in cases involving cochlear malformations (López Diez et al. 2023).

13.3.2 Image Segmentation for 3D Anatomical Modeling

Segmentation methods are used to characterize the shape of the anatomical structures. Semantic segmentation is a medical image technique that maps each point in an image to a label that determines to which anatomical part that point belongs. There are automated tools that allow for this segmentation process as well as creating the corresponding anatomical models of the unique geometry of each patient's cochlea and the relevant structures. These models provide deeper insights into individual anatomy and enable large-scale automated measurements, making it possible to assess variability in cochlear morphology across broad population groups. They also support applications such as cochlear shape characterization, optimizing the fit of diverse electrode arrays (Dhanasingh and Jolly 2017), and even producing physical replicas of the anatomy through 3D printing (Hrncirik et al. 2023).

Moreover, the insertion of the EA can sometimes be complicated by the access to the RW, using models that allow for intra-cochlear segmentation of the scala tympani, and calculating diameter of its cross-section can help assessing the CI therapy feasibility and optimal approach for better outcome. There are also other complications such as occlusion of the lumen due to fibrotic tissue growth, or the presence of critical geometrical features of the rollercoaster (Avci et al. 2014) that can be outlined with accurate anatomical 3D models.

Image segmentation of the inner ear helps to construct 3D anatomical models of relevant structures, such as the human bony labyrinth, the otic capsule, vestibule and semicircular canals, the scala tympani (ST) and scala vestibuli (SV), facial nerve

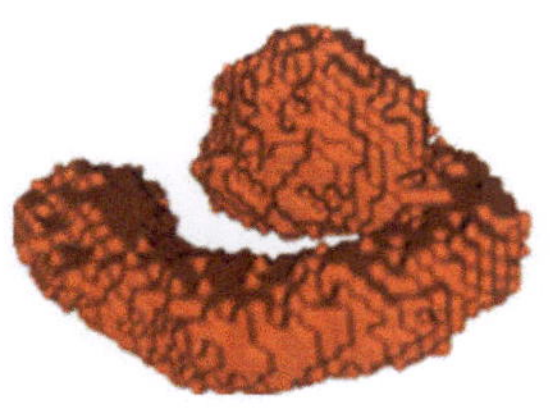

Fig. 13.11 Clinical CT imaging typically reveals only the otic capsule (shown in red), but not the boundaries of finer structures of the cochlea. Segmenting detailed components such as the scala tympani (blue) and scala vestibuli (yellow) can be achieved using knowledge-based approaches, supported by modeling and a priori information obtained from high-resolution modalities such as μCT

(FN), and cochlear nerve, or the auditory nerve (Li et al. 2025). Such modeling also allows to assist surgeons in preoperative planning of the cochlear implantation trajectories by providing accurate anatomical references (Abari et al. 2024; Al Marzouqi et al. 2007). Before the wide use of AI in medical image processing, traditional techniques such as thresholding, active contours, and snakes were commonly used. These methods are still sometimes combined with landmark detection approaches (López Diez et al. 2021) to segment thin neural structures or measure their diameters.

In many clinical imaging modalities, thin structures—such as the basilar membrane, which separates the scala tympani from the scala vestibuli—are difficult to visualize clearly. Shape modeling has been widely employed to segment such structures by leveraging anatomical similarities and variations across individuals. This approach often involves transferring knowledge from high-resolution imaging modalities (e.g., μCT) to guide and enhance segmentation quality, especially when target structures are not well-defined in standard, lower-resolution clinical images such as CT and MRI (see Fig. 13.11).

Parametric shape models represent anatomical structures using explicit mathematical equations—for example, modeling the cochlear duct as a logarithmic spiral. In these models, a small set of parameters controls various aspects of the shape. Patient-specific parameters can be estimated by identifying the best-fitting values, often using techniques such as Bayesian inference and numerical optimization (Demarcy et al. 2017; Wang et al. 2022), to determine the spiral parameters that best match the observed anatomical appearance (see Fig. 13.12).

Another approach to shape modeling involves the use of statistical shape models (SSMs), which are typically constructed using point-distribution representations of anatomical structures (Cootes et al. 1995). These models require point-to-point correspondences across all training shapes, allowing them to be aligned and analyzed for shape variability using principal component analysis (PCA) (Noble et al. 2011, 2012).

A patient-specific shape can then be reconstructed as a linear combination of the principal modes of shape variation, enabling the modeling of complex structures—such as nerves—that are difficult to represent with parametric models. SSMs are also

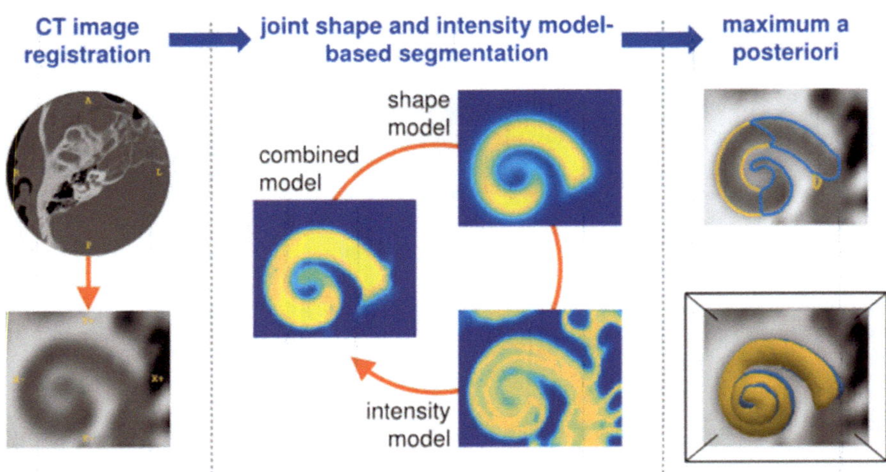

Fig. 13.12 A typical adaptation process of the parametric shape model, as described by Demarcy et al. (2017), involves recursively updating the intensity and shape models using input from the previous step. Overview of the method: Rigid registration is first performed to orient the cochlea within the cochlear coordinate system. Probabilities of intra-cochlear structures are then estimated by iteratively optimizing the shape and intensity model parameters. This optimization minimizes the discrepancy between the posterior probability of the label given: both shape and intensity (combined model), shape only (shape model), and intensity only (intensity model). Finally, hard segmentation is performed using the maximum a posteriori (MAP) estimate of intra-cochlear structure probabilities. The resulting isocontours and isosurfaces are then visualized

highly interpretable, as each principal component corresponds to a specific type of shape variation. Moreover, they perform particularly well even when the training dataset is relatively small.

In inner ear imaging, atlas-based techniques have been widely used for anatomical segmentation. This approach typically involves using one or more reference segmentations (atlases), which are transferred to a new patient's image through image registration techniques (Al-Dhamari et al. 2024; Gare et al. 2020; Powell et al. 2017).

In contrast, deep-learning-based segmentation is usually an end-to-end process, where a model learns to map input images directly to their corresponding label maps. These models are often implemented using fully convolutional neural networks, similar to those used for landmark detection (see Fig. 13.13). However, instead of predicting heatmaps, the network is trained to classify each pixel or voxel into a specific tissue class.

Hourglass convolutional network architectures are commonly used to extract the shape of the cochlea and other inner ear structures from medical images (Hussain et al. 2021; Lv et al. 2021; Margeta et al. 2022; Stebani et al. 2023; Wang et al. 2021a). These architectures are the most widely used segmentation model in deep learning. The model consists of a downward path (for feature extraction) and an upward path (for spatial reconstruction). The network operates by sliding a window across the image and predicting the probability that the center of the window belongs to a target structure (e.g., the cochlea).

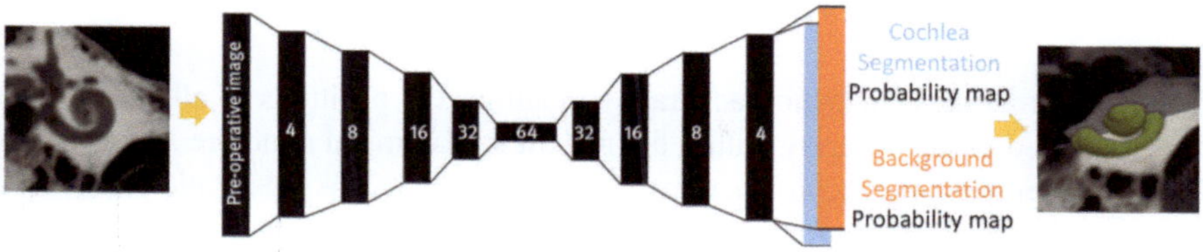

Fig. 13.13 Cochlear segmentation with U-Net network

Such architectures have been successfully applied to segment various anatomical structures of the inner ear from both CT and MRI images. A popular variation is the U-Net architecture (Ronneberger et al. 2015), which enhances the hourglass design with skip connections to preserve spatial detail. Unlike traditional methods, these deep learning approaches require less explicit modeling but typically depend on larger datasets or data augmentation techniques to effectively capture anatomical variability.

In these models, the network is trained to convert a CT image into a probability map distinguishing the cochlea from the background. Since standard CT imaging often lacks the resolution to clearly depict fine structures like the scalar separation, these shape models can be incorporated to enhance segmentation accuracy.

13.3.3 Image Registration Techniques

Image registration is an essential tool for aligning images from different modalities (Al-Dhamari et al. 2022; Bartling et al. 2005; Islam et al. 2021), evaluating preoperative and postoperative images together, or aligning images of different patients to a common reference frame for atlas construction. This enables visualization of the reconstructed electrode array from a postoperative CT image within a segmented cochlea from a preoperative CT or MRI.

This technique typically involves finding a transformation that maximizes the similarity between two input images, usually based on their intensity values. When comparing images of the same patient, the estimated transformation model is typically rigid (allowing only translation and rotation), as the anatomy remains consistent between images. In contrast, nonrigid transformations may be required when registering images from different patients, as demonstrated in (Wang et al. 2021b), or for atlas-based segmentation.

A common approach for initializing the registration algorithm, which often involves an iterative optimization process, is using previously estimated anatomical landmarks to estimate the initial registration (Margeta et al. 2022; Stritzel et al. 2024) followed by fine-tuning of the obtained transformation with more traditional optimization techniques.

Other approaches for localizing and aligning the cochlear region of interest are based on reinforcement learning techniques (Radutoiu et al. 2022) or adding fiducial markers (Regodic et al. 2020).

13.3.4 Electrode Array Detection and Localization

Electrode array (EA) detection accurately identifies the positions of all contacts on the implanted electrode array within the patient's anatomical structure and arranges them in order to enable reconstruction of a 3D EA model. This localization is crucial, as it provides valuable information about the precise placement of each electrode within the cochlea and helps assess whether the implantation was successful. Once EA detection is completed, distances to nearby anatomical structures can be measured, which can be used to estimate potential trauma resulting from the procedure (Eshraghi and Van De Water 2006). For assessing correct placement, outlining the electrode array offers visual support in identifying complications such as kinking, buckling, or tip fold-over (Ishiyama et al. 2020). Furthermore, the electrode locations within the cochlea can be utilized to assign patient-specific tonotopy and minimize place-frequency mismatch (Helpard et al. 2021). The platinum contacts of the array are captured on regular CT images with high-intensity values (see Fig. 13.2d). As such, simple thresholding techniques often cover most of the region of interest of the electrode array. This is often combined with a diverse set of ad hoc image processing filtering techniques (Schuman et al. 2010; Zhao 2018; Zhao et al. 2014) followed by graph-based techniques (Thormählen et al. 2024; Zhao et al. 2019) to reject invalid detections and add additional geometrical constraints.

To further simplify the ad hoc design of image filters, deep-learning-based techniques have been employed (Chi et al. 2019; Margeta et al. 2022) to directly locate electrodes from imaging data. These models can also estimate the position of the lead wire (see Fig. 13.14). Such approaches typically learn electrode heatmaps, similar to previously mentioned landmark detection techniques. After detection, ad hoc postprocessing steps are often applied to impose additional constraints on the detected electrode array or to identify abnormalities such as fold-overs or under-insertions. Once the detected electrode array is registered with the segmented image, the results can be used for anatomy-based CI fitting (Kurz et al. 2023; Li et al. 2021).

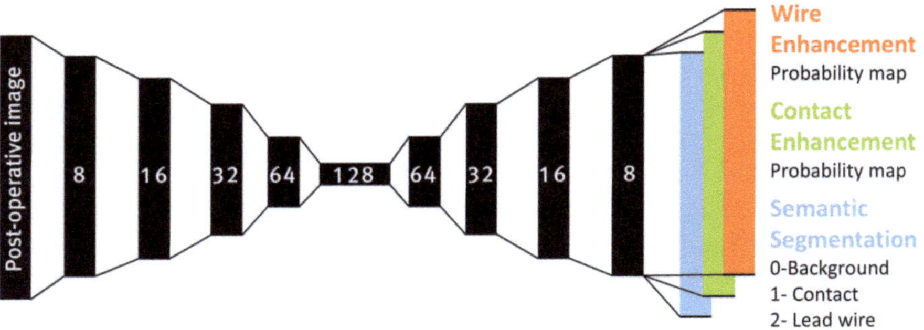

Fig. 13.14 Example network architecture for detection of electrode array (Margeta et al. 2022). Such approach should be followed by a contact ordering processing step

13.4 Emerging Technologies and Clinical Applications

Advances in medical imaging and AI are transforming cochlear implantation by enabling safer, more precise, and personalized procedures. This section highlights two emerging applications of AI in CI imaging: trajectory planning (Wimmer et al. 2014) and the detection of malformed cochleae (Lopez Diez 2024).

Trajectory planning now leverages high-resolution imaging and detailed anatomical segmentation to compute insertion paths that avoid critical structures, particularly in robot-assisted surgeries (Chap. 14). Similarly, AI techniques are being used to diagnose inner ear malformations and predict potential surgical complications.

To support these innovations, specialized software platforms are integrating AI into clinical workflows. Tools like OTOPLAN (Paouris et al. 2023), Nautilus (Margeta et al. 2022), and CImago (Geiger et al. 2023) streamline image analysis for both surgical planning and postoperative evaluation. As imaging data continues to grow, so does the need for robust AI models, ethical data handling, and standardized evaluation protocols. Public datasets such as OpenEar (Sieber et al. 2019) and Cross-Moda (Dorent et al. 2022) are helping to address these challenges, paving the way for scalable, data-driven cochlear care.

13.4.1 Surgical Trajectory Planning

Minimally invasive cochlear implant surgeries aim to provide safe and precise access to the inner ear while minimizing trauma (Abari et al. 2024; Wimmer et al. 2014). Segmentation of anatomical structures and identification of key landmarks are essential components for automating preoperative planning (see Fig. 13.15). This transforms surgical planning into a constrained trajectory computation that respects the patient's anatomy and avoids critical structures such as the chorda tympani and facial nerve, thereby enabling less traumatic insertions. Trajectories can be sampled and evaluated for safety based on their proximity to critical anatomical structures and their alignment with the scala tympani. This approach is particularly well-suited for determining optimal surgical trajectories in robot-assisted procedures.

13.4.2 Detection of Cochlear Malformations

Up to this point, the focus has been primarily on relatively normal cochlear anatomies. However, imaging assumes a significantly more critical role in clinical decision-making when dealing with inner ear malformations, which are present in approximately 20–30% of children with congenital hearing loss (Brotto et al. 2021). In cases where the pathology does not involve damage to the inner ear hair cells, diagnosis typically relies on radiological assessments—most commonly using CT or CBCT imaging modalities, as previously discussed (Lopez Diez 2024).

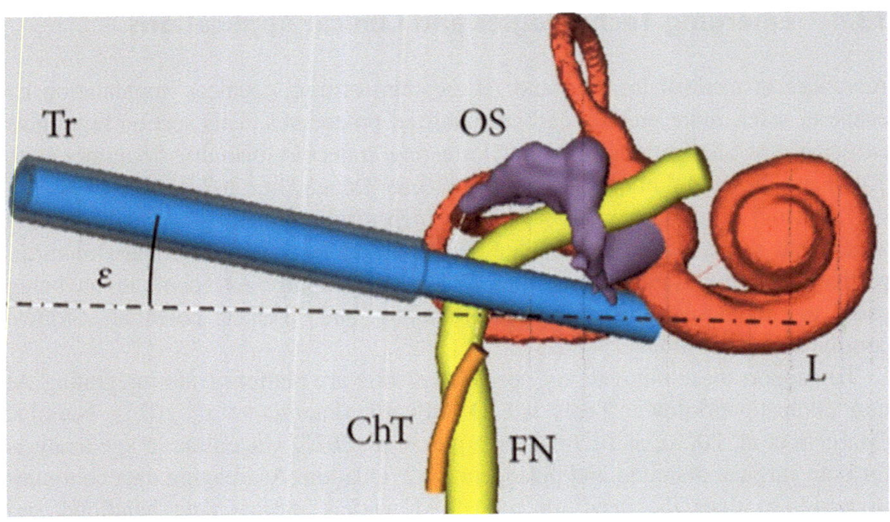

Fig. 13.15 Trajectory planning from high-resolution CT images constrained by Chorda tympani (ChT), Facial nerve (FN), Ossicles (OS), cochlear lumen (L). (Adapted from Wimmer et al. (2014) under the CC BY license)

Each inner ear malformation presents unique anatomical characteristics, which influence both the available treatment strategies and the feasibility of applying automated image processing techniques. These factors often vary not only between different malformation types but also among individual patients (Sennaroğlu and Demir Bajin 2017). One common clinical approach to identifying malformations involves extracting a set of anatomical measurements or descriptive features from imaging data (Dhanasingh et al. 2021). These features, manually annotated by clinicians, assist in classifying the specific type of malformation present.

From an artificial intelligence perspective, such malformations represent out-of-distribution cases, posing additional challenges for automated analysis pipelines. Effectively incorporating these anatomies requires specialized, tailored methodologies. The foundational step in this process is the accurate identification of malformations through medical imaging (see Fig. 13.16).

To detect such anomalies, various techniques have been explored. In particular, autoencoder-based methods have been applied to congenital inner ear malformations (López Diez et al. 2024).

This approach employs an autoencoder—architecturally similar to hourglass networks—trained in an unsupervised manner to compress and reconstruct 3D cochlear images. Since the model is trained exclusively on anatomically normal data, it learns to replicate healthy structures. When presented with abnormal anatomies during inference, the model attempts to reconstruct them as if they were normal, resulting in a measurable reconstruction error. This error serves as a proxy for detecting deviations from normal anatomy (see Fig. 13.17).

Cochlear aplasia | Common cavity | Incomplete partition I | Incomplete partition II | Incomplete partition III | Cochlear hypoplasia | Normal

Fig. 13.16 Examples of 3D cochlear shapes representing both normal and various congenital malformation types. The top row displays the 3D mesh overlaid on the original CT scan, while the bottom row shows the isolated 3D mesh. These cochlear models were generated using the 3D U-Net architecture described by Margeta et al. (2022). (Figure adapted from Lopez Diez 2024)

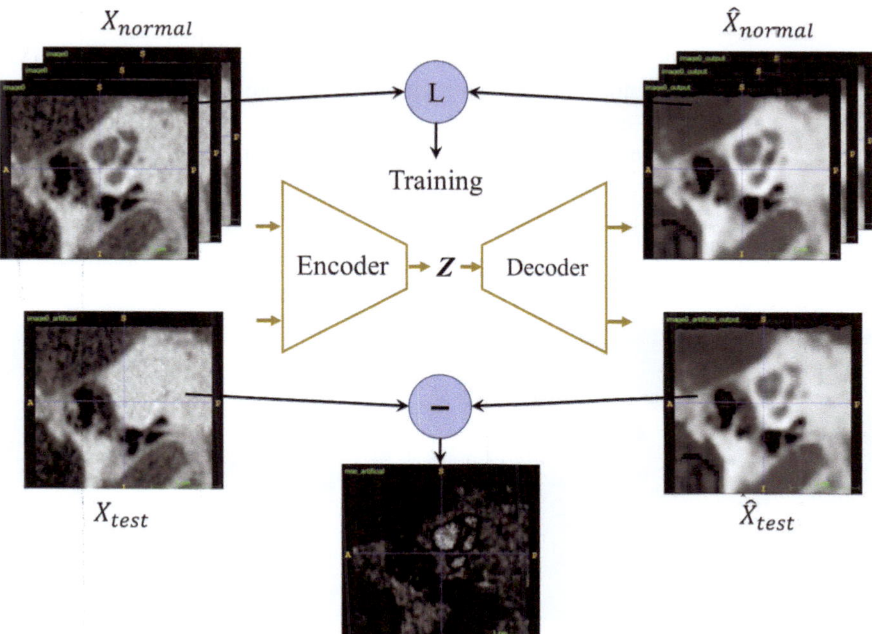

Fig. 13.17 An autoencoder-based anomaly detection framework. During training, the model learns to reconstruct normal cochlear anatomies. Anomalies are identified during inference by measuring the reconstruction error between the input and the output

In addition to voxel-based reconstruction errors, deviations in the spatial configuration of anatomical landmarks can also be analyzed to identify abnormalities (López Diez et al. 2021). More recently, the DeepDiffusion method (López Diez et al. 2023) has shown promise not only in detecting congenital malformations but also in categorizing them into distinct subtypes. This unsupervised technique encodes 3D cochlear shapes (as illustrated in Fig. 13.16) and clusters them based on morphological differences, enabling a more nuanced understanding of anatomical variability.

13.4.3 Software Platforms for Image Analysis and Planning

The rapid evolution of medical imaging technologies, combined with advances in modern machine learning, has opened new avenues for computer-assisted support in cochlear implantation. A growing number of software platforms are being developed to enable data-driven approaches for cochlear implant imaging and surgical planning (Al-Dhamari et al. 2024; Gatto et al. 2023; Margeta et al. 2022; Neves et al. 2021; Siebrecht et al. 2024; Stebani et al. 2023). These platforms integrate state-of-the-art image processing and artificial intelligence techniques, streamlining the workflow for anatomical analysis and clinical decision-making.

Examples include OTOPLAN® (Canfarotta et al. 2019; Paouris et al. 2023), which supports the entire patient journey—including robotic-assisted surgeries (Fourez et al. 2025)—as well as Nautilus™ (Margeta et al. 2022) and CImago Research Software (Geiger et al. 2023), both of which primarily focus on pre- and postoperative cochlear image analysis (Fig. 13.18).

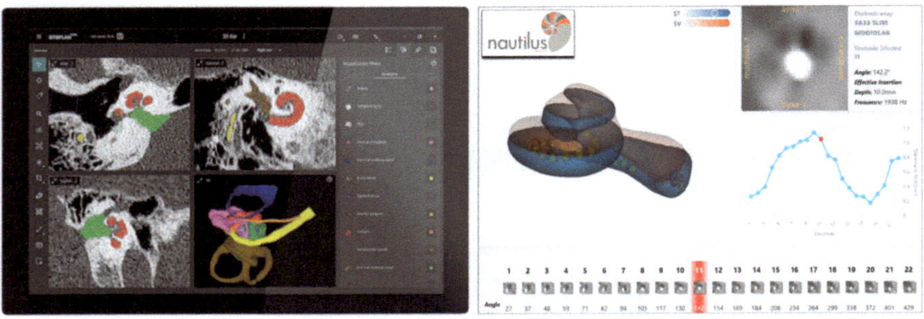

Fig. 13.18 Examples of CI image processing platforms: OTOPLAN® from CAScination AG and MED-EL GmbH (left) and Nautilus™ from Cochlear Limited (right). These platforms take input images and limited user input to generate 3D anatomical models for cochlear measurement an®d surgical planning. Images courtesy of the respective companies

13.4.4 Outlook and Research Prospects

Medical imaging is playing an increasingly important role in the journey of cochlear implant recipients. New imaging modalities are emerging, offering detailed insights into both anatomy and function. Automating the extraction of valuable information from these images enables large-scale analyses (Hussain et al. 2023), allowing us to challenge existing assumptions and gain new understanding of this intricate anatomical region.

The growing volume of imaging data necessitates the use of increasingly powerful AI techniques for analysis. However, this also introduces challenges related to data collection and handling. It is essential to ensure that all patient data—especially personal and sensitive information—is managed in compliance with current regulations, with robust safeguards to prevent data breaches. Additionally, significant effort must be made to ensure that AI models are free from intrinsic biases and that their evaluation follows best practices.

Detecting system failures and adapting to data distribution shifts over time is also critical. Comparing different AI approaches remains difficult, as state-of-the-art methods are often tested on private datasets, limiting direct comparison. Nevertheless, several publicly available datasets help mitigate this issue, including OpenEar (Sieber et al. 2019), which offers multimodal imaging; the Human Bony Labyrinth dataset (Wimmer et al. 2019), which includes co-registered CT and micro-CT images; and Cross-Moda (Dorent et al. 2022), which provides MRI and CT images of the head, often covering cochlear schwannomas.

The application of AI techniques now spans the entire spectrum of cochlear imaging. Thanks to advances in algorithms, increased availability of input data, and greater computational resources, the shift toward automated, data-driven methods is both ongoing and inevitable.

Acknowledgments The authors acknowledge the use of Microsoft Copilot for proofreading and enhancing the readability of the manuscript. All suggested changes were carefully reviewed and edited. The authors assume full responsibility for the content of the final publication.

References

Abari J, Al Saadi M, Van de Heyning P, Topsakal V (2024) Defining the ideal trajectory into the inner ear in image-guided cochlear implant surgery. Sci Rep 14(1):28426. https://doi.org/10.1038/s41598-024-79722-3

Al Marzouqi H, Noble J, Warren F, Labadie R, Fitzpatrick J, Dawant B (2007) Planning a safe drilling path for cochlear implantation surgery using image registration techniques. Proc SPIE Int Soc Opt Eng. https://doi.org/10.1117/12.710672

Alansary A, Oktay O, Li Y, Folgoc LL, Hou B, Vaillant G, Kamnitsas K, Vlontzos A, Glocker B, Kainz B, Rueckert D (2019) Evaluating reinforcement learning agents for anatomical landmark detection. Med Image Anal 53:156–164. https://doi.org/10.1016/j.media.2019.02.007

Al-Dhamari I, Helal R, Morozova O, Abdelaziz T, Jacob R, Paulus D, Waldeck S (2022) Automatic intra-subject registration and fusion of multimodal cochlea 3D clinical images. PLoS One 17(3): e0264449. https://doi.org/10.1371/journal.pone.0264449

Al-Dhamari I, Helal R, Abdelaziz T, Waldeck S, Paulus D (2024) Automatic cochlear multimodal 3D image segmentation and analysis using atlas–model-based method. Cochlear Implants Int. https://www.tandfonline.com/doi/abs/10.1080/14670100.2023.2274199

Alexiades G, Dhanasingh A, Jolly C (2015) Method to estimate the complete and two-turn cochlear duct length. Otol Neurotol 36(5):904–907. https://doi.org/10.1097/MAO.0000000000000620

Aschendorff A (2011) Imaging in cochlear implant patients. GMS Curr Top Otorhinolaryngol Head Neck Surg 10:Doc07. https://doi.org/10.3205/cto000080

Avci E, Nauwelaers T, Lenarz T, Hamacher V, Kral A (2014) Variations in microanatomy of the human cochlea. J Comp Neurol 522. https://doi.org/10.1002/cne.23594

Bartling SH, Peldschus K, Rodt T, Kral F, Matthies H, Kikinis R, Becker H (2005) Registration and fusion of CT and MRI of the temporal bone. J Comput Assist Tomogr 29(3):305–310. https://doi.org/10.1097/01.rct.0000160425.63884.5b

Batts S, Pham N, Tearney G, Stankovic KM (2025) The state of high-resolution imaging of the human inner ear: a look into the black box. Adv Sci 12(28):e00556. https://doi.org/10.1002/advs.202500556

Benson JC, Diehn F, Passe T, Guerin J, Silvera VM, Carlson ML, Lane J (2019) The forgotten second window: a pictorial review of round window pathologies. Am J Neuroradiol. https://doi.org/10.3174/ajnr.A6356

Braun K, Böhnke F, Stark T (2012) Three-dimensional representation of the human cochlea using micro-computed tomography data: presenting an anatomical model for further numerical calculations. Acta Otolaryngol 132(6):603–613. https://doi.org/10.3109/00016489.2011.653670

Brotto D, Sorrentino F, Cenedese R, Avato I, Bovo R, Trevisi P, Manara R (2021) Genetics of inner ear malformations: a review. Audiol Res 11(4):524–536. https://doi.org/10.3390/audiolres11040047

Canfarotta MW, Dillon MT, Buss E, Pillsbury HC, Brown KD, O'Connell BP (2019) Validating a new tablet-based tool in the determination of cochlear implant angular insertion depth. Otol Neurotol 40(8):1006–1010. https://doi.org/10.1097/MAO.0000000000002296

Chi Y, Wang J, Zhao Y, Noble JH, Dawant BM (2019) A deep-learning-based method for the localization of cochlear implant electrodes in CT images. In: 2019 IEEE 16th international symposium on biomedical imaging (ISBI 2019), pp 1141–1145. https://doi.org/10.1109/ISBI.2019.8759536

Cootes TF, Taylor CJ, Cooper DH, Graham J (1995) Active shape models-their training and application. Comput Vis Image Underst 61(1):38–59. https://doi.org/10.1006/cviu.1995.1004

Demarcy T, Vandersteen C, Guevara N, Raffaelli C, Gnansia D, Ayache N, Delingette H (2017) Automated analysis of human cochlea shape variability from segmented μCT images. Comput Med Imaging Graph 59:1–12. https://doi.org/10.1016/j.compmedimag.2017.04.002

Dhanasingh A, Jolly C (2017) An overview of cochlear implant electrode array designs. Hear Res 356:93–103. https://doi.org/10.1016/j.heares.2017.10.005

Dhanasingh A, Erpenbeck D, Assadi MZ, Doyle Ú, Roland P, Hagr A, Van Rompaey V, Van de Heyning P (2021) A novel method of identifying inner ear malformation types by pattern recognition in the mid modiolar section. Sci Rep 11(1):20868. https://doi.org/10.1038/s41598-021-00330-6

Dorent R, Kujawa A, Cornelissen S, Langenhuizen P, Shapey J, Vercauteren T (2022) Cross-modality domain adaptation challenge 2022 (crossMoDA) [Dataset]. Zenodo. https://doi.org/10.5281/zenodo.6504722

Escudé B, James C, Deguine O, Cochard N, Eter E, Fraysse B (2006) The size of the cochlea and predictions of insertion depth angles for cochlear implant electrodes. Audiol Neurotol 11(Suppl 1):27–33. https://doi.org/10.1159/000095611

Eser MB, Atalay B, Dogan MB, Gündüz N, Kalcioglu MT (2021) Measuring 3D cochlear duct length on MRI: is it accurate and reliable? AJNR Am J Neuroradiol 42(11):2016–2022. https://doi.org/10.3174/ajnr.A7287

Eshraghi AA, Van De Water TR (2006) Cochlear implantation trauma and noise-induced hearing loss: apoptosis and therapeutic strategies. Anat Rec A: Discov Mol Cell Evol Biol 288A(4):473–481. https://doi.org/10.1002/ar.a.20305

Essaid B, Kheddar H, Batel N, Chowdhury MEH, Lakas A (2024) Artificial intelligence for cochlear implants: review of strategies, challenges, and perspectives. IEEE Access 1–1. https://doi.org/10.1109/ACCESS.2024.3429524

Fedorov A, Beichel R, Kalpathy-Cramer J, Finet J, Fillion-Robin J-C, Pujol S, Bauer C, Jennings D, Fennessy F, Sonka M, Buatti J, Aylward S, Miller JV, Pieper S, Kikinis R (2012) 3D slicer as an image computing platform for the quantitative imaging network. Magn Reson Imaging 30(9): 1323–1341. https://doi.org/10.1016/j.mri.2012.05.001

Fourez A-L, Kaderbay A, Villerabel C, Korchagina J, Pean V, Mondain M, Venail F (2025) Implementation of robot-assisted minimally invasive cochlear implantation: a feasibility study. Otol Neurotol 46(7):809–815. https://doi.org/10.1097/MAO.0000000000004531

Gare BM, Hudson T, Rohani SA, Allen DG, Agrawal SK, Ladak HM (2020) Multi-atlas segmentation of the facial nerve from clinical CT for virtual reality simulators. Int J Comput Assist Radiol Surg 15(2):259–267. https://doi.org/10.1007/s11548-019-02091-0

Gatto A, Tofanelli M, Costariol L, Rizzo S, Borsetto D, Gardenal N, Uderzo F, Boscolo-Rizzo P, Tirelli G (2023) Otological planning software—OTOPLAN: a narrative literature review. Audiol Res 13(5):Article 5. https://doi.org/10.3390/audiolres13050070

Gee AH, Zhao Y, Treece GM, Bance ML (2021) Practicable assessment of cochlear size and shape from clinical CT images. Sci Rep 11(1):3448. https://doi.org/10.1038/s41598-021-83059-6

Geiger S, Iso-Mustajärvi M, Nauwelaers T, Avci E, Julkunen P, Linder P, Silvast T, Dietz A (2023) Automatic electrode scalar location assessment after cochlear implantation using a novel imaging software. Sci Rep 13(1):12416. https://doi.org/10.1038/s41598-023-39275-3

Gottfried TM, Dejaco D, Fischer N, Innerhofer V, Chacko LJ, Widmann G, Kremser C, Riechelmann H, Schmutzhard J (2022) Dimensions and forms of artefacts in 1.5 T and 3 T MRI caused by cochlear implants. Sci Rep 12(1):4884. https://doi.org/10.1038/s41598-022-08988-2

Helpard L, Li H, Rohani SA, Zhu N, Rask-Andersen H, Agrawal S, Ladak HM (2021) An approach for individualized cochlear frequency mapping determined from 3D synchrotron radiation phase-contrast imaging. IEEE Trans Biomed Eng 68(12):3602–3611. https://doi.org/10.1109/TBME.2021.3080116

Hermans R, Boomgaert L, Cockmartin L, Binst J, De Stefanis R, Bosmans H (2023) Photon-counting CT allows better visualization of temporal bone structures in comparison with current generation multi-detector CT. Insights Imaging 14(1):112. https://doi.org/10.1186/s13244-023-01467-w

Ho M-L, Juliano A, Eisenberg RL, Moonis G (2015) Anatomy and pathology of the facial nerve. Am J Roentgenol 204(6):W612–W619. https://doi.org/10.2214/AJR.14.13444

Hrncirik F, Roberts I, Sevgili I, Swords C, Bance M (2023) Models of cochlea used in cochlear implant research: a review. Ann Biomed Eng 51(7):1390–1407. https://doi.org/10.1007/s10439-023-03192-3

Hsieh SS, Leng S, Rajendran K, Tao S, McCollough CH (2021) Photon counting CT: clinical applications and future developments. IEEE Trans Radiat Plasma Med Sci 5(4):441–452. https://doi.org/10.1109/TRPMS.2020.3020212

Hussain R, Lalande A, Girum KB, Guigou C, Bozorg Grayeli A (2021) Automatic segmentation of inner ear on CT-scan using auto-context convolutional neural network. Sci Rep 11:4406. https://doi.org/10.1038/s41598-021-83955-x

Hussain R, Frater A, Calixto R, Karoui C, Margeta J, Wang Z, Hoen M, Delingette H, Patou F, Raffaelli C, Vandersteen C, Guevara N (2023) Anatomical variations of the human cochlea using an image analysis tool. J Clin Med 12(2):509. https://doi.org/10.3390/jcm12020509

Ishiyama A, Risi F, Boyd P (2020) Potential insertion complications with cochlear implant electrodes. Cochlear Implants Int 21(4):206–219. https://doi.org/10.1080/14670100.2020.1730066

Islam KT, Wijewickrema S, O'Leary S (2021) A deep learning based framework for the registration of three dimensional multi-modal medical images of the head. Sci Rep 11(1):1860. https://doi.org/10.1038/s41598-021-81044-7

Iyaniwura JE, Elfarnawany M, Riyahi-Alam S, Sharma M, Kassam Z, Bureau Y, Parnes LS, Ladak HM, Agrawal SK (2017) Intra- and interobserver variability of cochlear length measurements in clinical CT. Otol Neurotol 38(6):828. https://doi.org/10.1097/MAO.0000000000001411

Jerosha S, Subramonian SG, Mohanakrishnan A, Muralidharan Y, Natarajan P (2024) Preoperative assessment using CT and MRI scans of the temporal bone to determine the degree of difficulty in cochlear implant surgery. Cureus 16(7):e65196. https://doi.org/10.7759/cureus.65196

Koch RW, Ladak HM, Elfarnawany M, Agrawal SK (2017) Measuring cochlear duct length—a historical analysis of methods and results. J Otolaryngol Head Neck Surg 46(1):19. https://doi.org/10.1186/s40463-017-0194-2

Kurz A, Herrmann D, Hagen R, Rak K (2023) Using anatomy-based fitting to reduce frequency-to-place mismatch in experienced bilateral cochlear implant users: a promising concept. J Pers Med 13(7):1109. https://doi.org/10.3390/jpm13071109

Lesica NA, Mehta N, Manjaly JG, Deng L, Wilson BS, Zeng F-G (2021) Harnessing the power of artificial intelligence to transform hearing healthcare and research. Nat Mach Intell 3(10): 840–849. https://doi.org/10.1038/s42256-021-00394-z

Li H, Helpard L, Ekeroot J, Rohani SA, Zhu N, Rask-Andersen H, Ladak HM, Agrawal S (2021) Three-dimensional tonotopic mapping of the human cochlea based on synchrotron radiation phase-contrast imaging. Sci Rep 11(1):Article 1. https://doi.org/10.1038/s41598-021-83225-w

Li W, Wijewickrema S, Margeta J, Kamraoui R, Hussain R, Gerard J-M (2025) A systematic review of automated temporal bone segmentation methods. Biomed Eng Adv 10:100195. https://doi.org/10.1016/j.bea.2025.100195

Lopez Diez P (2024) Deep learning methods in 3D computed tomography images for implantable devices. PhD, Technical University of Denmark

López Diez P, Sundgaard JV, Patou F, Margeta J, Paulsen RR (2021) Facial and cochlear nerves characterization using deep reinforcement learning for landmark detection. In: de Bruijne M, Cattin PC, Cotin S, Padoy N, Speidel S, Zheng Y, Essert C (eds) Medical image computing and computer assisted intervention—MICCAI 2021. Springer, pp 519–528. https://doi.org/10.1007/978-3-030-87202-1_50

López Diez P, Margeta J, Diab K, Patou F, Paulsen RR (2023) Unsupervised classification of congenital inner ear malformations using deepdiffusion for latent space representation. In: Greenspan H, Madabhushi A, Mousavi P, Salcudean S, Duncan J, Syeda-Mahmood T, Taylor R (eds) Medical image computing and computer assisted intervention—MICCAI 2023. Springer, pp 652–662. https://doi.org/10.1007/978-3-031-43904-9_63

López Diez P, Sundgaard JV, Margeta J, Diab K, Patou F, Paulsen RR (2024) Deep reinforcement learning and convolutional autoencoders for anomaly detection of congenital inner ear malformations in clinical CT images. Comput Med Imaging Graph 113:102343. https://doi.org/10.1016/j.compmedimag.2024.102343

Lv Y, Ke J, Xu Y, Shen Y, Wang J, Wang J (2021) Automatic segmentation of temporal bone structures from clinical conventional CT using a CNN approach. Int J Med Robot Comput Assist Surg 17(2):e2229. https://doi.org/10.1002/rcs.2229

Margeta J, Hussain R, López Diez P, Morgenstern A, Demarcy T, Wang Z, Gnansia D, Martinez Manzanera O, Vandersteen C, Delingette H, Buechner A, Lenarz T, Patou F, Guevara N (2022) A web-based automated image processing research platform for cochlear implantation-related studies. J Clin Med 11(22):6640. https://doi.org/10.3390/jcm11226640

McClenaghan F, Nash R (2020) The modified Stenver's view for cochlear implants—what do the surgeons want to know? J Belg Soc Radiol 104(1):37. https://doi.org/10.5334/jbsr.2059

Mom T, Puechmaille M, El Yagoubi M, Lère A, Petersen J-E, Bécaud J, Saroul N, Gilain L, Mirafzal S, Chabrot P (2022) Robotized cochlear implantation under fluoroscopy: a preliminary series. J Clin Med 12(1):211. https://doi.org/10.3390/jcm12010211

Neves CA, Tran ED, Kessler IM, Blevins NH (2021) Fully automated preoperative segmentation of temporal bone structures from clinical CT scans. Sci Rep 11(1):116. https://doi.org/10.1038/s41598-020-80619-0

Noble JH, Labadie RF, Majdani O, Dawant BM (2011) Automatic segmentation of intra-cochlear anatomy in conventional CT. IEEE Trans Biomed Eng 58(9):2625–2632. https://doi.org/10.1109/TBME.2011.2160262

Noble JH, Gifford RH, Labadie RF, Dawant BM (2012) Statistical shape model segmentation and frequency mapping of cochlear implant stimulation targets in CT. Med Image Comput Comput Assist Interv 15(Pt 2):421–428. https://doi.org/10.1007/978-3-642-33418-4_52

Paouris D, Kunzo S, Goljerová I (2023) Validation of automatic cochlear measurements using OTOPLAN® software. J Pers Med 13(5):Article 5. https://doi.org/10.3390/jpm13050805

Perazzini C, Puechmaille M, Saroul N, Plainfossé O, Montrieul L, Bécaud J, Gilain L, Chabrot P, Boyer L, Mom T (2021) Fluoroscopy guided electrode-array insertion for cochlear implantation with straight electrode-arrays: a valuable tool in most cases. Eur Arch Otorrinolaringol 278(4): 965–975. https://doi.org/10.1007/s00405-020-06151-z

Powell KA, Liang T, Hittle B, Stredney D, Kerwin T, Wiet GJ (2017) Atlas-based segmentation of temporal bone anatomy. Int J Comput Assist Radiol Surg 12(11):1937–1944. https://doi.org/10.1007/s11548-017-1658-6

Puel U, Beysang A, Hossu G, Eliezer M, Assabah B, Ambarki K, Gondim Teixeira FA, Blum A, Parietti-Winkler C, Gillet R (2025) Comparison of CT-like MRI sequences for preoperative planning of cochlear implantation using super-high-resolution CT as a reference. Eur Radiol Exp 9(1):1. https://doi.org/10.1186/s41747-024-00538-x

Quatre R, Schmerber S, Attyé A (2024) Improving rehabilitation of deaf patients by advanced imaging before cochlear implantation. J Neuroradiol 51(2):145–154. https://doi.org/10.1016/j.neurad.2023.10.002

Radutoiu A-T, Patou F, Margeta J, Paulsen RR, López Diez P (2022) Accurate localization of inner ear regions of interests using deep reinforcement learning. In: Lian C, Cao X, Rekik I, Xu X, Cui Z (eds) Machine learning in medical imaging. Springer, pp 416–424. https://doi.org/10.1007/978-3-031-21014-3_43

Rak K, Spahn B, Müller-Graff F-T, Engert J, Voelker J, Hackenberg S, Hagen R, Petritsch B, Grunz J-P, Bley T, Neun T, Huflage H (2024) The photon-counting CT enters the field of Cochlear implantation: comparison to angiography DynaCT and conventional multislice CT. Otol Neurotol 45(6):662. https://doi.org/10.1097/MAO.0000000000004221

Regodic M, Bardosi Z, Freysinger W (2020) Automatic fiducial marker detection and localization in CT images: a combined approach. In: Medical imaging 2020: image-guided procedures, robotic interventions, and modeling, vol 11315, pp 507–514. https://doi.org/10.1117/12.2548852

Risbud A, Muhonen E, Abouzari M, Djalilian HR (2021) Symptoms: sudden hearing loss and facial paralysis. Hear J 74(5) https://journals.lww.com/thehearingjournal/fulltext/2021/05000/symptoms__sudden_hearing_loss_and_facial_paralysis.12.aspx

Risi F, Saldanha A, Leigh R, Gibson P (2004) Magnetic resonance imaging safety of Nucleus® 24 cochlear implants at 3.0 T. Int Congr Ser 1273:394–398. https://doi.org/10.1016/j.ics.2004.07.033

Ronneberger O, Fischer P, Brox T (2015) U-net: convolutional networks for biomedical image segmentation. In: Navab N, Hornegger J, Wells WM, Frangi AF (eds) Medical image computing and computer-assisted intervention—MICCAI 2015. Springer, pp 234–241. https://doi.org/10.1007/978-3-319-24574-4_28

Santek T, Hofmann E, Milewski C, Schwager K, Prescher A (2024) Clinical high-resolution imaging of the inner ear by using magnetic resonance imaging (MRI) and cone beam computed tomography (CBCT). J Pers Med 14(6):Article 6. https://doi.org/10.3390/jpm14060637

Schuman TA, Noble JH, Wright CG, Wanna GB, Dawant B, Labadie RF (2010) Anatomic verification of a novel method for precise intrascalar localization of cochlear implant electrodes in adult temporal bones using clinically available computed tomography. Laryngoscope 120(11):2277–2283. https://doi.org/10.1002/lary.21104

Schurzig D, Timm ME, Batsoulis C, Salcher R, Sieber D, Jolly C, Lenarz T, Zoka-Assadi M (2018) A novel method for clinical cochlear duct length estimation toward patient-specific cochlear implant selection. OTO Open 2(4):2473974X18800238. https://doi.org/10.1177/2473974X18800238

Sennaroğlu L, Demir Bajin M (2017) Classification and current management of inner ear malformations. Balkan Med J 34(5):397–411. https://doi.org/10.4274/balkanmedj.2017.0367

Shapey J, Kujawa A, Dorent R, Wang G, Dimitriadis A, Grishchuk D, Paddick I, Kitchen N, Bradford R, Saeed SR, Bisdas S, Ourselin S, Vercauteren T (2021) Segmentation of vestibular schwannoma from MRI, an open annotated dataset and baseline algorithm. Sci Data 8(1):286. https://doi.org/10.1038/s41597-021-01064-w

Sieber DM, Erfurt P, John S, Ribeiro dos Santos G, Schurzig D, Sørensen MS, Lenarz T (2019) The OpenEar library of 3D models of the human temporal bone based on computed tomography and micro-slicing [Dataset]. Zenodo. https://doi.org/10.5281/zenodo.1473724

Siebrecht M, Briaire JJ, Verbist BM, Kalkman RK, Frijns JHM (2024) Automated segmentation of clinical CT scans of the cochlea and analysis of the cochlea's vertical profile. Heliyon 10(16): e35737. https://doi.org/10.1016/j.heliyon.2024.e35737

Sismono F, Mancini L, Leblans M, Goyens J, De Greve G, Schneiders S, Beckers K, Dirckx J, De Foer B, Zarowski A (2021) Synchrotron radiation X-ray microtomography for the visualization of intra-cochlear anatomy in human temporal bones implanted with a perimodiolar cochlear implant electrode array. J Synchrotron Radiat 28(1):327–332. https://doi.org/10.1107/S1600577520014952

Spahn B, Müller-Graff F-T, Rak K, Engert J, Voelker J, Hackenberg S, Hagen R, Neun T, Petritsch B, Bley TA, Grunz J-P, Huflage H (2025) Pre- and postoperative imaging of cochlear implantation in cadaveric specimens using low-dose photon-counting detector CT. Am J Neuroradiol 46(2):362–371. https://doi.org/10.3174/ajnr.A8533

Starovoyt A, Pyka G, Putzeys T, Balcaen T, Wouters J, Kerckhofs G, Verhaert N (2023) Human cochlear microstructures at risk of electrode insertion trauma, elucidated in 3D with contrast-enhanced microCT. Sci Rep 13(1):2191. https://doi.org/10.1038/s41598-023-29401-6

Stebani J, Blaimer M, Zabler S, Neun T, Pelt DM, Rak K (2023) Towards fully automated inner ear analysis with deep-learning-based joint segmentation and landmark detection framework. Sci Rep 13(1):19057. https://doi.org/10.1038/s41598-023-45466-9

Stritzel J, Ebrahimzadeh AH, Büchner A, Lanfermann H, Marschollek M, Wolff D (2024) Landmark-based registration of a cochlear model to a human cochlea using conventional CT scans. Sci Rep 14(1):1115. https://doi.org/10.1038/s41598-023-50632-0

Swarup A, Karakkandy V, Chappity P, Naik S, Behera SK, Parida PK, Grover M, Gupta G, Giri PP, Sarkar S, Pradhan P, Samal DK, Kallyadan Veetil A, Adhikari A, Nayak S (2023) Comparing accuracy of cochlear measurements on magnetic resonance imaging and computed tomography: a step towards radiation-free cochlear implantation. J Otolaryngol 18(4):208–213. https://doi.org/10.1016/j.joto.2023.08.001

Thormählen J, Krüger B, Nogueira W (2024) Automatic localization of cochlear implant electrodes using cone beam computed tomography images. Biomed Eng Online 23(1):65. https://doi.org/10.1186/s12938-024-01249-5

Vaid S, Vaid N, Manikoth M, Zope A (2015) Role of HRCT and MRI of the temporal bone in predicting and grading the degree of difficulty of cochlear implant surgery. Indian J Otolaryngol Head Neck Surg 67(2):150–158. https://doi.org/10.1007/s12070-015-0858-z

Vogl T, Tawfik A, Emam A, Naguib N, Nour-Eldin A, Burck I, Stöver T (2015) Pre-, intra- and post-operative imaging of cochlear implants. RöFo 187(11):980–989. https://doi.org/10.1055/s-0035-1553413

Wagner F, Wimmer W, Leidolt L, Vischer M, Weder S, Wiest R, Mantokoudis G, Caversaccio MD (2015) Significant artifact reduction at 1.5T and 3T MRI by the use of a cochlear implant with removable magnet: an experimental human cadaver study. PLoS One 10(7):e0132483. https://doi.org/10.1371/journal.pone.0132483

Wang J, Lv Y, Wang J, Ma F, Du Y, Fan X, Wang M, Ke J (2021a) Fully automated segmentation in temporal bone CT with neural network: a preliminary assessment study. BMC Med Imaging 21(1):166. https://doi.org/10.1186/s12880-021-00698-x

Wang Z, Vandersteen C, Raffaelli C, Guevara N, Patou F, Delingette H (2021b) One-shot learning for landmarks detection. In: Engelhardt S, Oksuz I, Zhu D, Yuan Y, Mukhopadhyay A, Heller N, Huang SX, Nguyen H, Sznitman R, Xue Y (eds) Deep generative models, and data augmentation, labelling, and imperfections. Springer, pp 163–172. https://doi.org/10.1007/978-3-030-88210-5_15

Wang Z, Demarcy T, Vandersteen C, Gnansia D, Raffaelli C, Guevara N, Delingette H (2022) Bayesian logistic shape model inference: application to cochlear image segmentation. Med Image Anal 75:102268. https://doi.org/10.1016/j.media.2021.102268

Widmann G, Dejaco D, Luger A, Schmutzhard J (2020) Pre- and post-operative imaging of cochlear implants: a pictorial review. Insights Imaging 11(1):93. https://doi.org/10.1186/s13244-020-00902-6

Wimmer W, Venail F, Williamson T, Akkari M, Gerber N, Weber S, Caversaccio M, Uziel A, Bell B (2014) Semiautomatic cochleostomy target and insertion trajectory planning for minimally invasive cochlear implantation. Biomed Res Int 2014:e596498. https://doi.org/10.1155/2014/596498

Wimmer W, Anschuetz L, Weder S, Wagner F, Delingette H, Caversaccio M (2019) Human bony labyrinth: co-registered CT and micro-CT images, surface models and anatomical landmarks [Dataset]. Zenodo. https://doi.org/10.5281/zenodo.3355272

Yushkevich PA, Piven J, Hazlett HC, Smith RG, Ho S, Gee JC, Gerig G (2006) User-guided 3D active contour segmentation of anatomical structures: significantly improved efficiency and reliability. NeuroImage 31(3):1116–1128. https://doi.org/10.1016/j.neuroimage.2006.01.015

Zhao Y (2018) Automatic techniques for cochlear implant CT image analysis. Dissertation, Vanderbilt University

Zhao Y, Dawant BM, Labadie RF, Noble JH (2014) Automatic localization of cochlear implant electrodes in CT. In: Golland P, Hata N, Barillot C, Hornegger J, Howe R (eds) Medical image computing and computer-assisted intervention—MICCAI 2014, vol 8673. Springer, pp 331–338. https://doi.org/10.1007/978-3-319-10404-1_42

Zhao Y, Chakravorti S, Labadie RF, Dawant BM, Noble JH (2019) Automatic graph-based method for localization of cochlear implant electrode arrays in clinical CT with sub-voxel accuracy. Med Image Anal 52:1–12. https://doi.org/10.1016/j.media.2018.11.005

Advanced Surgical Techniques Using Robotics and Augmented Reality

14

Raabid Hussain, Shayan Alvansazyadi, Giuseppe Ricciardi, and Razeen Hussain

Abstract

The integration of advanced technologies such as virtual reality (VR), navigation systems, augmented reality (AR), and robotics aims to revolutionize cochlear implant (sCI) surgery. This chapter explores the transformative impact of these innovations on surgical precision, patient outcomes, and training methodologies. VR provides immersive simulations for preoperative planning and surgical training, enhancing the surgeon's spatial awareness and procedural confidence. Navigation systems offer real-time guidance, improving the accuracy of electrode placement and reducing the risk of complications. AR overlays critical anatomical information onto the surgical field, allowing for more informed decision-making during the procedure. Robotics introduces a new level of precision and control, enabling minimally invasive techniques and reducing surgeon fatigue. Despite these advancements, the full potential of these technologies has not yet been fully realized, and the advantages of their use are still under investigation. Together, these technologies are not only advancing the field of CI surgery but also setting new standards for surgical excellence and patient care.

Keywords

Augmented reality · Robotics · Cochlear implant surgery · Virtual reality · Navigation · Mixed reality

R. Hussain (✉) · G. Ricciardi
Advanced Innovation Department, Cochlear Ltd, Vallauris, France
e-mail: rhussain@cochlear.com

S. Alvansazyadi · R. Hussain
Department of Informatics, Bioengineering, Robotics, and Systems Engineering, University of Genoa, Genova, Italy
e-mail: razeen.hussain@edu.unige.it

B. Molaee-Ardekani, M. Segovia-Martinez (eds.), *Advancing Cochlear Implants*,
https://doi.org/10.1007/978-3-032-15719-5_14

14.1 Introduction

Cochlear implant (CI) surgery has had a transformative impact on individuals with profound hearing loss by enhancing auditory skills, speech, and language development, as well as academic achievement and quality of life (Cejas et al. 2023; Marschark et al. 2007; McRackan et al. 2018). Outcomes vary based on factors like age at implantation, prior language skills, hearing loss cause, time of hearing deprival, and implant use consistency (Carlson et al. 2023; Marschark et al. 2007). Recent advances in CI devices, electrodes, and surgical techniques have expanded eligibility and effectiveness, with a focus on preserving inner ear structures and minimizing intracochlear trauma (Eshraghi et al. 2012; Hamed et al. 2024).

Applying preservation principles benefits all recipients, including those with moderate and severe hearing loss, by protecting residual hearing and enhancing speech perception, while reducing complications like fibrosis and easing future reimplantation if needed (Finley et al. 2008; Mamelle et al. 2020; Schaefer et al. 2021). Future technologies such as cellular regeneration could also be enabled by limiting intracochlear injury (Devare et al. 2018).

Like any surgery, CI carries risks, including infection, bleeding, and potential facial nerve damage. Some patients may experience balance issues, worsened tinnitus, and other complications like fluid leaks, taste changes, or electrode migration (Binnetoglu et al. 2020). Due to the cochlea's proximity to sensitive structures, high precision is crucial in CI surgery, where technologies like robotics and augmented reality (AR) aid surgeons with submillimeter accuracy, minimizing trauma and reducing human error (De Seta et al. 2022; Hussain et al. 2019). Although currently adopted by only a few centers, these technologies hold promise for further improving CI surgery, but their full potential has yet to be realized.

Traditional CI Surgery and Alternatives

CI surgery is a complex procedure that involves several steps. The mastoidectomy posterior tympanotomy approach (MPTA) is the classic surgical approach for cochlear implantation, originally described by William House in 1961 (Fouad 2020). This approach is usually performed under general anesthesia. The MPTA process involves the following steps:

The surgery begins with the surgeon making a small incision behind the ear. This allows access to the mastoid bone, which is then drilled into to create a small opening into the cochlea, the spiral-shaped cavity of the inner ear. This is a delicate process, as the cochlea is filled with fluid that transmits sound impulses to the brain, and care must be taken not to damage it. Once the opening is made, the surgeon then inserts the electrode array of the CI into the cochlea. The body of the implant (i.e., the receiver/stimulator) is then secured to the skull. The receiver/stimulator receives signals from the speech processor and converts them into electrical impulses, which are sent along the electrode array in the cochlea. These impulses stimulate the auditory nerve, allowing the person to hear. The surgery typically takes between 1 and 3 h. Afterward, there is a recovery period before the device can be activated and programming can begin.

The MPTA approach has been the standard for cochlear implantation for more than half a century, with no major advancements or changes (Fouad 2020). However, several alternative approaches and technical innovations have emerged, each with relative advantages and disadvantages (Fouad 2020; Kiratzidis 2022; Nagaraj et al. 2023; Surmelioglu et al. 2016):

Suprameatal Approach: A less invasive technique that avoids mastoidectomy and posterior tympanotomy (Kronenberg et al. 2001). It aims to reduce the risk of facial nerve injury and cerebrospinal fluid leakage, which are potential complications of the traditional approach. It involves drilling a tunnel above the ear canal (suprameatal) to access the cochlea directly.

Transcanal Approach: This technique accesses the cochlea through the ear canal (Jang et al. 2012). It is less invasive than the traditional approach and may reduce the risk of complications such as facial nerve injury and cerebrospinal fluid leakage.

Canal Wall Down Approach: In this technique, the posterior and superior walls of the ear canal are removed to create a large cavity that provides easy access to the middle and inner ear (Carfrae and Foyt 2009). This approach is typically used in cases with extensive ear disease that cannot be managed with a less invasive approach.

Direct Access Approach (DCA): This is a recent advancement in minimally invasive cochlear implantation, often performed with robotic assistance. It replaces the standard mastoidectomy process with a small-diameter tunnel from the lateral bone surface to the cochlea for electrode array insertion (Weber et al. 2013). Robots can perform precise movements guided by preoperative imaging, potentially improving electrode placement accuracy and reducing complications (Fig. 14.1).

Conscious Sedation with Local Anesthesia (CS-LA): In certain cases, cochlear implantation can be performed under conscious sedation with local anesthesia instead of general anesthesia (Shabashev et al. 2017). This approach may reduce the risks associated with general anesthesia and may be preferred in certain patient populations.

Each of these techniques has its own relative advantages and disadvantages, and the choice of technique depends on the individual patient's anatomy and the surgeon's preference (Surmelioglu et al. 2016). The recovery process after CI surgery, regardless of the technique used, involves both physical healing from the surgery and auditory rehabilitation. Generally, physical recovery includes caring for the incision site, managing discomfort or dizziness, and allowing time for healing before the CI system is activated. Auditory rehabilitation involves programming the CI and learning to interpret the new sounds being perceived. This process is typically consistent across different surgical techniques.

Regardless of the technique, surgeons face several challenges during CI surgery. One of the main challenges is dealing with anatomical variations as every patient's inner ear structure is unique. This can make the surgical procedure complex and requires the surgeon to adapt their approach for each individual patient. Another

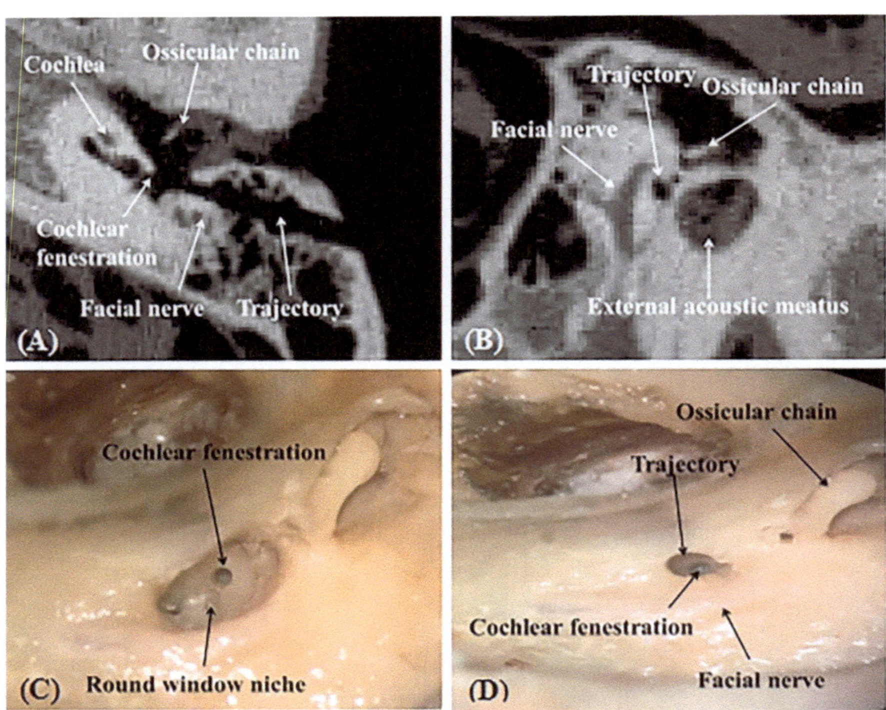

Fig. 14.1 Tunnel approach for cochlear implantation (Wang et al. 2020), licensed under CC BY

significant challenge is the risk of facial nerve injury. The facial nerve runs very close to the surgical site, and there is a risk of injury during the procedure, which can have serious consequences (Pedro and Anna 2022). Surgeons also need to manage the risk of cerebrospinal fluid (CSF) leakage during the surgery. This requires careful handling of the cochlea during the drilling and insertion process (Binnetoglu et al. 2020). Controlling bleeding is also a challenge as the surgeon needs to ensure that the surgical field is always clear and visible. Finally, securing the implant in the correct position is crucial for the success of the surgery and the subsequent functioning of the CI.

Ensuring precise electrode placement is regarded as the most difficult and crucial step for the success of the procedure and the subsequent functioning of the CI. The electrode array is placed with the help of an insertion tool, and the surgeon must avoid causing trauma to the cochlea and disturbing the fluid inside, which can affect residual hearing (Lawand et al. 2014). If the electrode insertion is very close to the basilar membrane, this can lead to structural damage and/or loss of residual hearing. Intraoperative imaging, such as plain radiography, can also be used to assess the number of electrodes inserted and their placement within the scala tympani. These measures help ensure that the electrode array is correctly and safely inserted into the cochlea, which is crucial for the CI to function properly and for the patient to achieve the best possible hearing outcome.

14.2 Virtual Reality-Based Training and Surgical Planning

In recent years, virtual reality (VR) has transformed medical education, particularly in fields where precision is critical. Traditional training methods, such as phantom and cadaver experiments, often fall short due to their lack of realism in replicating living tissues and physiological responses. These models also face ethical, legal, and cost-related challenges, limiting their availability and consistency. Additionally, they do not provide immediate feedback, making it harder for trainees to learn and improve in real time. VR bridges this gap by offering immersive simulations that accurately replicate human anatomy and surgical dynamics, allowing trainees to refine their skills and build confidence in a safe environment (Linke et al. 2013). VR also enhances preoperative planning by enabling surgeons to visualize and manipulate 3D models from advanced imaging, improving their understanding of complex cases and facilitating communication within surgical teams. Planning tools such as Nautilus (Cochlear Ltd.), Cimago (Advanced Bionics), and Otoplan (Cascination and MED-EL joint collaboration) further support surgical planning by providing accurate patient-specific 3D visualizations, significantly enhancing the effectiveness of otological procedures (Geiger et al. 2023; Hajr et al. 2023; Margeta et al. 2022).

Surgical VR simulators feature high-resolution VR headsets that provide a 360-degree view, immersing users in a realistic surgical environment. Integrated haptic feedback devices allow trainees to feel resistance when interacting with virtual anatomical structures, enhancing fine motor skills and muscle memory. Motion tracking captures user movements in real time, improving control and realism. Furthermore, VR simulators for CI use advanced 3D modeling to create detailed representations of the temporal bone, allowing for interactive navigation and practice of various procedures. Instructional elements guide users through techniques and provide immediate feedback. Comprehensive assessment algorithms track metrics like accuracy and timing, offering insights into a trainee's progress. Together, these features create a powerful training platform that significantly enhances surgical education and prepares healthcare professionals for complex procedures.

Recent years have seen the emergence of several VR-based surgical planning and training simulators, including the cochlear implant surgical assessment tool (CISAT) and the University of Melbourne Virtual Reality Temporal Bone Surgery Simulator (Fig. 14.2) (Frendø et al. 2021; Copson et al. 2017).

CISAT (Frendø et al. 2021, 2022a, b) is designed to enhance training and evaluation for CI procedures. It features detailed 3D models of ear anatomy, allowing surgeons to familiarize themselves with complex structures like the cochlea. Trainees can practice the entire surgical process in a controlled environment, benefiting from haptic feedback that mimics real surgical interactions. It tracks performance metrics such as precision and timing, providing valuable feedback and case-based scenarios to prepare users for real-world challenges. By enabling preoperative planning with patient-specific anatomy visualization, it enhances surgeons' skills and improves patient outcomes. Similarly, The University of Seville is also developing a training method using 3D-printed temporal bones, allowing trainees to

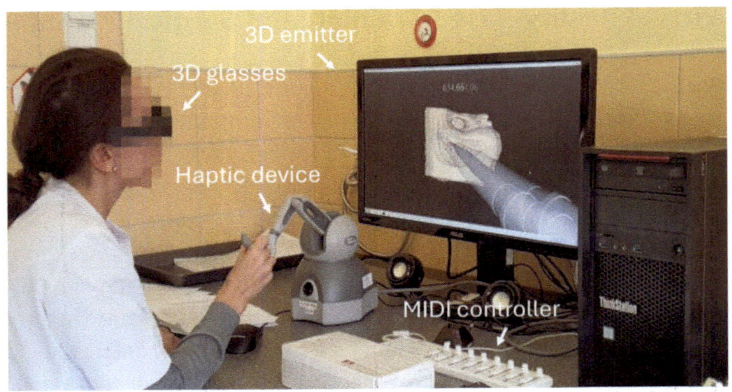

Fig. 14.2 Virtual reality temporal bone surgery simulator. A geomagic touch haptic device from 3D systems is used to capture input motions. Stereoscopic visualization of the virtual scene is enabled through a 3D emitter and 3D glasses. A MIDI controller allows the user to adjust parameters such as magnification level, burr size, and burr type (diamond or cutting). The software was developed by a team at the University of Melbourne. (Image adapted, with modifications, from Gawęcki et al. (2020), under the CC BY license)

work on a realistic structure. The drill is connected to a haptics system to simulate the resistance of different tissues, enhancing the training experience.

The University of Melbourne's simulator (Copson et al. 2017; Wijewickrema et al. 2017) enhances the skills of otolaryngology residents and surgeons through an immersive training experience. It combines advanced hardware and software to create a high-resolution VR environment with an expansive field of view. Haptic feedback devices simulate surgical sensations, while motion tracking technology captures hand movements for realistic technique practice. The software features sophisticated 3D modeling, allowing users to navigate surgical scenarios like mastoidectomy and tympanoplasty with step-by-step guidance. Real-time performance metrics provide immediate feedback, helping trainees assess their progress and identify areas for improvement.

A study by Jones et al. aimed to develop a VR simulator for CI surgery using the Unity3D (Unity Technologies, USA) game engine (Jones et al. 2019). High-resolution models of human cochleae were constructed from synchrotron-radiation phase-contrast imaging data, and the simulator included realistic representations of surgical instruments and procedures. Performance testing achieved a consistent frame rate of 25 frames per second, ensuring smooth operation under high computational demands. Validation indicated no significant differences between simulated and actual CI insertions, and expert surgeons provided positive qualitative feedback. The study concluded that VR simulators built with accessible game engines can enhance medical education by offering realistic, affordable training tools.

Advancements in technology have introduced ultrahigh-fidelity (UHF) graphics in VR training for temporal bone surgery, utilizing the eyepiece of a digital surgical microscope instead of a PC screen. These setups enhance depth perception and graphical realism compared to traditional screen-based VR, which often relies on anaglyph or shutter glasses for 3D effects. While high-fidelity simulations are believed to improve surgical skill acquisition, recent research indicates that UHF graphics may hinder novice performance during mastoidectomy training due to increased complexity and challenges in achieving stereovision for some users (Frithioff et al. 2022).

Overall, integrating VR into CI surgery training shows great promise for improving educational outcomes and surgical skills. VR allows trainees to practice complex procedures in immersive environments, enhancing their confidence and preoperative planning. However, challenges remain, including the high cost of equipment and varying effectiveness among users—particularly those struggling with stereovision. Ongoing validation of these simulators is essential. Future opportunities lie in enhancing simulation realism, expanding access in under-resourced areas, and incorporating artificial intelligence for personalized training (Chessa et al. 2024). As the field of VR advances, its potential to transform CI surgery training and planning will continue to grow, benefiting surgical practices and patient care.

14.3 Intraoperative Navigation Systems

Intraoperative navigation systems are essential tools in modern procedures, offering solutions to complex surgical challenges. These systems help surgeons accurately identify target anatomy, determine optimal implant and electrode placement, and track their current position, thereby enhancing safety and precision of CI procedures. Navigation systems act as advanced measurement and information tool, providing surgeons with real-time data crucial for surgical decision-making. Through superior accuracy and consistency, studies in different surgical domains have shown that dynamic navigation and surgical guides significantly improve implant placement precision, reducing errors and enhancing overall outcomes (Sun et al. 2020; Younis et al. 2024).

Torres et al. examined the variability in surgeons' mental representations of cochlear anatomy during CI surgery, especially across different experience levels (Torres et al. 2016). The study used artificial temporal bones, categorized surgeons by experience, and employed a DigiPointeur navigation system (Collin Medical, France) with a high accuracy threshold (0.11 mm and target accuracy <0.76 mm). Findings showed significant variability in cochlear entry points, with experienced surgeons demonstrating greater accuracy and efficiency in defining insertion axes compared to fellows and residents. These results emphasize the importance of navigation systems in supporting accurate cochlear anatomy navigation and enhancing surgical safety.

The O-Arm (Medtronic, USA) is another promising navigation system that may benefit CI procedures. It integrates intraoperative CT scanning with neuronavigation to provide real-time imaging and precise instrument tracking. In an experimental study by Duret et al., temporal bones were positioned on a plate equipped with an infrared sensor to ensure precise localization (Duret et al. 2020). Infrared markers placed on the bone and instruments enabled automatic pre-to-perioperative registration and accurate tracking, while an infrared tracker attached to the drill allowed for precise calibration. This setup, with high-resolution imaging, significantly enhanced the visibility of critical structures like the facial nerve. Despite the advantages, a notable drawback was the increased radiation dose (0.8 mSv per scan) compared to conventional CT scans (0.56 mSv), presenting a risk factor that CI teams must weigh against the benefits of enhanced surgical accuracy. Intraoperative CT-based navigation may be particularly beneficial for patients with complex ear anatomy (Stelter et al. 2012). Using infrared-based navigation, skull-mounted markers, and high-resolution CT scans, this approach improved electrode placement accuracy, minimized risks such as facial nerve damage, and provided perioperative feedback on final placement. However, drawbacks include high equipment costs and additional setup time, which may extend surgical duration.

Keyhole surgery, a minimally invasive surgery approach (MIS), involves making small incisions to access specific anatomical targets, allowing for precise procedures with reduced recovery times and fewer complications compared to traditional open surgery. This approach can be supported by advanced image-guided navigation systems. In a feasibility study, the use of high-resolution flat-panel volume computed tomography (fpVCT) to perform MIS CI procedures was explored (Majdani et al. 2008). Conducted on human temporal bones, the study aimed to achieve cochleostomy through a minimally invasive channel in the mastoid—traditionally accessed via mastoidectomy—to identify key landmarks such as the facial nerve, chorda tympani, and lateral semicircular canal from CT slices. In this chapter, four temporal bones with implanted fiducial markers were used for registration. Using VectorVision2 navigation system (Brainlab, Germany) with fiducial markers and iPlan2.0 software, the anatomical structures were identified and segmented with precision. The system's AutoPilot feature provided real-time drill guidance, enabling the surgeon to maintain the planned trajectory with a root mean square error (RMSE) of 0.2 mm. This approach demonstrated the potential of fpVCT-supported navigation in achieving the precision necessary for keyhole cochlear implantation.

Another increasingly used technology is fluoroscopy, which provides continuous X-ray visualization, allowing the surgeon to see the progress of the electrode array as it is inserted into the cochlea (Fig. 14.3). Fluoroscopy addresses common challenges in CI surgery, such as misrouting, kinking, and incomplete insertion, which can occur due to variations in cochlear anatomy. It is particularly valuable for cases with anatomical abnormalities, such as malformed cochleae or otosclerosis, where precise electrode placement is critical. The C-arm fluoroscopy setup enables surgeons to monitor the electrode array in real time, allowing for immediate adjustments if resistance or anatomical obstacles are encountered (Mom et al. 2022; Perazzini et al. 2021). This guidance helps prevent issues such as scala translocation, ensuring smooth insertion, reducing trauma to the cochlea, and supporting hearing preservation.

Intraoperative navigation systems play a critical role in CI surgery, offering enhanced precision and safety by providing surgeons with advanced imaging, tracking, and anatomical visualization capabilities. These systems address the unique challenges posed by cochlear anatomy variability, making CI procedures more efficient and outcomes more predictable for patients.

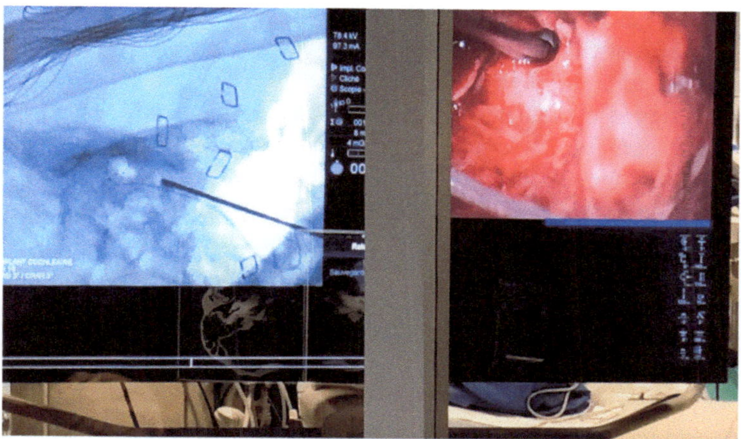

Fig. 14.3 Electrode array insertion using fluoroscopy imaging as feedback (Mom et al. 2022), licensed under CC BY

14.4 Advanced Non-robotic Electrode Insertion Techniques

Electrode insertion tools are increasingly recognized as essential in CI surgery due to the need for precision, control, and reduced trauma during the procedure. Manual insertion of electrode arrays—even by skilled surgeons—poses several challenges. Factors such as involuntary hand tremors and difficulty maintaining a steady, controlled insertion speed can compromise the precision of electrode placement (De Seta et al. 2022). Controlled and consistent insertion is key to minimizing trauma during surgery, which directly impacts hearing preservation.

A variety of manual and motorized tools have been proposed for electrode array insertion during CI surgery. The advanced bionics (AB) insertion tool is a notable commercial device designed for manual electrode array insertion in CI surgery. Initially paired with AB's HiFocus 1J and Helix arrays, it features a handle connected to a flexible shaft with a slide mechanism that ejects the electrode array from its insertion tube. AB now offers a similar tool for their Mid-Scala array, optimized for advance-off-stylet (AOS) insertion, ensuring more controlled and standardized placement. The guide assembly tool for AOS insertion represents a more complex manual option (Kratchman et al. 2012). It consists of a guide assembly, electrode array ejector, and stylet retractor, all integrated into a linear slide mechanism. This system, mounted on a micro-table, ensures precise intracochlear placement of the array, providing surgeons with enhanced control over the delicate insertion process.

Another simple yet effective tool is the two-piece titanium tube, designed for use during direct cochlear access (DCA) procedures (Caversaccio et al. 2017). This tool prevents the electrode array from deviating or kinking as it passes through the mastoid cells, and its two-piece construction allows for easy removal without disturbing the electrode array once inserted. Similar to the titanium tube, the polyimide tube assists in slow and controlled electrode array insertion during DCA by guiding the array through the drilled tunnel (Rojas et al. 2021). Its mirrored two-piece design also facilitates smooth post-insertion removal.

Motorized insertion tools offer several advantages over manual tools and techniques in CI surgery, including the ability to standardize the insertion process, produce smoother force profiles, eliminate the risk of human tremor, and achieve slow, continuous insertion speeds that minimize trauma to the inner ear (Nguyen et al. 2014). Various research groups have developed experimental motorized tools aimed at overcoming the limitations of manual insertion. One early example comes from the Hannover team, which developed an automated tool for AOS insertion (Hussong et al. 2008). They later incorporated a force-sensing unit, which—while producing slightly higher forces compared to manual insertion—exhibited greater reliability and fewer force peaks (Schurzig et al. 2010). The team's latest iteration includes a force sensor and can perform both traditional and AOS insertions. They also proposed a simpler tool using a syringe and infusion pump to control the forward motion of the electrode array, making it adaptable and sterile for clinical use (Fig. 14.4) (Rau et al. 2020).

Another development comes from the Paris team, which created a motorized insertion tool capable of slow, controlled insertion to analyze force profiles and assess the risk of trauma during surgery (Miroir et al. 2012). Later versions of the tool addressed visual field impairment and enabled compatibility with robotic arms, although the current design does not feature a force sensor for real-time feedback (Nguyen et al. 2012). Meanwhile, the University of Iowa developed a micro-mechanical control device that significantly reduces intracochlear pressure transients compared to manual insertion (Hartl et al. 2019). Their experimental motorized tool, with adjustable insertion speeds of 0.1, 0.5, and 1 mm/s, is currently available on the market as IotaSoft from IotaMotion, showing promising results in minimizing insertion-related trauma.

Additionally, electrocochleography (ECochG) and array coatings are two evolving techniques aimed at improving the safety and effectiveness of CI surgery. ECochG provides real-time feedback on cochlear function during electrode array insertion by recording cochlear microphonics (CM), helping surgeons monitor for signs of intracochlear damage and adjust electrode placement accordingly. While the prognostic value of intraoperative ECochG is still under investigation—with some studies showing a correlation between CM recordings and hearing preservation—its potential to improve surgical precision remains promising (Koka et al. 2018). Visualizing ECochG data through digital imaging, using tools such as SmartNav (Cochlear Ltd., Australia) and AIM (Advanced Bionics, USA), further enhances the ability to make real-time adjustments, potentially leading to better hearing preservation outcomes (Arweiler-Harbeck et al. 2021). Additionally, SmartNav enables radiation-free electrode array placement checks perioperatively. This innovation enhances confidence in surgical outcomes and provides data on insertion speed, depth, impedance, and other objective assessments.

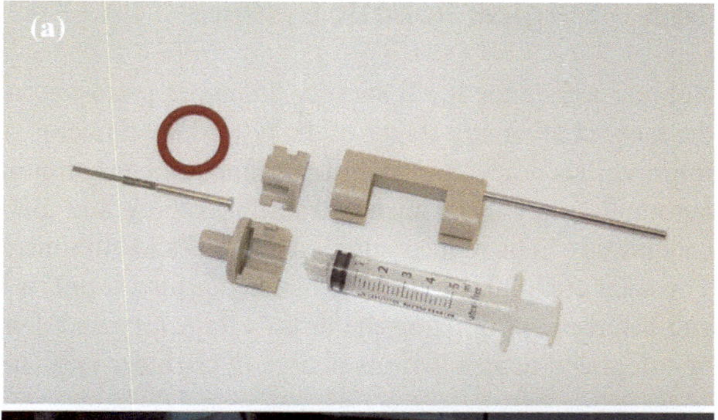

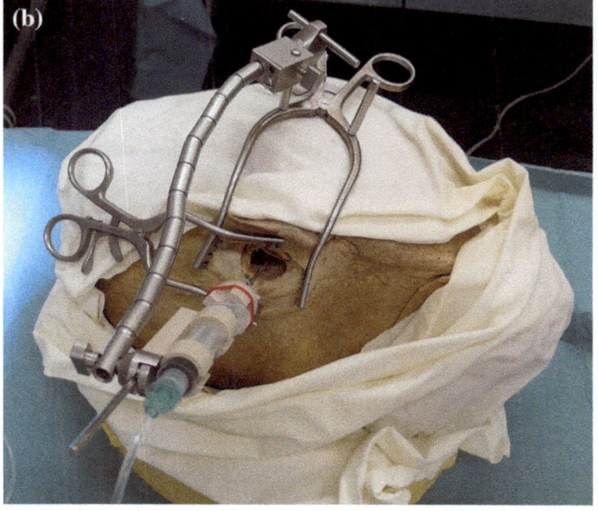

Fig. 14.4 Hydraulic insertions of CI electrode arrays (Geraldine Zuniga et al. 2022), licensed under CC BY

Array coatings, on the other hand, aim to reduce the forces exerted during electrode insertion and minimize trauma to the inner ear's delicate structures. Traditional arrays, made of platinum and iridium wires encased in silicone, can cause friction and stiffness during insertion. Newer materials and coatings have been developed to address these issues. For instance, electrode arrays made from carbon nanotube bundles have been shown to require significantly less insertion force (Choi et al. 2019). Additionally, coatings such as zwitterionic-sulfobetaine-methacrylate have demonstrated reduced friction coefficients in cadaveric studies, lowering insertion forces and reducing the risk of trauma (Hugl et al. 2019). Beyond mechanical benefits, certain coatings are also being explored for drug delivery, enhancing implant functionality and further underscoring their potential to improve outcomes in CI surgery.

Electrode insertion tools are critical for improving both the accuracy and safety of CI surgery. They address the limitations of manual insertion, support minimally invasive surgical techniques, and reduce the likelihood of trauma to the cochlea—essential for preserving hearing. By offering better control, precision, and consistency, these tools help surgeons achieve more predictable and successful outcomes, ultimately benefiting the patient's hearing experience.

14.5 Surgical Robotic Systems

Robotic assistance in CI surgery enhances precision and control, especially during the more challenging stages of the procedure. Systems such as RobOtol and HEARO improve accuracy by stabilizing movements, mitigating natural tremors, and automating key steps such as mastoidectomy and electrode array insertion. These capabilities help reduce surgeon fatigue and minimize errors (Miroir et al. 2010; Topsakal et al. 2022). Robotic assistance also supports efficient electrode alignment and insertion, optimizing the array's trajectory and depth within the cochlea. It can be integrated across various phases of cochlear implantation, including:

- Performing a robot-assisted mastoidectomy and posterior tympanotomy
- Performing a minimally invasive cochleostomy with robot-guided drilling
- Inserting the electrode array

The OTOBOT robotic system, proposed in 2011, utilized a Mitsubishi RV-3S industrial robot controlled by custom software to perform mastoidectomy (Danichelko et al. 2011; Dillon et al. 2015). This system provided real-time movement feedback through the Polaris Spectra optical tracking system, ensuring precision even in the presence of patient movement. Titanium fiducial markers implanted around the mastoid enabled accurate alignment between CT imaging and the physical space, allowing the surgeon to define safe milling boundaries. OTOBOT achieved successful milling on temporal bone specimens, with postoperative CT scans showing no damage to critical structures and a maximum error of 0.6 mm.

The ROSA robotic system (Zimmer Biomet, USA) represents another advancement, tested on five patients undergoing CI surgery (Klopp-Dutote et al. 2021). This system combines rigid skull fixation with a robotic arm offering six degrees of freedom and 3D planning via Rosanna® software. Intraoperative imaging with a CBCT scanner maintains a precise milling trajectory, ensuring safety margins from critical structures. The procedure involves staged drilling and robotic-guided electrode insertion, with real-time verification through facial nerve neuromonitoring, endoscopic checks, and CBCT scans. Despite requiring extended anesthesia (6 h) and surgical duration (3 h), the ROSA system demonstrated high accuracy with a median radial error of 0.67 mm.

The Da Vinci Si robotic system (Intuitive Surgical, USA) has also been explored in cadaveric studies, focusing on augmented visual navigation to enhance precision (Liu et al. 2014). This system employs a master-slave configuration for cochleostomy and mastoidectomy, integrating augmented reality with CBCT guidance. The augmented reality overlay provides real-time visualization of vital anatomy, such as the facial nerve, aiding in complex anatomical challenges. However, limitations include the lack of haptic feedback and the need for refined instruments specifically suited for otological procedures.

RobOtol (Collin Medical, France) was specifically designed for the constrained space of the middle ear (Miroir et al. 2010). While it is not suitable for the heavy drilling required during the mastoidectomy phase, its primary application has

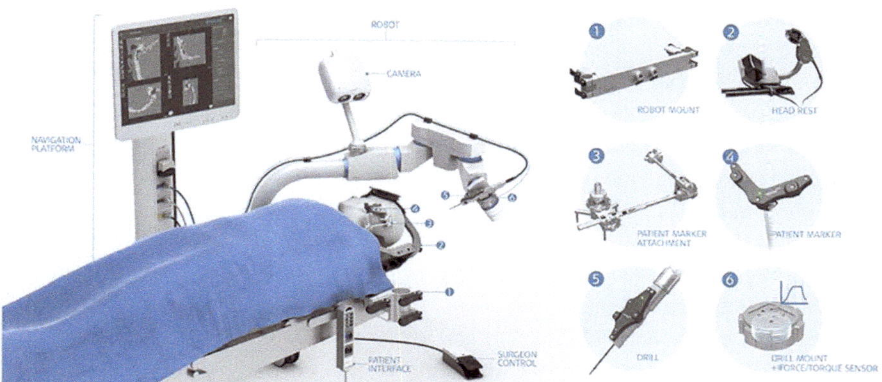

Fig. 14.5 HEARO robotic setup and system for CI procedures (Topsakal et al. 2022), licensed under CC BY

recently shifted to CI electrode array insertion. This shift is due to its ability to perform precise, slow, and stable movements, resulting in reduced translocation rates and fewer misplaced electrodes compared to manual methods (De Seta et al. 2021). Currently, RobOtol is designed for lateral wall electrode arrays, but efforts are underway to develop electrode holders for modiolar-hugging arrays. Despite longer surgery times, robotic assistance with RobOtol potentially offers less traumatic insertions and the ability to modify insertion paths based on real-time feedback, such as electrocochleography. It can also be directly linked to surgical navigation systems, providing precise trajectory planning and localization capabilities. In comparison, the recently introduced OTODRIVE system (Cascination AG, Switzerland) offers similar but somewhat less advanced capabilities. It has demonstrated precise and safe electrode insertion, minimizing trauma and improving hearing thresholds post-surgery (Abari et al. 2024).

The HEARO system (Cascination AG, Switzerland) (Fig. 14.5) further advanced robotic CI surgery by enabling automated inner ear access and electrode insertion with force-feedback monitoring and six-axis force-torque sensors (Topsakal et al. 2020). Studies comparing robotic versus conventional CI surgery showed significant improvements in audiological outcomes for HEARO patients, particularly in pure-tone and speech perception (Heuninck et al. 2023). Following similar principles, various startups like Eindhoven Medical Robotics (Netherlands), OtoJig (Germany), and others promise to further revolutionize the robotic-assisted CI landscape.

14.6 Surgical Augmented Reality Systems

AR is a transformative technology that merges the real and digital worlds by overlaying digital content on users' view. This approach is utilized across various fields, including healthcare, education, engineering, and entertainment. Unlike traditional displays, AR enhances user interactions with real-world objects. Since its development in the mid-20th century, AR has advanced significantly, particularly in optics and sensors. A key goal is to ensure seamless interaction between virtual and real elements without discomfort, meeting visual requirements like field of view (FOV) and image resolution. The human eye typically has an FOV of about 120° for binocular vision. AR has become increasingly important in medicine, especially for surgical planning and navigation, providing real-time, 3D overlays that enhance precision (Yin et al. 2021).

A typical AR system for temporal bone procedures involves calibration, registration, and tracking methodologies to visualize preoperative information about target and other critical structures (Hussain et al. 2019). Navigational guidance information may also be augmented to the surgical view using microscope, head-mounted displays (HMDs), or a monitor (Guigou et al. 2022). Several studies have highlighted AR's capacity to enhance visualization and navigation, offering real-time overlays of critical structures for increased accuracy and safety.

One such study integrated AR with the da Vinci Si surgical system, combining CBCT and stereo video augmentation to help surgeons visualize essential structures like the facial nerve, cochlea, and round window during operations (Liu et al. 2014). The AR system guided precise cochleostomy and electrode placement, with overlays ensuring accuracy in critical areas, achieving a mean registration error of under 0.5 mm for vital structures such as the facial nerve. In a related approach, AR technology was used to merge 2D microscope video with 3D preoperative CT scans, offering a solution that eliminates the need for external tracking systems. By mapping the visible portion of the incus to preoperative CT data, the AR system provided surgeons with real-time, precise guidance during electrode insertion (Zhang et al. 2024). This setup achieved average rotation errors of less than 25 degrees and translation errors below 2 mm across the x, y, and z axes, underscoring AR's growing impact on improving accuracy in complex anatomical navigation.

Further exploring AR's role in cochlear implantation, another study applied AR to transmodiolar CI surgery by overlaying preoperative CT scan data onto the live surgical view for real-time navigation (Guigou et al. 2022). This allowed for precise visualization of the modiolar axis and the entry point for implant placement (Fig. 14.6). The use of 3D navigation cues enabled surgeons to achieve an average 2D registration error of 0.41 mm and an angular error of 13.5°, which was within the surgical specification of 15° (Hussain et al. 2020). A visual patient movement tracking strategy was employed based on keypoint feature matching. However, visual tracking procedures are limited by drift errors over time, and a drift-correction algorithm often needs to be integrated for longer procedures (Gu et al. 2023). Another similar application utilized a stereo camera pair attached to the auxiliary lenses. Using deep reinforcement learning to detect anatomical landmarks, the system enabled precise registration of the cochlear apex and modiolus (the central axis). With a target registration error of 0.31 mm and an angular error of 15.10°, this AR setup provided a high level of precision, particularly valuable in transmodiolar cochlear implant procedures (Hussain et al. 2020).

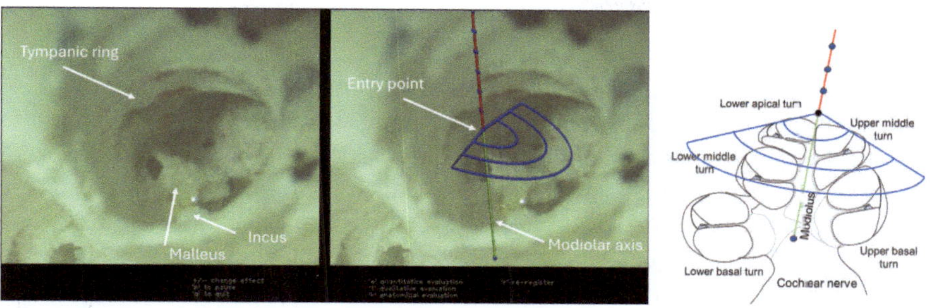

Fig. 14.6 AR display for transmodiolar electrode insertion guidance via external auditory canal. The augmented reality displayed the entry point (intersection of red, green, and blue lines), the modiolar axis (green line), the extension of the modiolar axis outside the promontory (red line), and semicircular lines in a perpendicular plane to the modiolar axis (blue) guiding the surgeon to the right angle of approach. A straight pick was placed on the promontory at the entry point, and its equivalent markers were aligned with the semicircular lines. Then the pick position was progressively aligned with the red line, and when the markers intersected with the points on the straight line, the alignment in the 3D space was insured. During the progression, the pick allowed to check for the correct axis. The modiolar depth was indicated by a marker on the burr shaft

The introduction of Twin-S, a digital twin framework developed for temporal bone procedures, showcases how AR can enhance visualization and precision in other complex procedures (Shu et al. 2023). Twin-S integrates real-time simulation with high-precision optical tracking to offer AR-based guidance. Surgeons benefit from continuously updated overlays of the patient's anatomy during surgery, which reflect real-time interactions between the tools and the anatomy. During mastoidectomy, where protecting delicate nerves and vessels is critical, Twin-S achieved an average error of 1.39 mm in the drilling process, demonstrating its value in improving surgical precision.

For AR-based CI surgery, precise and ergonomic visualization is essential to navigate the complex anatomy of the ear and avoid critical structures like the facial nerve. Digital microscopes, Brainlab-integrated systems and HMDs, play key roles in achieving this level of precision (Hussain et al. 2019). Digital microscopes, like Arriscope (Munich Surgical Imaging, Germany) and RoboticScope (BHS Technologies, Germany), combined with AR overlays enable real-time 3d imaging that enhances a surgeon's ability to assess delicate structures in the middle ear during electrode insertion, improving spatial awareness and reducing risks associated with misalignment. Brainlab systems (Brainlab, Germany), when integrated with surgical microscopes (Carl Zeiss NTS Ltd., Germany; Leica Microsystems, Germany), provide continuous imaging updates and augmented anatomical overlays, allowing surgeons to focus directly on the surgical field with real-time feedback, thereby minimizing potential complications from tool misplacement.

14.7 Discussion

The integration of AR and robotic systems in CI surgery is a rapidly evolving field that holds significant promise for improving surgical outcomes. It is a testament to the rapid advancements in medical technology. Table 14.1 presents an overview of different commercial and research systems available to the CI community. These technologies are being harnessed to improve surgical precision, reduce risks, and ultimately enhance patient outcomes (Panara et al. 2021; Seetohul et al. 2023; Shoman 2022). The need for such tools becomes even more evident when considering the importance of electrode placement in achieving optimal hearing outcomes. The success of a CI largely depends on the proximity of the electrodes to the auditory nerve's ganglion cells, the depth of insertion, and the array's stability within the cochlea (Chakravorti et al. 2019). Precise positioning ensures better auditory performance while minimizing damage to residual hearing. Without specialized tools, maintaining this level of accuracy is much more difficult, particularly in such a delicate and confined anatomical space.

Electrode insertion tools are particularly beneficial when performing minimally invasive techniques, such as DCA which requires creating a precise path through the temporal bone to reach the cochlea. This procedure demands an extremely high level of accuracy to avoid damaging surrounding structures like the facial nerve. Specialized tools and robotic systems play a critical role in this process by guiding the surgeon's movements and ensuring the drilling trajectory is precise, thereby reducing the risks inherent in manual approaches (Riojas et al. 2021).

Robotic CI surgery significantly differs from traditional CI surgery in terms of the surgical approach, precision, and potential for hearing preservation. Robotic systems, with their high precision and stability, are being increasingly utilized in various stages of CI surgery. These stages include:

- Drilling a keyhole approach to the middle ear
- Gaining access to the inner ear
- Electrode insertion into the cochlea

While this minimizes trauma and recovery time, the lack of direct visualization raises safety concerns, prompting the use of multipolar facial nerve monitoring during the procedure (Auinger et al. 2021; De Seta et al. 2022; Panara et al. 2021).

In terms of precision and control, traditional CI surgery relies on the surgeon's skill to manually insert the electrode array using specialized tools (Daoudi et al. 2021; De Seta et al. 2022). The precision of the insertion is subject to variability due to the surgeon's experience (Daoudi et al. 2021). Robotic systems, however, provide enhanced precision by eliminating tremors, standardizing the insertion speed, and offering real-time feedback on insertion forces (De Seta et al. 2022; Panara et al. 2021). This control reduces the likelihood of trauma to the cochlea's delicate structures, such as the basilar membrane, and studies have highlighted that robotic assistance may result in more accurate electrode placement with fewer cases of electrode translocation.

Table 14.1 Overview of some commercial or research systems available for CI procedures

Procedure	Problem/solution	Examples of tools available	Main adaptability challenges
Surgical training	Enhanced virtual platforms mimicking real environments to allow easy access to training especially for complex procedures	Visible Ear Simulator/ CISAT (VES, Denmark) UMelb simulator (University of Melbourne, Australia)	Latency, complex computations, and haptic feedback integration
Surgery planning and assessment	Use of pre- and postoperative imaging to analyze biomarkers and implant location	Nautilus (Cochlear Ltd., Australia) OtoPlan (MED-El, Austria; Cascination AG, Switzerland) CImago (Advanced Bionics, USA)	Needed for additional imaging and added workflow for typical anatomy patients
Digital microscopes	Use of digital imaging to enable augmented surgical scene visualizations and system integration	Arriscope (Munich Surgical Imaging, Germany) RoboticScope (BHS Technologies, Germany)	Lack of contract in deep workspaces
Navigation and visualization	Use of electromagnetic, optical, or AR systems for surgical guidance	Collin Navigation System/ DigiPointeur (Collin Medical, France) Brainlab microscopes (Carl Zeiss NTS Ltd., Germany; Leica Microsystems, Germany) Polaris (NDI, Canada) FusionTrack (Atracsys LLC)	Ergomics, electromagnetic artifacts, line-of-sight requirements, setup time and complexity, immersive visualization, accuracy requirements
Automatic drilling	Use of preoperative planning, navigation systems and robotics to define minimally invasive drilling trajectories and achieve them with sub-millimetric precision	HEARO (Cascination AG, Switzerland) EMR (Eindhoven Medical Robotics) OtoJig (OtoJig, Germany)	Cost, setup time and complexity, visual feedback
Electrode array insertions	Slow and steady movements through robotized or motorized insertion devices	Robotol HEARO (Cascination AG, Switzerland) Otodrive (Cascination AG, Switzerland) IotaSoft (IotaMotion, USA) Hydrodrive (MHH, Germany) OtoJig (OtoJig, Germany)	Additional surgical workflow/time, setup time and complexity, maneuverability
	Use of real-time X-ray systems or objective measurements to provide insertion feedback and placement check	Fluoroscopy imaging Smartnav (Cochlear, Australia) AIM (Advanced Bionics, USA)	Additional surgical workflow/time, radiation exposure or real-time feedback association

Hearing preservation is a major goal of CI surgery, particularly for patients with residual hearing. In traditional surgery, hearing loss is often linked to trauma caused by drilling near the cochlea and the insertion of the electrode array (Panara et al. 2021). Robotic systems aim to reduce this risk by offering a more controlled insertion process, which has been shown to lead to better preservation of residual hearing in some studies (Douadi et al. 2021). Although the initial data is promising, more clinical evidence is needed to determine the long-term efficacy of robotic systems in preserving hearing. This is also true for motorized insertion tools, particularly when combined with robotic systems, as no clinical benefits in terms of residual hearing preservation or speech performance have been demonstrated thus far (Maheo et al. 2023).

Despite these advancements, there are notable differences between the two approaches beyond surgical outcomes. For instance, robotic CI surgery tends to take longer than traditional methods, largely due to the time required to set up and register the robotic system (Daoudi et al. 2021). Additionally, robotic systems are significantly more expensive, which poses a barrier to widespread adoption, particularly in the absence of definitive evidence showing superior outcomes compared to traditional surgery (De Seta et al. 2022). Surgeons also face a learning curve when transitioning to robotic systems, as these require specialized training (Guigou et al. 2022; Panara et al. 2021). This additional complexity could initially increase the duration of procedures and potentially impact patient outcomes as surgeons gain proficiency. However, robotic surgery offers the potential to overcome limitations inherent in manual techniques, such as tremor and force variability.

In addition to robotics, mixed-reality systems have emerged as a promising tool in CI surgery, with potential to enhance visualization, precision, and accuracy. AR systems overlay computer-generated images onto the surgeon's view of the surgical field, allowing for enhanced visualization of critical structures like the cochlea and facial recess during surgery. This can aid in surgical planning, reduce the risk of complications, and improve the accuracy of electrode array insertion (Hussain et al. 2019, 2020). Real-time guidance provided by AR systems helps surgeons navigate complex anatomy and optimize the surgical approach (Guigou et al. 2022; Hussain et al. 2020). AR systems also improve the precision of CI surgery by offering feedback on instrument position and depth, minimizing trauma to surrounding tissues (Hussain et al. 2019). This level of precision can potentially enhance hearing preservation outcomes, though more clinical evidence is needed to confirm its efficacy. AR further supports minimally invasive approaches by providing better visualization during drilling, potentially reducing trauma, recovery times, and scarring.

However, integrating AR into CI surgery also presents challenges. Accurate alignment of preoperative imaging data with the patient's anatomy in real time is technically complex, particularly when dealing with delicate structures like the temporal bone (Hussain et al. 2019). Ensuring an ergonomic and user-friendly interface is also critical to avoid distracting surgeons during procedures (Katic et al. 2015; Thoranaghatte et al. 2008). Additionally, the cost of AR systems and their integration into surgical workflows pose barriers to widespread adoption.

In conclusion, while robotic and AR systems represent significant technological advancements in CI surgery, further research is needed to establish their long-term benefits, safety, and cost effectiveness. Both technologies hold potential to improve surgical precision, reduce trauma, and enhance hearing preservation, but challenges remain in terms of cost, complexity, and clinical validation. Currently, only a handful of centers worldwide have these advanced systems in routine practice. As these technologies evolve, they may become standard tools in CI procedures, providing real-time feedback and assurance to surgeons, enabling less-experienced surgeons to perform like experienced surgeons, and offering improved outcomes for patients.

Acknowledgments The authors acknowledge the use of Microsoft Copilot for proofreading and enhancing the readability of the manuscript. All suggested changes were carefully reviewed and edited. The authors assume full responsibility for the content of the final publication.

References

Abari J, Heuninck E, Topsakal V (2024) Entirely robotic cochlear implant surgery. Am J Otolaryngol 45(5):104360. https://doi.org/10.1016/j.amjoto.2024.104360

Arweiler-Harbeck D, D'heygere V, Meyer M, Hans S, Waschkies L, Lang S, Anton K, Hessel H, Schneider A, Heiler T, Höing B (2021) Digital live imaging of intraoperative electrocochleography–first description of feasibility and hearing preservation during cochlear implantation. Otol Neurotol 42(9):1342–1346. https://doi.org/10.1097/MAO.0000000000003256

Auinger AB, Dahm V, Liepins R, Riss D, Baumgartner WD, Arnoldner C (2021) Robotic cochlear implant surgery: imaging-based evaluation of feasibility in clinical routine. Front Surg 8:742219. https://doi.org/10.3389/fsurg.2021.742219

Binnetoglu A, Demir B, Batman C (2020) Surgical complications of cochlear implantation: a 25-year retrospective analysis of cases in a tertiary academic center. Eur Arch Otorrinolaringol 277:1917–1923. https://doi.org/10.1007/s00405-020-05916-w

Carfrae MJ, Foyt D (2009) Intact meatal skin, canal wall down approach for difficult cochlear implantation. J Laryngol Otol 123(8):903–906. https://doi.org/10.1017/S0022215108004039

Carlson RJ, Walsh T, Mandell JB, Aburayyan A, Lee MK, Gulsuner S et al (2023) Association of genetic diagnoses for childhood-onset hearing loss with cochlear implant outcomes. JAMA Otolaryngol Head Neck Surg 149(3):212–222. https://doi.org/10.1001/jamaoto.2022.4463

Caversaccio M, Gavaghan K, Wimmer W, Williamson T, Ansò J, Mantokoudis G, Gerber N, Rathgeb C, Feldmann A, Wagner F, Scheidegger O (2017) Robotic cochlear implantation: surgical procedure and first clinical experience. Acta Otolaryngol 137(4):447–454. https://doi.org/10.1080/00016489.2017.1278573

Cejas I, Barker DH, Petruzzello E, Sarangoulis CM, Quittner AL (2023) Cochlear implantation and educational and quality-of-life outcomes in adolescence. JAMA Otolaryngol Head Neck Surg 149(8):708–715. https://doi.org/10.1001/jamaoto.2023.1327

Chakravorti S, Noble JH, Gifford RH, Dawant BM, O'Connell BP, Wang J, Labadie RF (2019) Further evidence of the relationship between cochlear implant electrode positioning and hearing outcomes. Otol Neurotol 40(5):617–624. https://doi.org/10.1097/MAO.0000000000002204

Chessa M, Gerini L, Hussain R, Martini M, Pizzo M, Solari F, Viola E (2024) Extended reality and artificial intelligence for exergaming: opportunities and open challenges for rehabilitation and cognitive training. In: 2024 IEEE 8th Forum on Research and Technologies for Society and Industry Innovation (RTSI). IEEE, pp 357–362

Choi GJ, Gwon TM, Kim DH, Park J, Kim SM, Oh SH, Lim Y, Jun SB, Kim SJ (2019) CNT bundle 944 based thin intracochlear electrode array. Biomed Microdevices 21(27):945. https://doi.org/10.1007/s10544-019-0384-y

Copson B, Wijewickrema S, Zhou Y, Piromchai P, Briggs R, Bailey J, Kennedy G, O'Leary S (2017) Supporting skill acquisition in cochlear implant surgery through virtual reality simulation. Cochlear Implants Int 18(2):89–96. https://doi.org/10.1080/14670100.2017.1289299

Danilchenko A, Balachandran R, Toennies JL, Baron S, Munske B, Fitzpatrick JM, Withrow TJ, Webster RJ III, Labadie RF (2011) Robotic mastoidectomy. Otol Neurotol 32(1):11–16. https://doi.org/10.1097/MAO.0b013e3181fcee9e

Daoudi H, Lahlou G, Torres R, Sterkers O, Lefeuvre V, Ferrary E, Mosnier I, Nguyen Y (2021) Robot-assisted cochlear implant electrode array insertion in adults: a comparative study with manual insertion. Otol Neurotol 42(4):e438–e444. https://doi.org/10.1097/MAO.0000000000003002

De Seta D, Nguyen Y, Daoudi H, Torres R, Mosnier I, Sterkers O (2021) Robot-assisted cochlear implantation: the Robotol® system experience. ENT Updates, 11

De Seta D, Daoudi H, Torres R, Ferrary E, Sterkers O, Nguyen Y (2022) Robotics, automation, active electrode arrays, and new devices for cochlear implantation: a contemporary review. Hear Res 414:108425. https://doi.org/10.1016/j.heares.2021.108425

Devare J, Gubbels S, Raphael Y (2018) Outlook and future of inner ear therapy. Hear Res 368:127–135. https://doi.org/10.1016/j.heares.2018.05.009

Dillon NP, Balachandran R, Fitzpatrick JM, Siebold MA, Labadie RF, Wanna GB, Withrow TJ, Webster RJ III (2015) A compact, bone-attached robot for mastoidectomy. J Med Devices 9(3):031003. https://doi.org/10.1115/1.4030083

Duret S, Guigou C, Grelat M, Bozorg-Grayeli A (2020) Minimally invasive cochlear implantation assisted by intraoperative CT scan combined to neuronavigation. Otol Neurotol 41(4):e441–e448. https://doi.org/10.1097/MAO.0000000000002577

Eshraghi AA, Nazarian R, Telischi FF, Rajguru SM, Truy E, Gupta C (2012) The cochlear implant: historical aspects and future prospects. Anat Rec 295(11):1967–1980.7. https://doi.org/10.1002/ar.22580

Finley CC, Holden TA, Holden LK, Whiting BR, Chole RA, Neely GJ et al (2008) Role of electrode placement as a contributor to variability in cochlear implant outcomes. Otol Neurotol 29(7):920–928. https://doi.org/10.1097/MAO.0b013e318184f492

Fouad YA (2020) Advances in surgical and anesthetic techniques for cochlear implantation. In: Advances in rehabilitation of hearing loss. IntechOpen, London

Frendø M, Frithioff A, Konge L, Sørensen MS, Andersen SA (2021) Cochlear implant surgery: learning curve in virtual reality simulation training and transfer of skills to a 3D-printed temporal bone—a prospective trial. Cochlear Implants Int 22(6):330–337. https://doi.org/10.1080/14670100.2021.1940629

Frendø M, Frithioff A, Konge L, Cayé-Thomasen P, Sørensen MS, Andersen SAW (2022a) Cochlear implant surgery: virtual reality simulation training and transfer of skills to cadaver dissection—a randomized, controlled trial. J Int Adv Otol 18(3):219. https://doi.org/10.5152/iao.2022.21429

Frendø M, Frithioff A, Konge L, Foghsgaard S, Mikkelsen PT, Sørensen MS, Cayé-Thomasen P, Andersen SAW (2022b) Assessing competence in cochlear implant surgery using the newly developed cochlear implant surgery assessment tool. Eur Arch Otorrinolaringol 279:127–136. https://doi.org/10.1007/s00405-021-06632-9

Frithioff A, Frendø M, Mikkelsen PT, Sørensen MS, Andersen SAW (2022) Cochlear implantation: exploring the effects of 3D stereovision in a digital microscope for virtual reality simulation training—a randomized controlled trial. Cochlear Implants Int 23(2):80–86. https://doi.org/10.1080/14670100.2021.1997026

Gawęcki W, Węgrzyniak M, Mickiewicz P, Gawłowska MB, Talar M, Wierzbicka M (2020) The impact of virtual reality training on the quality of real antromastoidectomy performance. J Clin Med 9(10):3197. https://doi.org/10.3390/jcm9103197

Geiger S, Iso-Mustajärvi M, Nauwelaers T, Avci E, Julkunen P, Linder P et al (2023) Automatic electrode scalar location assessment after cochlear implantation using a novel imaging software. Sci Rep 13(1):12416. https://doi.org/10.1038/s41598-023-39275-3

Geraldine Zuniga M, Lenarz T, Rau TS (2022) Hydraulic insertions of cochlear implant electrode arrays into the human cadaver cochlea: preliminary findings. Eur Arch Otorrinolaringol 279(6): 2827–2835. https://doi.org/10.1007/s00405-021-06979-z

Gu W, Knopf J, Cast J, Higgins LD, Knopf D, Unberath M (2023) Nail it! Vision-based drift correction for accurate mixed reality surgical guidance. Int J Comput Assist Radiol Surg 18(7): 1235–1243. https://doi.org/10.1007/s11548-023-02950-x

Guigou C, Hussain R, Lalande A, Grayeli AB (2022) Augmented reality based transmodiolar cochlear implantation. Otol Neurotol 43(2):190–198. https://doi.org/10.1097/MAO.0000000000003437

Hajr E, Abdelsamad Y, Almuhawas F, Alashour A, Hagr A (2023) Cochlear implantation: the use of OTOPLAN reconstructed images in trajectory identification. Ear Nose Throat J 104(10): 01455613221134742. https://doi.org/10.1177/01455613221134742

Hamed N, Alajmi N, Alkoblan FI, Alghtani YA, Abdelsamad Y, Alhussien A et al (2024) The chronological evolution of cochlear implant contraindications: a comprehensive review. J Clin Med 13(8):2337. https://doi.org/10.3390/jcm13082337

Hartl RMB, Kaufmann C, Hansen MR, Tollin DJ (2019) Intracochlear pressure transients during 1010 cochlear implant electrode insertion: effect of micro-mechanical control on limiting pressure 1011 trauma. Otol Neurotol 40:736–744. https://doi.org/10.1097/MAO.0000000000002164

Heuninck E, Van de Heyning P, Van Rompaey V, Mertens G, Topsakal V (2023) Audiological outcomes of robot-assisted cochlear implant surgery. Eur Arch Otorrinolaringol 280(10): 4433–4444. https://doi.org/10.1007/s00405-023-07961-7

Hügl S, Scheper V, Gepp MM, Lenarz T, Rau TS, Schwieger J (2019) Coating stability and insertion forces of an alginate-cell-based drug delivery implant system for the inner ear. J Mech Behav Biomed Mater 97:90–98. https://doi.org/10.1016/j.jmbbm.2019.05.007

Hussain R, Lalande A, Guigou C, Bozorg-Grayeli A (2019) Contribution of augmented reality to minimally invasive computer-assisted cranial base surgery. IEEE J Biomed Health Inform 24(7):2093–2106. https://doi.org/10.1109/JBHI.2019.2954003

Hussain R, Lalande A, Berihu Girum K, Guigou C, Grayeli AB (2020) Augmented reality for inner ear procedures: visualization of the cochlear central axis in microscopic videos. Int J Comput Assist Radiol Surg 15(10):1703–1711. https://doi.org/10.1007/s11548-020-02240-w

Hussong A, Rau T, Eilers H, Baron S, Heimann B, Leinung M, Lenarz T, Majdani O (2008) Conception and design of an automated insertion tool for cochlear implants. In: 2008 30th annual international conference of the IEEE Engineering in Medicine and Biology Society. IEEE, pp 5593–5596. https://doi.org/10.1109/IEMBS.2008.4650482

Jang JH, Song JJ, Yoo JC, Lee JH, Oh SH, Chang SO (2012) An alternative procedure for cochlear implantation: transcanal approach. Acta Otolaryngol 132(8):845–849. https://doi.org/10.3109/00016489.2012.667880

Jones B, Rohani SA, Ong N, Tayeh T, Chalabi A, Agrawal SK, Ladak HM (2019) A virtual-reality training simulator for cochlear implant surgery. Simul Gaming 50(2):243–258

Katić D, Spengler P, Bodenstedt S, Castrillon-Oberndorfer G, Seeberger R, Hoffmann J, Dillmann R, Speidel S (2015) A system for context-aware intraoperative augmented reality in dental implant surgery. Int J Comput Assist Radiol Surg 10:101–108. https://doi.org/10.1007/s11548-014-1005-0

Kiratzidis T (2022) Cochlear implant surgery using the Veria technique. In: Cochlear implants: new and future directions. Springer Nature Singapore, Singapore, pp 215–235

Klopp-Dutote N, Lefranc M, Strunski V, Page C (2021) Minimally invasive fully ROBOT-assisted cochlear implantation in humans: preliminary results in five consecutive patients. Clin Otolaryngol 46(6):1326–1330. https://doi.org/10.1111/coa.13840

Koka K, Riggs WJ, Dwyer R, Holder JT, Noble JH, Dawant BM, Ortmann A, Valenzuela CV, Mattingly JK, Harris MM, O'Connell BP (2018) Intra-cochlear electrocochleography during cochlear implant electrode insertion is predictive of final scalar location. Otol Neurotol 39(8): e654–e659. https://doi.org/10.1097/MAO.0000000000001906

Kratchman LB, Schurzig D, McRackan TR, Balachandran R, Noble JH, Webster RJ III, Labadie RF (2012) A manually operated, advance off-stylet insertion tool for minimally invasive cochlear implantation surgery. IEEE Trans Biomed Eng 59(10):2792–2800. https://doi.org/10.1109/TBME.2012.2210220

Kronenberg J, Migirov L, Dagan T (2001) Suprameatal approach: new surgical approach for cochlear implantation. J Laryngol Otol 115(4):283–285. https://doi.org/10.1258/0022215011907451

Lawand NS, French PJ, Driel JV, Briaire JJ, Frijns JHM (2014) Cochlear implant electrode improvement for stimulation and sensing. In: Sensing technology: current status and future trends II, pp 1–25. https://doi.org/10.1007/978-3-319-02315-1_1

Linke R, Leichtle A, Sheikh F, Schmidt C, Frenzel H, Graefe H, Wollenberg B, Meyer JE (2013) Assessment of skills using a virtual reality temporal bone surgery simulator. Acta Otorhinolaryngol Ital 33(4):273

Liu WP, Azizian M, Sorger J, Taylor RH, Reilly BK, Cleary K, Preciado D (2014) Cadaveric feasibility study of da Vinci Si–assisted cochlear implant with augmented visual navigation for otologic surgery. JAMA Otolaryngol Head Neck Surg 140(3):208–214. https://doi.org/10.1001/jamaoto.2013.6443

Maheo C, Marie A, Torres R, Archutick J, Leclère JC, Marianowski R (2023) Robot-assisted and manual cochlear implantation: an intra-individual study of speech recognition. J Clin Med 12(20):6580. https://doi.org/10.3390/jcm12206580

Majdani O, Bartling SH, Leinung M, Stöver T, Lenarz M, Dullin C, Lenarz T (2008) A true minimally invasive approach for cochlear implantation: high accuracy in cranial base navigation through flat-panel-based volume computed tomography. Otol Neurotol 29(2):120–123. https://doi.org/10.1097/mao.0b013e318157f7d8

Mamelle E, Granger B, Sterkers O, Lahlou G, Ferrary E, Nguyen Y, Mosnier I (2020) Long-term residual hearing in cochlear implanted adult patients who were candidates for electro-acoustic stimulation. Eur Arch Otorrinolaringol 277(3):705–713. https://doi.org/10.1007/s00405-019-05745-6

Margeta J, Hussain R, López Diez P, Morgenstern A, Demarcy T, Wang Z et al (2022) A web-based automated image processing research platform for cochlear implantation-related studies. J Clin Med 11(22):6640. https://doi.org/10.3390/jcm11226640

Marschark M, Rhoten C, Fabich M (2007) Effects of cochlear implants on children's reading and academic achievement. J Deaf Stud Deaf Educ 12(3):269–282. https://doi.org/10.1093/deafed/enm013

McRackan TR, Bauschard M, Hatch JL, Franko-Tobin E, Droghini HR, Nguyen SA, Dubno JR (2018) Meta-analysis of quality-of-life improvement after cochlear implantation and associations with speech recognition abilities. Laryngoscope 128(4):982–990. https://doi.org/10.1002/lary.26738

Miroir M, Nguyen Y, Szewczyk J, Mazalaigue S, Ferrary E, Sterkers O, Grayeli AB (2010) RobOtol: from design to evaluation of a robot for middle ear surgery. In: 2010 IEEE/RSJ international conference on intelligent robots and systems. IEEE, pp 850–856

Miroir M, Nguyen Y, Kazmitcheff G, Ferrary E, Sterkers O, Grayeli AB (2012) Friction force measurement during cochlear implant insertion: application to a force-controlled insertion tool design. Otol Neurotol 33(6):1092–1100. https://doi.org/10.1097/MAO.0b013e31825f24de

Mom T, Puechmaille M, El Yagoubi M, Lère A, Petersen JE, Bécaud J, Saroul N, Gilain L, Mirafzal S, Chabrot P (2022) Robotized cochlear implantation under fluoroscopy: a preliminary series. J Clin Med 12(1):211. https://doi.org/10.3390/jcm12010211

Nagaraj S, Singhal P, Singh AS, Sharma S, Sharma AK, Jat KS et al (2023) A comparative study of surgical outcomes in children with cochlear implantation using posterior tympanotomy &

modified Veria technique. Indian J Otolaryngol Head Neck Surg 75(Suppl 1):255–260. https://doi.org/10.1007/s12070-022-03399-1

Nguyen Y, Miroir M, Kazmitcheff G, Sutter J, Bensidhoum M, Ferrary E, Sterkers O, Bozorg Grayeli A (2012) Cochlear implant insertion forces in microdissected human cochlea to evaluate a prototype array. Audiol Neurotol 17(5):290–298. https://doi.org/10.1159/000338406

Nguyen Y, Kazmitcheff G, De Seta D, Miroir M, Ferrary E, Sterkers O (2014) Definition of metrics to evaluate cochlear array insertion forces performed with forceps, insertion tool, or motorized tool in temporal bone specimens. Biomed Res Int 2014(1):532570. https://doi.org/10.1155/2014/532570

Panara K, Shahal D, Mittal R, Eshraghi AA (2021) Robotics for cochlear implantation surgery: challenges and opportunities. Otol Neurotol 42(7):e825–e835. https://doi.org/10.1097/MAO.0000000000003165

Pedro C, Anna K (2022) Explantation and reimplantation of cochlear implants. In: Cochlear implants: new and future directions. Springer Nature Singapore, Singapore, pp 403–425

Perazzini C, Puechmaille M, Saroul N, Plainfossé O, Montrieul L, Bécaud J, Gilain L, Chabrot P, Boyer L, Mom T (2021) Fluoroscopy guided electrode-array insertion for cochlear implantation with straight electrode-arrays: a valuable tool in most cases. Eur Arch Otorrinolaringol 278: 965–975. https://doi.org/10.1007/s00405-020-06151-z

Rau TS, Zuniga MG, Salcher R, Lenarz T (2020) A simple tool to automate the insertion process in cochlear implant surgery. Int J Comput Assist Radiol Surg 15:1931–1939. https://doi.org/10.1007/s11548-020-02243-7

Riojas KE, Tran ET, Freeman MH, Noble JH, Webster RJ III, Labadie RF (2021) Clinical translation of an insertion tool for minimally invasive cochlear implant surgery. J Med Devices 15(3):031001. https://doi.org/10.1115/1.4050203

Schaefer S, Sahwan M, Metryka A, Kluk K, Bruce IA (2021) The benefits of preserving residual hearing following cochlear implantation: a systematic review. Int J Audiol 60(8):561–577. https://doi.org/10.1080/14992027.2020.1863484

Schurzig D, Webster RJ III, Dietrich MS, Labadie RF (2010) Force of cochlear implant electrode insertion performed by a robotic insertion tool: comparison of traditional versus Advance Off-Stylet techniques. Otol Neurotol 31(8):1207–1210. https://doi.org/10.1097/MAO.0b013e3181f2ebc3

Seetohul J, Shafiee M, Sirlantzis K (2023) Augmented reality (AR) for surgical robotic and autonomous systems: state of the art, challenges, and solutions. Sensors 23(13):6202. https://doi.org/10.3390/s23136202

Shabashev S, Fouad Y, Huncke TK, Roland JT (2017) Cochlear implantation under conscious sedation with local anesthesia; safety, efficacy, costs, and satisfaction. Cochlear Implants Int 18(6):297–303. https://doi.org/10.1080/14670100.2017.1376423

Shoman NM (2022) Robotics and cochlear implant surgery: goals and developments. Curr Opin Otolaryngol Head Neck Surg 30(5):314–319. https://doi.org/10.1097/MOO.0000000000000837

Shu H, Liang R, Li Z, Goodridge A, Zhang X, Ding H, Nagururu N, Sahu M, Creighton FX, Taylor RH, Munawar A (2023) Twin-S: a digital twin for skull base surgery. Int J Comput Assist Radiol Surg 18(6):1077–1084. https://doi.org/10.1007/s11548-023-02863-9

Stelter K, Ledderose G, Hempel JM, Morhard DF, Flatz W, Krause E, Mueller J (2012) Image guided navigation by intraoperative CT scan for cochlear implantation. Comput Aided Surg 17(3):153–160. https://doi.org/10.3109/10929088.2012.668937

Sun TM, Lee HE, Lan TH (2020) Comparing accuracy of implant installation with a navigation system (NS), a laboratory guide (LG), NS with LG, and freehand drilling. Int J Environ Res Public Health 17(6):2107. https://doi.org/10.3390/ijerph17062107

Sürmelioğlu Ö, Özdemir S, Tarkan Ö, Tuncer Ü, Çetik F, Kara K, Kıroğlu M (2016) Alternative techniques in cochlear implantation. J Int Adv Otol 12(1):109–112. https://doi.org/10.5152/iao.2015.1022

Thoranaghatte RU, Giraldez JG, Zheng G (2008) Landmark based augmented reality endoscope system for sinus and skull-base surgeries. In: 2008 30th annual international conference of the IEEE Engineering in Medicine and Biology Society. IEEE, pp 74–77. https://doi.org/10.1109/IEMBS.2008.4649094

Topsakal, V., Matulic M, Assadi MZ, Mertens G, Van Rompaey V, Van de Heyning P (2020) Comparison of the surgical techniques and robotic techniques for cochlear implantation in terms of the trajectories toward the inner ear. The journal of international advanced otology 16(1):3–7. https://doi.org/10.5152/iao.2020.8113

Topsakal V, Heuninck E, Matulic M, Tekin AM, Mertens G, Van Rompaey V, Galeazzi P, Zoka-Assadi M, Van de Heyning P (2022) First study in men evaluating a surgical robotic tool providing autonomous inner ear access for cochlear implantation. Front Neurol 13:804507. https://doi.org/10.3389/fneur.2022.804507

Torres R, Kazmitcheff G, Bernardeschi D, De Seta D, Bensimon JL, Ferrary E, Sterkers O, Nguyen Y (2016) Variability of the mental representation of the cochlear anatomy during cochlear implantation. Eur Arch Otorrinolaringol 273:2009–2018. https://doi.org/10.1007/s00405-015-3763-x

Wang J, Liu H, Ke J, Hu L, Zhang S, Yang B, Sun S, Guo N, Ma F (2020) Image-guided cochlear access by non-invasive registration: a cadaveric feasibility study. Sci Rep 10(1):18318. https://doi.org/10.1038/s41598-020-75530-7

Weber S, Gerber N, Gavaghan KA, Williamson T, Wimmer W, Anso J, Brogna-Salas L, Chen D, Weisstanner C, Caversaccio M, Bell B (2013) Image guided and robotic assisted minimally invasive cochlear implantation. In: The Hamlyn symposium on medical robotics, pp 17–18

Wijewickrema S, Copson B, Zhou Y, Ma X, Briggs R, Bailey J, Kennedy G, O'Leary S (2017) Design and evaluation of a virtual reality simulation module for training advanced temporal bone surgery. In: 2017 IEEE 30th international symposium on computer-based medical systems (CBMS). IEEE, pp 7–12

Yin K, He Z, Xiong J, Zou J, Li K, Wu ST (2021) Virtual reality and augmented reality displays: advances and future perspectives. J Phys Photonics 3(2):022010. https://doi.org/10.1088/2515-7647/abf02e

Younis H, Lv C, Xu B, Zhou H, Du L, Liao L, Zhao N, Long W, Elayah SA, Chang X, He L (2024) Accuracy of dynamic navigation compared to static surgical guides and the freehand approach in implant placement: a prospective clinical study. Head Face Med 20(1):30. https://doi.org/10.1186/s13005-024-00433-1

Zhang Y, Anaya ED, Su D, Lou A, Noble JH (2024) Monocular microscope to CT registration using pose estimation of the incus for augmented reality cochlear implant surgery. In: Medical imaging 2024: image-guided procedures, robotic interventions, and modeling, vol 12928. SPIE, pp 556–562

Future Optical Cochlear Implants 15

Antoine Huet

Abstract

The electrical cochlear implant (eCI) has restored speech understanding to the majority of the one million patients worldwide who would otherwise be deaf. However, hearing with an eCI is far from normal and patients report an unmet medical need to understand speech in the presence of background noise. This is due to the wide spread of excitation around each electrode contact, which limits the transfer of sound frequency information. Future optical cochlear implants (oCI) will use light instead of electricity because light can be spatially confined and will enable hearing restoration with finer tonotopic resolution. Photosensitization of the cochlea will be achieved by optogenetic modification of spiral ganglion neurons using an optogene therapy. This chapter discusses the last decade of development of an oCI and outlines the remaining challenges for it to become a new clinical standard in hearing rehabilitation.

Keywords

Optical cochlear implants · Optogene therapy · Channelrhodopsin · Spiral ganglion neurons · Hearing restoration · Sensory neuroprosthetics

A. Huet (✉)
Institute for Neurosciences of Montpellier (INM), University Montpellier, INSERM, UMR 1298, Montpellier, France

Institute for Auditory Neuroscience and InnerEarLab, University Medical Center Göttingen, Göttingen, Germany

Cluster of Excellence "Multiscale Bioimaging: From Molecular Machines to Networks of Excitable Cells" (MBExC), University of Göttingen, Göttingen, Germany

Auditory Circuit Lab, Institute for Auditory Neuroscience, University Medical Center Göttingen, Göttingen, Germany
e-mail: antoine.huet@inserm.fr

B. Molaee-Ardekani, M. Segovia-Martinez (eds.), *Advancing Cochlear Implants*, https://doi.org/10.1007/978-3-032-15719-5_15

15.1 Introduction

One in two people will experience hearing loss (HL) during their lifetime, which, if left untreated, will most likely lead to a reduction in daily activities and quality of life (WHO 2021). The yearly cost of untreated HL is high and was evaluated at 980 billion US$ (WHO 2021). The causes of HL are diverse (e.g., genetic disorders, noise exposure, aging), but mainly result from cochlear disorders associated with a loss or dysfunction of the mechano-transduction of sound into a neural activation of the spiral ganglion neurons (SGN).

The last decade has seen the development of cell and gene therapies that target the underlying cause of HL, as exemplified by the recent clinical reports of otoferlin-associated HL (Akil et al. 2019; Rankovic et al. 2021; Lv et al. 2024; Qi et al. 2024; Moser et al. 2024). However, these approaches are etiology specific and therefore inherently limited in the number of patients they can cure. Alternatively, hearing aids and cochlear implants are one-size-fits-all solutions and will therefore remain the main strategy for hearing rehabilitation. In general terms, hearing aids are mainly used to rehabilitate moderate HL where speech understanding is impaired, whereas cochlear implants are used to restore hearing in severely hearing-impaired patients where speech understanding is not possible. Hearing aids and cochlear implants differ fundamentally in the way they work. Hearing aids are nonlinear sound amplifiers designed to compensate for the impaired mechano-transduction of the sound. In contrast, cochlear implants bypass the dysfunctional mechano-transduction of the sound and electrically stimulate the SGNs directly (Fig. 15.1a).

Both the acoustic and electrical modalities rely on the concept of cochlear tonotopy to encode the sound frequency. High-frequency sounds are transduced at the base of the cochlea, whereas low-frequency sounds are transduced at the apex. In acoustic hearing, the decomposition of sound according to its frequency is achieved by the resonance property of the basilar membrane, which varies from base to apex (Békésy and Wever 1960). In electrical hearing, the frequency decomposition of the

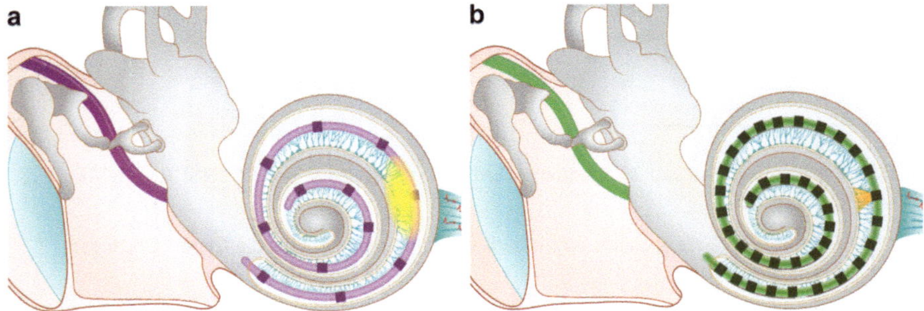

Fig. 15.1 Schematic representations of the internal component of an electrical (**a**) and optical (**b**) cochlear implant. It can be observed that, in the case of the electrical implant, there is a broad spread of activation around the electrode contact, which limits the number of independent stimulation channels. In contrast, the optical cochlear implant allows for a more focused activation. (Figure modified from Dieter et al. (2020a), under the CC BY license)

sound is achieved by the sound processor, which sequentially activates the implanted electrodes at specific tonotopic positions (Zeng et al. 2008).

The success of the electrical cochlear implant (eCI) is represented by its more than one million users, the majority of whom have regained open speech understanding in quiet (Zeng 2022). Remarkably, if diagnosed and implanted soon after the onset of HL, children with a prelingual hearing loss can acquire almost normal speech and language development.

Nevertheless, hearing with an eCI is far from normal, and patients report an unmet medical need to understand speech in the presence of background noise (Wolf et al. 2022; Hunniford et al. 2023). This complaint is far from anecdotal, as most everyday social interactions take place in the presence of background sounds. This deficiency is caused by the wide distribution of the electric current in the saline-filled cochlea, around each electrode contact, which leads to the recruitment of a large population of SGNs and thus results in poor frequency resolution (Kral et al. 1998). For 25 years, researchers tried to reduce the electric cochlear electrical spread of activation by developing multipolar stimulation (Bierer et al. 2010) or by improving the interface between the electrode and the neural tissue: use of intra-scalar electrodes (Cords et al. 2000) or intraneural electrodes (Middlebrooks and Snyder 2007), by promoting neurite growth toward the electrodes (Pinyon et al. 2014). Where tested clinically, the benefits have remained limited, and further efforts are warranted.

An alternative to reduce the cochlear spread of stimulation is to use light instead of electricity, as light can be spatially confined (Fig. 15.1b). The concept of optical cochlear implants (oCI) was first introduced by Richter and colleagues who demonstrated more precise photoactivation of the SGNs using cochlear near-infrared stimulation (NIS) compared to electrical stimulation (Izzo et al. 2007). Unfortunately, the mechanism by which SGNs are activated by NIS has since been questioned, and it is quite unlikely that future oCIs will rely on this mode of stimulation (Teudt et al. 2011; Thompson et al. 2015; Kallweit et al. 2016; Baumhoff et al. 2019; Shapiro et al. 2012). The concept of oCI was then further developed by Moser and colleagues, who rely on optogenetics to photocontrol the SGNs, i.e., the expression of photosensitive proteins in targeted cells. The latter approach will be discussed further in this chapter.

15.2 Fundamentals of Optical Cochlear Implants

15.2.1 Roadmap to Success: The Rise of Optogenetics in Biomedical Research

The emergence of optogenetics is a striking example of how basic science can lead to profound advances in applied research. It all started with scientists studying phototaxis (i.e., the ability to move toward or away from a light source) in green algae. First, channelrhodopsin (ChR) was identified as the light-sensitive protein responsible for the phototaxis in green algae (Nagel et al. 2002, 2003). Later on, hippocampal neurons genetically engineered to express channelrhodopsin-2 (ChR-2) were shown to be light-drivable (Boyden et al. 2005). This series of groundbreaking discoveries led to a scientific revolution by providing new tools for understanding the brain and its circuitry (Emiliani et al. 2022; Kim et al. 2017), as well as for developing new therapies. Multiple proof-of-principle studies demonstrated how optogenetics can restore or improve the function of diseased organs such as the eye, ear, brain, heart, larynx and stomach (Arrenberg et al. 2010; Bi et al. 2006; Bruegmann et al. 2010; Hernandez et al. 2014; Kim et al. 2021; Chen et al. 2021; Lagali et al. 2008; Vedam-Mai et al. 2021; Vogt et al. 2021; Walsh et al. 2018; Wykes et al. 2012). The year 2021 marked a milestone in the advancement of optogenetics in biomedical research, as Sahel and colleagues demonstrated for the first time the use of optogenetics to partially restore vision in human retinas (Sahel et al. 2021).

15.2.2 Proof-of-Concept Studies of Optical Hearing Restoration

The feasibility of optogenetic activation of the auditory pathway was first shown using mice whose SGNs were engineered to express ChR-2. Following microLED (µLED) stimulation delivered to the cochlea, activation of the auditory pathway was demonstrated in vivo by recording (*i*) optically evoked auditory brainstem response (oABR); (*ii*) optically evoked firing in single SGN; and (*iii*) optically evoked local field potential in the inferior colliculus (IC) (Hernandez et al. 2014). In addition, the potential of optical hearing restoration was first assessed using congenitally deaf mice in which optical electrophysiological responses were recorded in the absence of acoustic responses.

In order for the approach to be translational, SGNs need to be optogenetically modified on demand at a post-developmental stage. This was first achieved using an engineered adeno-associated virus (AAV), carrying ChR, which was injected into the modiolus of adult gerbils (Wrobel et al. 2018). A few weeks after injection, ~40% of the injected cochleae contained SGNs expressing ChR, and the SGN activation was demonstrated in vivo by recording their unitary responses to light (i.e., optically evoked single SGN recordings). Furthermore, the optical cochlear activation was shown to propagate from the auditory nerve to the auditory cortex, where single A1 neuron-evoked responses were recorded.

Can cochlear optogenetic stimulation restore auditory perception in deafened subjects? Pharmacologically deafened gerbils implanted with wired single-channel oCIs (Wrobel et al. 2018) and rats implanted with autonomous complete multichannel LED-based oCIs (Keppeler et al. 2020) performed auditory-driven behavior following activation of the implant, and thus optogenetic activation of their SGNs.

Finally, the original hypothesis that optogenetic stimulation would allow more spatially restricted cochlear activation was demonstrated. Multielectrode array recordings from the IC allowed mapping of the cochlear spread of activation. Using this approach, acoustic, electrical and optogenetic cochlear stimulation were compared, revealing near-physiological spread of activation for single optical fiber

(Dieter et al. 2019) and μLED-based oCI (Dieter et al. 2020b; Azees et al. 2023) stimulation.

The aforementioned proof-of-principle studies collectively endorse the advancement of an oCI toward clinical application. In comparison to an eCI, an oCI necessitates a dual technological development: first, a medical device, the light-emitting device, and second, an optogene therapy to photosensitize the SGNs.

15.3 Design of the Optical Medical Device

It is anticipated that the oCI will possess numerous common features with the eCI technology. Such features include the speech processor, the transmission coil, the titanium package and some of the implant electronics (Fig. 15.2a). Two different implementations are envisioned for the light-emitting device that differ by the place where light will be generated: (i) an active design where light is generated by the implant in the cochlea (Fig. 15.2b) and (ii) a passive design where light is generated outside of the cochlea and brought inside by an array of waveguides (Fig. 15.2c).

The initial oCI prototypes consisted of implantable arrays of blue light-emitting diodes (LEDs). In this active approach, the size of the LEDs can be reduced to a range of approximately 100 μm (i.e., μLEDs), and a large density of μLEDs can be transferred to the flexible array through a wafer-based process (Dieter et al. 2020b; Klein et al. 2018). The high number of μLEDs precludes the possibility of addressing each optrode by an individual line, as is the case for the electrodes of the eCI. This limitation can be circumvented by matrix-like addressing schemes, whereby the n- and p-contacts of the μLEDs are reached by independent lines ideally placed in two isolated planes. In this approach, achieving good passivation of the electronics, while preserving bending properties, is of prime importance to avoid direct electrical

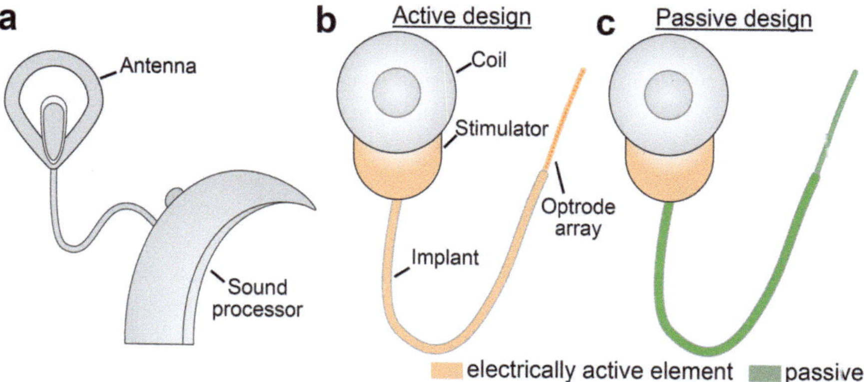

Fig. 15.2 Schematic representation of future optical cochlear implants: (**a**) external behind-the-ear sound processor and antenna; (**b**, **c**) internal component with active (**a**) or passive (**b**) optrodes. The electrically active components of the implant are highlighted in orange. (Figure inspired by Khurana et al. 2023)

stimulation of the cochlea (Keppeler et al. 2020). Unfortunately, current LED technology does not allow for medically safe light emission of orange/red light that might be required to activate some of the ChRs considered for optical hearing restoration (details in the next section).

The second generation of oCI prototypes will be based on a passive design, with the light emitters situated outside of the cochlea and a waveguide array used to conduct the light into the cochlea (Huet et al. 2024). It is anticipated that the optoelectronic device will be integrated into the titanium package, alongside the electronic material responsible for generating the light. In order for the light to exit the titanium housing while maintaining the hermetic sealing, a sapphire-based transparent window is a potential candidate for interfacing the light emitters and the waveguide array. In this approach, a multitude of light beams, generated by an array of light emitters, will undergo continuous interleaved scanning of the interfacing point of the waveguides. Further details concerning the properties of the light emitter array and waveguides are soon to be expected.

15.4 Design of the Optogene Therapy

15.4.1 Optimal ChR for Hearing Restoration

The fundamental component of the optogene therapy is the light-activated protein, channelrhodopsin, that is expressed in the SGN. In response to light, opening of ChR induces a photocurrent that triggers the SGNs to fire. The optogenetic activation of SGNs should support high-rate firing with high temporal precision, similar to physiological sound encoding. Achieving such photocontrol is far from trivial as it requires matching the biophysical properties of the ChR with those of the SGNs and the physical properties of the light emitters. A first constraint arises from the SGNs, as most of them only fire a few action potentials in response to millisecond-long current injections (Markowitz and Kalluri 2020). Thus, continuous firing can only be achieved by pulsed light stimulation. Similarly, active and passive optrodes require interleaved stimulation to accommodate the large number of stimulation channels. Consequently, the ideal ChR for optical hearing restoration must support fast optical stimulation. Biophysically, this translates into a ChR with a short closing kinetics (Gunaydin et al. 2010; Berndt et al. 2010; Klapoetke et al. 2014): the shorter the channel closing kinetics when the light is turned off, the higher the stimulation rate that can be used. The second constraint relates to the energy consumption of an oCI. Indeed, the increased number of stimulation channels is directly associated with an increase in energy consumption. Therefore, the optimal ChR must mediate large photocurrents when expressed in the SGNs. The third limitation relates to the safety of optical stimulation, since high-frequency and high-intensity optical stimulation carries the risk of phototoxicity. This risk can be reduced by using less phototoxic red light instead of blue light for stimulation. This is achieved by employing a ChR that responds within the orange/red spectrum.

To this end, various naturally occurring and engineered ChRs were screened for their ability to mediate activation of the auditory pathway (ChR2 (Hernandez et al. 2014); CatCh (Wrobel et al. 2018); ChR2-H134R (Ajay et al. 2023); f-Chrimson (Mager et al. 2018; Bali et al. 2021); vf-Chrimson (Bali et al. 2021); Chronos (Keppeler et al. 2018; Mittring et al. 2023); ChReef (Alekseev et al. 2025)). Light-evoked firing properties were assessed using in vivo single SGN recordings in response to optical stimulation delivered into the cochlea via laser-coupled optical fiber (Fig. 15.3a). For all tested ChRs, the SGNs fire one to multiple spikes per light pulse with high temporal precision (vector strength ~1; first spike latency jitter ~0.1 ms) in response to slow (i.e., <50 Hz) light pulse trains (Fig. 15.3b, c). In response to faster light pulse trains, the reliability of firing declines rapidly above 100 Hz stimulation rate for all tested ChRs, except for f-Chrimson, which mediates reliable firing up to 200 Hz stimulation rate (Fig. 15.3c–e). Surprisingly, f-Chrimson closing kinetics is slower (i.e., closing kinetic ~3 ms at body temperature) than vf-Chrimson and Chronos (~1.5 ms and ~1 ms, respectively), suggesting that other properties such as the level of ChR membrane expression (Keppeler et al. 2018; Meng et al. 2019) and the single channel conductance may influence the maximum stimulation rates that can be used.

Next, identification of the ChR mediating the activation of the auditory pathway at the lowest intensity was conducted. To this end, optically evoked auditory brainstem responses (oABR) were measured in response to 10 Hz light pulse trains delivered into the cochlea via a laser-coupled optical fiber (Table 15.1) at various light intensities. The lowest activation thresholds were measured for ChRmine, and its engineered variant, ChReef, in response to green light (wavelength = 522 nm) amounted to 0.17 and 0.27 μJ, respectively. Such thresholds are one order of magnitude lower than previously reported for f-Chrimson (~1 μJ) in response to orange light (wavelength = 594 nm). Interestingly, ChRmine and ChReef activation spectrum enable activation by orange light, which resulted in activation thresholds of 1.25 and 2.45 μJ, respectively.

Considering the previously described data from oABRs and single SGN recordings, f-Chrimson appears to be an appropriate ChR for optical hearing restoration as it (i) supports stimulation with high-rate light pulse trains; (ii) can be activated by low light pulses; and (iii) has an orange/red-shifted activation spectrum.

15.4.2 Virus-Mediated Expression of ChR in the SGN

The objective of the optogene therapy is to drive safe and long-lasting ChR expression in the spiral ganglion neurons of mature cochlea. Performing gene therapy in the cochlea is not new as attested by the recent clinical trials aiming at (i) inducing trans-differentiation of cochlear supporting cells into sensory hair cells for regenerative hearing restoration (ClinicalTrials.gov ID: NCT02132130); (ii) replacing the *OTOF* gene in inner hair cells lacking functional otoferlin (Lv et al. 2024; Qi et al. 2024). Classically, inner ear gene therapy is performed using adeno-associated virus (AAV) as viral vectors, to deliver the gene in the cell type of interest. The optogene therapy

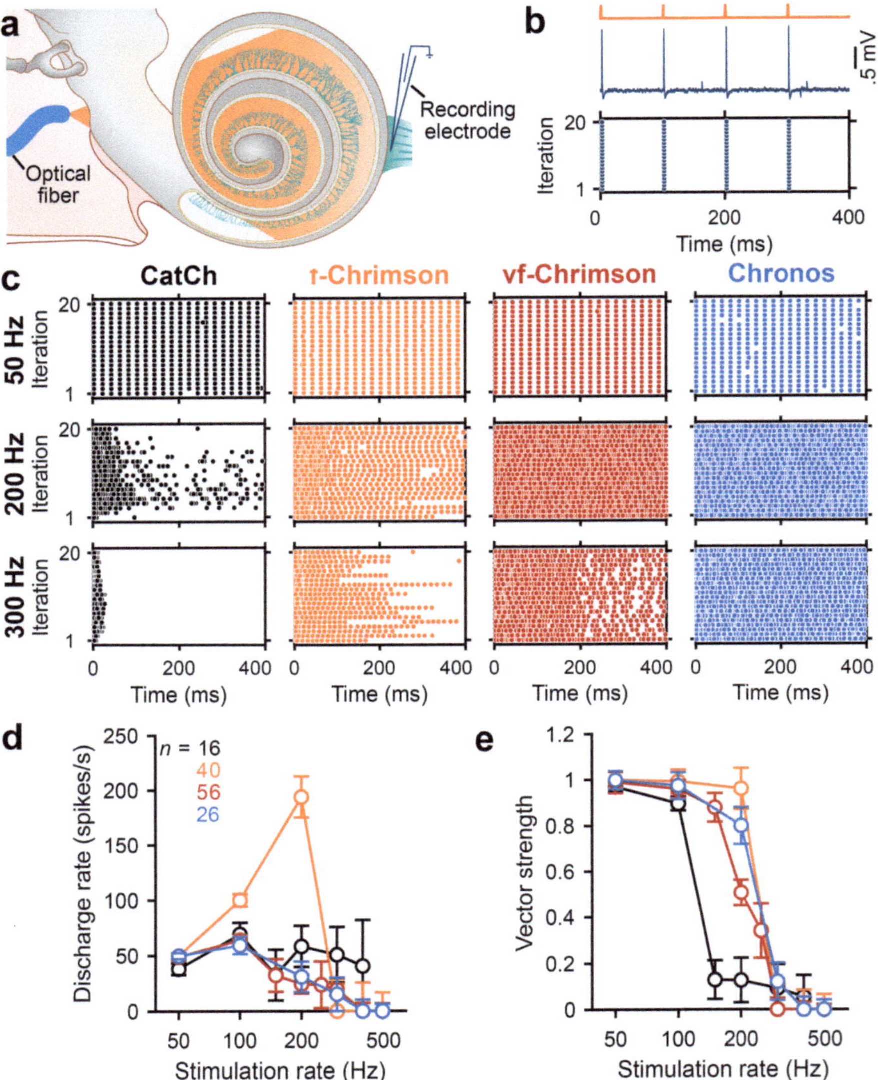

Fig. 15.3 Screening of the optimal channelrhodopsin for hearing restoration. (**a**) Single recording of optogenetically modified SGNs in response to light is the gold standard approach to assess optogenetically evoked firing. Light is delivered into the cochlea via a laser-coupled optical fiber. (The cochlea cartoon was modified from Dieter et al. 2020a). (**b**) Exemplary recording from a CatCh-expressing SGN in response to a 10 Hz light pulse train. (Top) light signal used as a stimulus; (middle) juxtacellular spikes in response to one iteration of the stimulus; (bottom) raster plots built from 20 presentations of the stimulus. (**c**) Raster plots from the best recorded SGNs expressing CatCh (black), f-Chrimson (orange), vf-Chrimson (red) and Chronos (blue) in response to light pulse trains of 50, 200 and 300 Hz. (**d**, **e**) Quantification of the discharge rate (**d**) and vector strength (**e**) measured from SGNs expressing CatCh (Wrobel et al. 2018), f-Chrimson (Mager et al. 2018), vf-Chrimson (Bali et al. 2021) and Chronos (Keppeler et al. 2018)

Table 15.1 oABR threshold measured for various ChRs expressed in the SGNs

ChR	Virus injection model	oABR threshold (stimulation wavelength)			Reference
		470 nm	522 nm	594 nm	
CatCh	Adult gerbils	4.5 μJ	N/A	N/A	Wrobel et al. (2018)
Chronos	p8 mice	4.5 μJ	N/A	N/A	Keppeler et al. (2018)
f-Chrimson	p8 mice	N/A	N/A	1 μJ	Mager et al. (2018)
	p8 gerbils	N/A	N/A	1 μJ	Huet et al. (2021)
	Adult gerbils	N/A	N/A	5 μJ	Huet et al. (2021)
vf-Chrimson	p8 mice	12 μJ	N/A	N/A	Bali et al. (2021)
ChRmine	p8 mice	N/A	0.27 μJ	2.45 μJ	Alekseev et al. (2025)
ChReef	p8 mice	N/A	0.17 μJ	1.25 μJ	Alekseev et al. (2025)

The virus administration model refers to the species and the age at which the intracochlear injection was performed
N/A Not applicable

planned for optical hearing restoration differs from previously mentioned works in that it targets the SGN somas, which are housed in Rosenthal's canal. The location of the SGN in a narrow and bony structure makes them more difficult to reach than the cells of the organ of Corti, which are situated in the fluid-filled scala tympani where viral vectors can be directly administered.

The optogene therapy is based on three fundamental elements: (*i*) the ChR and auxiliary sequences that compose the so-called transgene; (*ii*) a viral vector and promoter for the specific transduction and expression of the transgene in the SGNs; and (*iii*) the administration approach of the viral suspension to the inner ear.

Recombinant adeno-associated viruses (rAAV) represent the most prevalent viral vector utilized in contemporary gene therapy. These viruses are regarded as safe due to their replication-deficient nature and were modified to prevent integration into the host genome. To date, a broad range of rAAVs have been evaluated for their ability to mediate ChR transduction in the SGNs. When directly injected into the cochlea of early postnatal rodents (i.e., 3–8 days after birth, prior to the onset of hearing function; see (Huet and Rankovic 2021) for detailed methodology), the majority of tested rAAVs were able to mediate ChR expression in at least 50% of the SGNs using the human synapsin promoter (Hernandez et al. 2014; Mager et al. 2018; Huet et al. 2021; Wrobel et al. 2018; Bali et al. 2021; Keppeler et al. 2018; Mittring et al. 2023; Zerche et al. 2023; Duarte et al. 2018; Richardson et al. 2021). A longitudinal study demonstrated that ChR expression in the SGNs persisted and remained functional for the lifetime of the mice (i.e., 24 months after injection (Bali et al. 2022)). However, the transduction of some ChRs in the immature SGN does not necessarily result in optical hearing. This is exemplified by the case of the ultrafast blue light-activated Chronos, which necessitated the use of highly potent rAAVs to support the optical activation of the SGNs (Duarte et al. 2018; Keppeler et al. 2018).

Nevertheless, when the same viral vectors were injected into the mature cochlea, a substantially lower transduction rate of the SGNs was observed (Huet et al. 2021; Wrobel et al. 2018; Michael et al. 2023). In the mature cochlea, the injection of a small volume (approximately 3 μL) of viral vector solution into the Rosenthal's canal, where the SGN somas are housed, resulted in an SGN transduction rate of ~30% in ~40% of the treated cochlea (Huet et al. 2021; Wrobel et al. 2018; Michael et al. 2023). In contrast, injection into the scala tympani or the semicircular canal did not result in SGN transduction (Richardson et al. 2021). The precise mechanism underlying this reduction in efficacy between the immature and mature cochlea injection remains unclear. However, it has been recently demonstrated that increasing the number of virus particles administered into the cochlea enables partial correction of this effect. Indeed, the slow administration of a volume of 5 μL via a pump-controlled microcatheter inserted into the scala tympani in conjunction with an evacuation vent (μ-catheter + vent) of the AAV-PHP.S enabled to optogenetically modify ~30% of the SGN in 80% of the treated cochlea (Thirumalai et al. 2025). In this approach, the cochlear perilymph is partially replaced by the viral vector solution. The μ-catheter + vent approach has been successfully employed to optogenetically modify the SGNs of pharmacologically deafened gerbils (Thirumalai et al. 2025) and nonhuman primates (Alekseev et al. 2025).

15.5 Physiology of Optical Hearing (eCI vs oCI: Spatial Resolution, Intensity Coding, Timing and Hybrid Stimulation)

In order to restore hearing, the auditory nerve must be provided with three characteristics of sound: sound frequency, sound intensity and timing.

15.5.1 Improved Spatial Resolution of Optical Hearing

Optogenetic SGN stimulation would result in an increased number of independent stimulation channels and thus increase the amount of spectral information available to the CI user. Experimentally, the spread of cochlear activation can be measured from the IC using multielectrode array recording (Fig. 15.4a). In the IC, the cochlear tonotopy is conserved with the dorsal part of the IC coding low-frequency sounds (i.e., apical cochlear activation) and the ventral part coding for the high-frequency ones (i.e., basal cochlear activation).

Using this approach in ChR-2 transgenic mice, a pioneering study demonstrated an indistinguishable small cochlear activation spread between optical cochlear base stimulation and pure tone acoustic stimulation, in contrast to the significantly larger activation spread elicited by electrical stimulation (Hernandez et al. 2014). It was later confirmed, using Mongolian gerbils in which the SGNs were optogenetically modified at adult age, that optical stimulation leads to a near-physiological spread of cochlear activation at multiple cochlear locations (basal, middle and apical turns, Fig. 15.4b, Dieter et al. 2019). With regard to spectral selectivity, optogenetic

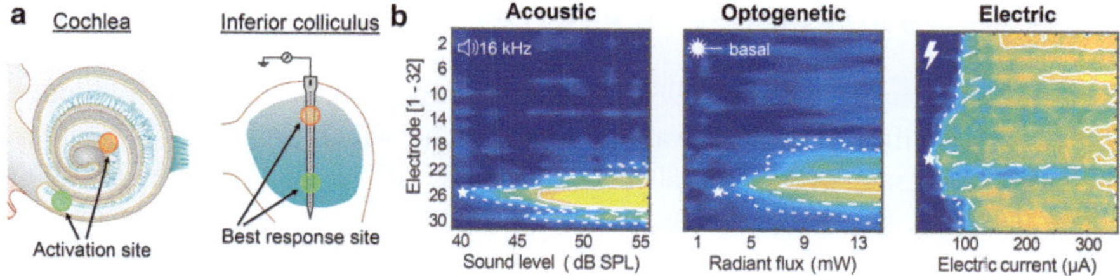

Fig. 15.4 Near-physiological spread of cochlear activation of optical hearing. (**a**) The distribution of cochlear activation can be recorded from the IC using a multielectrode array. This measure is based on the concept of tonotopy, i.e., the spatial distribution of cochlear and IC activation as a function of sound frequency. (**b**) Exemplary multi-unit frequency response area measured in response to acoustic, optogenetic and electric stimulus delivered at similar cochlear position. (Figure modified from Dieter et al. (2020a), under the CC BY license)

stimulation demonstrated superior efficacy compared to both monopolar and bipolar electrical stimulation, exhibiting an almost twofold improvement at medium and high IC activation strength (Fig. 15.4b, Dieter et al. 2019, 2020b). Since that time, certain aspects of these findings have been replicated in ChR-2 transgenic mice, albeit with a diminished spectral selectivity enhancement (Richardson et al. 2021; Azees et al. 2023; Thompson et al. 2020). This is likely attributable to the fact that the cochleae of these mice are approximately four times smaller than those of gerbils. Computational simulations of light propagation in the gerbil cochlea, conducted using the Monte Carlo ray tracing method, corroborated these findings, thereby confirming the potential for even greater spectral selectivity in cochlear optogenetics with intracochlear emitters (Dieter et al. 2019; Wrobel et al. 2018). Performing a similar computational approach on the human cochlea with both active and passive oCI designs demonstrated once more greater spatial selectivity compared to the eCI (Khurana et al. 2022). The measured spatial spread of excitation was found to be 4 to 12 times smaller for all tested tonotopic positions, depending on the light source properties. Furthermore, this spread was found to be robust to different scenarios of oCI location in the cochlea and scar tissue development. The narrowest spatial spread of excitation was obtained using a passive oCI design with a small beam opening angle. However, the greatest light directionality comes at the cost of the potential for not illuminating the SGNs, for example, during insertion or upon later rotation of the oCI. Therefore, future oCI designs may rely on an intermediate beam opening angle to achieve good spatial selectivity while ensuring stable neural activation. The same framework predicted that the first-generation oCI, envisioned for clinical trial, could mediate 32 to 64 independent stimulation channels (Khurana et al. 2022). The provision of additional spectral information via refined optogenetic stimulation may facilitate more naturalistic auditory perceptions. In particular, by enhancing the comprehension of speech in complex acoustic environments, such as those characterized by background noise, competing speakers and reverberations. Furthermore, the promised enhanced spectral coding may contribute to facilitate the perception of melodies in speech and music by CI users. However, the estimation of these perceptual benefits remains to be demonstrated upon psychophysical testing of the oCI in preclinical studies and clinical trials.

15.5.2 Optogenetic Encoding of the Sound Level

In acoustic hearing, the sound level is encoded by the joint activation of individual SGNs, which differ in their activation threshold and which progressively increase their discharge rate over a 10–60 dB range (Fig. 15.5a, d, Winter et al. 1990; Liberman 1978; Sachs and Abbas 1974; Ohlemiller et al. 1991). Furthermore, the sound level is encoded by the proportion of SGNs recruited along the tonotopic axis. Thus, the louder the sound at a given frequency, the larger the proportion of SGNs recruited at and beyond the corresponding tonotopic place. This results in a perceptual dynamic range of ~120 dB for the sound level. The sound level encoding is primarily governed by the cochlear micromechanics (Yates et al. 1990) and the heterogenous synapses between the IHCs and the SGNs (for review, see Moser et al. 2023). In electrical hearing, the previously described mechanisms are bypassed by the direct stimulation of the SGNs, resulting in a diminished dynamic range of ~1 dB at the level of the individual SGN and an audible perceptual range of 10–20 dB (Rubinstein 2004; Zeng 2004).

In optogenetic hearing, the intensity of the stimulus can be encoded in two ways: either by controlling the amplitude of the light pulse (Fig. 15.5b, e), which is the most direct approach, or by controlling the duration of the light pulse (Fig. 15.5c, f) to adapt to laser diodes, which are best controlled when operated with strong currents that are modulated in duration. Amplitude-based control of stimulus intensity was investigated in SGNs expressing f-Chrimson, vf-Chrimson and Chronos (Bali et al. 2021; Mittring et al. 2023). Similar to what is observed acoustically, single SGNs gradually increased their discharge rate with light pulse amplitude until saturation was reached (Fig. 15.5e). The dynamic range over which the discharge rate increased appeared to be ChR specific, ranging from 1 dB for vf-Chrimson (Bali et al. 2021) to 3 dB for f-Chrimson (Bali et al. 2021) and Chronos (Mittring et al. 2023). Interestingly, for all tested ChRs, the activation threshold of the SGNs was heterogeneously distributed over 15 dB, resulting in a wider dynamic range of intensity encoding (Bali et al. 2021; Mittring et al. 2023). Duration-based control of stimulus intensity was investigated in SGNs expressing Chronos (Mittring et al. 2023). Similarly, a gradual activation of single SGNs could be achieved over an estimated dynamic range of 3 dB (Fig. 15.5f). Interestingly, the dynamic range measured from a population of SGNs was similar and amounted to ~7 dB for the two modes of control (Fig. 15.5e, f).

The studies discussed in the above paragraph focused on how SGNs encode light intensity, but for the brain to perceive the loudness of the stimulus, the optically driven activation of the SGNs has to be recoded by various populations of neurons localized between the cochlea and the auditory cortex. Using IC multielectrode array recordings from Mongolian gerbils with the CatCh-expressing SGNs, a dynamic range of activation of at least 10 dB was measured (Dieter et al. 2019). More recently, an IC activity level-dependent comparison between optogenetic and acoustic stimulation was conducted, and the dynamic range of optical hearing to 26 dB (peak equivalent) was approximated, which roughly corresponds to half of the dynamic range measured for acoustic hearing (~50 $dB_{\text{peak equivalent}}$, Michael et al.

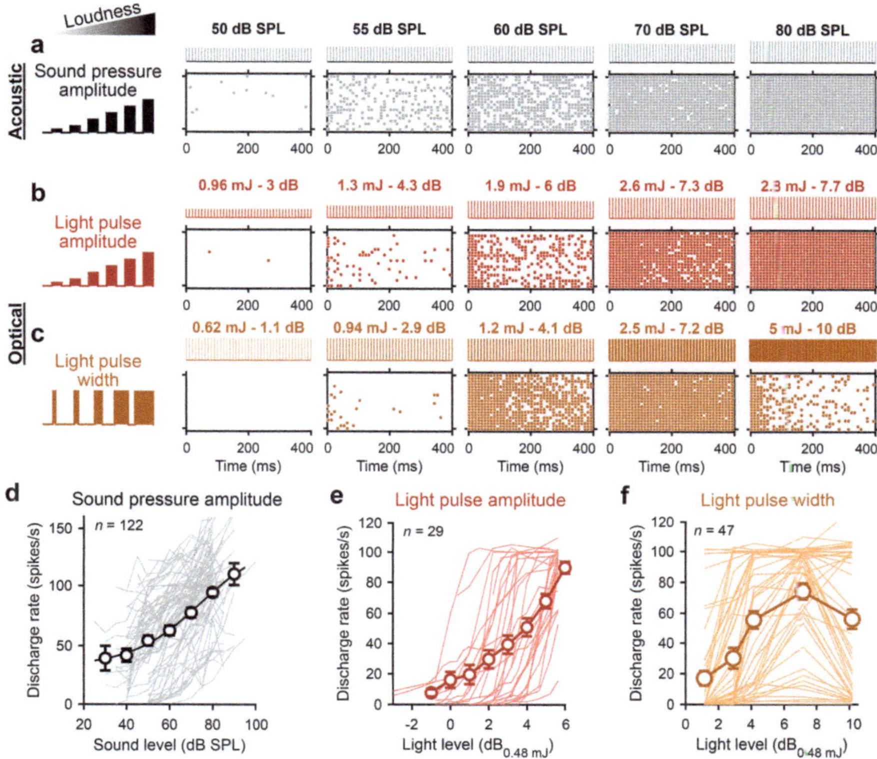

Fig. 15.5 The optical gradual activation of the SGNs can be achieved through the control of the light pulse amplitude and width. (**a**–**c**) The figure illustrates the representative acoustically (**a**) and optogenetically (**b**, **c**) evoked responses of a single SGN in response to varying stimulation strengths. In order to achieve gradual activation of the SGNs via an acoustic method, the sound pressure amplitude must be increased. The sound level is displayed superimposed on each raster plot. In optogenetic experiments, the gradual activation of SGNs can be achieved by controlling the light pulse amplitude (**b**) or the light pulse width (**c**) independently. The energy content of the 400 ms light pulse train is displayed superimposed on each raster plot. For comparison, the light level relative to a 100 Hz, 1 ms, 12 mW light pulse train was calculated and is also displayed on top of each raster plot. D-F. Quantification of the discharge rate as a function of the sound level (**d**, non-optogenetically modified SGNs) and the light level controlled either by the light pulse amplitude (**e**, Chronos-expressing SGNs) or light pulse width (**f**, Chronos-expressing SGNs). (Data were reanalyzed and replotted from Mittring et al. 2023)

2023). However, estimating the loudness perception of optical hearing remains to be characterized by psychophysical testing of the oCI in preclinical and clinical studies.

15.5.3 Temporally Precise Optogenetic Control of the Spiral Ganglion Neurons

Sound is composed of two essential temporal cues: the temporal envelope and the temporal fine structure (TFS). The envelope captures slow amplitude fluctuations, typically in the range of 1–50 Hz, while the TFS corresponds to the rapid oscillations at higher frequencies. In a quiet background, the envelope is critical for speech perception, while the TFS is most important for pitch perception and sound localization (Smith et al. 2002). In electrical hearing, most sound coding strategies focus on providing the auditory system with envelope information carried by high stimulation rate (~1 kHz) biphasic pulse trains (Zeng 2004). The rationale for using such a high carrier frequency remains a topic of debate in the community, but is discussed to ensure accurate representation of sound stimuli and to induce desynchronization of the SGNs by bringing them into relative refractoriness. In both quiet and noisy environments, several studies have shown that stimulation rates higher than 300 Hz do not significantly improve speech intelligibility (Friesen et al. 2005; Shannon et al. 2011). In optical hearing, desynchronization of SGNs does not appear to be necessary, as the temporal distribution of activation across the population of SGNs is similar to that measured with acoustic stimulation (Mittring et al. 2023). This could reflect the heterogeneous expression of ChR between neighboring SGNs, resulting in more variable light-evoked spike timing between neurons. Additionally, the spike statistics evoked by light supports enhancement of the response in the cochlear nucleus neurons (Mittring et al. 2023), reminiscent of what was described acoustically (Joris et al. 1994). Collectively, these findings suggest that utilizing a stimulation rate within the range of 100–300 Hz would allow for optimal recruitment of SGNs and faithful encoding of the stimulation by most ChRs. Future studies will confirm the sound coding strategies of the oCIs to faithfully encode the speech envelope using electrophysiological and behavioral approaches.

15.6 Outlook and Conclusion

This chapter has discussed the current development of optogenetic spiral ganglion neurons stimulation for hearing restoration with an optical cochlear implant. In order to become a new clinical standard, future investigations will be necessary to confirm and validate the findings discussed above. First, optimize the optogenetic tools and their delivery method to improve the efficacy and safety of optogenetic control to ensure precise targeting and efficient stimulation of SGNs. Second, develop a suitable processor and sound encoding strategy to effectively control the oCIs and generate optimal optical signals to mimic natural hearing. Third, validate the technological feasibility of multichannel oCIs and compare their efficacy with eCIs using preclinical electrophysiological and behavioral studies in vivo. Fourth, conduct a thorough preclinical evaluation of the biosafety and long-term stability of optogenetic SGN stimulation.

Acknowledgments I would like to thank Tobias Moser and colleagues from the Institute for Auditory Neuroscience for the vibrant years spent together working toward the development of an optical cochlear implant. This chapter provides an insight into the fantastic science taking place there. The author declares the use of DeepL to increase readability and to correct language mistakes of the author's chapter. The author carefully checked all suggested changes and reviewed and edited the content. The author takes full responsibility for the content of the final publication.

References

Ajay EA, Trang EP, Thompson AC, Wise AK, Grayden DB, Fallon JB, Richardson RT (2023) Auditory nerve responses to combined optogenetic and electrical stimulation in chronically deaf mice. J Neural Eng 20:026035. https://doi.org/10.1088/1741-2552/acc75f

Akil O, Dyka F, Calvet C, Emptoz A, Lahlou G, Nouaille S, Monvel JB de, Hardelin J-P, Hauswirth WW, Avan P, et al (2019) Dual AAV-mediated gene therapy restores hearing in a DFNB9 mouse model. Proc Natl Acad Sci 116: 4496–4501. https://doi.org/10.1073/pnas.1817537116

Alekseev A, Hunniford V, Zerche M, Jeschke M, May FE, Vavakou A, Siegenthaler D, Hüser MA, Kiehn SM, Garrido-Charles A et al (2025) Efficient and sustained optogenetic control of sensory and cardiac systems. Nat Biomed Eng. https://doi.org/10.1038/s41551-025-01461-1

Arrenberg AB, Stainier DYR, Baier H, Huisken J (2010) Optogenetic control of cardiac function. Science 330:971–974. https://doi.org/10.1126/science.1195929

Azees AA, Thompson AC, Thomas R, Zhou J, Ruther P, Wise AK, Ajay EA, Garrett DJ, Quigley A, Fallon JB et al (2023) Spread of activation and interaction between channels with multi-channel optogenetic stimulation in the mouse cochlea. Hear Res 440:108911. https://doi.org/10.1016/j.heares.2023.108911

Bali B, Lopez de la Morena D, Mittring A, Mager T, Rankovic V, Huet AT, Moser T (2021) Utility of red-light ultrafast optogenetic stimulation of the auditory pathway. EMBO Mol Med 13: e13391. https://doi.org/10.15252/emmm.202013391

Bali B, Gruber-Dujardin E, Kusch K, Rankovic V, Moser T (2022) Analyzing efficacy, stability, and safety of AAV-mediated optogenetic hearing restoration in mice. Life Sci Alliance 5: e202101338. https://doi.org/10.26508/lsa.202101338

Baumhoff P, Kallweit N, Kral A (2019) Intracochlear near infrared stimulation: feasibility of optoacoustic stimulation in vivo. Hear Res 371:40–52. https://doi.org/10.1016/j.heares.2018.11.003

Berndt A, Prigge M, Gradmann D, Hegemann P (2010) Two open states with progressive proton selectivities in the branched channelrhodopsin-2 photocycle. Biophys J 98:753. https://doi.org/10.1016/j.bpj.2009.10.052

Bi A, Cui J, Ma Y-P, Olshevskaya E, Pu M, Dizhoor AM, Pan Z-H (2006) Ectopic expression of a microbial-type rhodopsin restores visual responses in mice with photoreceptor degeneration. Neuron 50:23–33. https://doi.org/10.1016/j.neuron.2006.02.026

Bierer JA, Bierer SM, Middlebrooks JC (2010) Partial tripolar cochlear implant stimulation: spread of excitation and forward masking in the inferior colliculus. Hear Res 270:134–142. https://doi.org/10.1016/j.heares.2010.08.006

Boyden ES, Zhang F, Bamberg E, Nagel G, Deisseroth K (2005) Millisecond-timescale, genetically targeted optical control of neural activity. Nat Neurosci 8:1263–1268. https://doi.org/10.1038/nn1525

Bruegmann T, Malan D, Hesse M, Beiert T, Fuegemann CJ, Fleischmann BK, Sasse P (2010) Optogenetic control of heart muscle in vitro and in vivo. Nat Methods 7:897–900. https://doi.org/10.1038/nmeth.1512

Chen R, Gore F, Nguyen Q-A, Ramakrishnan C, Patel S, Kim SH, Raffiee M, Kim YS, Hsueh B, Krook-Magnusson E et al (2021) Deep brain optogenetics without intracranial surgery. Nat Biotechnol 39:161–164. https://doi.org/10.1038/s41587-020-0679-9

Cords SM, Reuter G, Issing PR, Sommer A, Kuzma J, Lenarz T (2000) A silastic positioner for a modiolus-hugging position of intracochlear electrodes: electrophysiologic effects. Am J Otolaryngol 21:212–217

Dieter A, Duque-Afonso CJ, Rankovic V, Jeschke M, Moser T (2019) Near physiological spectral selectivity of cochlear optogenetics. Nat Commun 10:1962. https://doi.org/10.1038/s41467-019-09980-7

Dieter A, Keppeler D, Moser T (2020a) Towards the optical cochlear implant: optogenetic approaches for hearing restoration. EMBO Mol Med 12:e11618. https://doi.org/10.15252/emmm.201911618

Dieter A, Klein E, Keppeler D, Jablonski L, Harczos T, Hoch G, Rankovic V, Paul O, Jeschke M, Ruther P et al (2020b) μLED-based optical cochlear implants for spectrally selective activation of the auditory nerve. EMBO Mol Med 12(8):e12387. https://doi.org/10.15252/emmm.202012387

Duarte MJ, Kanumuri VV, Landegger LD, Tarabichi O, Sinha S, Meng X, Hight AE, Kozin ED, Stankovic KM, Brown MC et al (2018) Ancestral adeno-associated virus vector delivery of opsins to spiral ganglion neurons: implications for optogenetic cochlear implants. Mol Ther J Am Soc Gene Ther 26:1931–1939. https://doi.org/10.1016/j.ymthe.2018.05.023

Emiliani V, Entcheva E, Hedrich R, Hegemann P, Konrad KR, Lüscher C, Mahn M, Pan Z-H, Sims RR, Vierock J et al (2022) Optogenetics for light control of biological systems. Nat Rev Methods Primer 2:1–25. https://doi.org/10.1038/s43586-022-00136-4

Friesen LM, Shannon RV, Cruz RJ (2005) Effects of stimulation rate on speech recognition with cochlear implants. Audiol Neurotol 10:169–184. https://doi.org/10.1159/000084027

Gunaydin LA, Yizhar O, Berndt A, Sohal VS, Deisseroth K, Hegemann P (2010) Ultrafast optogenetic control. Nat Neurosci 13:387–392. https://doi.org/10.1038/nn.2495

Hernandez VH, Gehrt A, Reuter K, Jing Z, Jeschke M, Mendoza Schulz A, Hoch G, Bartels M, Vogt G, Garnham CW et al (2014) Optogenetic stimulation of the auditory pathway. J Clin Invest 124:1114–1129. https://doi.org/10.1172/JCI69050

Huet AT, Rankovic V (2021) Application of targeting-optimized chronos for stimulation of the auditory pathway. Methods Mol Biol (Clifton, NJ) 2191:261–285. https://doi.org/10.1007/978-1-0716-0830-2_16

Huet AT, Dombrowski T, Rankovic V, Thirumalai A, Moser T (2021) Developing fast, red-light optogenetic stimulation of spiral ganglion neurons for future optical cochlear implants. Front Mol Neurosci 14:635897. https://doi.org/10.3389/fnmol.2021.635897

Huet A, Mager T, Gossler C, Moser T (2024) Toward optogenetic hearing restoration. Annu Rev Neurosci 47(1):103–121. https://doi.org/10.1146/annurev-neuro-070623-103247

Hunniford V, Kühler R, Wolf B, Keppeler D, Strenzke N, Moser T (2023) Patient perspectives on the need for improved hearing rehabilitation: a qualitative survey study of German cochlear implant users. Front Neurosci 17:1105562. https://doi.org/10.3389/fnins.2023.1105562

Izzo AD, Suh E, Pathria J, Walsh JT, Whitlon DS, Richter C-P (2007) Selectivity of neural stimulation in the auditory system: a comparison of optic and electric stimuli. J Biomed Opt 12:021008. https://doi.org/10.1117/1.2714296

Joris PX, Carney LH, Smith PH, Yin TC (1994) Enhancement of neural synchronization in the anteroventral cochlear nucleus. I. Responses to tones at the characteristic frequency. J Neurophysiol 71:1022–1036

Kallweit N, Baumhoff P, Krueger A, Tinne N, Kral A, Ripken T, Maier H (2016) Optoacoustic effect is responsible for laser-induced cochlear responses. Sci Rep 6:28141. https://doi.org/10.1038/srep28141

Keppeler D, Merino RM, Lopez De La Morena D, Bali B, Huet AT, Gehrt A, Wrobel C, Subramanian S, Dombrowski T, Wolf F et al (2018) Ultrafast optogenetic stimulation of the auditory pathway by targeting-optimized Chronos. EMBO J 37:e99649. https://doi.org/10.15252/embj.201899649

Keppeler D, Schwaerzle M, Harczos T, Jablonski L, Dieter A, Wolf B, Ayub S, Vogl C, Wrobel C, Hoch G et al (2020) Multichannel optogenetic stimulation of the auditory pathway using microfabricated LED cochlear implants in rodents. Sci Transl Med 12:eabb8086. https://doi.org/10.1126/scitranslmed.abb8086

Khurana L, Keppeler D, Jablonski L, Moser T (2022) Model-based prediction of optogenetic sound encoding in the human cochlea by future optical cochlear implants. Comput Struct Biotechnol J 20:3621–3629. https://doi.org/10.1016/j.csbj.2022.06.061

Khurana L, Harczos T, Moser T, Jablonski L (2023) En route to sound coding strategies for optical cochlear implants. iScience 26:107725. https://doi.org/10.1016/j.isci.2023.107725

Kim CK, Adhikari A, Deisseroth K (2017) Integration of optogenetics with complementary methodologies in systems neuroscience. Nat Rev Neurosci 18:222–235. https://doi.org/10.1038/nrn.2017.15

Kim WS, Hong S, Park SI (2021) Robust, wireless gastric optogenetic implants for the study of peripheral pathways and applications in obesity. Annu Int Conf IEEE Eng Med Biol Soc IEEE Eng Med Biol Soc Annu Int Conf 2021:5742–5746

Klapoetke NC, Murata Y, Kim SS, Pulver SR, Birdsey-Benson A, Cho YK, Morimoto TK, Chuong AS, Carpenter EJ, Tian Z et al (2014) Independent optical excitation of distinct neural populations. Nat Methods 11:338–346. https://doi.org/10.1038/nmeth.2836

Klein E, Gossler C, Paul O, Ruther P (2018) High-density μLED-based optical cochlear implant with improved thermomechanical behavior. Front Neurosci 12:659. https://doi.org/10.3389/fnins.2018.00659

Kral A, Hartmann R, Mortazavi D, Klinke R (1998) Spatial resolution of cochlear implants: the electrical field and excitation of auditory afferents. Hear Res 121:11–28

Lagali PS, Balya D, Awatramani GB, Münch TA, Kim DS, Busskamp V, Cepko CL, Roska B (2008) Light-activated channels targeted to ON bipolar cells restore visual function in retinal degeneration. Nat Neurosci 11:667–675. https://doi.org/10.1038/nn.2117

Liberman MC (1978) Auditory-nerve response from cats raised in a low-noise chamber. J Acoust Soc Am 63:442–455

Lv J, Wang H, Cheng X, Chen Y, Wang D, Zhang L, Cao Q, Tang H, Hu S, Gao K et al (2024) AAV1-hOTOF gene therapy for autosomal recessive deafness 9: a single-arm trial. Lancet 403: 2317–2325. https://doi.org/10.1016/S0140-6736(23)02874-X

Mager T, Lopez de la Morena D, Senn V, Schlotte J, Errico AD, Feldbauer K, Wrobel C, Jung S, Bodensiek K, Rankovic V et al (2018) High frequency neural spiking and auditory signaling by ultrafast red-shifted optogenetics. Nat Commun 9:1750. https://doi.org/10.1038/s41467-018-04146-3

Markowitz AL, Kalluri R (2020) Gradients in the biophysical properties of neonatal auditory neurons align with synaptic contact position and the intensity coding map of inner hair cells. elife 9:e55378. https://doi.org/10.7554/eLife.55378

Meng X, Murali S, Cheng Y-F, Lu J, Hight AE, Kanumuri VV, Brown MC, Holt JR, Lee DJ, Edge ASB (2019) Increasing the expression level of ChR2 enhances the optogenetic excitability of cochlear neurons. J Neurophysiol 122:1962–1974. https://doi.org/10.1152/jn.00828.2018

Michael M, Wolf BJ, Klinge-Strahl A, Jeschke M, Moser T, Dieter A (2023) Devising a framework of optogenetic coding in the auditory pathway: insights from auditory midbrain recordings. Brain Stimul 16:1486–1500

Middlebrooks JC, Snyder RL (2007) Auditory prosthesis with a penetrating nerve array. J Assoc Res Otolaryngol 8:258–279. https://doi.org/10.1007/s10162-007-0070-2

Mittring A, Moser T, Huet AT (2023) Graded optogenetic activation of the auditory pathway for hearing restoration. Brain Stimul 16:466–483. https://doi.org/10.1016/j.brs.2023.01.1671

Moser T, Karagulyan N, Neef J, Jaime Tobón LM (2023) Diversity matters—extending sound intensity coding by inner hair cells via heterogeneous synapses. EMBO J 42:e114587. https://doi.org/10.15252/embj.2023114587

Moser T, Chen H, Kusch K, Behr R, Vona B (2024) Gene therapy for deafness: are we there now? EMBO Mol Med 16:675–677. https://doi.org/10.1038/s44321-024-00058-6

Nagel G, Ollig D, Fuhrmann M, Kateriya S, Musti AM, Bamberg E, Hegemann P (2002) Channelrhodopsin-1: a light-gated proton channel in green algae. Science 296:2395–2398. https://doi.org/10.1126/science.1072068

Nagel G, Szellas T, Huhn W, Kateriya S, Adeishvili N, Berthold P, Ollig D, Hegemann P, Bamberg E (2003) Channelrhodopsin-2, a directly light-gated cation-selective membrane channel. Proc Natl Acad Sci 100:13940–13945. https://doi.org/10.1073/pnas.1936192100

Ohlemiller KK, Echteler SM, Siegel JH (1991) Factors that influence rate-versus-intensity relations in single cochlear nerve fibers of the gerbil. J Acoust Soc Am 90:274–287

Pinyon JL, Tadros SF, Froud KE, Wong ACY, Tompson IT, Crawford EN, Ko M, Morris R, Klugmann M, Housley GD (2014) Close-field electroporation gene delivery using the cochlear implant electrode array enhances the bionic ear. Sci Transl Med 6:233ra54. https://doi.org/10.1126/scitranslmed.3008177

Qi J, Tan F, Zhang L, Lu L, Zhang S, Zhai Y, Lu Y, Qian X, Dong W, Zhou Y et al (2024) AAV-mediated gene therapy restores hearing in patients with DFNB9 deafness. Adv Sci Weinh Baden-Wurtt Ger 11(11):e2306788. https://doi.org/10.1002/advs.202306788

Rankovic V, Vogl C, Dörje NM, Bahader I, Duque-Afonso CJ, Thirumalai A, Weber T, Kusch K, Strenzke N, Moser T (2021) Overloaded adeno-associated virus as a novel gene therapeutic tool for otoferlin-related deafness. Front Mol Neurosci 13:6000051. https://doi.org/10.3389/fnmol.2020.600051

Richardson RT, Thompson AC, Wise AK, Ajay EA, Gunewardene N, O'Leary SJ, Stoddart PR, Fallon JB (2021) Viral-mediated transduction of auditory neurons with opsins for optical and hybrid activation. Sci Rep 11:11229. https://doi.org/10.1038/s41598-021-90764-9

Rubinstein JT (2004) How cochlear implants encode speech. Curr Opin Otolaryngol Head Neck Surg 12:444–448

Sachs MB, Abbas PJ (1974) Rate versus level functions for auditory-nerve fibers in cats: tone-burst stimuli. J Acoust Soc Am 56:1835–1847

Sahel J-A, Boulanger-Scemama E, Pagot C, Arleo A, Galluppi F, Martel JN, Esposti SD, Delaux A, de Saint Aubert J-B, de Montleau C et al (2021) Partial recovery of visual function in a blind patient after optogenetic therapy. Nat Med 27:1223–1229. https://doi.org/10.1038/s41591-021-01351-4

Shannon RV, Cruz RJ, Galvin JJ (2011) Effect of stimulation rate on cochlear implant users' phoneme, word and sentence recognition in quiet and in noise. Audiol Neurotol 16:113–123. https://doi.org/10.1159/000315115

Shapiro MG, Homma K, Villarreal S, Richter C-P, Bezanilla F (2012) Infrared light excites cells by changing their electrical capacitance. Nat Commun 3:736. https://doi.org/10.1038/ncomms1742

Smith ZM, Delgutte B, Oxenham AJ (2002) Chimaeric sounds reveal dichotomies in auditory perception. Nature 416:87–90. https://doi.org/10.1038/416087a

Teudt IU, Maier H, Richter C-P, Kral A (2011) Acoustic events and 'optophonic' cochlear responses induced by pulsed near-infrared laser. IEEE Trans Biomed Eng 58:1648–1655. https://doi.org/10.1109/TBME.2011.2108297

Thirumalai A, Henseler J, Enayati M, Kusch K, Hessler R, Moser T, Huet AT (2025) Improved optogenetic modification of the spiral ganglion neurons for future optical cochlear implants. Theranostics 15(10):4270–4286

Thompson AC, Fallon JB, Wise AK, Wade SA, Shepherd RK, Stoddart PR (2015) Infrared neural stimulation fails to evoke neural activity in the deaf guinea pig cochlea. Hear Res 324:46–53. https://doi.org/10.1016/j.heares.2015.03.005

Thompson AC, Wise AK, Hart WL, Needham K, Fallon JB, Gunewardene N, Stoddart PR, Richardson RT (2020) Hybrid optogenetic and electrical stimulation for greater spatial resolution and temporal fidelity of cochlear activation. J Neural Eng 17:056046. https://doi.org/10.1088/1741-2552/abbff0

Vedam-Mai V, Deisseroth K, Giordano J, Lazaro-Munoz G, Chiong W, Suthana N, Langevin J-P, Gill J, Goodman W, Provenza NR et al (2021) Proceedings of the eighth annual deep brain stimulation think tank: advances in optogenetics, ethical issues affecting DBS research, neuromodulatory approaches for depression, adaptive neurostimulation, and emerging DBS technologies. Front Hum Neurosci 15:644593. https://doi.org/10.3389/fnhum.2021.644593

Vogt M, Schulz B, Wagdi A, Lebert J, van Belle GJ, Christoph J, Bruegmann T, Patejdl R (2021) Direct optogenetic stimulation of smooth muscle cells to control gastric contractility. Theranostics 11:5569–5584. https://doi.org/10.7150/thno.53883

von Békésy G, Wever EG (1960) Experiments in hearing. McGraw-Hill, New York

Walsh JJ, Christoffel DJ, Heifets BD, Ben-Dor GA, Selimbeyoglu A, Hung LW, Deisseroth K, Malenka RC (2018) 5-HT release in nucleus accumbens rescues social deficits in mouse autism model. Nature 560:589–594

WHO (2021) Deafness and hearing loss. Geneva: World Health Organization (WHO); 2021. https://www.who.int/news-room/fact-sheets/detail/deafness-and-hearing-loss

Winter IM, Robertson D, Yates GK (1990) Diversity of characteristic frequency rate-intensity functions in guinea pig auditory nerve fibres. Hear Res 45:191–202

Wolf BJ, Kusch K, Hunniford V, Vona B, Kühler R, Keppeler D, Strenzke N, Moser T (2022) Is there an unmet medical need for improved hearing restoration? EMBO Mol Med 14:e15798

Wrobel C, Dieter A, Huet A, Keppeler D, Duque-Afonso CJ, Vogl C, Hoch G, Jeschke M, Moser T (2018) Optogenetic stimulation of cochlear neurons activates the auditory pathway and restores auditory-driven behavior in deaf adult gerbils. Sci Transl Med 10:eaao0540. https://doi.org/10.1126/scitranslmed.aao0540

Wykes RC, Heeroma JH, Mantoan L, Zheng K, MacDonald DC, Deisseroth K, Hashemi KS, Walker MC, Schorge S, Kullmann DM (2012) Optogenetic and potassium channel gene therapy in a rodent model of focal neocortical epilepsy. Sci Transl Med 4:161ra152. https://doi.org/10.1126/scitranslmed.3004190

Yates GK, Winter IM, Robertson D (1990) Basilar membrane nonlinearity determines auditory nerve rate-intensity functions and cochlear dynamic range. Hear Res 45:203–219. https://doi.org/10.1016/0378-5955(90)90121-5

Zeng F-G (2004) Trends in cochlear implants. Trends Amplif 8:1–34

Zeng F-G (2022) Celebrating the one millionth cochlear implant. JASA Express Lett 2:077201. https://doi.org/10.1121/10.0012341

Zeng F-G, Rebscher S, Harrison WV, Sun X, Feng H (2008) Cochlear implants: system design, integration and evaluation. IEEE Rev Biomed Eng 1:115–142. https://doi.org/10.1109/RBME.2008.2008250

Zerche M, Wrobel C, Kusch K, Moser T, Mager T (2023) Channelrhodopsin fluorescent tag replacement for clinical translation of optogenetic hearing restoration. Mol Ther Methods Clin Dev 29:202–212. https://doi.org/10.1016/j.omtm.2023.03.009

MIX
Papier aus verantwortungsvollen Quellen
Paper from responsible sources
FSC® C105338

If you have any concerns about our products,
you can contact us on
ProductSafety@springernature.com

In case Publisher is established outside the EU,
the EU authorized representative is:
Springer Nature Customer Service Center GmbH
Europaplatz 3, 69115 Heidelberg, Germany

Printed by Libri Plureos GmbH
in Hamburg, Germany